Atlas of the Visible Human Male

National Library of Medicine

Atlas of the Visible Human Male

Reverse Engineering of the Human Body

VICTOR M. SPITZER, PH.D.
DAVID G. WHITLOCK, M.D., PH.D.
University of Colorado Health Sciences Center

JONES AND BARTLETT PUBLISHERS
Sudbury, Massachusetts

BOSTON   LONDON   SINGAPORE

Editorial, Sales, and Customer Service Offices

Jones and Bartlett Publishers, Inc.
40 Tall Pine Drive
Sudbury, MA 01776
508-443-5000
800-832-0034
info@jbpub.com
http://www.jbpub.com

Jones and Bartlett Publishers International
Barb House, Barb Mews
London W6 7PA
UK

Library of Congress Cataloging-in-Publication Data

Spitzer, Victor M.
    Atlas of the visible human male : reverse engineering of the human
body / Victor M. Spitzer , David G. Whitlock.
      p. cm.
    At head of title: National Library of Medicine.
    Includes index.
    ISBN 0-7637-0273-0 (hardcover : alk. paper). — ISBN 0-7637-0347-8
(pbk. : alk. paper)
    1. Human anatomy—Atlases.   2. Tomography—Atlases. I. Whitlock,
David G., 1924-   .   II. National Library of Medicine (U.S.)
III. Title.
    [DNLM: 1. Visible Human Project (National Library of Medicine
(U.S.))   2. Anatomy, Cross-Sectional—atlases.   3. Image Processing,
Computer-Assisted.   QS 17 S761a 1997]
QM25.S74   1997
611' .0022'2—dc21
DNLM/DLC
for Library of Congress   97-16572
                                             CIP

Production Editor: Sue Michener
Editorial Production Service: Tower Graphics
Text Design, Cover Design, and Production: Marshall Henrichs
Text Printing: World Color
Cover Printing: Henry N. Sawyer Company
Binding: World Color (paperback), Acme Bookbinding (hardcover)

Printed in the United States of America
01 00 99 98 97   10 9 8 7 6 5 4 3 2 1

*Our foremost debt and highest acknowledgment are
to all people who have given their last perfect gift to society—their bodies.
Without these great gifts medical and health-care professionals cannot obtain
the rich understanding of the human body that underpins
all areas of medical knowledge.*

◆

To these generous people we dedicate this work.

◆

*To my father, one of these generous people,
I dedicate my small part in realizing his large dreams.*

VMS

Contents

ACKNOWLEDGMENTS

We are grateful to our University of Colorado department chairmen, Karl Pfenninger, MD, of Cellular and Structural Biology, and Michael Manco-Johnson, MD, of Radiology, for their collective vision in the late 1980s, which led to the teaming of the authors. They recognized that a superlative knowledge of anatomy was a crucial basis for medicine and that radiology had techniques and technology that could enormously benefit the computerization of anatomical knowledge. Their vision has been realized in the creation of the National Library of Medicine's The Visible Human Project™.

We are also deeply indebted to the National Library of Medicine (NLM) for providing the leadership and funds for the development of the Visible Human male database. Especially the Director, Donald Lindberg, MD, for his support and encouragement, and the NLM Boards of Directors, who approved this project and continue to support its development and use.

Michael Ackerman, the NLM project officer, deserves special commendation for his dedication to the project from its inception. We especially appreciate his efforts in facilitating special approvals for procedures not covered by government regulations, such as the expedient transport of dead bodies, and his insistence that review committees understand the pictures included in our proposals. We also gratefully acknowledge the contract work of Sharon Cummings and take solace in her concern that "her Library" was going to start cataloging corpses in the basement.

The acquisition, management, and preparation of the cadavers were also critical to the success of this project and for this, we offer our thanks to the State Anatomical Boards (SAB) of Colorado (Director, Michael Carry, PhD), Texas (Director, Andrew Payer, PhD), and Maryland (Director, Ronald S. Wade) for their cooperation and interest in the project. A special word of appreciation is also extended to Ed Molock and Mark Blatchley of the Colorado SAB for their round-the-clock efforts each time a cadaver was acquired. Also to Allen Tyler of the Texas SAB, thank you for spending many hours with one of the authors (VMS) waiting for the arrival of cadavers, during the embalming process, and even on a Lear Jet flight from Texas to Denver with one of the cadavers. Alan's companionship and experience in this uncommon situation was comforting at all times.

Radiological images, which were the basis for selection of the cadavers and later part of the Visible Human dataset, were acquired with expediency and expertise. The "emergency" procedures used to obtain radiographs, MR, and CT images, all within 24 to 36 hours and all with state-of-the-art equipment, required expert planning, cooperation, and hard work. We are grateful to Ann Scherzinger, PhD, for design and coordination of all the radiological imaging. Charles Ahrens, RT, provided radiographic imaging expertise and procedure development. His efforts and cooperation are especially noted. David Rubinstein, MD, put in many long evenings and early mornings to help with positioning of the cadavers and the acquisition and interpretation of the images. Paul Russ, MD, provided interpretations of the body images.

The long, night imaging hours, between regular, busy clinical days, for both the MR and CT imaging will always be remembered and appreciated. Cathy Gustafson, RT, Donna Callan, RT, and Deborah Singel, RT, put in many long nights on the MR scanner. Teresa Bonner, RT, supervised the CT imaging for all the cadavers and Thomas Hogan, RT, imaged the Visible Human Male. We are grateful to all these imagers for their last-minute response and dedication to the goals of the project.

Martha Pelster and Tim Butzer directed the support team and cut most of the 7,000 slices of the male and female specimens themselves. This team not only devised and tested freezing chambers and techniques and numerous cadaver handling systems, but also demonstrated the tenacity to perform the same function day after day. They were assisted by Helen Pelster, James Heath, Brian McNevin, Kate Finger, and Gregory Spitzer.

Helen Pelster, assisted by Gregory Spitzer, directed the image management, archiving, registration, and cataloging of the image database. Photography was overseen by James Heath. Control mechanisms for both the cryomacrotome and the cameras were designed and built by Charles Rush, MS, in collaboration with Karl Reinig, PhD. Numerous designs and redesigns of the equipment and protocol were successful because of their willingness, interest, and involvement along with support staff Bruce Burkhart and Dick Kennedy.

We also gratefully acknowledge Deborah Lehman, John Blackwell, and Renee Torgler for their work in the early portions of the contract and acknowledge that Renee was certainly willing, and in fact did, give the shirt off her back for the glory and goals of this work.

Of course, our gratitude for the participants in the Visible Human Male Project is quite different from our appreciation for the heroic efforts in bringing this book to fruition. Although numerous publishers showed little interest, Clayton Jones recognized the value of and believed in multimedia enough to not ignore the old and trusted media, the printed page, even for this entirely digital subject. It was his snap decision to commit (in less than one hour from presentation of our concept) to publish this book that has spurred our dedication to this effort. We also gratefully acknowledge Niel Patterson who shared confidence in our work and introduced us to Clayton. The constant facilitation and occasional badgering by our editors, Dave Phanco and Paul Lembo, could not have been done with more professionalism. The production staff of Jones and Bartlett, and the very special design influence of Marshall Henrichs, has been superb after they got over the normally encountered hurdles of what it was that these two crazy people in Colorado actually did.

For the detailed layout of all 250 labeled plates we are grateful to Raechel O'Kelly, Paul Sommers, Amy Schilling, and once again to Helen Pelster and Greg Spitzer for image management and database development. The 3D rendered images in the "Image Gallery" are only possible because of the teamwork of Zhonjun Fu and a team of 25 students and staff who have classified each and every visible structure in the body over the past 18 months. A few examples of this work are shown in the 3D pictures in the last part of the book. For these ray-traced renderings, which reflect both the color and texture of the original voxel data, we are indebted to Karl Reinig and Helen Pelster.

The images of simulated procedures, in the last part, reflect the exciting new directions that these fully segmented and classified data are facilitating. These images reflect thousands of hours of teamwork and cooperation from members of the University of Colorado Center for Human Simulation, in our effort to more fully utilize the potential of this extraordinary national resource. Karl Reinig, PhD, has led this simulation effort with the dedicated support of Tom Mahalik, PhD, Travis Johnson, and again Helen Pelster.

Our participation in this book certainly depended on previous support that laid the foundation for developing and assembling the technology necessary for the successful completion of the project. That includes a pivotal grant from the Frost Foundation, publishing support from Mosby-Yearbook, and equipment support from Hewlett Packard, Intel Corporation, Intergraph Computer Systems, Digital Equipment Corporation, Exabyte, Kodak, and Apple computer. Funding for all classification and segmentation of the Visible Human Male, which facilitates the 3D rendered images in the last part of this book, was provided by Gold Standard Multimedia. Research Systems Inc. generously provided the image processing software (IDL), which allowed rapid prototyping and development of image processing and handling of the Visible Human dataset.

Second to last, and most important, we express our continual appreciation of our wives and families for their support and understanding during the long hours involved in the preparation of this work. We also appreciate their technical assistance and occasional dominance. Last of all, we acknowledge all the people who have significantly contributed to the Visible Human Male Project at the University of Colorado and to this book but for some strange reason have slipped our collective minds.

The hand colored title page to the first edition of the "De Humani Corporis Fabrica," 1543

INTRODUCTION

The Visible Human Male

The Visible Human Male is the most complete computerized database of the human body ever assembled. This book is your tour guide—taking you through the 1,878, one-millimeter thick transverse slices of an adult human male as well as sagittal and coronal reconstructions and representative clinical scans and 3D reconstructions. Called "the greatest contribution to anatomy since Vesalius's 1543 publication of De Humani Corporis Fabrica," the first collection of faithfully rendered drawings of human anatomy, these images, constituting 15 gigabytes of information, are the seeds for a growing medical revolution. With this atlas those already familiar with anatomy, such as radiologists and other health-care professionals, can follow structures through different views of the same body, seeing how an organ or muscle changes in space. Medical students can supplement their knowledge of gross anatomy by studying the labeled images. For the lay audience it is a window into the unknown, a way to visualize the body's mysterious machinery, and see what the professionals see. This dataset should change the way doctors are trained and surgery is practiced; it may help cut the cost of health care and allow physicians and other professionals to perform procedures unthinkable before. The National Library of Medicine's The Visible Human Project™ is an anatomy lesson for the twenty-first century.

Cadavers have been dissected since the thirteenth century, but it was not until the Renaissance that anatomists stopped searching for the seat of emotion or temperament locked within our organs and began seeing the body as a miraculous machine whose inner workings should be understood and mastered by physicians to help heal the sick. In 1405 the University of Bologna allowed up to twenty students to observe the dissection of a male corpse; thirty were permitted at a female's dissection. The pool of bodies for dissection consisted mainly of executed criminals who sometimes numbered only three or four a year, so simply put, there were never enough cadavers to go around. In the sixteenth century demand became so great that amphitheaters were built to accommodate the large audiences. Physicians, prominent members of the community, intellectuals, and the inevitable idle and rich snobs, in addition to medical students, gathered for the spectacle, which would last three to four days—a time frame determined by the lack of refrigeration. Later, audiences were made up of lay people as well as professionals, making it difficult to keep a sense of decorum, especially when people gathered in costume and sometimes became so rowdy that they would grab specimens right off the dissecting table.

Since then, anatomy has been studied in exquisite detail. Bodies have been dissected, cross-sectioned, photographed, and described ad infinitum. Yet those who have first-hand knowledge of the inside of a human body are still a highly select group. Most medical students only get to explore one cadaver, usually shared with at least three classmates. With The Visible

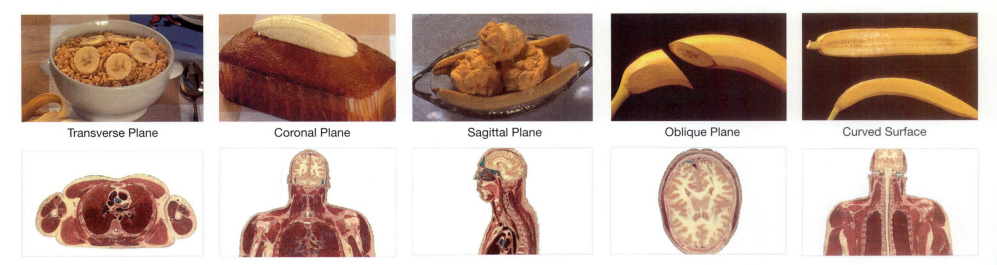

| Transverse Plane | Coronal Plane | Sagittal Plane | Oblique Plane | Curved Surface |

These views and their symmetry properties (though maybe not their names) are very familiar outside the world of medicine.

Human Project we expect to change that, enhancing the experience of cadaver dissection in medical education and providing virtual cadavers to anyone interested in the human body. The images in this book are today's version of a dissection in a public amphitheater, but ours is a global amphitheater able to serve an unlimited audience. The pieces we present can be reassembled in sections, just the brain, for example, or as a complete unit. It is what we like to call reverse engineering. Our man in the machine allows interested citizens of any age or profession to travel through and explore a real human body in its entirety on their own computer screens.

The Visible Human Project was conceived by the National Library of Medicine (NLM) in 1988 after bringing together representatives of eight medical centers working in the area of 3D anatomical visualization. Collectively these scientists decided that the NLM, until then a text-dominated repository of medical information, should create an image database of human anatomy. The idea, the dream really, was to obtain a uniform set of photographs and scans from one optimum specimen. Over one hundred medical schools vied for the job, but a proposal submitted by our school, the University of Colorado School of Medicine, won the contract which ultimately totaled $1.4 million for work on both the man and the woman. We inaugurated the project on September 1, 1991. The success of the Colorado proposal, it was later revealed, was not based on the superior writing skills of the authors, but rather on twenty previously captured pictures of a human abdomen that were submitted with the proposal. The pictures, each 7 megabytes or 3.5 million words, must have been worth a few million words.

To be of maximum clinical and educational value, it was decided that our virtual couple needed to be documented in several different formats commonly used by radiologists and other physicians. We decided on six: traditional X-rays and CT (computerized tomography) scans to optimally visualize bone, MRI (magnetic resonance imaging) for soft tissue, and three types of color photographs for definitive resolution. These would make up the multispectral database of images that would be captured by the most clinically useful imaging techniques currently available and one technique—photography—that would provide the ultimate standard of comparison.

Our first and most formidable task was to locate a specimen that could be taken as "normal" or representative of a large population. To achieve this goal, the NLM decided we should obtain a sample set of three cadavers and let a blue ribbon panel of experts select the "winner" based on radiological images of their remains. The success of this search for a healthy-appearing cadaver depended on resources (numbers of cadavers per year) beyond that of the State Anatomical Board of Colorado and we were very fortunate to be joined by the State Anatomical Board of Texas, in our original proposal, and later also by Maryland. State Anatomical Boards are organizations through which citizens can donate their bodies for medical research and training.

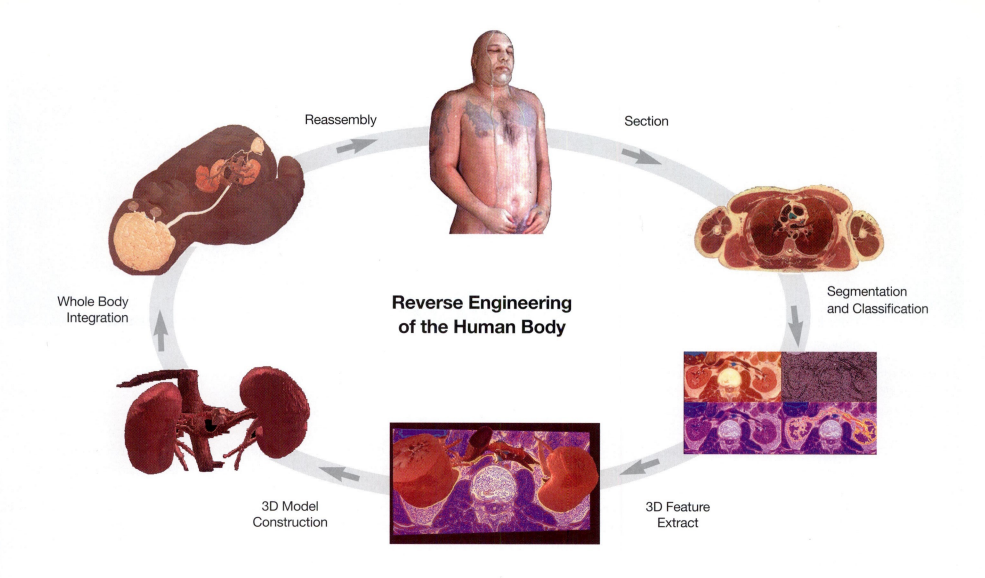

Reassembly

Section

Whole Body
Integration

**Reverse Engineering
of the Human Body**

Segmentation
and Classification

3D Model
Construction

3D Feature
Extract

What we were searching for was someone 21- to 60-years of age who died without traumatic injuries or invasive or infectious disease. We got lucky. Some inmates on death row in Texas had decided to donate their bodies to science. They were young, relatively healthy men whose organs, tainted by lethal injection, were rendered unsuitable for transplant. Through screening of cadavers such as these as well as individuals obtained from other donations, our panel selected the body of a recently executed 38-year-old male. It was not lost on us that victims of execution, a population that taught anatomist's centuries ago, would be teaching us once again, perhaps in some way repaying society for their crimes.

Aside from some minor imperfections—no appendix, Number 14 tooth, and left testicle—this specimen fit all our criteria, which were that the body be: less than 6 feet tall, less than 22-inches wide, less than 14-inches from front to back, and normal height for weight. What accounted for these "ideal" human dimensions? Most were based on the limitations of equipment we were using either to image the body or slice it. Initially,

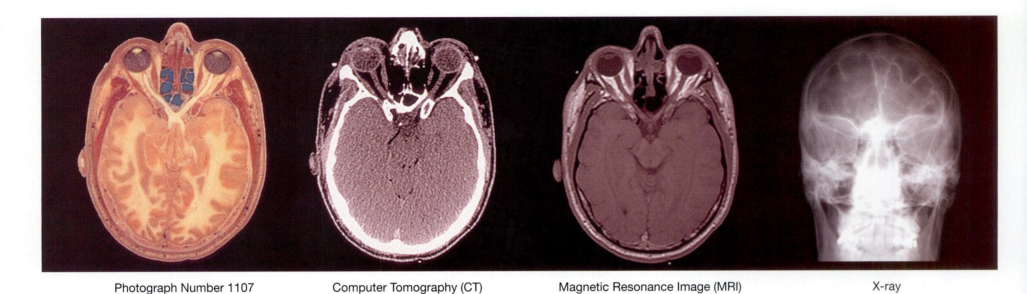

| Photograph Number 1107 | Computer Tomography (CT) | Magnetic Resonance Image (MRI) | X-ray |

The right-left symmetry is indicated by the bilateral structures in the head. The tilt of the head (proximity of the chin to the chest) is indicated by the length of the optic nerve in this single plane.

the ideal position was to be achieved by suspending the cadaver vertically while freezing it to preserve a "life-like" human form in the standing position. This required constructing a special freezing chamber for which the most practical height was no more than six feet. The process worked well for positioning internal viscera, because the body was suspended by the skull, underarms, and pelvis. But there were undesirable distortions in the areas of the cheeks, arms, and lower abdomen. After three or four attempts, this method was abandoned.

Other dimension limits had to do with the size of the cryomacrotome and other tools we used to slice the body. We were breaking new ground. According to published literature, what we did was unfeasible. Fortunately, no one told us that, so we experimented, sometimes erring, but improving our techniques every step of the way. Ultimately we opted to obtain the CT scans of the cadaver in the supine position and to hold the cadaver in that exact same position throughout the physical sectioning process. We did this by positioning the cadaver in a plywood box, adapted to mount on the CT gantry, and then filling all extraneous space in that confining box with a foaming agent

called Alpha Cradle™ that hardens in 20 minutes. This is a product used in radiation therapy to construct a custom-fit cradle for patients to lie on during successive treatments. The foaming agent is not much different from foaming agents used for filling space in electronics packaging, with the exception that it generates so little heat that it does not burn the patient. A stable, low temperature was also an important issue for the cadaver because we were trying to preserve the tissue.

The standard technique for positioning a patient in the CT scanner includes a vertical positioning of the median sagittal plane which is obtained by positioning the ears in the same horizontal plane. This positioning is generally done on the CT scanner but for our cadaver the positioning was done in the laboratory. The success of the "positioning" can be judged by the symmetry of the body, in particular the head in the slices taken transverse to the long axis of the body. A particularly good example of this positioning is demonstrated by the symmetry of the optic nerves in Slice Number 1107. The length and symmetry of the optic nerves in the same slice also demonstrates a desirable tilt of the head from front to back.

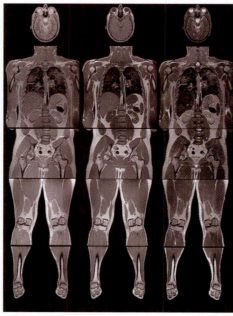

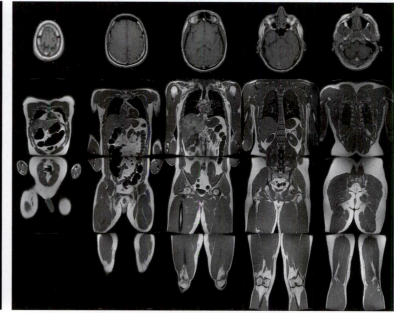

Proton density   T1   T2
weighted   weighted   weighted

Five different depths

Five of the 1871 transverse CT cross-sections

Magnetic resonance images (MRI) were obtained in transverse planes of the head and in coronal planes of the body.

When the NLM contract for the Visible Human was awarded, MR images of the body were not routine clinical practice, so the original contract called for MR images only of the head. By the time the Visible Human Male was acquired, body imaging was of clinical importance and we added it to our protocol. The initial contract also called for injection of the major vessels of the body with a colored latex (red in the arteries, blue in the veins) and iodine solution that could be seen in the MRI, CT scans, and physically cut sections. We injected five cadavers, and although this gave good results in the CT images, the solution leaked from the vessels in some areas of the body in three of the five cadavers. There is a long history of preparations using these kinds on injections (including latex, microfil, India ink, barium, etc.)—but they are rarely used on whole cadavers. Because it was unreasonable to assume that we would be able to prepare a leak-free specimen, vessel injection was eliminated from the protocol. Since MRI provided good delineation of the large vessels, these images were substituted for marking them by injec-

tion. The major drawback for MR images is simply that the body was not in the same position as used for CT scanning.

A $5,000 "backsaw" was commissioned from a local designer to partition the intact cadaver. We used it to section the cadaver into four blocks each less than 20 inches in height, the maximum height our cryomacrotome could handle. Concerned with future integration of data, we wanted our cuts to be in the exact same plane as the CT slices taken earlier and to make the saw kerf (the width of the saw cut) as thin as possible. The missing images in the Dataset represent anatomy destroyed by the saw. We will avoid this compromise in the future by expansion of the cryomacrotome to permit cutting the entire cadaver intact. The saw was manually operated with a single straight blade at 30,000 lb. of tension facilitating as straight a cut as possible. The major difficulty we had with the saw was when the blade encountered tissues of different resistance, in particular bone/soft tissue interfaces. We noted the problem and tried to overcome it by freezing the

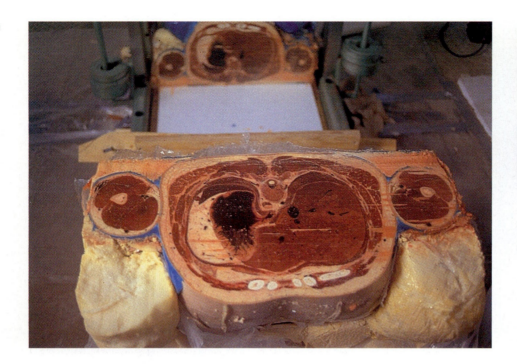

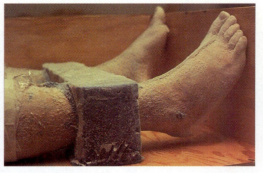

The legs of the cadaver were glued together to maintain their position during the cutting process. The "glue" was the same gelatin mixture later used to form the block to be sectioned.

cadaver to a lower temperature ($-70°$ C) to equalize the tissue characteristics. Backsaw cuts made after packing the body in dry ice for two days prior to sectioning were most successful. We also carefully positioned the cut so that it went directly through the hard bone of the spine and not between vertebrae, which often deflected the blade and caused irregularities in the cut. The most difficult section to cut was that through the diaphragm and arms. What do you do with the arms once they are cut off? Cutting them separately and repositioning them at the correct locations in the computer database would have been extremely difficult, so we "glued" them to the abdomen and pelvis using colored gelatin, straight from our cafeteria, prior to sectioning the cadaver. To keep the legs in their relative positions we also used a "bridge" of gelatin between them before sectioning.

Six months after receiving the body, and after sectioning it into the blocks described above, we began imaging the 1-mm thick slices. These images were obtained by photographing the top surface of the block after removal of each successive millimeter by the cryomacrotome. No actual slices were retained, however, since we destroyed each documented slice by milling it down to get to the one underneath. A key element during the process was to keep all factors affecting the photography of the newly exposed surface constant, especially the lighting and specimen temperature. Uniformity of the lighting field was achieved with four polarized strobe lights. We optimized the consistency of the color of the top surface of the specimen by continuous cutting. We had two teams of two people each, cutting in shifts for as many as 12 hours a day. In a lab with no air conditioning during a hot Colorado summer, however, some cutting sessions were necessarily restricted to less than eight hours. Despite our progress, we were still somewhat limited by refrigeration, one of the same constraints encountered by fifteenth-century anatomists. Each slice, including cutting, preparation, photography, refreezing, and repositioning took anywhere from four to ten minutes.

The photography included image capture with three cameras, one digital and two conventional film, one 35-mm and one 70-mm. Refreezing the specimen was a "hand" operation. It entailed setting a tray of dry ice on the block surface for 20 seconds. The variable in the process was the surface preparation time. This included removing any debris from the surface, checking for tissue that was not cleanly cut (e.g., some tendons and fascia), spraying the surface with ethanol, applying a black mask over the bright white dry ice packing the block, attaching a Kodak photographic color strip (for

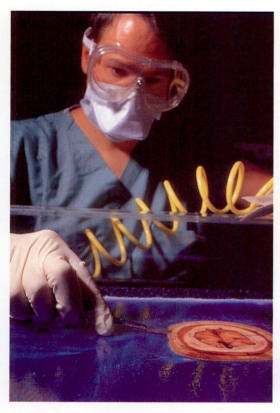

Surface preparation after removing each 1-mm slice includes removing all debris and checking for any "uncut" connective tissue.

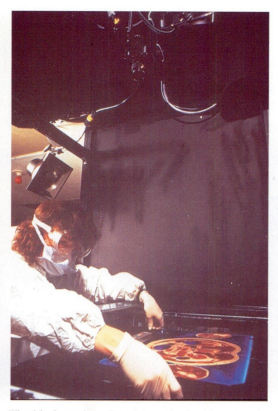

The black mask is placed over the freshly cut and prepared surface for photography. The three cameras used for image capture are at the top of the picture.

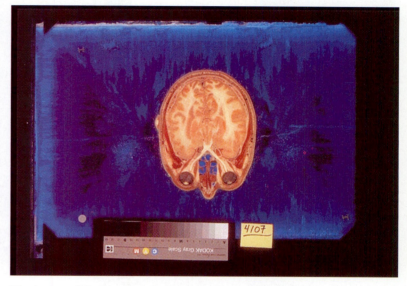

The image of the block surface as the camera viewed it. The color normalization strip is included at the bottom of each picture. The slice identifier indicates this was the 107th-mm from the top of the fourth block that was cut.

image color normalization), and assigning an internal slice number (written on a yellow Post-it Note™) indicating the block and the millimeter that each photograph represented (i.e., Slice 3105 is the 105th-mm cut in the third block). There were few typical days. Most components broke down at least once. A typical log entry:

> The first day is pretty good—It kinda looked like we would never get finished. First slice—power cord on 70-mm camera disconnected—took picture manually. Third slice—image capture computer stuck at "blue" image—reboot. Fifth slice—film counter on 70-mm camera malfunctions—fix programmed for the day. Sixth slice—computer stuck again— reboot. Slice 31—computer crashed again. Then things went well to slice 97.

The entire process took nearly nine months—four months of cutting time and the remainder for specimen preparation of each block.

The top and bottom few slices of each block were usually partial slices because the backsaw surface was not a perfect plane or the specimen section was not perfectly parallel to the cutting plane of the cryomacrotome. These partial slices and the 1.3-mm gap attributed to the saw kerf have been eliminated from the final, photographic image database. These missing slices account for the gaps at the diaphragm, upper thigh and just below the knee.

Our virtual male made his debut on November 28, 1994, at a meeting of the Radiological Society of North America in Chicago. Aware that the press might seize the opportunity to focus on the data's past—as housing for a convicted killer—and not its future—a revolution in the study of anatomy and surgical practice, we rehearsed our announcement carefully. Ironically the next day's headlines read, "Cannibal Dahmer Killed in Jail." By comparison, our news was tame. We stopped worrying.

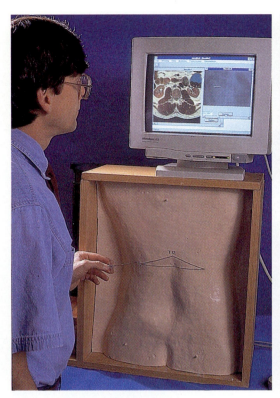

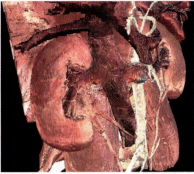

A celiac plexus nerve block is practiced with the plastic back, built from the data of the visible human (left). The student feels each tissue the needle encounters (upper right), through forces controlled by the Phantom, a haptic feedback interface from Sensable Technologies (lower right), and confirms the proper placement of the needle with a simulated radiograph showing its current position, as shown on the left side of monitor (upper left).

The student that just cut through the thigh of the Visible Human felt the cut as it was made. The cut reveals tissue defined by and taken from the Visible Human database.

With the release, over the Internet, of the *The Visible Human Project* by the National Library of Medicine, the images are now in the hands of medical researchers and technology designers all over the world. This dataset will facilitate the planning and rehearsal of medical interventional procedures prior to carrying them out in "real" operating or procedure rooms. Just as the aircraft industry provides simulators for its pilots, human body simulations are being constructed for training health-care personnel, including physicians, dentists, nurses, technicians, and EMTs. Col. Richard Satava, MD has predicted the impact of this new industry as "The use of the Visible Human database will fundamentally change the practice of medicine."

At the University of Colorado Center for Human Simulation, anesthesiologists use a haptic feedback device coupled to the classified data of the Visible Human abdomen to practice a celiac plexus block— an injection into a nerve center in the middle of the abdomen. With the haptic feedback device providing the interface to the computer, the student can "feel" the entire procedure. Before this simulator based on the Visible Human data was available, the procedure was performed by the student for the first time on a real patient. With the same haptic feedback device, this time coupled to the anatomy of the eye, we have developed a simulator for ophthalmologists to practice radial keratotomy; as they touch or cut the cornea, they feel and see the deformation they are causing. By coupling the same device to anatomical images of a tooth, dentists can experience, first-hand, the feeling and appearance of a tooth with carries and calculus. We can also mimic different stages in the progression of the disease. Simulators allow unlimited practice and rehearsal. These programs have value for the public sector as well, including patient education, general interest, and entertainment. Progress in this area is updated on our web site http://www.uchsc.edu/sm/chs.

We have embarked on other uses of this database in order to empower ourselves to more rapidly simulate the "new models" of humans, be the differences ever so slight, that keep coming off the assembly line. Our ultimate goal is to reverse engineer a single, or small number of models, of the human body in each fundamentally different category. We can then develop variations of these standard models to more closely match an individual of interest. This may mean a specific patient for the doctor, a small population of virtual heads for the helmet or headgear designer, an endless army of Size 6 bodies for the garment designer, or the ultimate stuntman for the movie producer. Our first model, presented here, is a middle-aged, Caucasian male. The second model, already sliced and imaged, is a middle-aged Caucasian female. We hope to expand on these categories with a variety of ages, shapes, and races. It is not unreasonable to extend this reverse engineering process to different species.

Available from the Internet (www.nlm.nih.gov/research/visible/visible_human.html) free of charge by agreeing to credit the NLM and send them a copy of any product produced, *The Visible Human Project* offers the medical and lay communities unlimited and previously unimaginable possibilities. We now have a renewable cadaver, a standardized patient, and a basis for digital populations of the future. Not only can we dissect it, we can put it back together again and start all over. Currently more than 700 licenses to use the dataset have been granted in more than 26 countries and applications are proliferating. The real future of medical education is not in the Visible Human Dataset itself but rather in the manipulation, distortion, and modification of the data to produce whole populations of virtual humans of every age, race, and pathology. In the past, we have provided visions of our theories and concepts through medical illustrations, photographs, and models, but each of these was a single representation in time. The Visible Human computerized database of the body, defined in three- or four-dimensional space, permits the viewing of our ideas in time—present, past, and future.

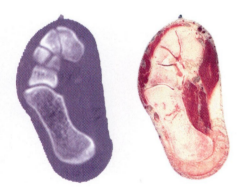

Transverse CT and corresponding anatomy planes through the ankle.

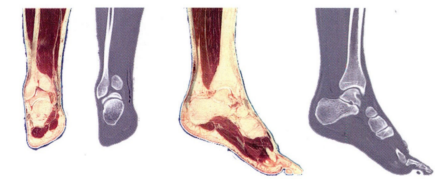

Coronal and sagittal reconstructions of anatomical and CT images through the ankle.

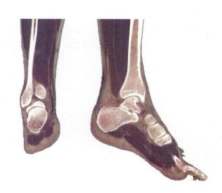

Overlay of the CT on the anatomical images showing alignment of the two modalities.

PART I

ORIGINAL IMAGES
Transverse

The transverse images are the original, digital pictures of the National Library of Medicine's The Visible Human Dataset™. Every picture (every millimeter) is included, from the head through the feet. These transverse planes are particularly useful when comparing bilateral anatomical structures. All images are presented in the standard radiological format, as though the viewer was standing at the foot of the bed and looking toward the patient's head.

The maximum resolution of images in the transverse plane is determined by the camera and field of view. All images in this part were acquired with a LEAF CCD camera. The pixel size is 0.32 millimeters square and the anatomy is contained in a matrix of 1760 x 1024 pixels. This pixel size and density is of similar resolution to the printing process used for the twelve-image arrays on the right-hand pages of this book.

The single labeled transverse image, on each left-hand page, was chosen from one of the twelve images, on the facing right-hand page. No attempt was made to keep labels in the same location from slice to slice. Bilateral structures are labeled on both sides when space permits. Labeling of the images generally follows naming conventions in *Gray's Anatomy,* 30th Amer-

ican edition (C. D. Clemente, editor, Lea & Febiger, 1985). Tissue classes are included when appropriate and abbreviations are used only when necessary. The word "bone" is used only with the carpal, tarsal, and skull bones. The word muscle is attached to all labeled muscles. The abbreviations used include: M., muscle; N., nerve; Lig., ligament; Trans., transverse; Post., posterior; and Ant., anterior. The twelve-image array permits anatomical structures to be followed easily, millimeter by millimeter, vertically through the body.

Every acquired transverse slice is included in this image collection. The only missing images are the three saw kerfs. An image relating to each saw kerf has been inserted in the image collection at these locations. The magnification used for the labeled images changes multiple times to accomidate the labels and to maximize the image size. The twelve images on the right-hand pages, change scale only once—at the base of the neck. Another image modification, in the labeled diagrams, occurs in the lower extremities where space has been removed between the legs and feet to bring them closer together. The original leg spacing is maintained for the twelve-image array pages.

1

anterior

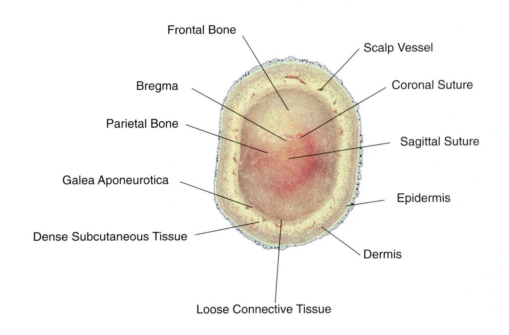

Frontal Bone

Scalp Vessel

Bregma

Coronal Suture

Parietal Bone

Sagittal Suture

Galea Aponeurotica

Epidermis

Dense Subcutaneous Tissue

Dermis

Loose Connective Tissue

right

left

posterior

a_vm1001

a_vm1002

a_vm1003

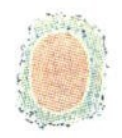

a_vm1004

a_vm1005

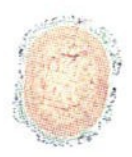

a_vm1006

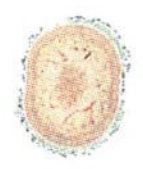

a_vm1007

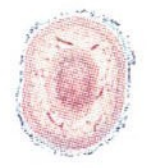

a_vm1008

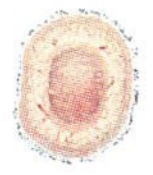

a_vm1009

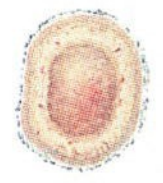

a_vm1010

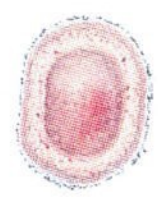

a_vm1011

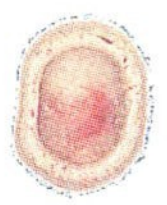

a_vm1012

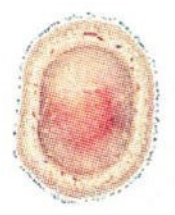

anterior

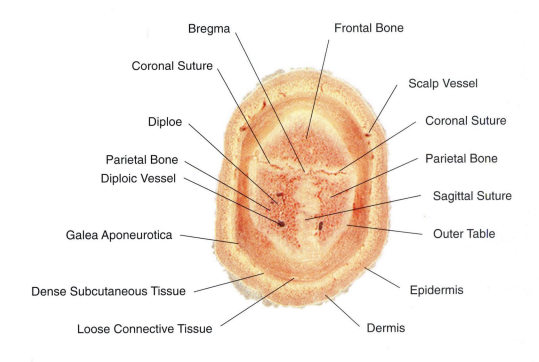

Bregma

Frontal Bone

Coronal Suture

Scalp Vessel

Coronal Suture

Diploe

Parietal Bone

Parietal Bone

Diploic Vessel

Sagittal Suture

Galea Aponeurotica

Outer Table

Dense Subcutaneous Tissue

Epidermis

Loose Connective Tissue

Dermis

right

left

posterior

4

a_vm1013	a_vm1014	a_vm1015
a_vm1016	a_vm1017	**a_vm1018**
a_vm1019	a_vm1020	a_vm1021
a_vm1022	a_vm1023	a_vm1024

anterior

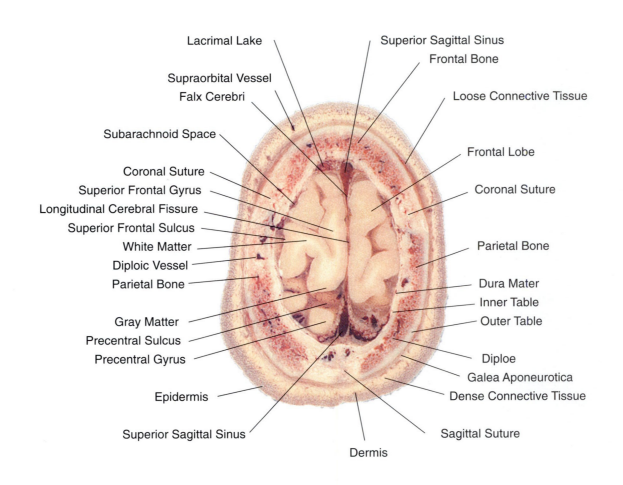

Lacrimal Lake

Supraorbital Vessel

Falx Cerebri

Subarachnoid Space

Coronal Suture

Superior Frontal Gyrus

Longitudinal Cerebral Fissure

Superior Frontal Sulcus

White Matter

Diploic Vessel

Parietal Bone

Gray Matter

Precentral Sulcus

Precentral Gyrus

Epidermis

Superior Sagittal Sinus

Dermis

Superior Sagittal Sinus

Frontal Bone

Loose Connective Tissue

Frontal Lobe

Coronal Suture

Parietal Bone

Dura Mater

Inner Table

Outer Table

Diploe

Galea Aponeurotica

Dense Connective Tissue

Sagittal Suture

right

left

posterior

a_vm1025

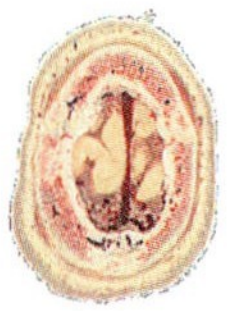

a_vm1026

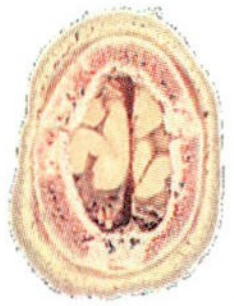

a_vm1027

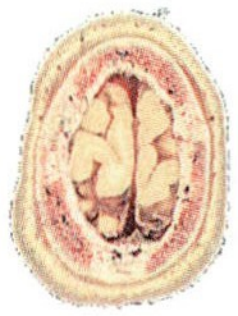

a_vm1028

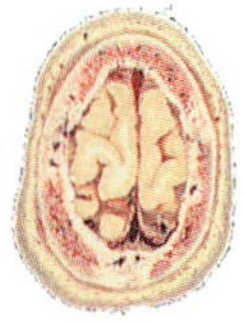

a_vm1029

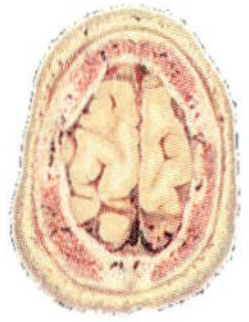

a_vm1030

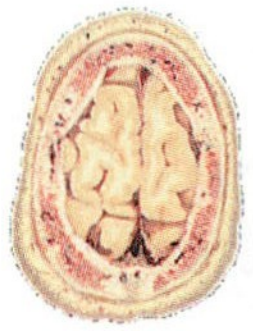

a_vm1031

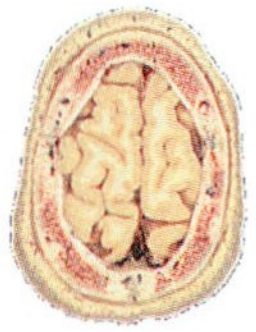

a_vm1032

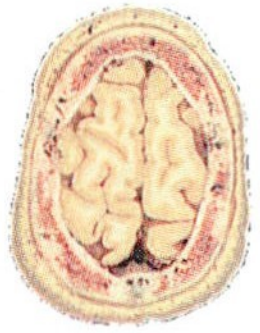

a_vm1033

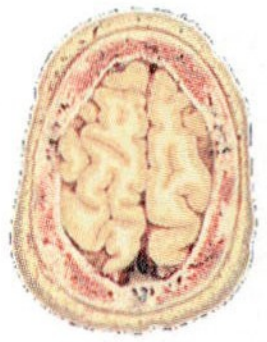

a_vm1034

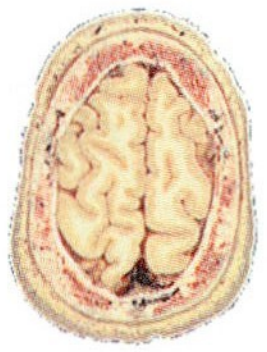

a_vm1035

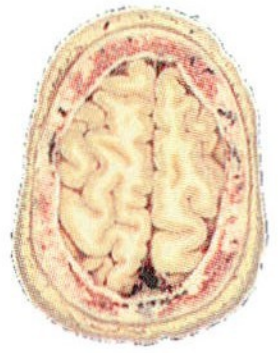

a_vm1036

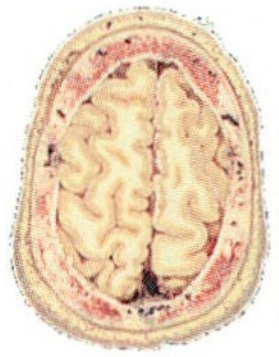

anterior

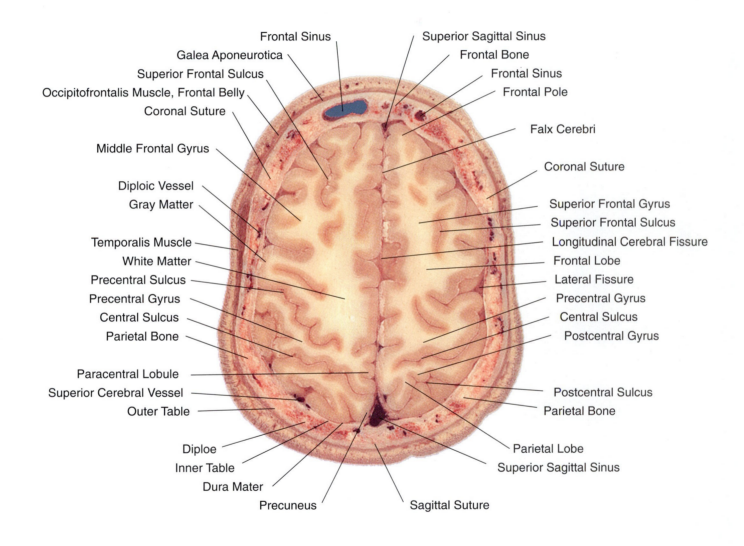

Frontal Sinus

Galea Aponeurotica

Superior Frontal Sulcus

Occipitofrontalis Muscle, Frontal Belly

Coronal Suture

Middle Frontal Gyrus

Diploic Vessel

Gray Matter

Temporalis Muscle

White Matter

Precentral Sulcus

Precentral Gyrus

Central Sulcus

Parietal Bone

Paracentral Lobule

Superior Cerebral Vessel

Outer Table

Diploe

Inner Table

Dura Mater

Precuneus

Superior Sagittal Sinus

Frontal Bone

Frontal Sinus

Frontal Pole

Falx Cerebri

Coronal Suture

Superior Frontal Gyrus

Superior Frontal Sulcus

Longitudinal Cerebral Fissure

Frontal Lobe

Lateral Fissure

Precentral Gyrus

Central Sulcus

Postcentral Gyrus

Postcentral Sulcus

Parietal Bone

Parietal Lobe

Superior Sagittal Sinus

Sagittal Suture

right

left

posterior

a_vm1037

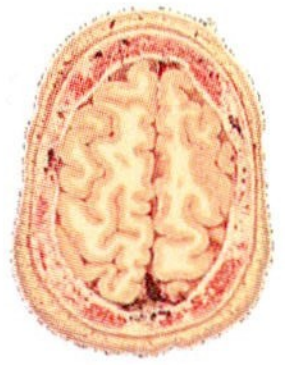

a_vm1038

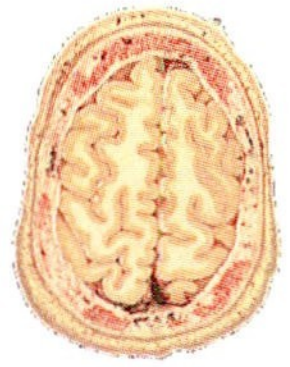

a_vm1039

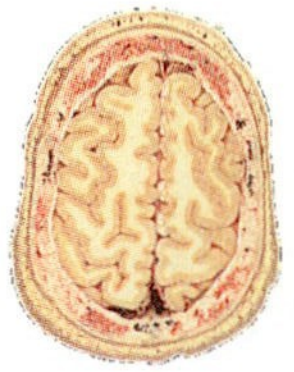

a_vm1040

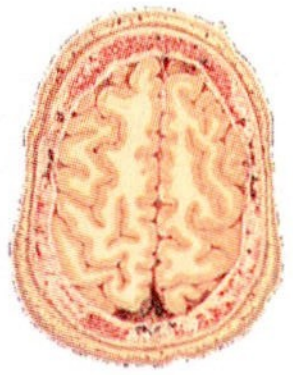

a_vm1041

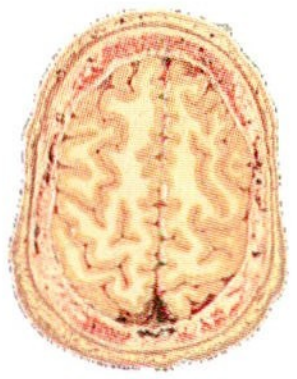

a_vm1042

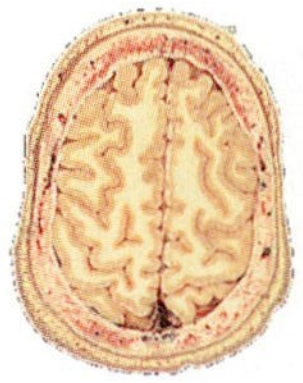

a_vm1043

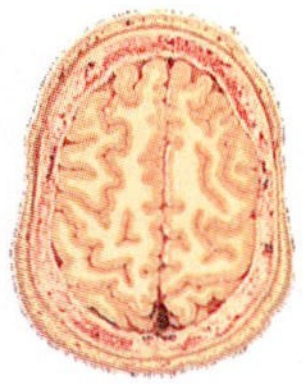

a_vm1044

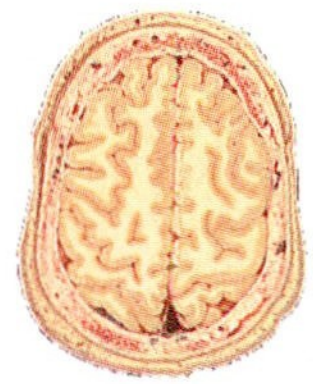

a_vm1045

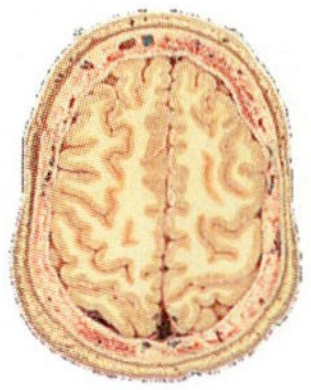

a_vm1046

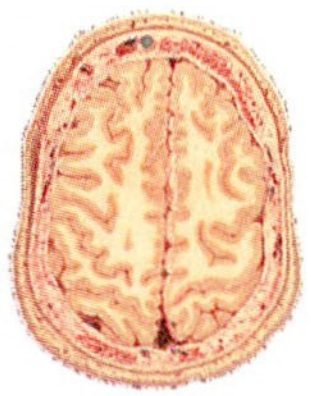

a_vm1047

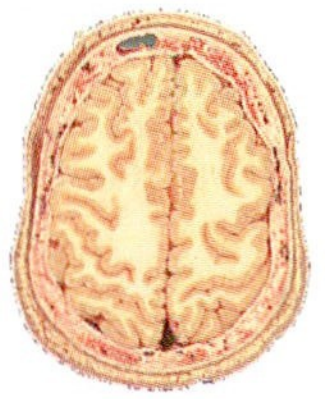

a_vm1048

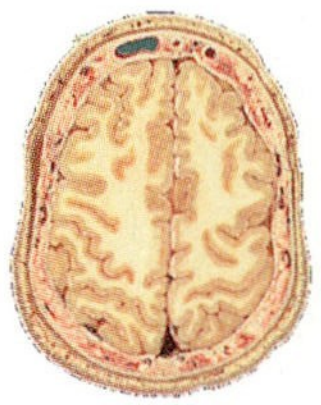

anterior

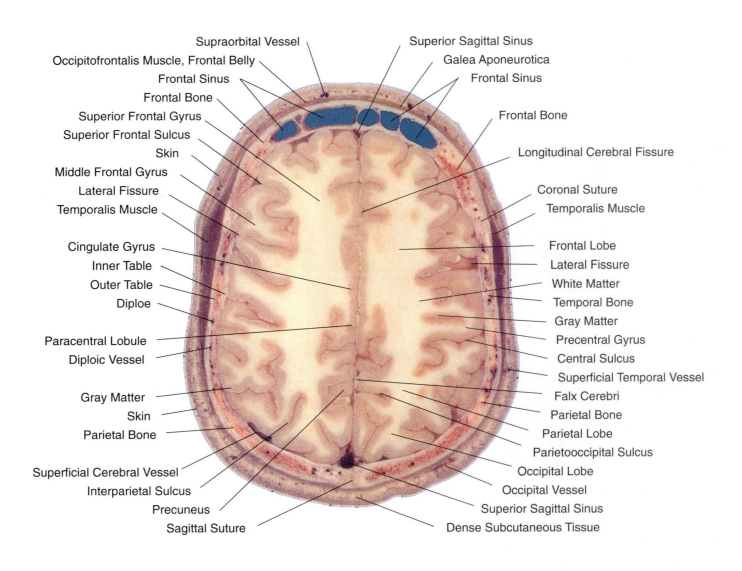

Supraorbital Vessel

Superior Sagittal Sinus

Occipitofrontalis Muscle, Frontal Belly

Galea Aponeurotica

Frontal Sinus

Frontal Sinus

Frontal Bone

Superior Frontal Gyrus

Frontal Bone

Superior Frontal Sulcus

Skin

Longitudinal Cerebral Fissure

Middle Frontal Gyrus

Lateral Fissure

Coronal Suture

Temporalis Muscle

Temporalis Muscle

Cingulate Gyrus

Frontal Lobe

Inner Table

Lateral Fissure

Outer Table

White Matter

Diploe

Temporal Bone

Gray Matter

Paracentral Lobule

Precentral Gyrus

Diploic Vessel

Central Sulcus

Superficial Temporal Vessel

Gray Matter

Falx Cerebri

Skin

Parietal Bone

Parietal Bone

Parietal Lobe

Parietooccipital Sulcus

Superficial Cerebral Vessel

Occipital Lobe

Interparietal Sulcus

Occipital Vessel

Precuneus

Superior Sagittal Sinus

Sagittal Suture

Dense Subcutaneous Tissue

posterior

right

left

a_vm1049

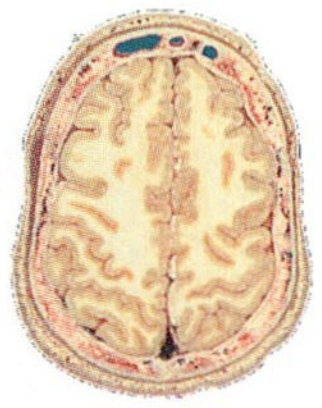

a_vm1050

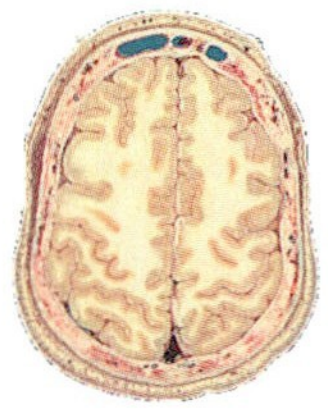

a_vm1051

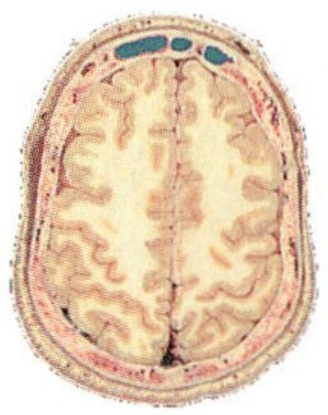

a_vm1052

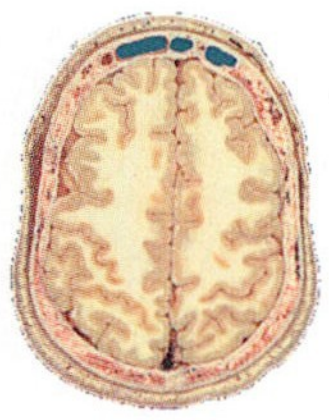

a_vm1053

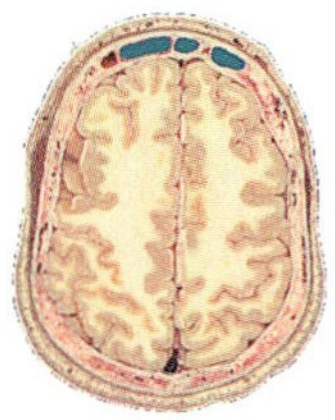

a_vm1054

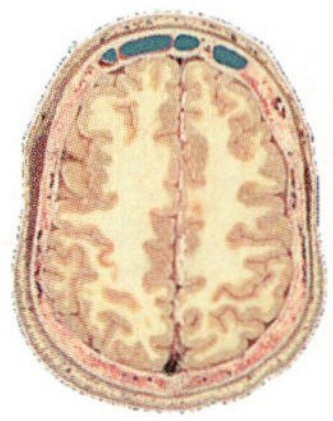

a_vm1055

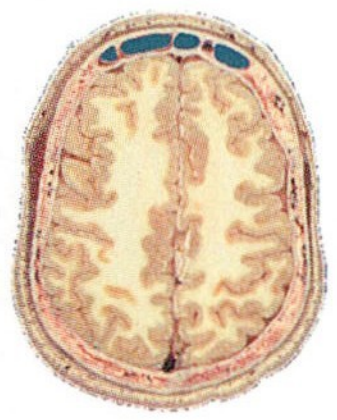

a_vm1056

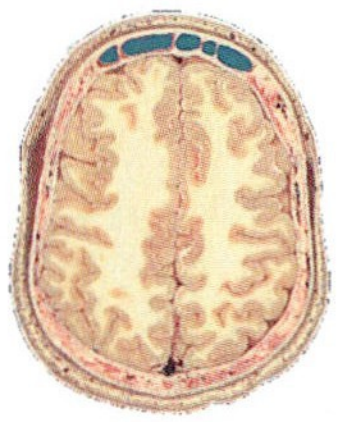

a_vm1057

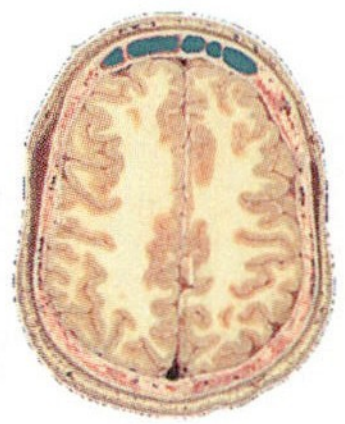

a_vm1058

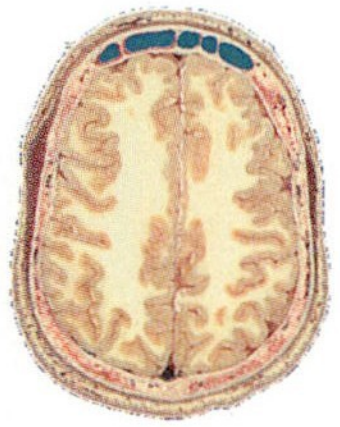

a_vm1059

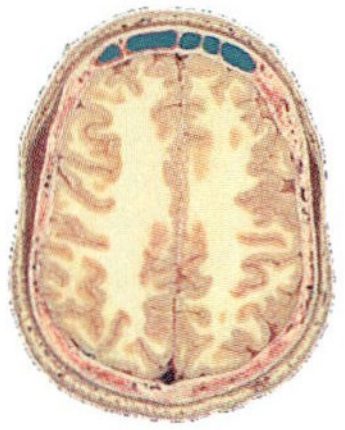

a_vm1060

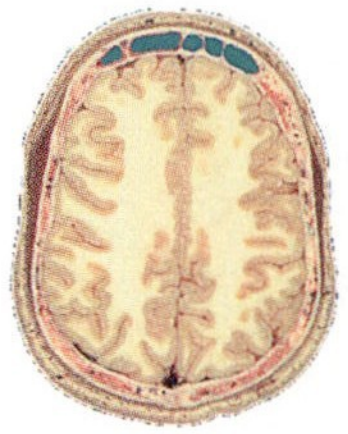

Transverse
a_vm1072

anterior

Supraorbital Vessel | Frontal Sinus
Medial Frontal Gyrus
Occipitofrontalis Muscle, Frontal Belly | Superior Sagittal Sinus
Superior Frontal Gyrus | Cingulate Sulcus
Middle Frontal Gyrus | Cingulate Gyrus
Frontal Bone
Inferior Frontal Sulcus | Frontal Lobe
Lateral Fissure | Inferior Frontal Sulcus
Temporalis Muscle | Lateral Ventricle
Inferior Frontal Gyrus | Inferior Frontal Gyrus
Superficial Temporal Vessel | Temporalis Muscle
Diploic Vessel | Coronal Suture
Outer Table | Caudate Nucleus
Diploe | Corona Radiata
Inner Table | Corpus Callosum

Corona Radiata | White Matter
Lateral Ventricle | Cingulate Gyrus
Dura Mater | Gray Matter
Skin | Parietal Bone
Longitudinal Cerebral Fissure | Precuneus
Parietal Lobe
Occipitofrontalis Muscle, Occipital Belly | Parietal Lobe
Parietooccipital Sulcus
Falx Cerebri | Occipital Vessel
Lateral Occipital Gyrus | Occipital Lobe
Superior Sagittal Sinus | Sagittal Suture

right   **left**

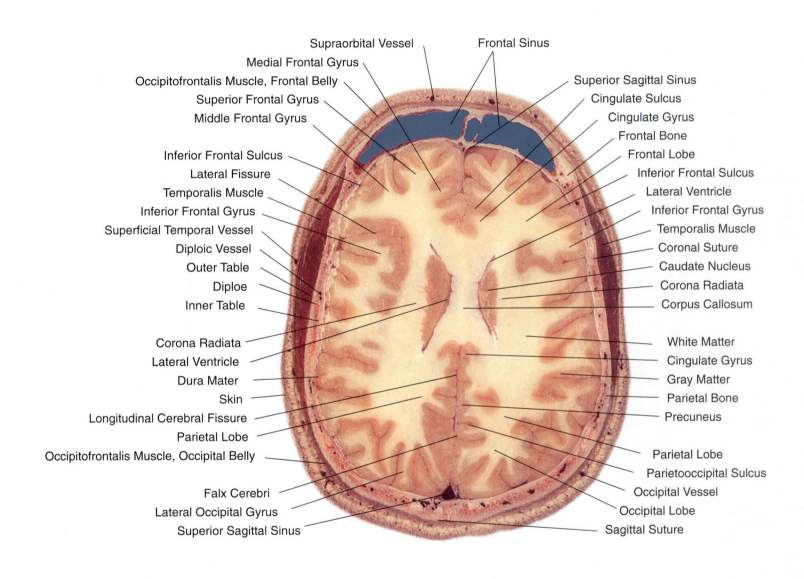

posterior

12

a_vm1061

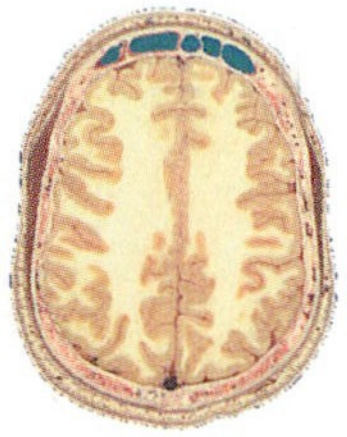

a_vm1062

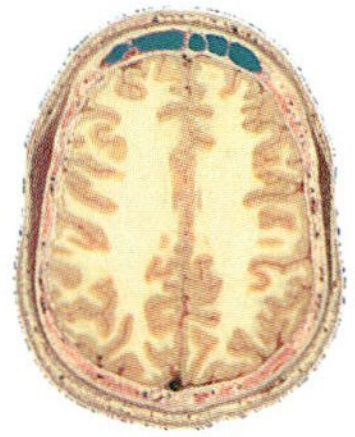

a_vm1063

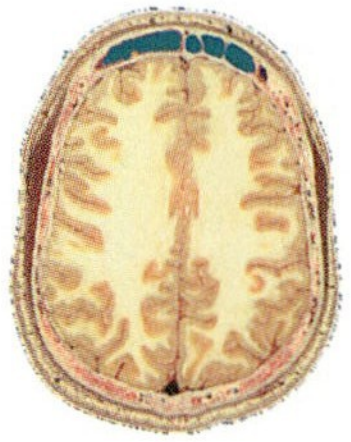

a_vm1064

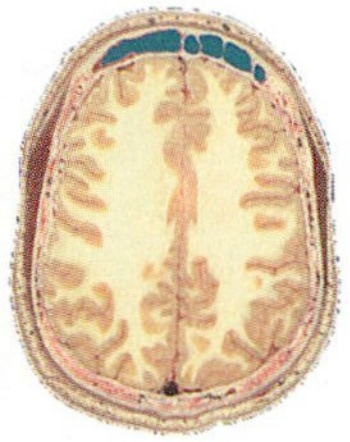

a_vm1065

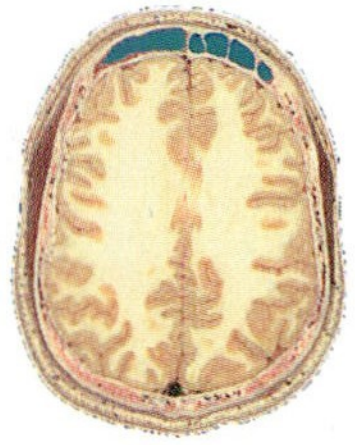

a_vm1066

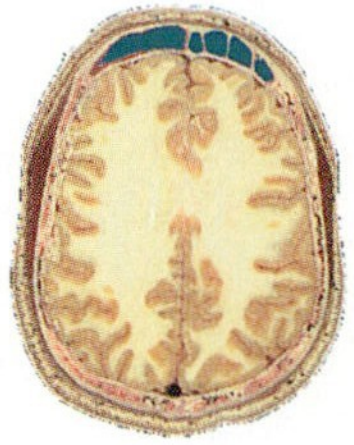

a_vm1067

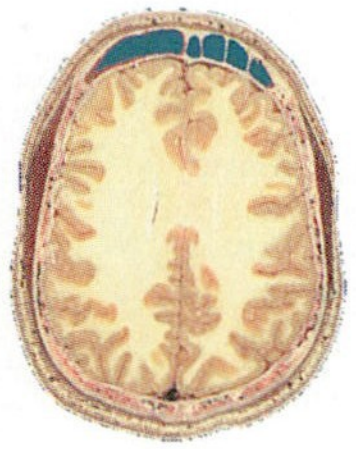

a_vm1068

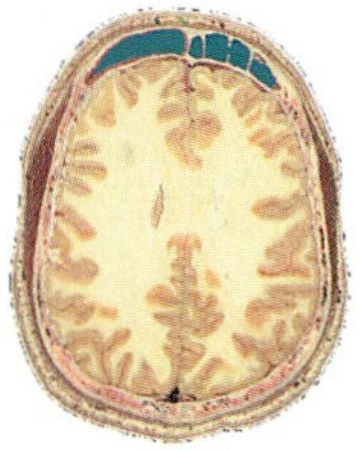

a_vm1069

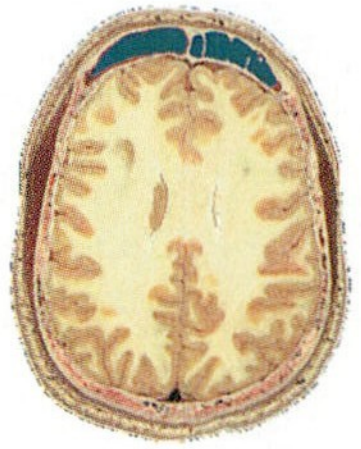

a_vm1070

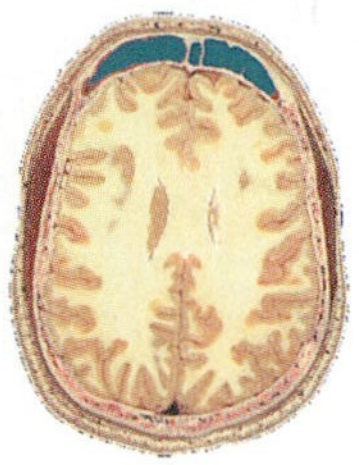

a_vm1071

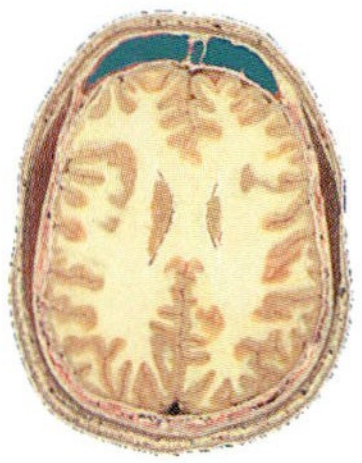

a_vm1072

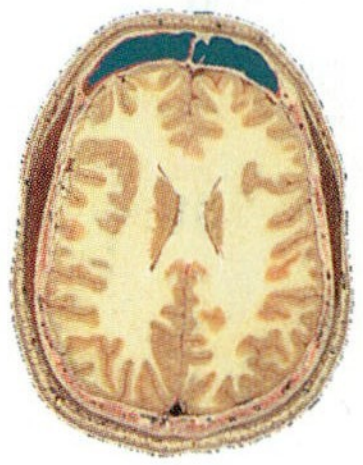

anterior

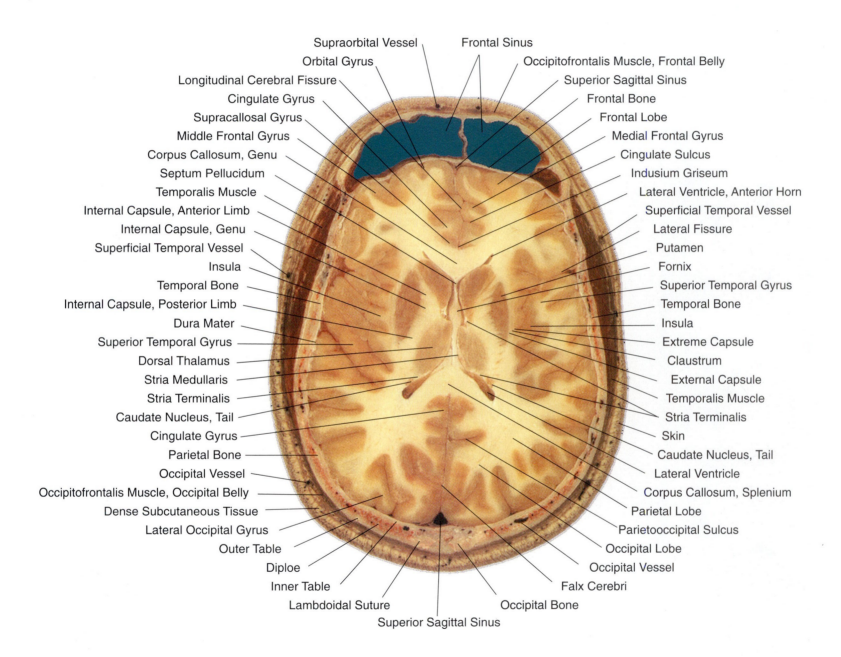

Supraorbital Vessel

Frontal Sinus

Orbital Gyrus

Occipitofrontalis Muscle, Frontal Belly

Longitudinal Cerebral Fissure

Superior Sagittal Sinus

Cingulate Gyrus

Frontal Bone

Supracallosal Gyrus

Frontal Lobe

Middle Frontal Gyrus

Medial Frontal Gyrus

Corpus Callosum, Genu

Cingulate Sulcus

Septum Pellucidum

Indusium Griseum

Temporalis Muscle

Lateral Ventricle, Anterior Horn

Internal Capsule, Anterior Limb

Superficial Temporal Vessel

Internal Capsule, Genu

Lateral Fissure

Superficial Temporal Vessel

Putamen

Insula

Fornix

Temporal Bone

Superior Temporal Gyrus

Internal Capsule, Posterior Limb

Temporal Bone

Dura Mater

Insula

Superior Temporal Gyrus

Extreme Capsule

Dorsal Thalamus

Claustrum

Stria Medullaris

External Capsule

Stria Terminalis

Temporalis Muscle

Caudate Nucleus, Tail

Stria Terminalis

Cingulate Gyrus

Skin

Parietal Bone

Caudate Nucleus, Tail

Occipital Vessel

Lateral Ventricle

Occipitofrontalis Muscle, Occipital Belly

Corpus Callosum, Splenium

Dense Subcutaneous Tissue

Parietal Lobe

Lateral Occipital Gyrus

Parietooccipital Sulcus

Outer Table

Occipital Lobe

Diploe

Occipital Vessel

Inner Table

Falx Cerebri

Lambdoidal Suture

Occipital Bone

Superior Sagittal Sinus

right

left

posterior

a_vm1073

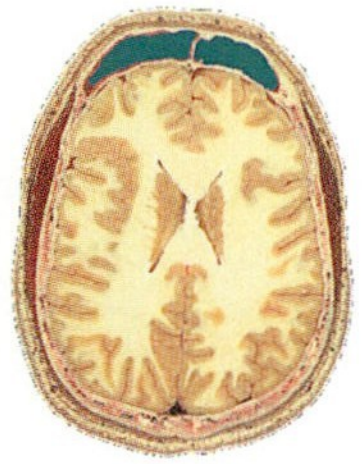

a_vm1074

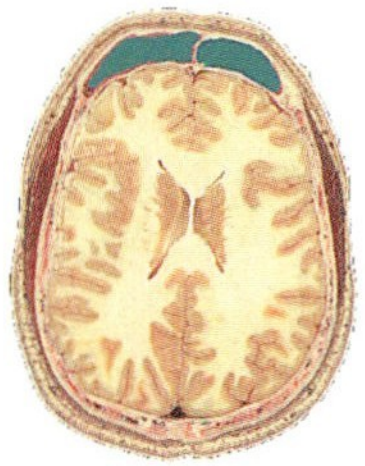

a_vm1075

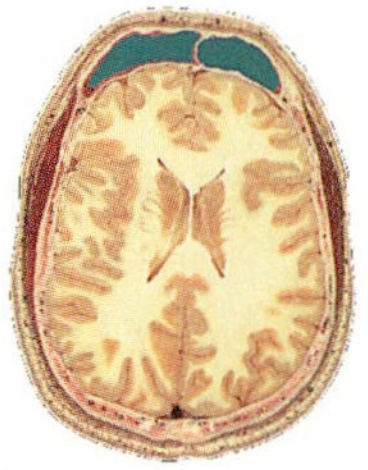

a_vm1076

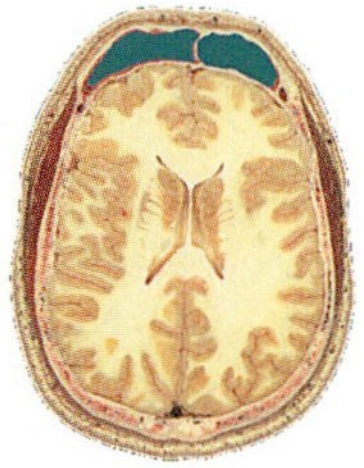

a_vm1077

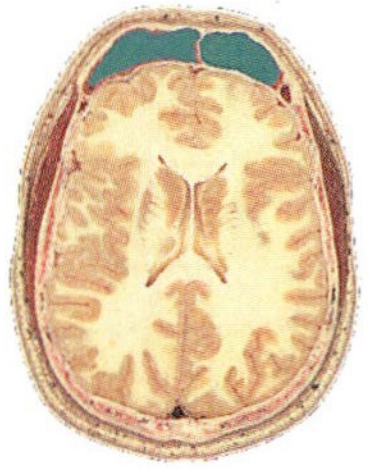

a_vm1078

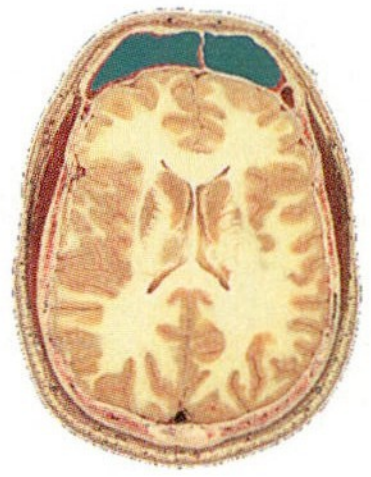

a_vm1079

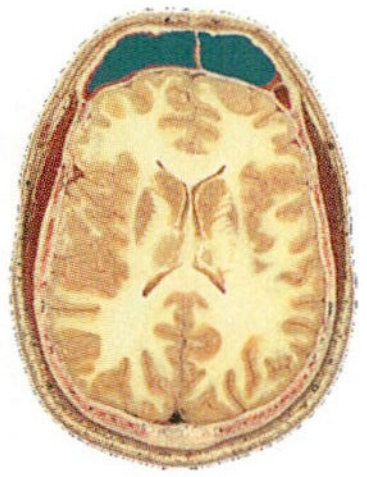

a_vm1080

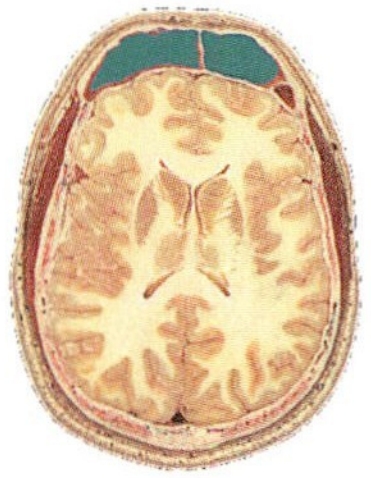

a_vm1081

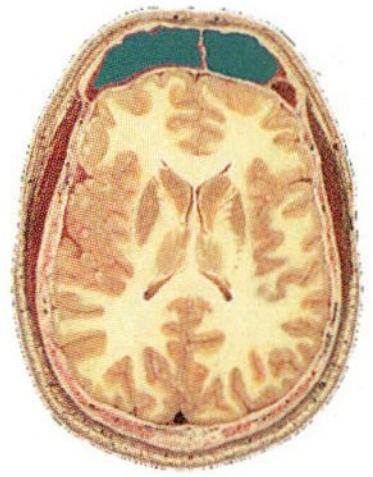

a_vm1082

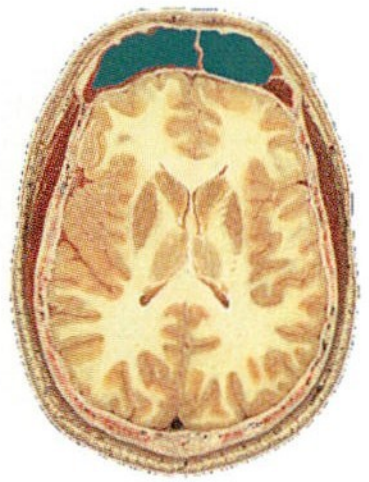

a_vm1083

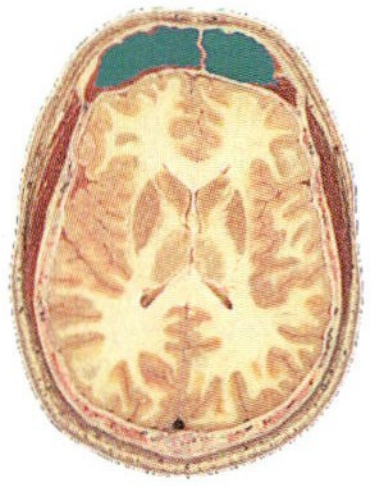

a_vm1084

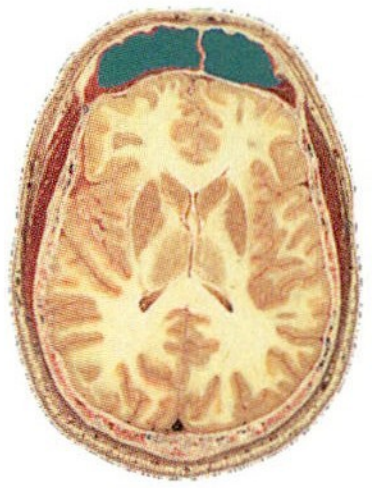

anterior

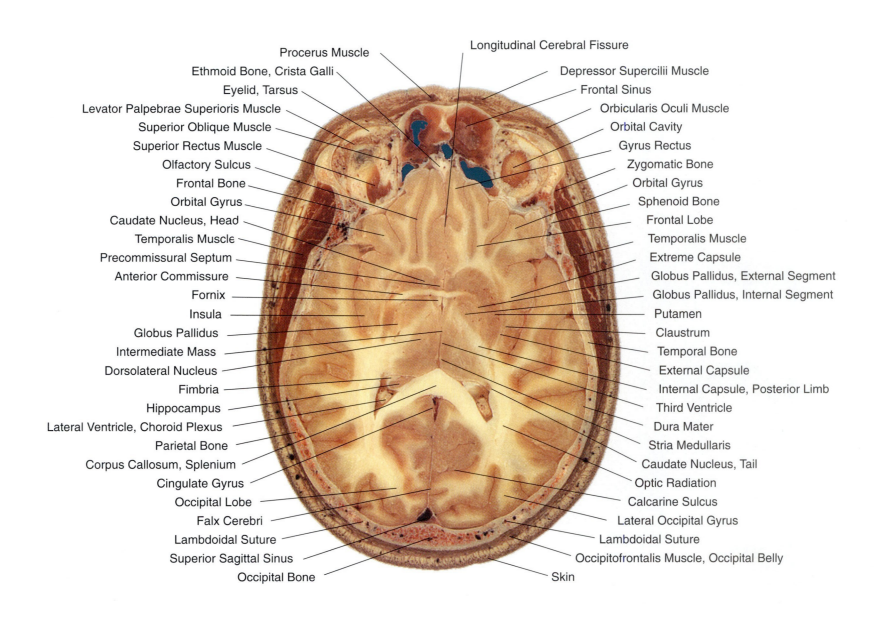

Procerus Muscle

Ethmoid Bone, Crista Galli

Eyelid, Tarsus

Levator Palpebrae Superioris Muscle

Superior Oblique Muscle

Superior Rectus Muscle

Olfactory Sulcus

Frontal Bone

Orbital Gyrus

Caudate Nucleus, Head

Temporalis Muscle

Precommissural Septum

Anterior Commissure

Fornix

Insula

Globus Pallidus

Intermediate Mass

Dorsolateral Nucleus

Fimbria

Hippocampus

Lateral Ventricle, Choroid Plexus

Parietal Bone

Corpus Callosum, Splenium

Cingulate Gyrus

Occipital Lobe

Falx Cerebri

Lambdoidal Suture

Superior Sagittal Sinus

Occipital Bone

Longitudinal Cerebral Fissure

Depressor Supercilii Muscle

Frontal Sinus

Orbicularis Oculi Muscle

Orbital Cavity

Gyrus Rectus

Zygomatic Bone

Orbital Gyrus

Sphenoid Bone

Frontal Lobe

Temporalis Muscle

Extreme Capsule

Globus Pallidus, External Segment

Globus Pallidus, Internal Segment

Putamen

Claustrum

Temporal Bone

External Capsule

Internal Capsule, Posterior Limb

Third Ventricle

Dura Mater

Stria Medullaris

Caudate Nucleus, Tail

Optic Radiation

Calcarine Sulcus

Lateral Occipital Gyrus

Lambdoidal Suture

Occipitofrontalis Muscle, Occipital Belly

Skin

right

left

posterior

16

a_vm1085

a_vm1086

a_vm1087

a_vm1088

a_vm1089

a_vm1090

a_vm1091

a_vm1092

a_vm1093

a_vm1094

a_vm1095

a_vm1096

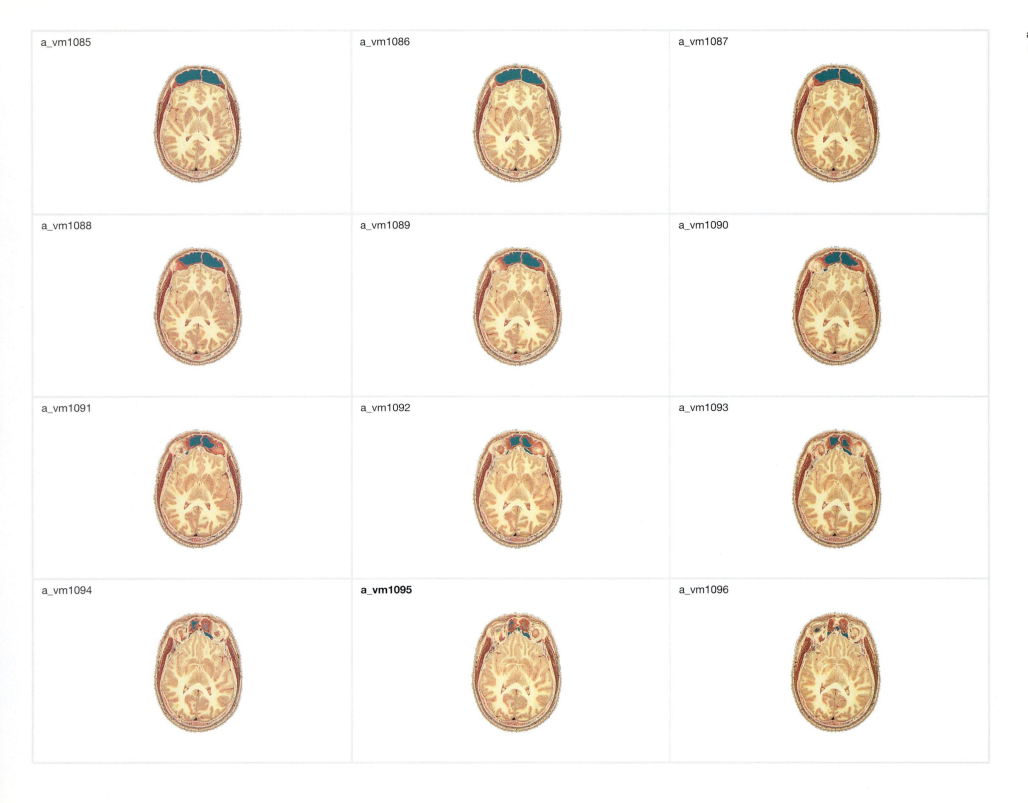

anterior

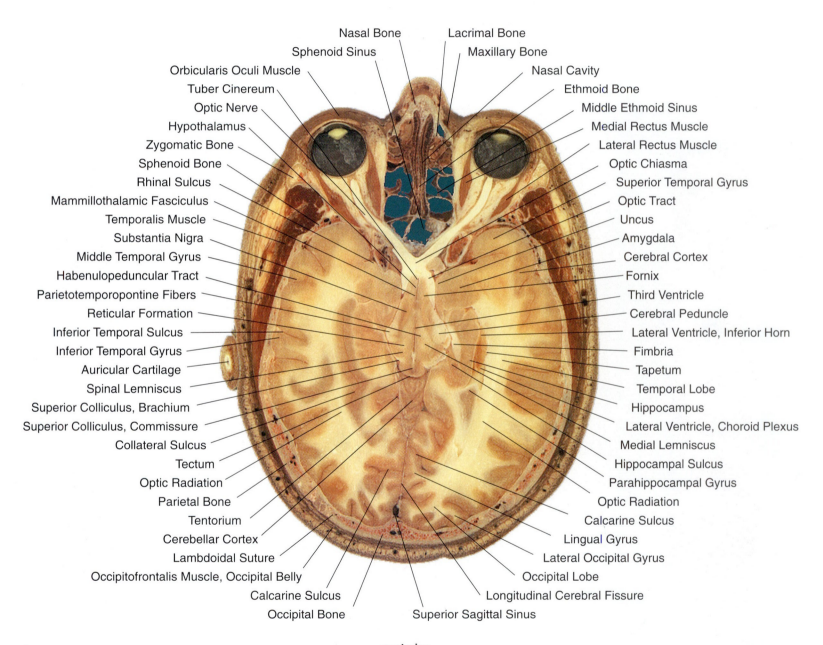

Nasal Bone
Lacrimal Bone
Sphenoid Sinus
Maxillary Bone
Orbicularis Oculi Muscle
Nasal Cavity
Tuber Cinereum
Ethmoid Bone
Optic Nerve
Middle Ethmoid Sinus
Hypothalamus
Medial Rectus Muscle
Zygomatic Bone
Lateral Rectus Muscle
Sphenoid Bone
Optic Chiasma
Rhinal Sulcus
Superior Temporal Gyrus
Mammillothalamic Fasciculus
Optic Tract
Temporalis Muscle
Uncus
Substantia Nigra
Amygdala
Middle Temporal Gyrus
Cerebral Cortex
Habenulopeduncular Tract
Fornix
Parietotemporopontine Fibers
Third Ventricle
Reticular Formation
Cerebral Peduncle
Inferior Temporal Sulcus
Lateral Ventricle, Inferior Horn
Inferior Temporal Gyrus
Fimbria
Auricular Cartilage
Tapetum
Spinal Lemniscus
Temporal Lobe
Superior Colliculus, Brachium
Hippocampus
Superior Colliculus, Commissure
Lateral Ventricle, Choroid Plexus
Collateral Sulcus
Medial Lemniscus
Tectum
Hippocampal Sulcus
Optic Radiation
Parahippocampal Gyrus
Parietal Bone
Optic Radiation
Tentorium
Calcarine Sulcus
Cerebellar Cortex
Lingual Gyrus
Lambdoidal Suture
Lateral Occipital Gyrus
Occipitofrontalis Muscle, Occipital Belly
Occipital Lobe
Calcarine Sulcus
Longitudinal Cerebral Fissure
Occipital Bone
Superior Sagittal Sinus

right

left

posterior

18

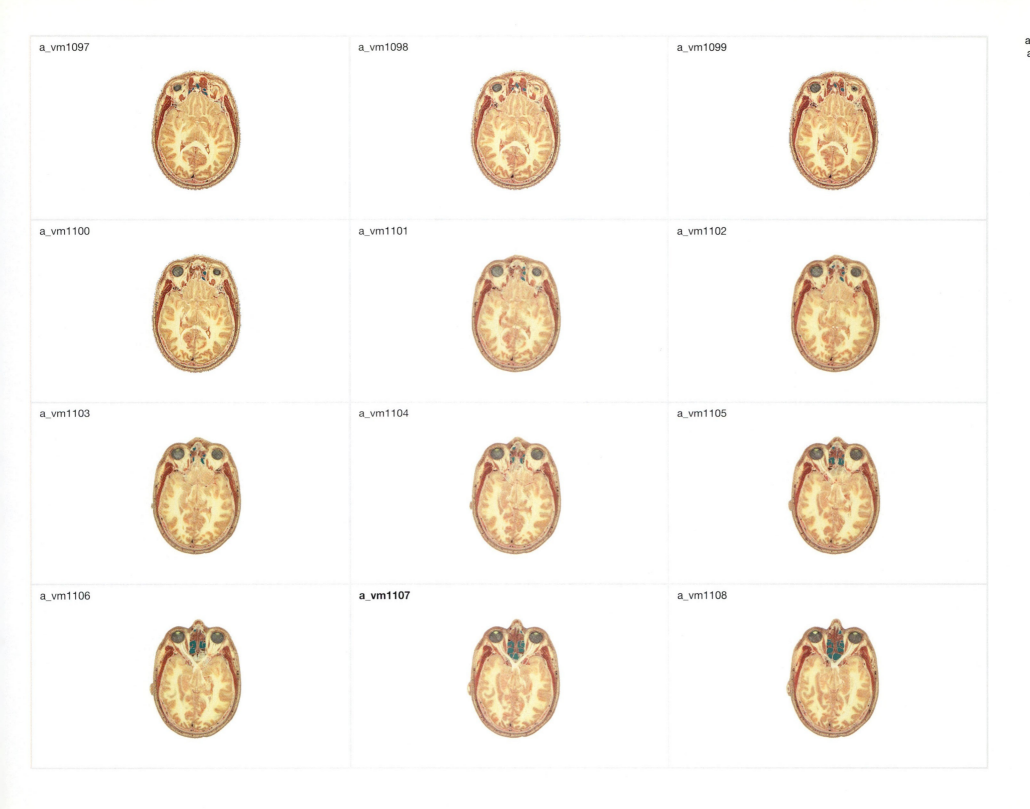

a_vm1097

a_vm1098

a_vm1099

a_vm1100

a_vm1101

a_vm1102

a_vm1103

a_vm1104

a_vm1105

a_vm1106

a_vm1107

a_vm1108

19

anterior

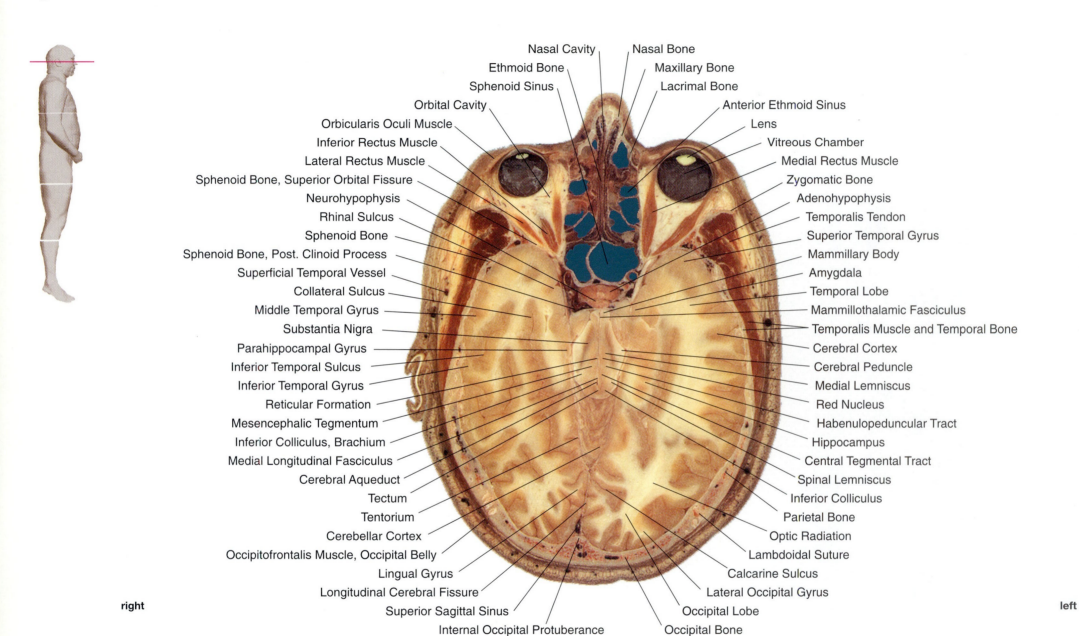

Nasal Cavity

Nasal Bone

Ethmoid Bone

Maxillary Bone

Sphenoid Sinus

Lacrimal Bone

Orbital Cavity

Anterior Ethmoid Sinus

Orbicularis Oculi Muscle

Lens

Inferior Rectus Muscle

Vitreous Chamber

Lateral Rectus Muscle

Medial Rectus Muscle

Sphenoid Bone, Superior Orbital Fissure

Zygomatic Bone

Neurohypophysis

Adenohypophysis

Rhinal Sulcus

Temporalis Tendon

Sphenoid Bone

Superior Temporal Gyrus

Sphenoid Bone, Post. Clinoid Process

Mammillary Body

Superficial Temporal Vessel

Amygdala

Collateral Sulcus

Temporal Lobe

Middle Temporal Gyrus

Mammillothalamic Fasciculus

Substantia Nigra

Temporalis Muscle and Temporal Bone

Parahippocampal Gyrus

Cerebral Cortex

Inferior Temporal Sulcus

Cerebral Peduncle

Inferior Temporal Gyrus

Medial Lemniscus

Reticular Formation

Red Nucleus

Mesencephalic Tegmentum

Habenulopeduncular Tract

Inferior Colliculus, Brachium

Hippocampus

Medial Longitudinal Fasciculus

Central Tegmental Tract

Cerebral Aqueduct

Spinal Lemniscus

Tectum

Inferior Colliculus

Tentorium

Parietal Bone

Cerebellar Cortex

Optic Radiation

Occipitofrontalis Muscle, Occipital Belly

Lambdoidal Suture

Lingual Gyrus

Calcarine Sulcus

Longitudinal Cerebral Fissure

Lateral Occipital Gyrus

Superior Sagittal Sinus

Occipital Lobe

Internal Occipital Protuberance

Occipital Bone

right

left

posterior

a_vm1109

a_vm1110

a_vm1111

a_vm1112

a_vm1113

a_vm1114

a_vm1115

a_vm1116

a_vm1117

a_vm1118

a_vm1119

a_vm1120

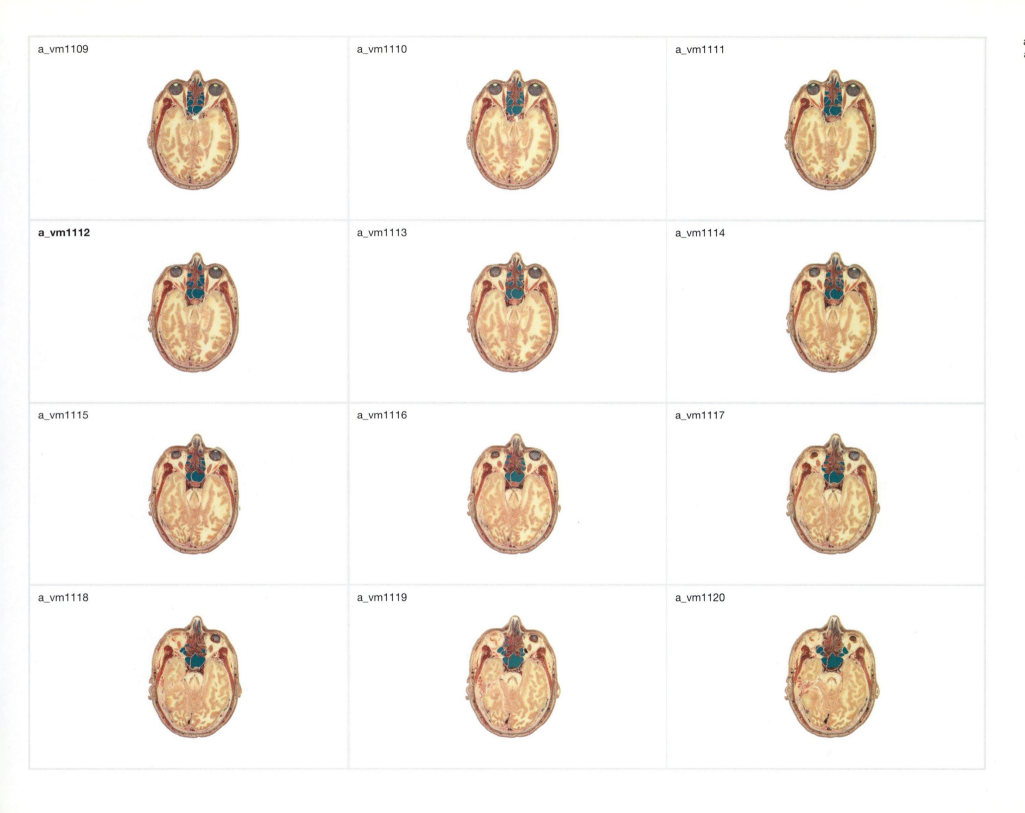

anterior

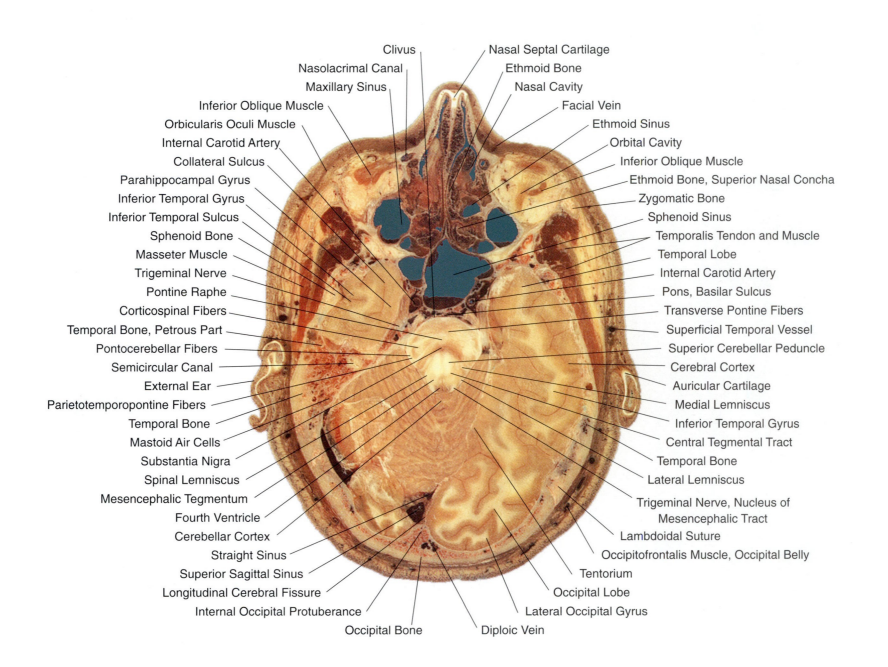

Clivus
Nasolacrimal Canal
Maxillary Sinus
Inferior Oblique Muscle
Orbicularis Oculi Muscle
Internal Carotid Artery
Collateral Sulcus
Parahippocampal Gyrus
Inferior Temporal Gyrus
Inferior Temporal Sulcus
Sphenoid Bone
Masseter Muscle
Trigeminal Nerve
Pontine Raphe
Corticospinal Fibers
Temporal Bone, Petrous Part
Pontocerebellar Fibers
Semicircular Canal
External Ear
Parietotemporopontine Fibers
Temporal Bone
Mastoid Air Cells
Substantia Nigra
Spinal Lemniscus
Mesencephalic Tegmentum
Fourth Ventricle
Cerebellar Cortex
Straight Sinus
Superior Sagittal Sinus
Longitudinal Cerebral Fissure
Internal Occipital Protuberance

Nasal Septal Cartilage
Ethmoid Bone
Nasal Cavity
Facial Vein
Ethmoid Sinus
Orbital Cavity
Inferior Oblique Muscle
Ethmoid Bone, Superior Nasal Concha
Zygomatic Bone
Sphenoid Sinus
Temporalis Tendon and Muscle
Temporal Lobe
Internal Carotid Artery
Pons, Basilar Sulcus
Transverse Pontine Fibers
Superficial Temporal Vessel
Superior Cerebellar Peduncle
Cerebral Cortex
Auricular Cartilage
Medial Lemniscus
Inferior Temporal Gyrus
Central Tegmental Tract
Temporal Bone
Lateral Lemniscus
Trigeminal Nerve, Nucleus of Mesencephalic Tract
Lambdoidal Suture
Occipitofrontalis Muscle, Occipital Belly
Tentorium
Occipital Lobe
Lateral Occipital Gyrus

Occipital Bone
Diploic Vein

right

left

posterior

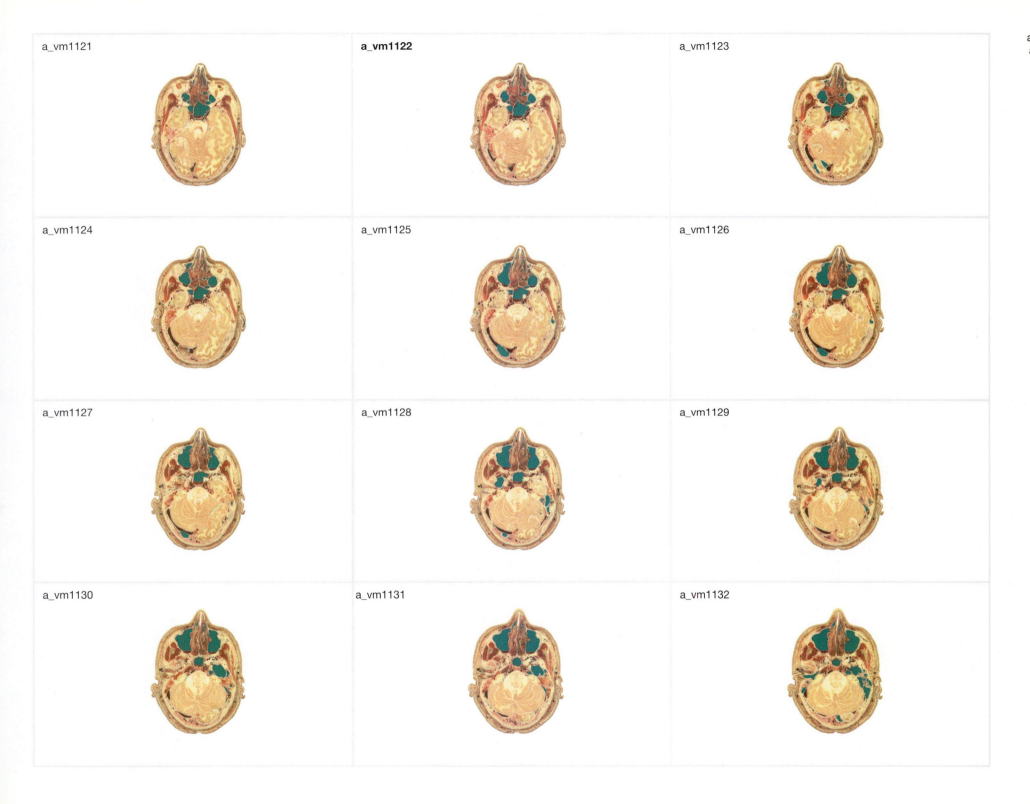

a_vm1121

a_vm1122

a_vm1123

a_vm1124

a_vm1125

a_vm1126

a_vm1127

a_vm1128

a_vm1129

a_vm1130

a_vm1131

a_vm1132

anterior

Nasal Septal Cartilage

Nasal Cavity

Nasolacrimal Canal

Pontine Nuclei

Sphenoid Bone

Corticospinal Fibers

Temporalis Muscle

Zygomaticus Major Muscle

Masseter Muscle

Pontocerebellar Fibers

Temporal Bone, Auditory Tube

Temporal Bone, Tubercle

Temporomandibular Joint

Mandible, Condyle

Sphenoid Bone, Foramen Spinosum

Superficial Temporal Vessel

Carotid Canal

External Acoustic Meatus

Middle Cerebellar Peduncle

Temporal Bone

Locus Ceruleus

External Ear

Mastoid Air Cells

Transverse Sinus

Superior Cerebellar Peduncle

Cerebellum, Fastigial Nucleus

Cerebellum, Emboliform Nucleus

Cerebellar Cortex

Lateral Cerebellar Hemisphere

Diploic Vessel

Semispinalis Capitis Muscle

Loose Connective Tissue

Occipital Bone, Outer Table

Cerebellum, Vermis

Ethmoid Bone, Middle Nasal Concha

Maxillary Bone

Levator Labii Superioris Muscle

Occipital Bone, Basal Portion

Maxillary Sinus

Orbicularis Oculi Muscle

Pons, Basilar Sulcus

Lateral Pterygoid Muscle

Mandible, Coronoid Process

Internal Carotid Artery

Pontine Raphe

Transverse Pontine Fibers

Masseter Muscle

Retromandibular Vein

Medial Lemniscus

Superficial Temporal Vein

Tympanic Cavity

Cochlea

Auricular Cartilage

Temporal Bone

Lateral Lemniscus

Pons, Tegmentum

Central Tegmental Tract

Mastoid Air Cells

Spinal Lemniscus

Medial Longitudinal Fasciculus

Lambdoidal Suture

Fourth Ventricle

Cerebellum, Globose Nucleus

Artifact

Internal Occipital Protuberance

Dense Connective Tissue

Occipital Bone

External Occipital Protuberance

right

left

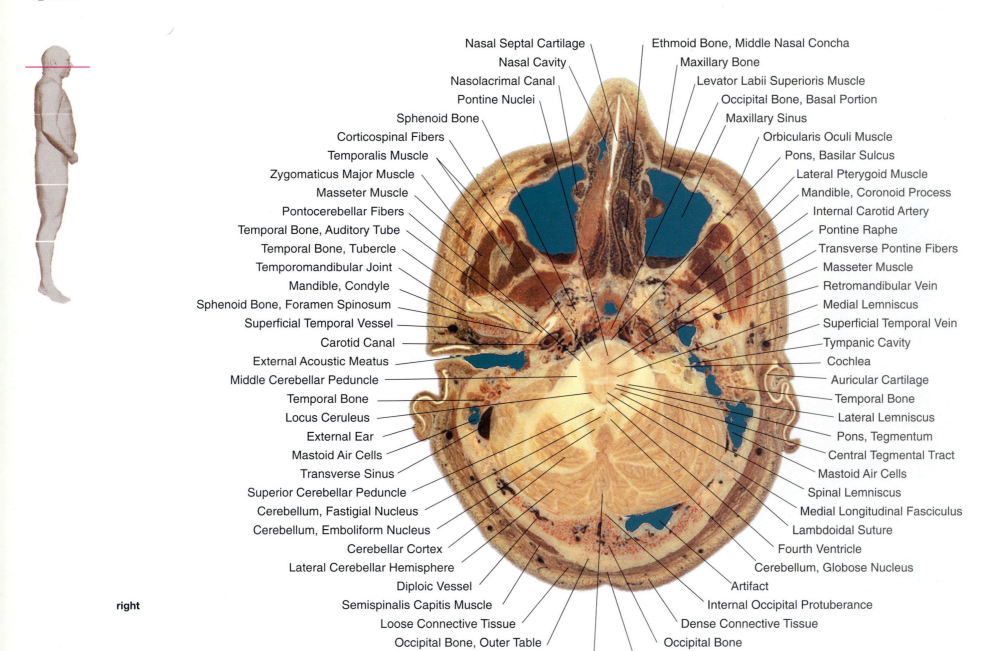

posterior

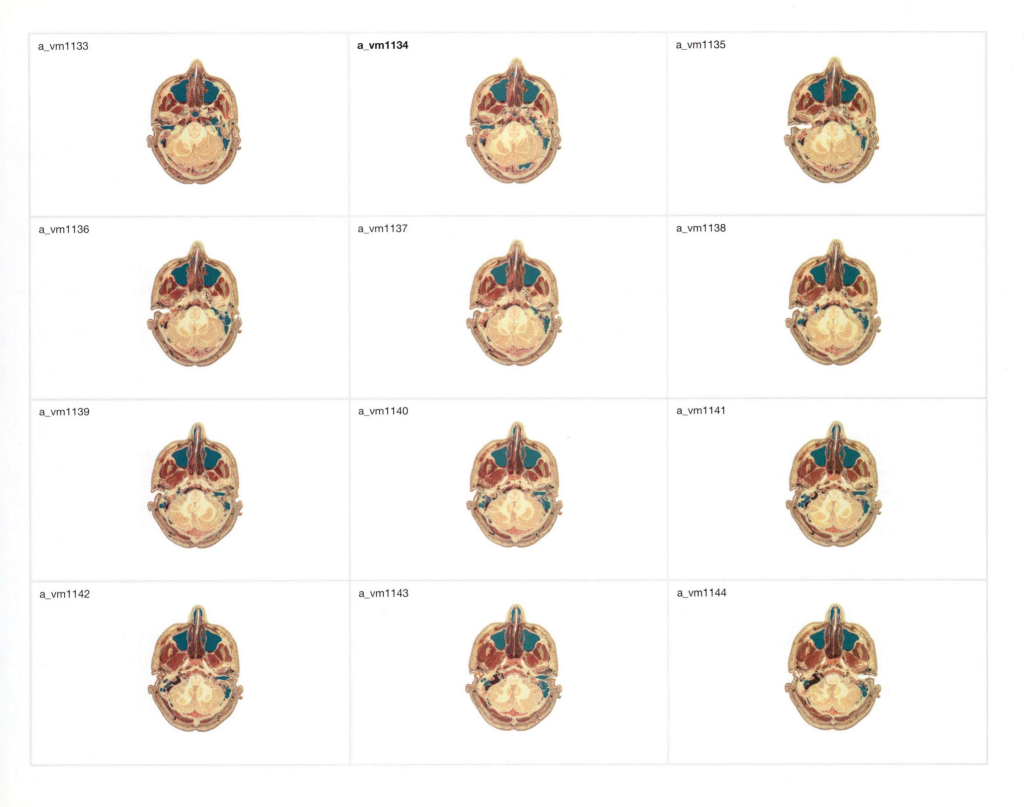

a_vm1133

a_vm1134

a_vm1135

a_vm1136

a_vm1137

a_vm1138

a_vm1139

a_vm1140

a_vm1141

a_vm1142

a_vm1143

a_vm1144

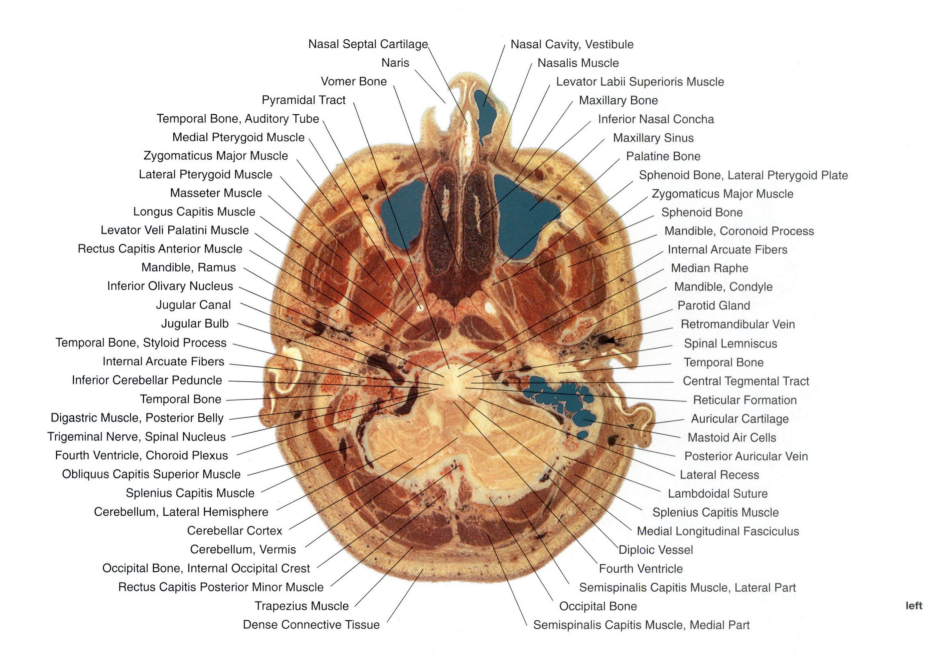

anterior

Nasal Septal Cartilage
Naris
Vomer Bone
Pyramidal Tract
Temporal Bone, Auditory Tube
Medial Pterygoid Muscle
Zygomaticus Major Muscle
Lateral Pterygoid Muscle
Masseter Muscle
Longus Capitis Muscle
Levator Veli Palatini Muscle
Rectus Capitis Anterior Muscle
Mandible, Ramus
Inferior Olivary Nucleus
Jugular Canal
Jugular Bulb
Temporal Bone, Styloid Process
Internal Arcuate Fibers
Inferior Cerebellar Peduncle
Temporal Bone
Digastric Muscle, Posterior Belly
Trigeminal Nerve, Spinal Nucleus
Fourth Ventricle, Choroid Plexus
Obliquus Capitis Superior Muscle
Splenius Capitis Muscle
Cerebellum, Lateral Hemisphere
Cerebellar Cortex
Cerebellum, Vermis
Occipital Bone, Internal Occipital Crest
Rectus Capitis Posterior Minor Muscle
Trapezius Muscle
Dense Connective Tissue

Nasal Cavity, Vestibule
Nasalis Muscle
Levator Labii Superioris Muscle
Maxillary Bone
Inferior Nasal Concha
Maxillary Sinus
Palatine Bone
Sphenoid Bone, Lateral Pterygoid Plate
Zygomaticus Major Muscle
Sphenoid Bone
Mandible, Coronoid Process
Internal Arcuate Fibers
Median Raphe
Mandible, Condyle
Parotid Gland
Retromandibular Vein
Spinal Lemniscus
Temporal Bone
Central Tegmental Tract
Reticular Formation
Auricular Cartilage
Mastoid Air Cells
Posterior Auricular Vein
Lateral Recess
Lambdoidal Suture
Splenius Capitis Muscle
Medial Longitudinal Fasciculus
Diploic Vessel
Fourth Ventricle
Semispinalis Capitis Muscle, Lateral Part
Occipital Bone
Semispinalis Capitis Muscle, Medial Part

right

left

posterior

26

a_vm1145

a_vm1146

a_vm1147

a_vm1148

a_vm1149

a_vm1150

a_vm1151

a_vm1152

a_vm1153

a_vm1154

a_vm1155

a_vm1156

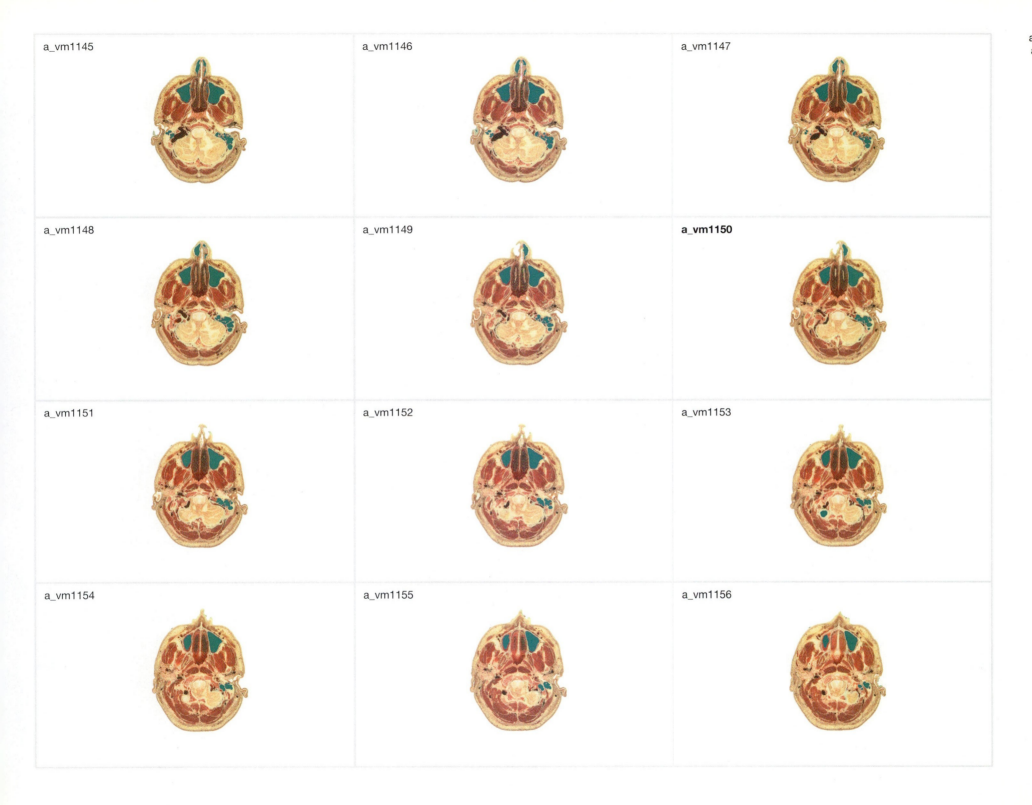

anterior

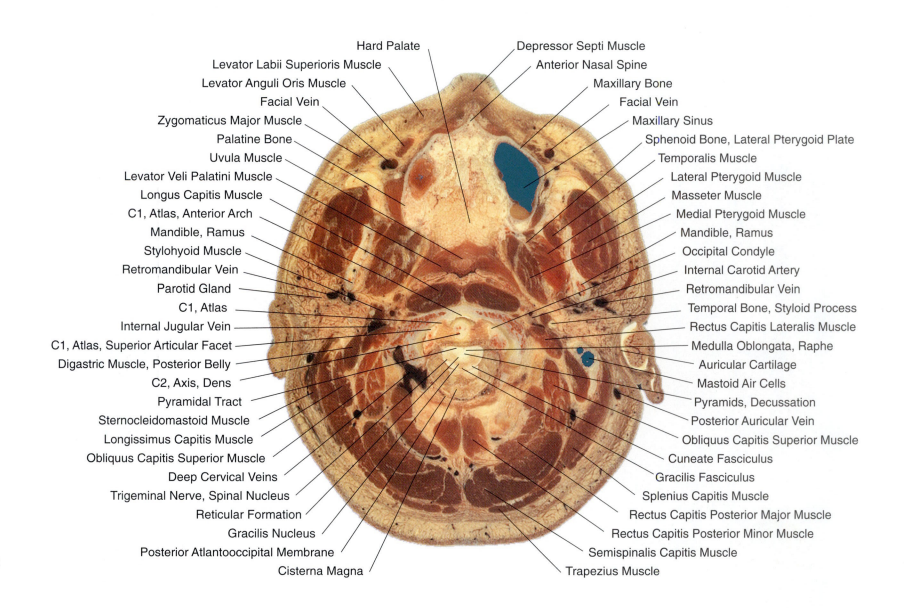

Hard Palate

Levator Labii Superioris Muscle

Levator Anguli Oris Muscle

Facial Vein

Zygomaticus Major Muscle

Palatine Bone

Uvula Muscle

Levator Veli Palatini Muscle

Longus Capitis Muscle

C1, Atlas, Anterior Arch

Mandible, Ramus

Stylohyoid Muscle

Retromandibular Vein

Parotid Gland

C1, Atlas

Internal Jugular Vein

C1, Atlas, Superior Articular Facet

Digastric Muscle, Posterior Belly

C2, Axis, Dens

Pyramidal Tract

Sternocleidomastoid Muscle

Longissimus Capitis Muscle

Obliquus Capitis Superior Muscle

Deep Cervical Veins

Trigeminal Nerve, Spinal Nucleus

Reticular Formation

Gracilis Nucleus

Posterior Atlantooccipital Membrane

Cisterna Magna

Depressor Septi Muscle

Anterior Nasal Spine

Maxillary Bone

Facial Vein

Maxillary Sinus

Sphenoid Bone, Lateral Pterygoid Plate

Temporalis Muscle

Lateral Pterygoid Muscle

Masseter Muscle

Medial Pterygoid Muscle

Mandible, Ramus

Occipital Condyle

Internal Carotid Artery

Retromandibular Vein

Temporal Bone, Styloid Process

Rectus Capitis Lateralis Muscle

Medulla Oblongata, Raphe

Auricular Cartilage

Mastoid Air Cells

Pyramids, Decussation

Posterior Auricular Vein

Obliquus Capitis Superior Muscle

Cuneate Fasciculus

Gracilis Fasciculus

Splenius Capitis Muscle

Rectus Capitis Posterior Major Muscle

Rectus Capitis Posterior Minor Muscle

Semispinalis Capitis Muscle

Trapezius Muscle

right

left

posterior

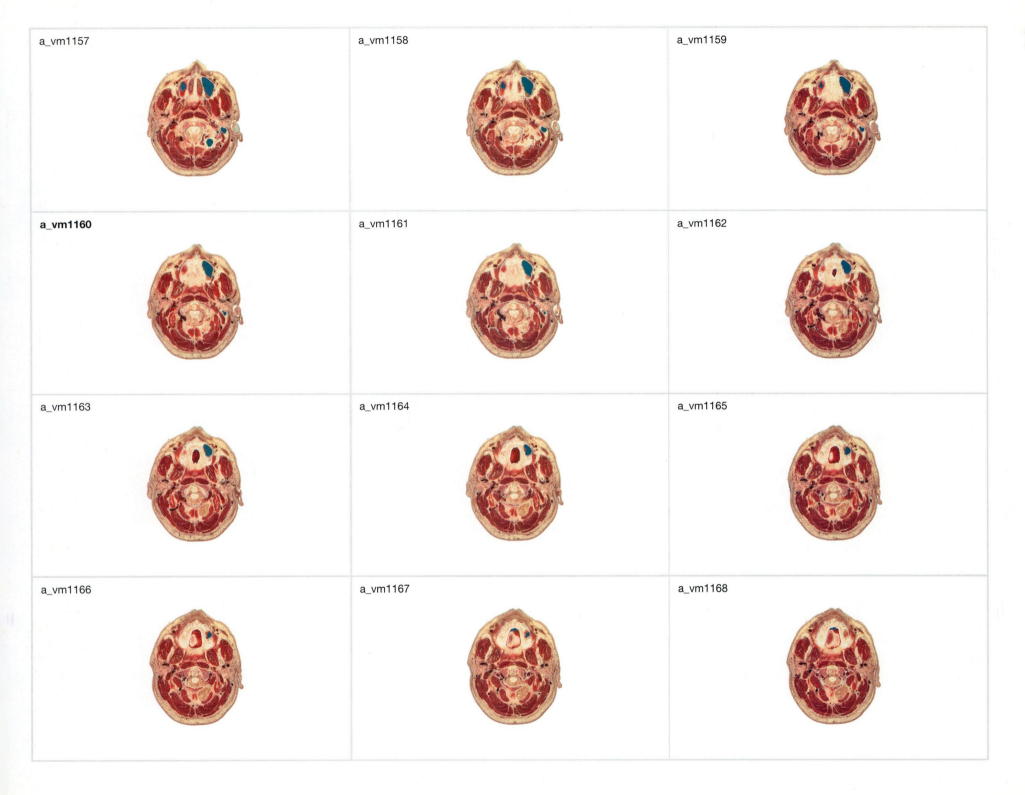

a_vm1157

a_vm1158

a_vm1159

a_vm1160

a_vm1161

a_vm1162

a_vm1163

a_vm1164

a_vm1165

a_vm1166

a_vm1167

a_vm1168

anterior

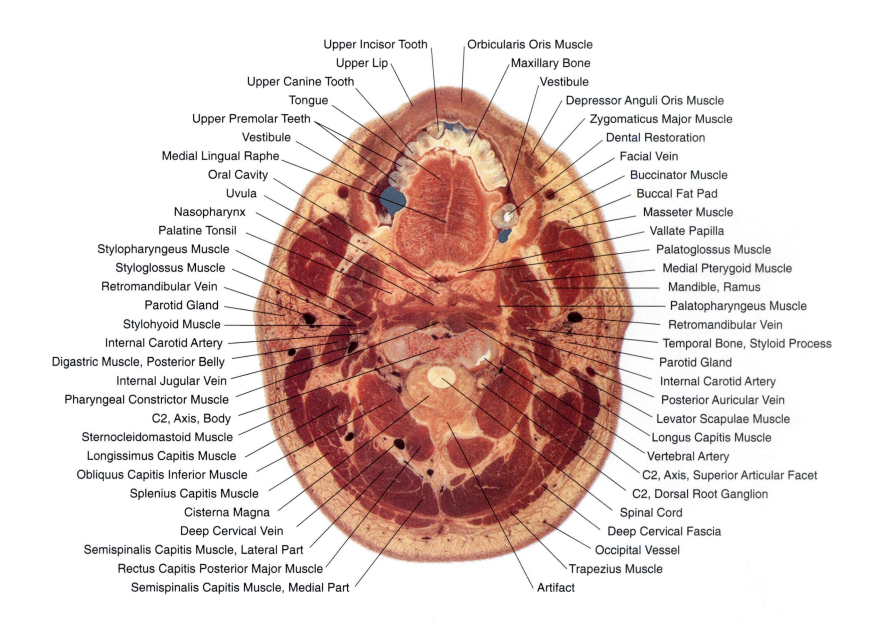

Upper Incisor Tooth

Orbicularis Oris Muscle

Upper Lip

Maxillary Bone

Upper Canine Tooth

Vestibule

Tongue

Depressor Anguli Oris Muscle

Upper Premolar Teeth

Zygomaticus Major Muscle

Vestibule

Dental Restoration

Medial Lingual Raphe

Facial Vein

Oral Cavity

Buccinator Muscle

Uvula

Buccal Fat Pad

Nasopharynx

Masseter Muscle

Palatine Tonsil

Vallate Papilla

Stylopharyngeus Muscle

Palatoglossus Muscle

Styloglossus Muscle

Medial Pterygoid Muscle

Retromandibular Vein

Mandible, Ramus

Parotid Gland

Palatopharyngeus Muscle

Stylohyoid Muscle

Retromandibular Vein

Internal Carotid Artery

Temporal Bone, Styloid Process

Digastric Muscle, Posterior Belly

Parotid Gland

Internal Jugular Vein

Internal Carotid Artery

Pharyngeal Constrictor Muscle

Posterior Auricular Vein

C2, Axis, Body

Levator Scapulae Muscle

Sternocleidomastoid Muscle

Longus Capitis Muscle

Longissimus Capitis Muscle

Vertebral Artery

Obliquus Capitis Inferior Muscle

C2, Axis, Superior Articular Facet

Splenius Capitis Muscle

C2, Dorsal Root Ganglion

Cisterna Magna

Spinal Cord

Deep Cervical Vein

Deep Cervical Fascia

Semispinalis Capitis Muscle, Lateral Part

Occipital Vessel

Rectus Capitis Posterior Major Muscle

Trapezius Muscle

Semispinalis Capitis Muscle, Medial Part

Artifact

right

left

posterior

30

a_vm1169

a_vm1170

a_vm1171

a_vm1172

a_vm1173

a_vm1174

a_vm1175

a_vm1176

a_vm1177

a_vm1178

a_vm1179

a_vm1180

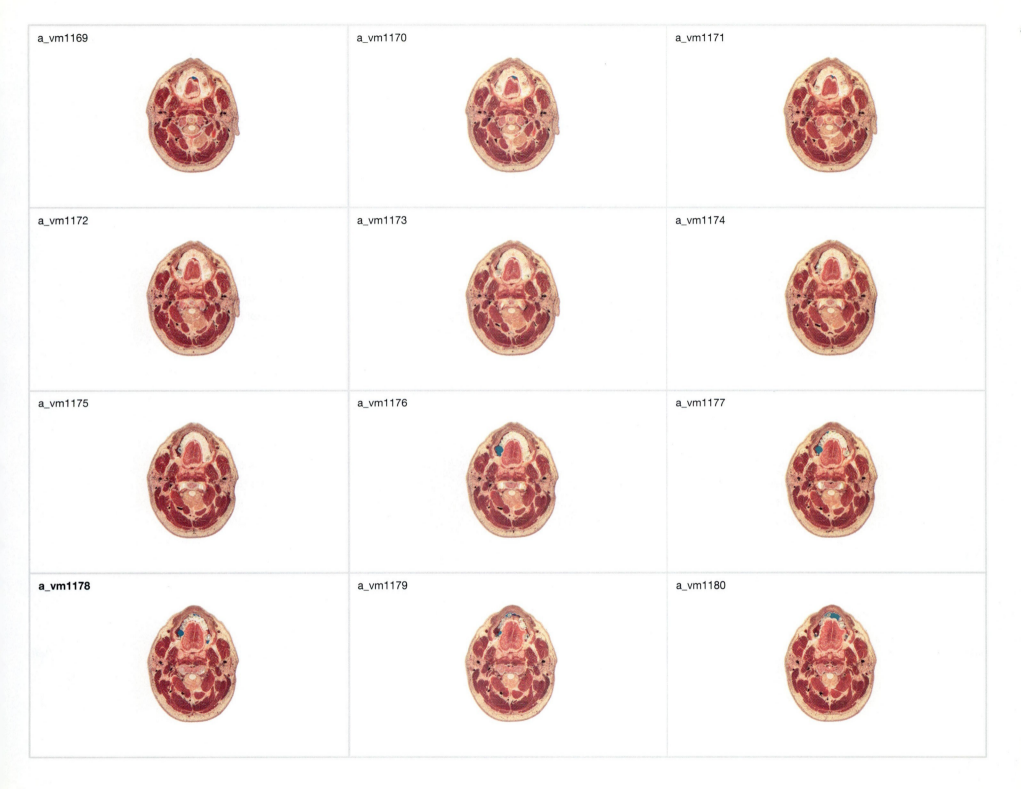

anterior

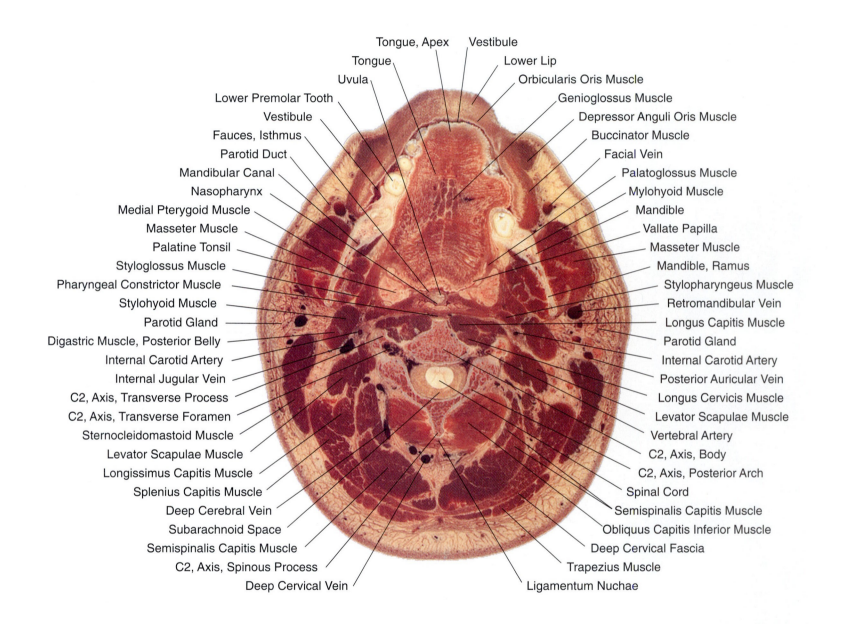

Tongue, Apex

Vestibule

Tongue

Lower Lip

Uvula

Orbicularis Oris Muscle

Lower Premolar Tooth

Genioglossus Muscle

Vestibule

Depressor Anguli Oris Muscle

Fauces, Isthmus

Buccinator Muscle

Parotid Duct

Facial Vein

Mandibular Canal

Palatoglossus Muscle

Nasopharynx

Mylohyoid Muscle

Medial Pterygoid Muscle

Mandible

Masseter Muscle

Vallate Papilla

Palatine Tonsil

Masseter Muscle

Styloglossus Muscle

Mandible, Ramus

Pharyngeal Constrictor Muscle

Stylopharyngeus Muscle

Stylohyoid Muscle

Retromandibular Vein

Parotid Gland

Longus Capitis Muscle

Digastric Muscle, Posterior Belly

Parotid Gland

Internal Carotid Artery

Internal Carotid Artery

Internal Jugular Vein

Posterior Auricular Vein

C2, Axis, Transverse Process

Longus Cervicis Muscle

C2, Axis, Transverse Foramen

Levator Scapulae Muscle

Sternocleidomastoid Muscle

Vertebral Artery

Levator Scapulae Muscle

C2, Axis, Body

Longissimus Capitis Muscle

C2, Axis, Posterior Arch

Splenius Capitis Muscle

Spinal Cord

Deep Cerebral Vein

Semispinalis Capitis Muscle

Subarachnoid Space

Obliquus Capitis Inferior Muscle

Semispinalis Capitis Muscle

Deep Cervical Fascia

C2, Axis, Spinous Process

Trapezius Muscle

Deep Cervical Vein

Ligamentum Nuchae

right

left

posterior

32

a_vm1181

a_vm1182

a_vm1183

a_vm1184

a_vm1185

a_vm1186

a_vm1187

a_vm1188

a_vm1189

a_vm1190

a_vm1191

a_vm1192

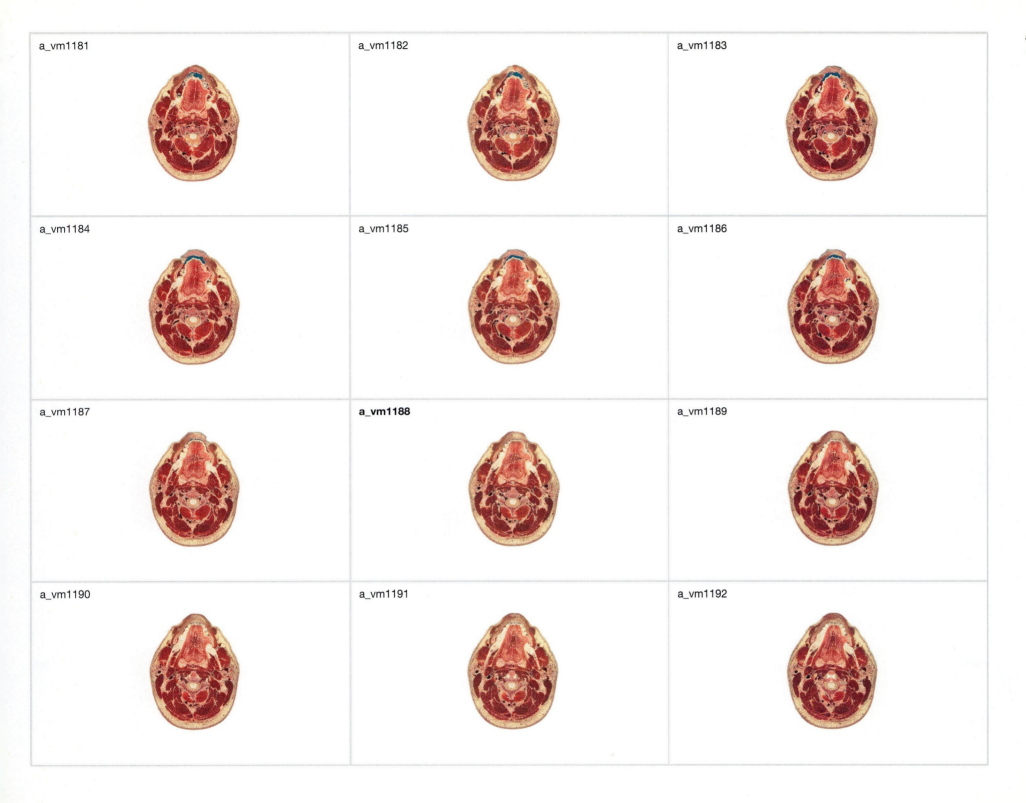

anterior

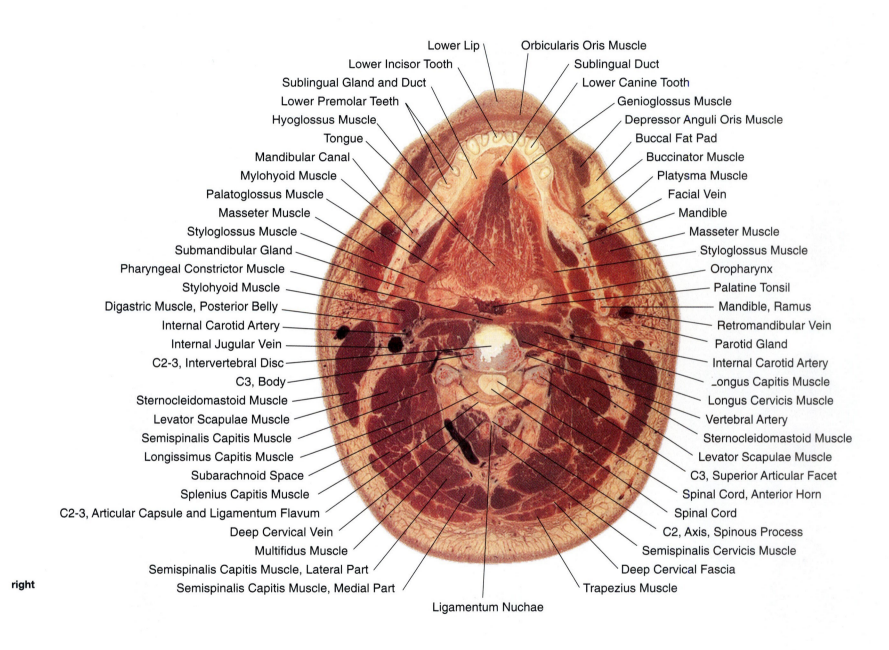

Lower Lip

Orbicularis Oris Muscle

Lower Incisor Tooth

Sublingual Duct

Sublingual Gland and Duct

Lower Canine Tooth

Lower Premolar Teeth

Genioglossus Muscle

Hyoglossus Muscle

Depressor Anguli Oris Muscle

Tongue

Buccal Fat Pad

Mandibular Canal

Buccinator Muscle

Mylohyoid Muscle

Platysma Muscle

Palatoglossus Muscle

Facial Vein

Masseter Muscle

Mandible

Styloglossus Muscle

Masseter Muscle

Submandibular Gland

Styloglossus Muscle

Pharyngeal Constrictor Muscle

Oropharynx

Stylohyoid Muscle

Palatine Tonsil

Digastric Muscle, Posterior Belly

Mandible, Ramus

Internal Carotid Artery

Retromandibular Vein

Internal Jugular Vein

Parotid Gland

C2-3, Intervertebral Disc

Internal Carotid Artery

C3, Body

Longus Capitis Muscle

Sternocleidomastoid Muscle

Longus Cervicis Muscle

Levator Scapulae Muscle

Vertebral Artery

Semispinalis Capitis Muscle

Sternocleidomastoid Muscle

Longissimus Capitis Muscle

Levator Scapulae Muscle

Subarachnoid Space

C3, Superior Articular Facet

Splenius Capitis Muscle

Spinal Cord, Anterior Horn

C2-3, Articular Capsule and Ligamentum Flavum

Spinal Cord

Deep Cervical Vein

C2, Axis, Spinous Process

Multifidus Muscle

Semispinalis Cervicis Muscle

Semispinalis Capitis Muscle, Lateral Part

Deep Cervical Fascia

Semispinalis Capitis Muscle, Medial Part

Trapezius Muscle

Ligamentum Nuchae

right

left

posterior

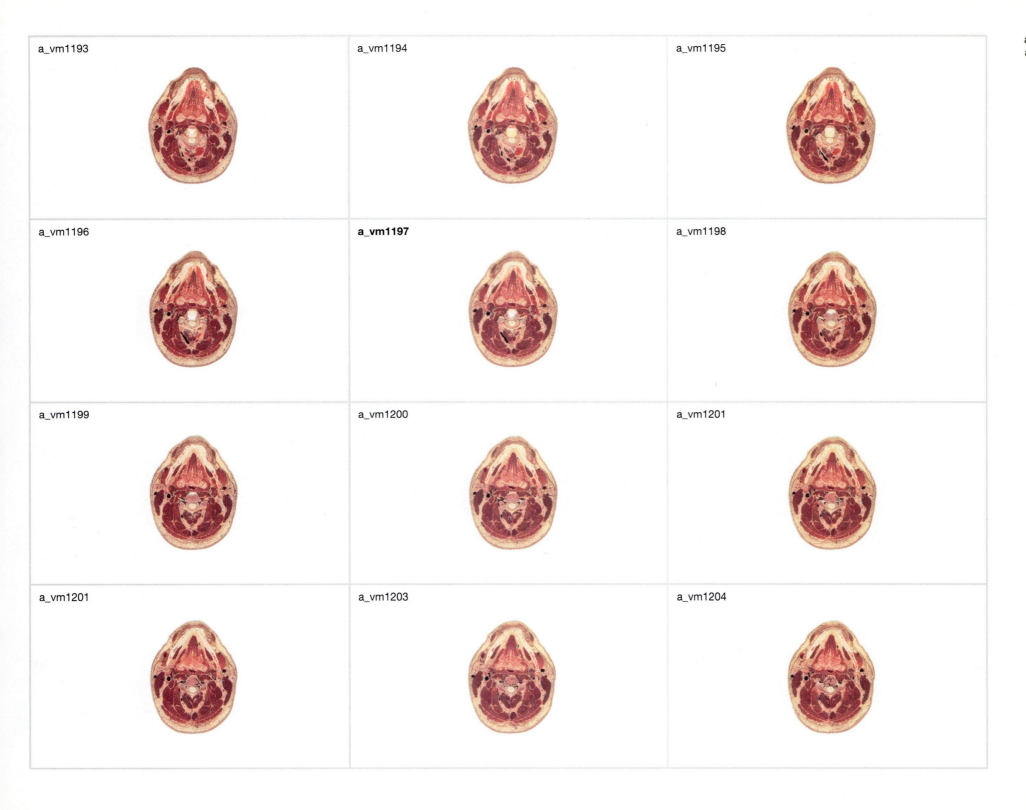

a_vm1193

a_vm1194

a_vm1195

a_vm1196

a_vm1197

a_vm1198

a_vm1199

a_vm1200

a_vm1201

a_vm1201

a_vm1203

a_vm1204

anterior

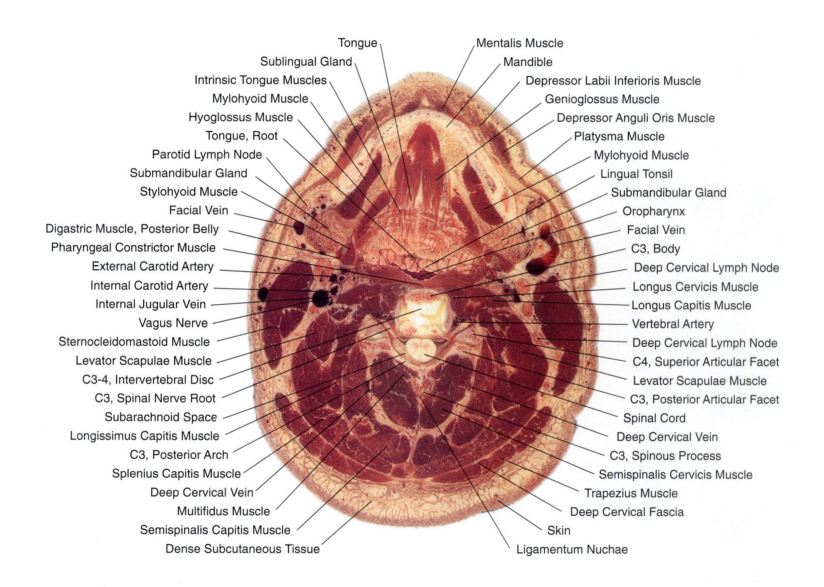

Tongue

Sublingual Gland

Intrinsic Tongue Muscles

Mylohyoid Muscle

Hyoglossus Muscle

Tongue, Root

Parotid Lymph Node

Submandibular Gland

Stylohyoid Muscle

Facial Vein

Digastric Muscle, Posterior Belly

Pharyngeal Constrictor Muscle

External Carotid Artery

Internal Carotid Artery

Internal Jugular Vein

Vagus Nerve

Sternocleidomastoid Muscle

Levator Scapulae Muscle

C3-4, Intervertebral Disc

C3, Spinal Nerve Root

Subarachnoid Space

Longissimus Capitis Muscle

C3, Posterior Arch

Splenius Capitis Muscle

Deep Cervical Vein

Multifidus Muscle

Semispinalis Capitis Muscle

Dense Subcutaneous Tissue

Mentalis Muscle

Mandible

Depressor Labii Inferioris Muscle

Genioglossus Muscle

Depressor Anguli Oris Muscle

Platysma Muscle

Mylohyoid Muscle

Lingual Tonsil

Submandibular Gland

Oropharynx

Facial Vein

C3, Body

Deep Cervical Lymph Node

Longus Cervicis Muscle

Longus Capitis Muscle

Vertebral Artery

Deep Cervical Lymph Node

C4, Superior Articular Facet

Levator Scapulae Muscle

C3, Posterior Articular Facet

Spinal Cord

Deep Cervical Vein

C3, Spinous Process

Semispinalis Cervicis Muscle

Trapezius Muscle

Deep Cervical Fascia

Skin

Ligamentum Nuchae

right

left

posterior

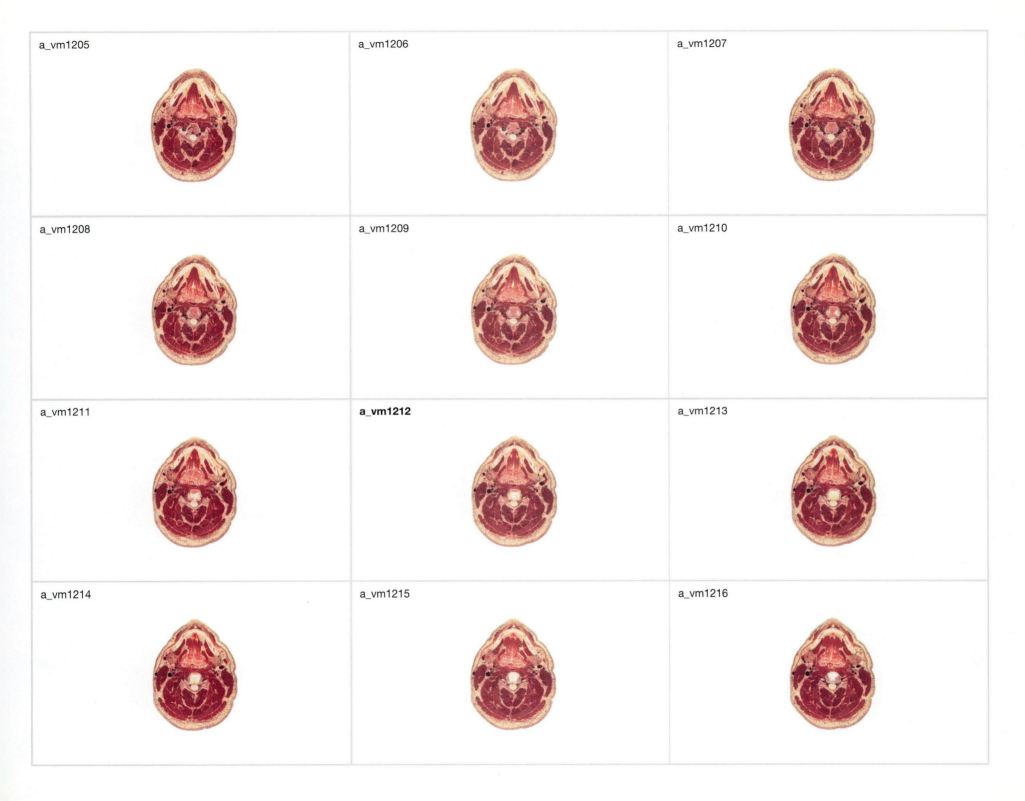

a_vm1205

a_vm1206

a_vm1207

a_vm1208

a_vm1209

a_vm1210

a_vm1211

a_vm1212

a_vm1213

a_vm1214

a_vm1215

a_vm1216

anterior

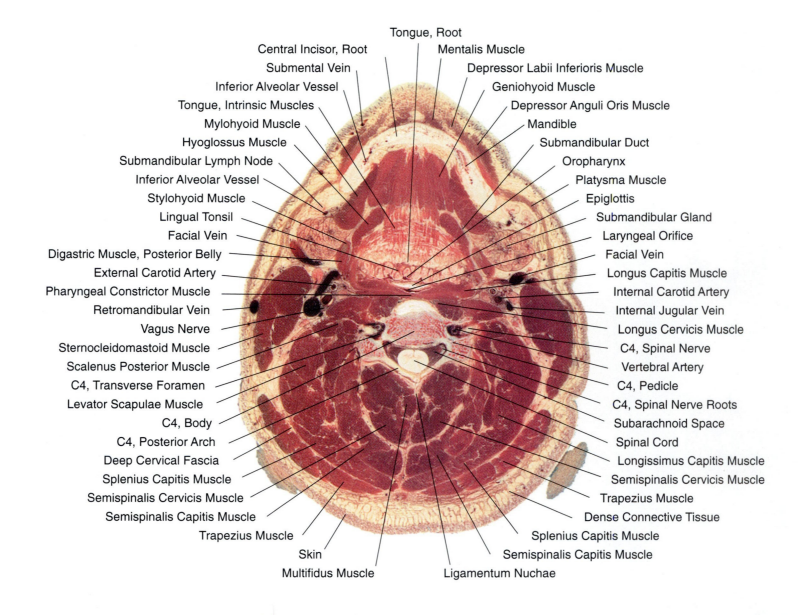

Tongue, Root

Central Incisor, Root
Mentalis Muscle

Submental Vein
Depressor Labii Inferioris Muscle

Inferior Alveolar Vessel
Geniohyoid Muscle

Tongue, Intrinsic Muscles
Depressor Anguli Oris Muscle

Mylohyoid Muscle
Mandible

Hyoglossus Muscle
Submandibular Duct

Submandibular Lymph Node
Oropharynx

Inferior Alveolar Vessel
Platysma Muscle

Stylohyoid Muscle
Epiglottis

Lingual Tonsil
Submandibular Gland

Facial Vein
Laryngeal Orifice

Digastric Muscle, Posterior Belly
Facial Vein

External Carotid Artery
Longus Capitis Muscle

Pharyngeal Constrictor Muscle
Internal Carotid Artery

Retromandibular Vein
Internal Jugular Vein

Vagus Nerve
Longus Cervicis Muscle

Sternocleidomastoid Muscle
C4, Spinal Nerve

Scalenus Posterior Muscle
Vertebral Artery

C4, Transverse Foramen
C4, Pedicle

Levator Scapulae Muscle
C4, Spinal Nerve Roots

C4, Body
Subarachnoid Space

C4, Posterior Arch
Spinal Cord

Deep Cervical Fascia
Longissimus Capitis Muscle

Splenius Capitis Muscle
Semispinalis Cervicis Muscle

Semispinalis Cervicis Muscle
Trapezius Muscle

Semispinalis Capitis Muscle
Dense Connective Tissue

Trapezius Muscle
Splenius Capitis Muscle

Skin
Semispinalis Capitis Muscle

Multifidus Muscle
Ligamentum Nuchae

right

left

posterior

a_vm1217

a_vm1218

a_vm1219

a_vm1220

a_vm1221

a_vm1222

a_vm1223

a_vm1224

a_vm1225

a_vm1226

a_vm1227

a_vm1228

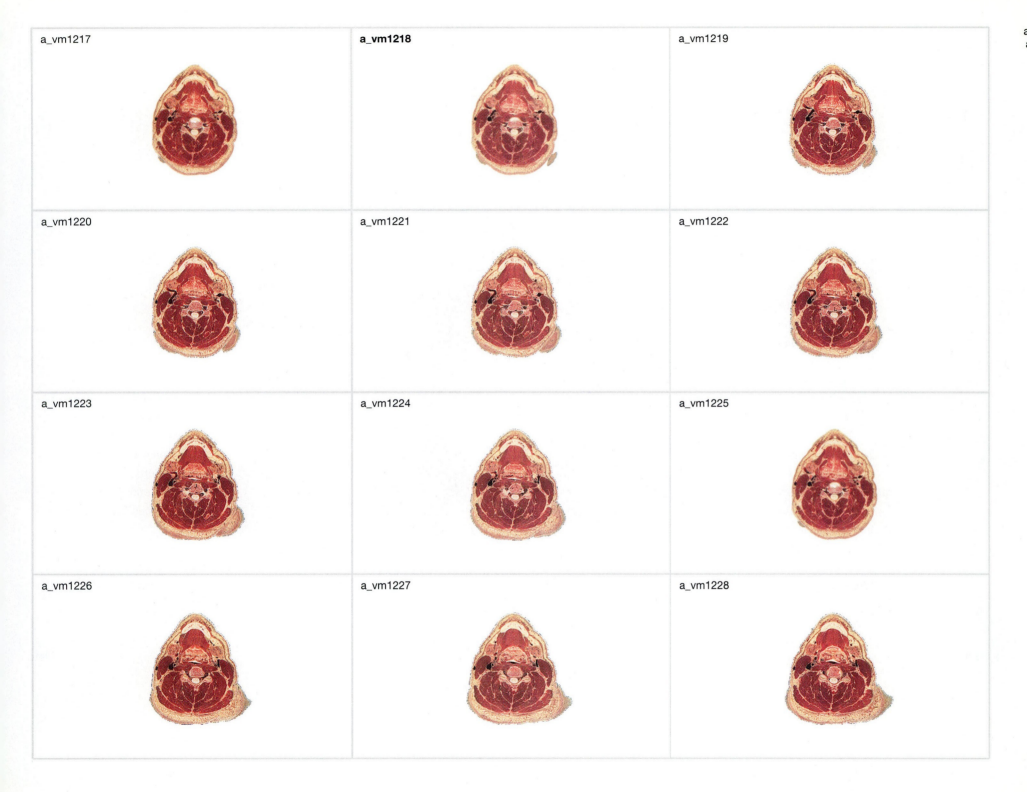

anterior

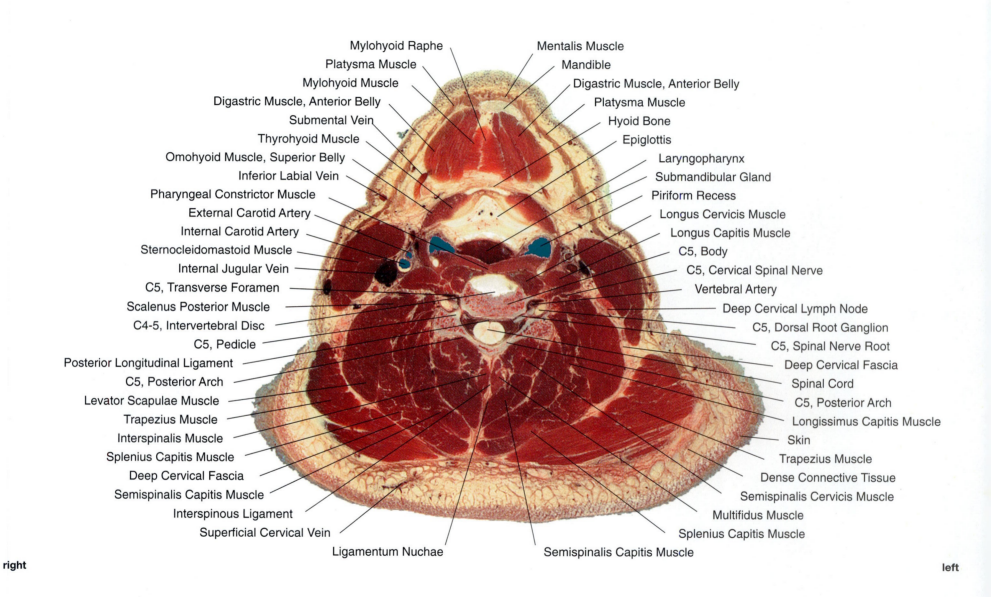

Mylohyoid Raphe

Mentalis Muscle

Platysma Muscle

Mandible

Mylohyoid Muscle

Digastric Muscle, Anterior Belly

Digastric Muscle, Anterior Belly

Platysma Muscle

Submental Vein

Hyoid Bone

Thyrohyoid Muscle

Epiglottis

Omohyoid Muscle, Superior Belly

Laryngopharynx

Inferior Labial Vein

Submandibular Gland

Pharyngeal Constrictor Muscle

Piriform Recess

External Carotid Artery

Longus Cervicis Muscle

Internal Carotid Artery

Longus Capitis Muscle

Sternocleidomastoid Muscle

C5, Body

Internal Jugular Vein

C5, Cervical Spinal Nerve

C5, Transverse Foramen

Vertebral Artery

Scalenus Posterior Muscle

Deep Cervical Lymph Node

C4-5, Intervertebral Disc

C5, Dorsal Root Ganglion

C5, Pedicle

C5, Spinal Nerve Root

Posterior Longitudinal Ligament

Deep Cervical Fascia

C5, Posterior Arch

Spinal Cord

Levator Scapulae Muscle

C5, Posterior Arch

Trapezius Muscle

Longissimus Capitis Muscle

Interspinalis Muscle

Skin

Splenius Capitis Muscle

Trapezius Muscle

Deep Cervical Fascia

Dense Connective Tissue

Semispinalis Capitis Muscle

Semispinalis Cervicis Muscle

Interspinous Ligament

Multifidus Muscle

Superficial Cervical Vein

Splenius Capitis Muscle

Ligamentum Nuchae

Semispinalis Capitis Muscle

right

left

posterior

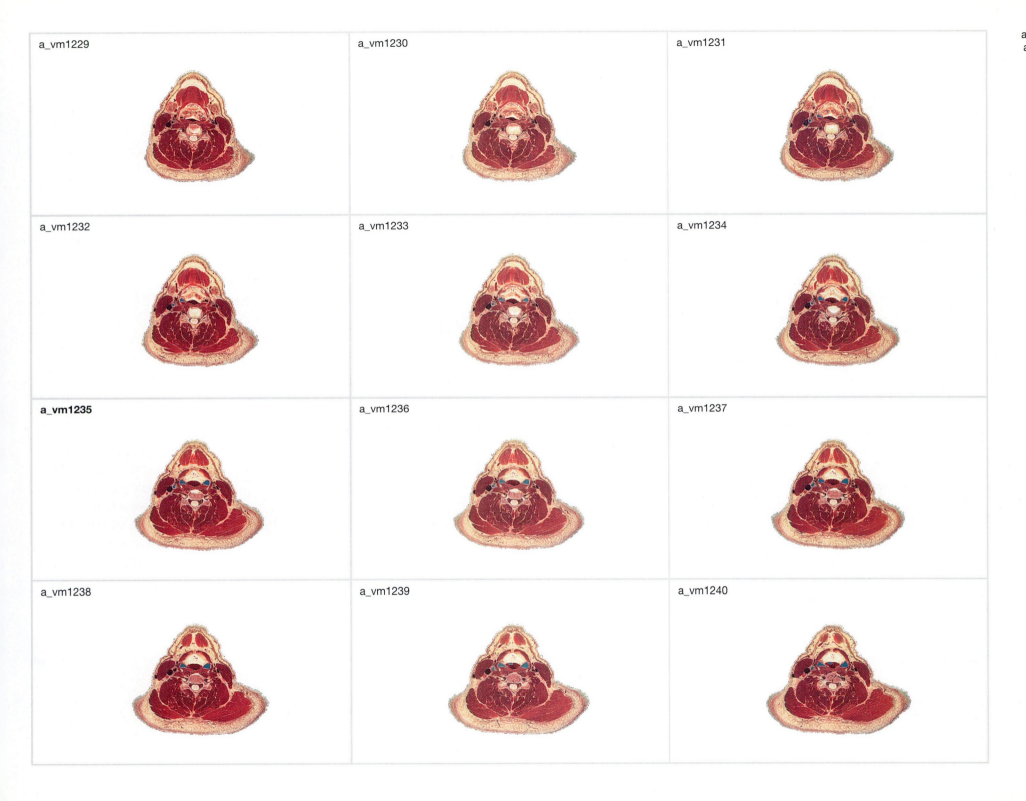

a_vm1229

a_vm1230

a_vm1231

a_vm1232

a_vm1233

a_vm1234

a_vm1235

a_vm1236

a_vm1237

a_vm1238

a_vm1239

a_vm1240

anterior

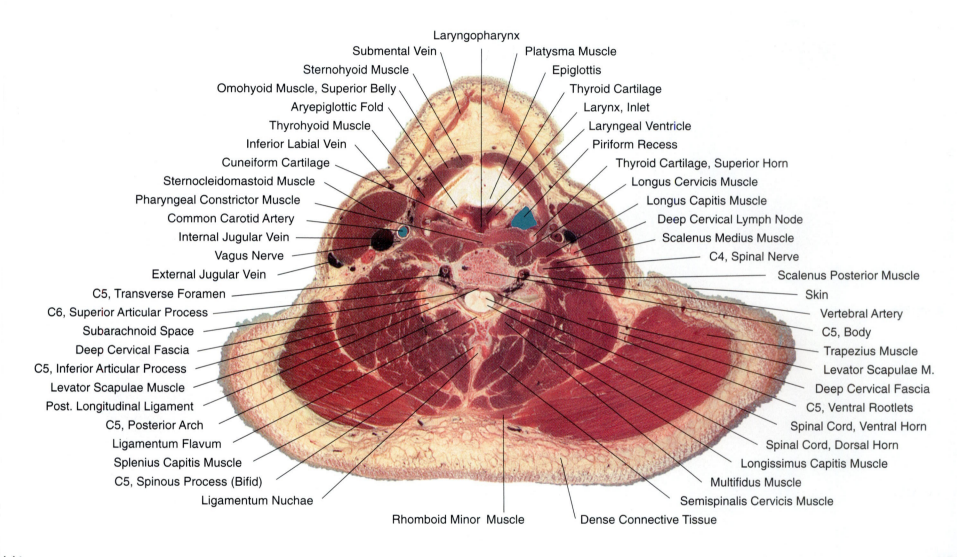

Laryngopharynx

Submental Vein

Sternohyoid Muscle

Omohyoid Muscle, Superior Belly

Aryepiglottic Fold

Thyrohyoid Muscle

Inferior Labial Vein

Cuneiform Cartilage

Sternocleidomastoid Muscle

Pharyngeal Constrictor Muscle

Common Carotid Artery

Internal Jugular Vein

Vagus Nerve

External Jugular Vein

C5, Transverse Foramen

C6, Superior Articular Process

Subarachnoid Space

Deep Cervical Fascia

C5, Inferior Articular Process

Levator Scapulae Muscle

Post. Longitudinal Ligament

C5, Posterior Arch

Ligamentum Flavum

Splenius Capitis Muscle

C5, Spinous Process (Bifid)

Ligamentum Nuchae

Platysma Muscle

Epiglottis

Thyroid Cartilage

Larynx, Inlet

Laryngeal Ventricle

Piriform Recess

Thyroid Cartilage, Superior Horn

Longus Cervicis Muscle

Longus Capitis Muscle

Deep Cervical Lymph Node

Scalenus Medius Muscle

C4, Spinal Nerve

Scalenus Posterior Muscle

Skin

Vertebral Artery

C5, Body

Trapezius Muscle

Levator Scapulae M.

Deep Cervical Fascia

C5, Ventral Rootlets

Spinal Cord, Ventral Horn

Spinal Cord, Dorsal Horn

Longissimus Capitis Muscle

Multifidus Muscle

Semispinalis Cervicis Muscle

Rhomboid Minor Muscle

Dense Connective Tissue

right

left

posterior

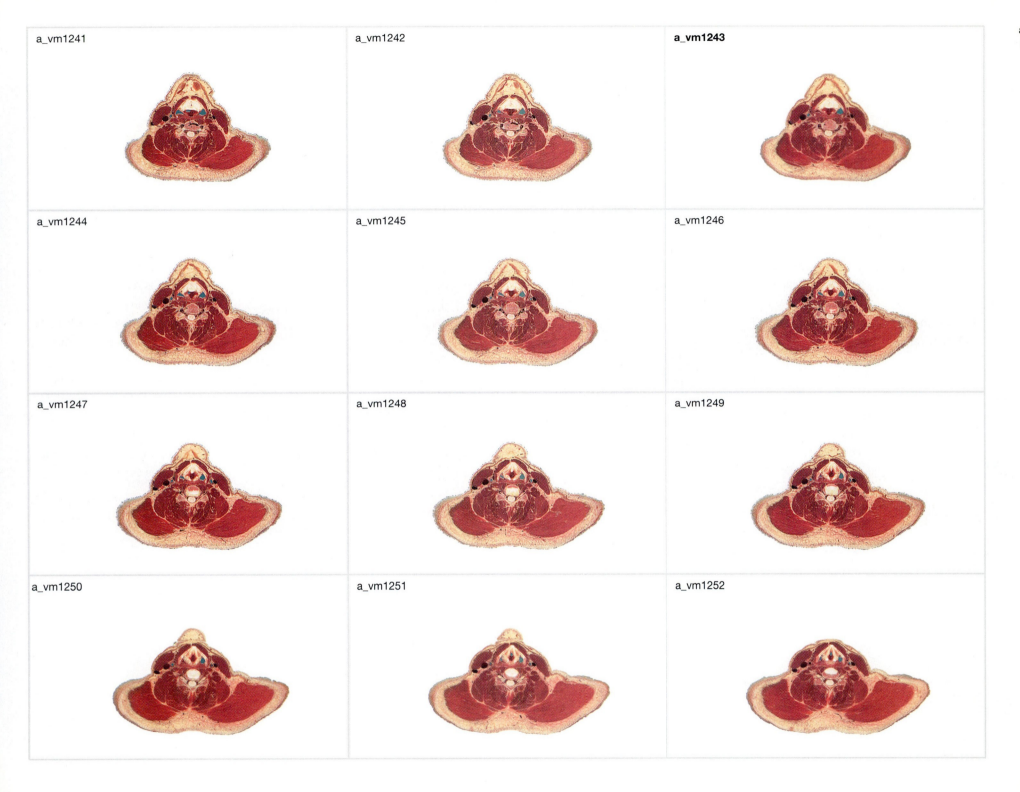

a_vm1241

a_vm1242

a_vm1243

a_vm1244

a_vm1245

a_vm1246

a_vm1247

a_vm1248

a_vm1249

a_vm1250

a_vm1251

a_vm1252

anterior

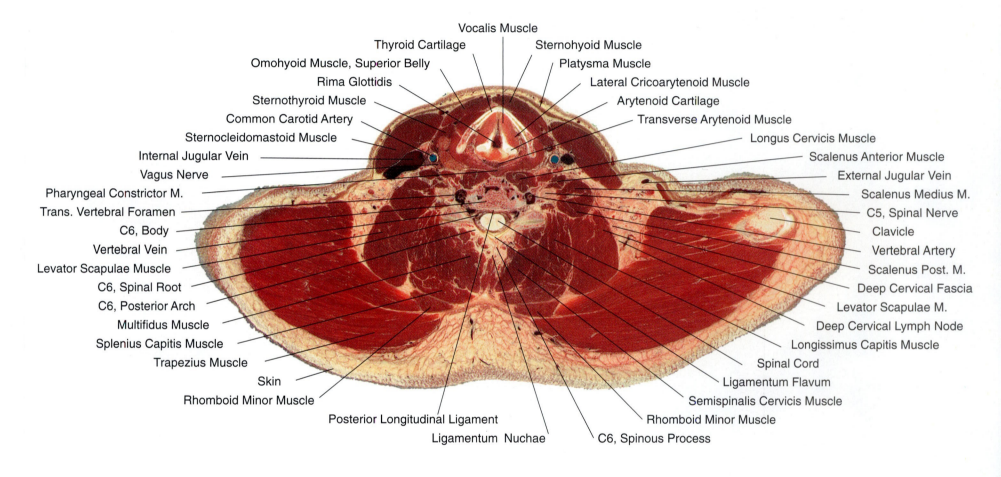

Vocalis Muscle

Thyroid Cartilage

Sternohyoid Muscle

Omohyoid Muscle, Superior Belly

Platysma Muscle

Rima Glottidis

Lateral Cricoarytenoid Muscle

Sternothyroid Muscle

Arytenoid Cartilage

Common Carotid Artery

Transverse Arytenoid Muscle

Sternocleidomastoid Muscle

Longus Cervicis Muscle

Internal Jugular Vein

Scalenus Anterior Muscle

Vagus Nerve

External Jugular Vein

Pharyngeal Constrictor M.

Scalenus Medius M.

Trans. Vertebral Foramen

C5, Spinal Nerve

C6, Body

Clavicle

Vertebral Vein

Vertebral Artery

Levator Scapulae Muscle

Scalenus Post. M.

C6, Spinal Root

Deep Cervical Fascia

C6, Posterior Arch

Levator Scapulae M.

Multifidus Muscle

Deep Cervical Lymph Node

Splenius Capitis Muscle

Longissimus Capitis Muscle

Trapezius Muscle

Spinal Cord

Skin

Ligamentum Flavum

Rhomboid Minor Muscle

Semispinalis Cervicis Muscle

Posterior Longitudinal Ligament

Rhomboid Minor Muscle

Ligamentum Nuchae

C6, Spinous Process

right

left

posterior

44

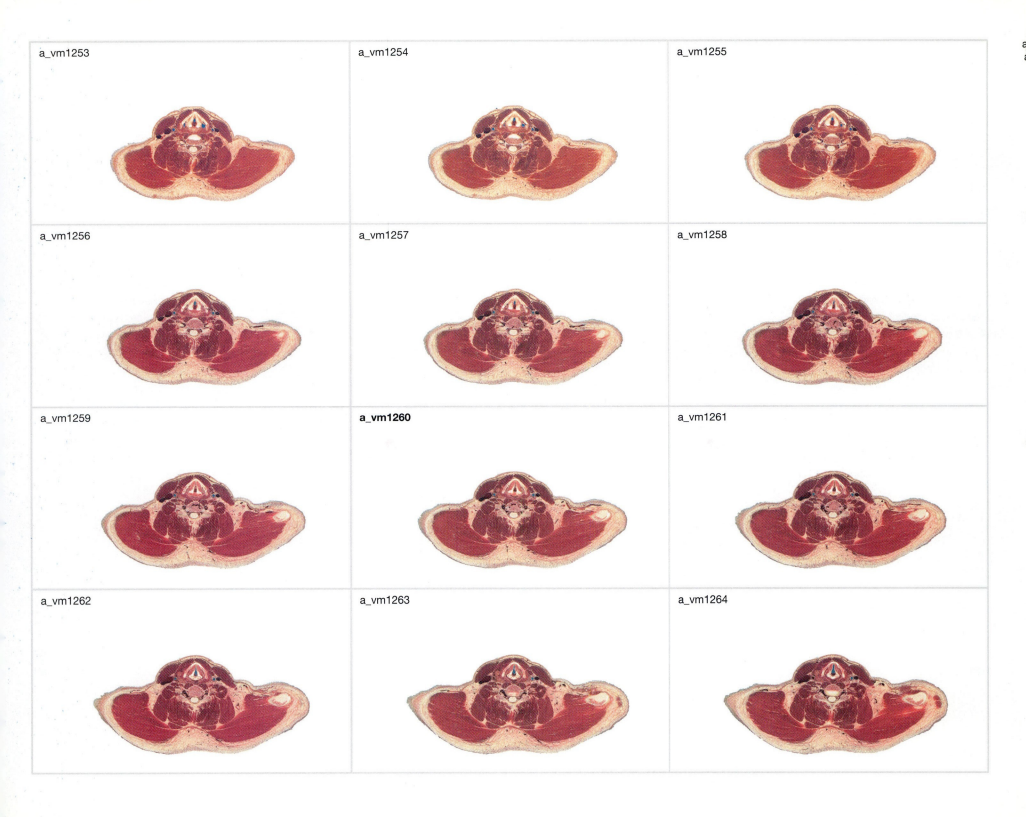

a_vm1253

a_vm1254

a_vm1255

a_vm1256

a_vm1257

a_vm1258

a_vm1259

a_vm1260

a_vm1261

a_vm1262

a_vm1263

a_vm1264

anterior

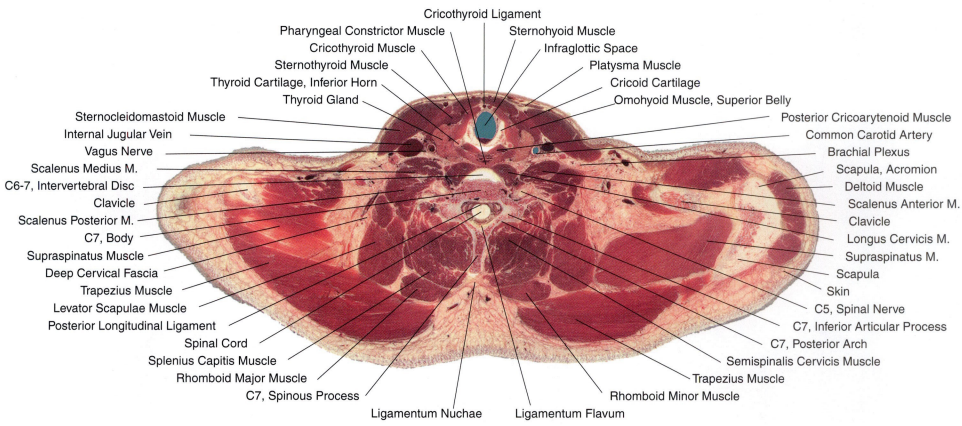

Cricothyroid Ligament

Pharyngeal Constrictor Muscle
Sternohyoid Muscle

Cricothyroid Muscle
Infraglottic Space

Sternothyroid Muscle
Platysma Muscle

Thyroid Cartilage, Inferior Horn
Cricoid Cartilage

Thyroid Gland
Omohyoid Muscle, Superior Belly

Sternocleidomastoid Muscle
Posterior Cricoarytenoid Muscle

Internal Jugular Vein
Common Carotid Artery

Vagus Nerve
Brachial Plexus

Scalenus Medius M.
Scapula, Acromion

C6-7, Intervertebral Disc
Deltoid Muscle

Clavicle
Scalenus Anterior M.

Scalenus Posterior M.
Clavicle

C7, Body
Longus Cervicis M.

Supraspinatus Muscle
Supraspinatus M.

Deep Cervical Fascia
Scapula

Trapezius Muscle
Skin

Levator Scapulae Muscle
C5, Spinal Nerve

Posterior Longitudinal Ligament
C7, Inferior Articular Process

Spinal Cord
C7, Posterior Arch

Splenius Capitis Muscle
Semispinalis Cervicis Muscle

Rhomboid Major Muscle
Trapezius Muscle

C7, Spinous Process
Rhomboid Minor Muscle

Ligamentum Nuchae
Ligamentum Flavum

right

left

posterior

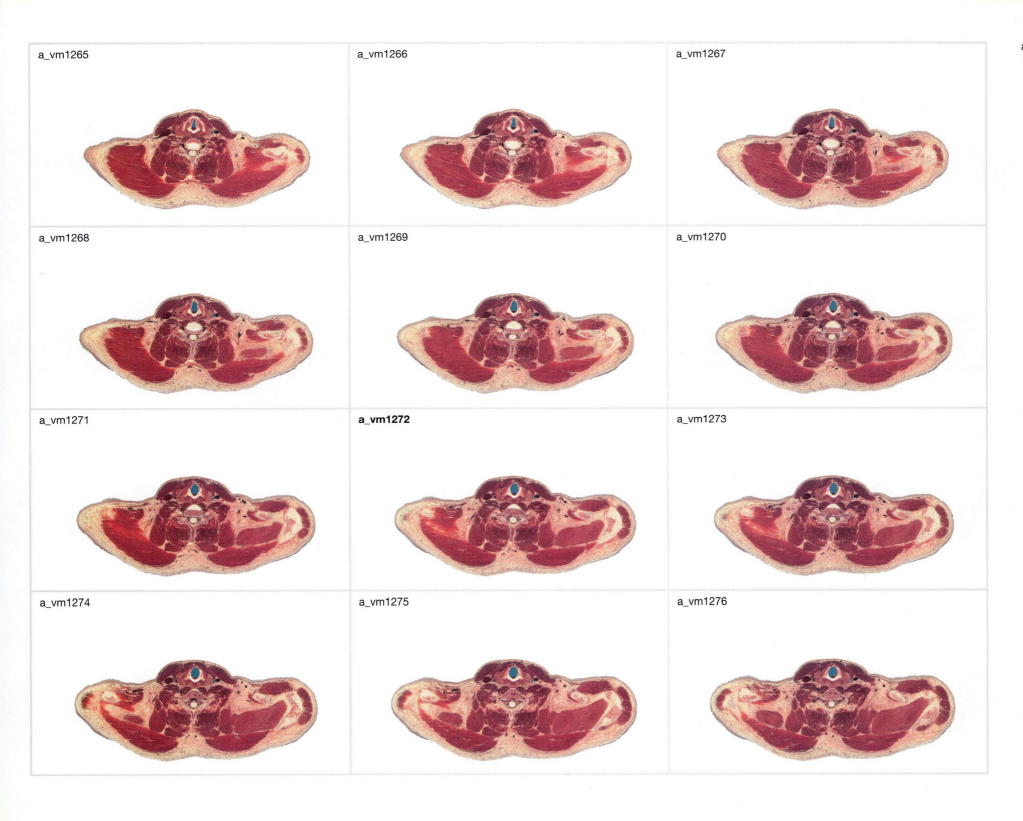

a_vm1265

a_vm1266

a_vm1267

a_vm1268

a_vm1269

a_vm1270

a_vm1271

a_vm1272

a_vm1273

a_vm1274

a_vm1275

a_vm1276

47

anterior

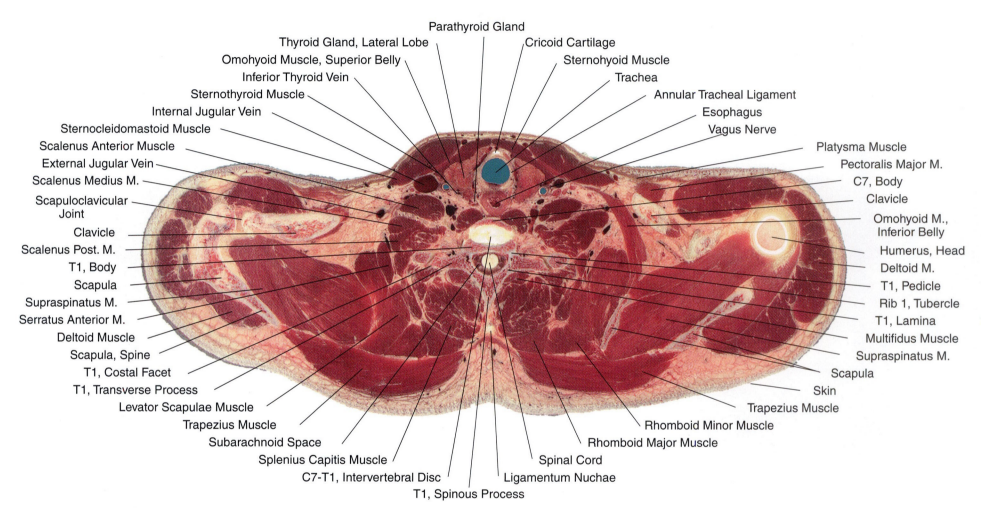

Parathyroid Gland

Thyroid Gland, Lateral Lobe

Cricoid Cartilage

Omohyoid Muscle, Superior Belly

Sternohyoid Muscle

Inferior Thyroid Vein

Trachea

Sternothyroid Muscle

Annular Tracheal Ligament

Internal Jugular Vein

Esophagus

Sternocleidomastoid Muscle

Vagus Nerve

Scalenus Anterior Muscle

Platysma Muscle

External Jugular Vein

Pectoralis Major M.

Scalenus Medius M.

C7, Body

Scapuloclavicular Joint

Clavicle

Clavicle

Omohyoid M., Inferior Belly

Scalenus Post. M.

Humerus, Head

T1, Body

Deltoid M.

Scapula

T1, Pedicle

Supraspinatus M.

Rib 1, Tubercle

Serratus Anterior M.

T1, Lamina

Deltoid Muscle

Multifidus Muscle

Scapula, Spine

Supraspinatus M.

T1, Costal Facet

Scapula

T1, Transverse Process

Skin

Levator Scapulae Muscle

Trapezius Muscle

Trapezius Muscle

Rhomboid Minor Muscle

Subarachnoid Space

Rhomboid Major Muscle

Splenius Capitis Muscle

Spinal Cord

C7-T1, Intervertebral Disc

Ligamentum Nuchae

T1, Spinous Process

right

left

posterior

a_vm1277

a_vm1278

a_vm1279

a_vm1280

a_vm1281

a_vm1282

a_vm1283

a_vm1284

a_vm1285

a_vm1286

a_vm1287

a_vm1288

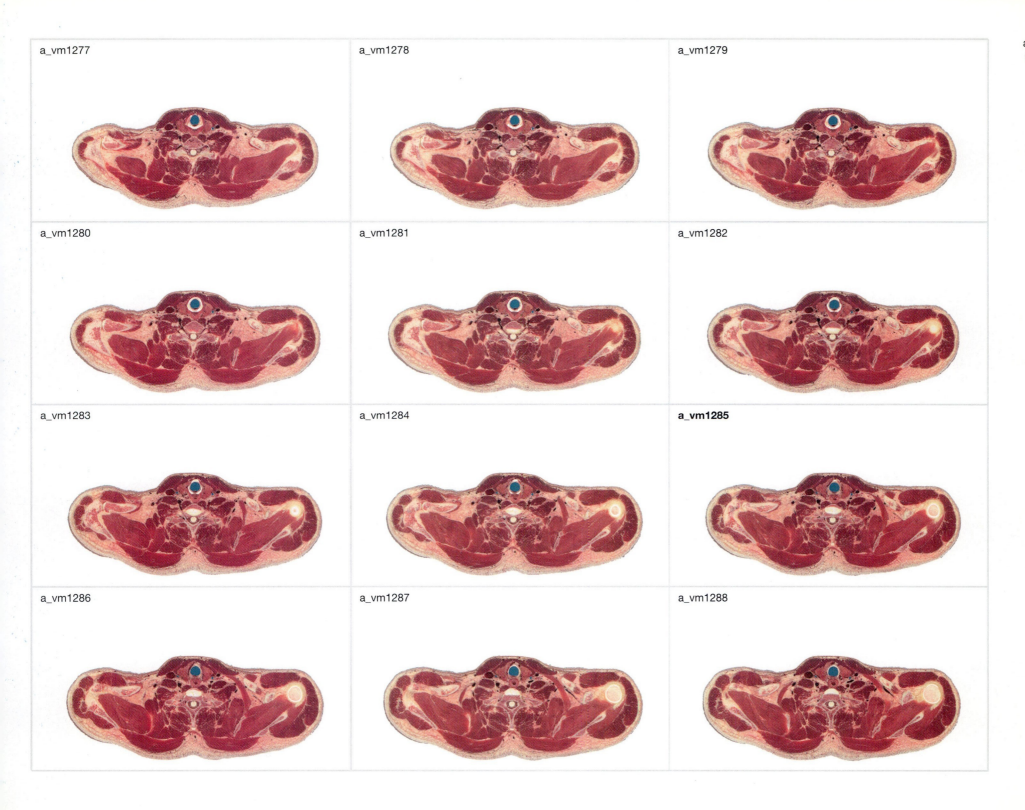

anterior

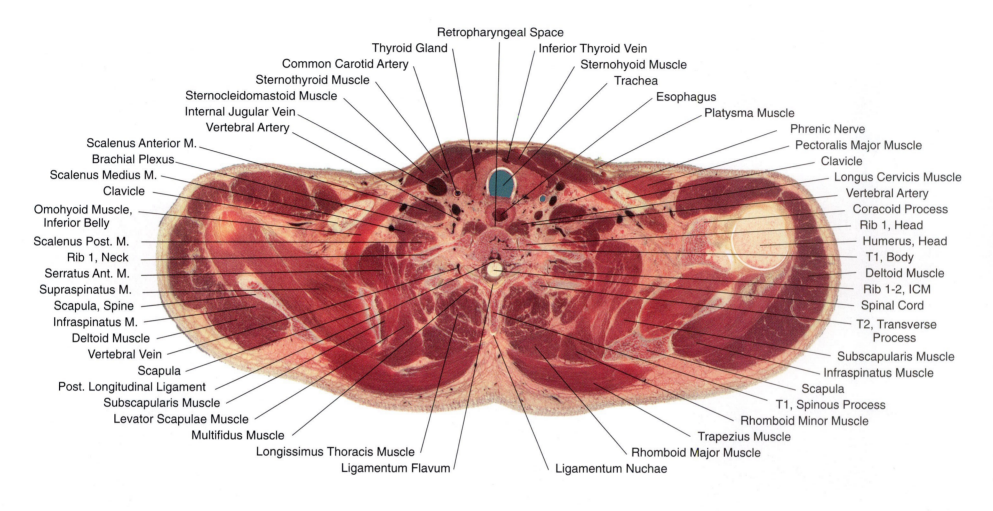

Retropharyngeal Space
Thyroid Gland
Inferior Thyroid Vein
Common Carotid Artery
Sternohyoid Muscle
Sternothyroid Muscle
Trachea
Sternocleidomastoid Muscle
Esophagus
Internal Jugular Vein
Platysma Muscle
Vertebral Artery
Phrenic Nerve
Scalenus Anterior M.
Pectoralis Major Muscle
Brachial Plexus
Clavicle
Scalenus Medius M.
Longus Cervicis Muscle
Clavicle
Vertebral Artery
Omohyoid Muscle,
Coracoid Process
Inferior Belly
Rib 1, Head
Scalenus Post. M.
Humerus, Head
Rib 1, Neck
T1, Body
Serratus Ant. M.
Deltoid Muscle
Supraspinatus M.
Rib 1-2, ICM
Scapula, Spine
Spinal Cord
Infraspinatus M.
T2, Transverse
Deltoid Muscle
Process
Vertebral Vein
Subscapularis Muscle
Scapula
Infraspinatus Muscle
Post. Longitudinal Ligament
Scapula
Subscapularis Muscle
T1, Spinous Process
Levator Scapulae Muscle
Rhomboid Minor Muscle
Multifidus Muscle
Trapezius Muscle
Longissimus Thoracis Muscle
Rhomboid Major Muscle
Ligamentum Flavum
Ligamentum Nuchae

right

left

ICM=Intercostal Muscles

posterior

50

a_vm1289

a_vm1290

a_vm1291

a_vm1292

a_vm1293

a_vm1294

a_vm1295

a_vm1296

a_vm1297

a_vm1298

a_vm1299

a_vm1300

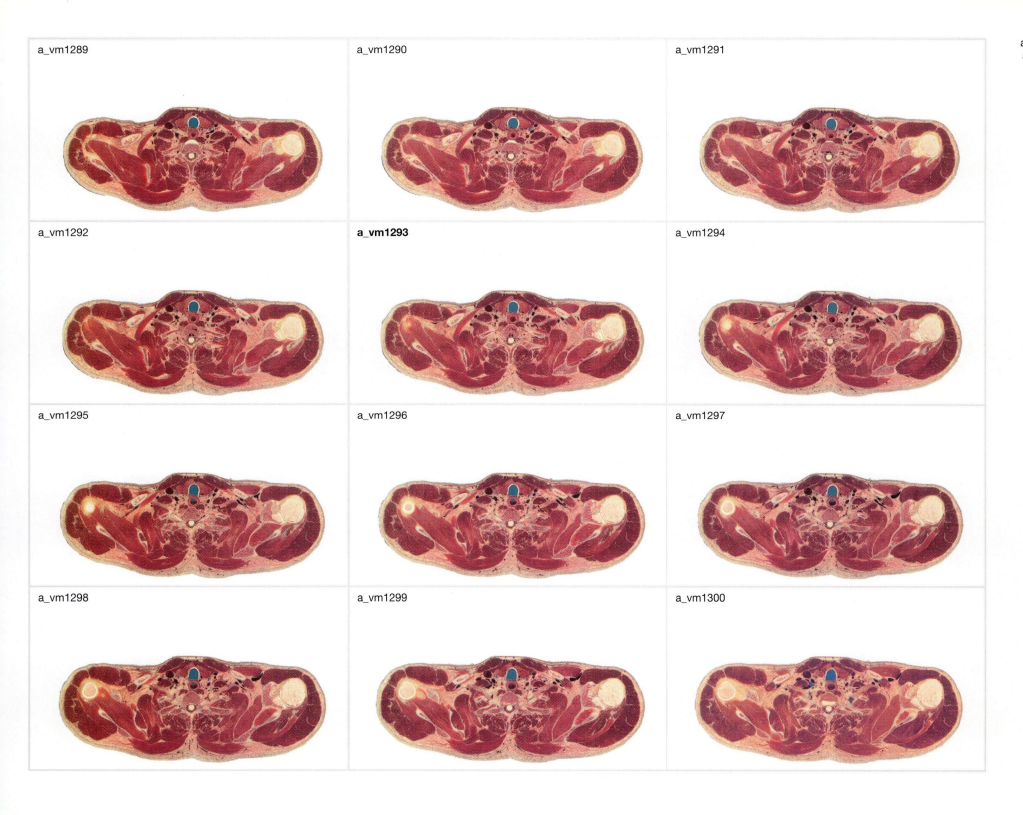

Transverse
a_vm1307

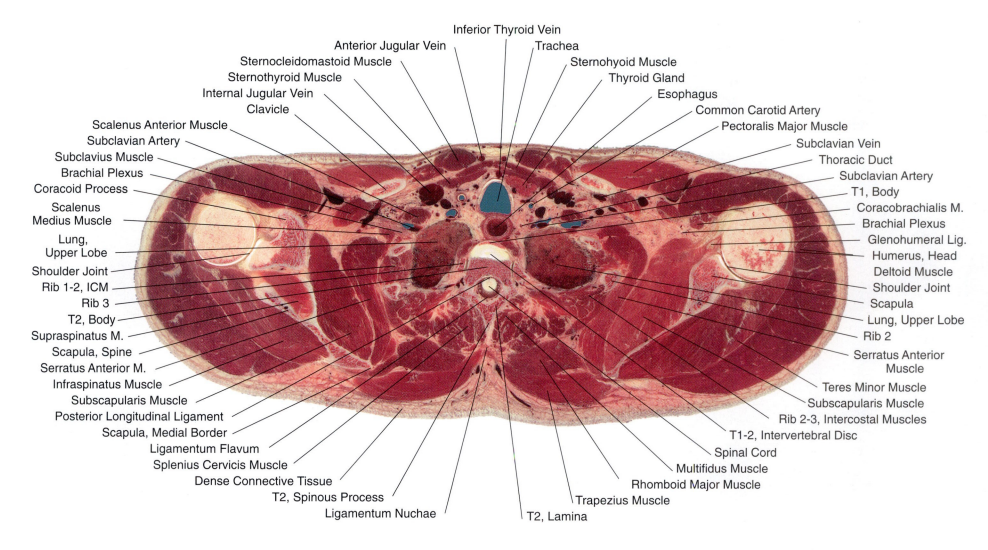

Inferior Thyroid Vein

Anterior Jugular Vein
Sternocleidomastoid Muscle
Sternothyroid Muscle
Internal Jugular Vein
Clavicle

Trachea
Sternohyoid Muscle
Thyroid Gland
Esophagus
Common Carotid Artery
Pectoralis Major Muscle

Scalenus Anterior Muscle
Subclavian Artery
Subclavius Muscle
Brachial Plexus
Coracoid Process
Scalenus
Medius Muscle
Lung,
Upper Lobe
Shoulder Joint
Rib 1-2, ICM
Rib 3
T2, Body
Supraspinatus M.
Scapula, Spine
Serratus Anterior M.
Infraspinatus Muscle
Subscapularis Muscle
Posterior Longitudinal Ligament
Scapula, Medial Border
Ligamentum Flavum
Splenius Cervicis Muscle
Dense Connective Tissue
T2, Spinous Process
Ligamentum Nuchae

Subclavian Vein
Thoracic Duct
Subclavian Artery
T1, Body
Coracobrachialis M.
Brachial Plexus
Glenohumeral Lig.
Humerus, Head
Deltoid Muscle
Shoulder Joint
Scapula
Lung, Upper Lobe
Rib 2
Serratus Anterior
Muscle
Teres Minor Muscle
Subscapularis Muscle
Rib 2-3, Intercostal Muscles
T1-2, Intervertebral Disc
Spinal Cord
Multifidus Muscle
Rhomboid Major Muscle
Trapezius Muscle
T2, Lamina

right

left

ICM=Intercostal Muscles

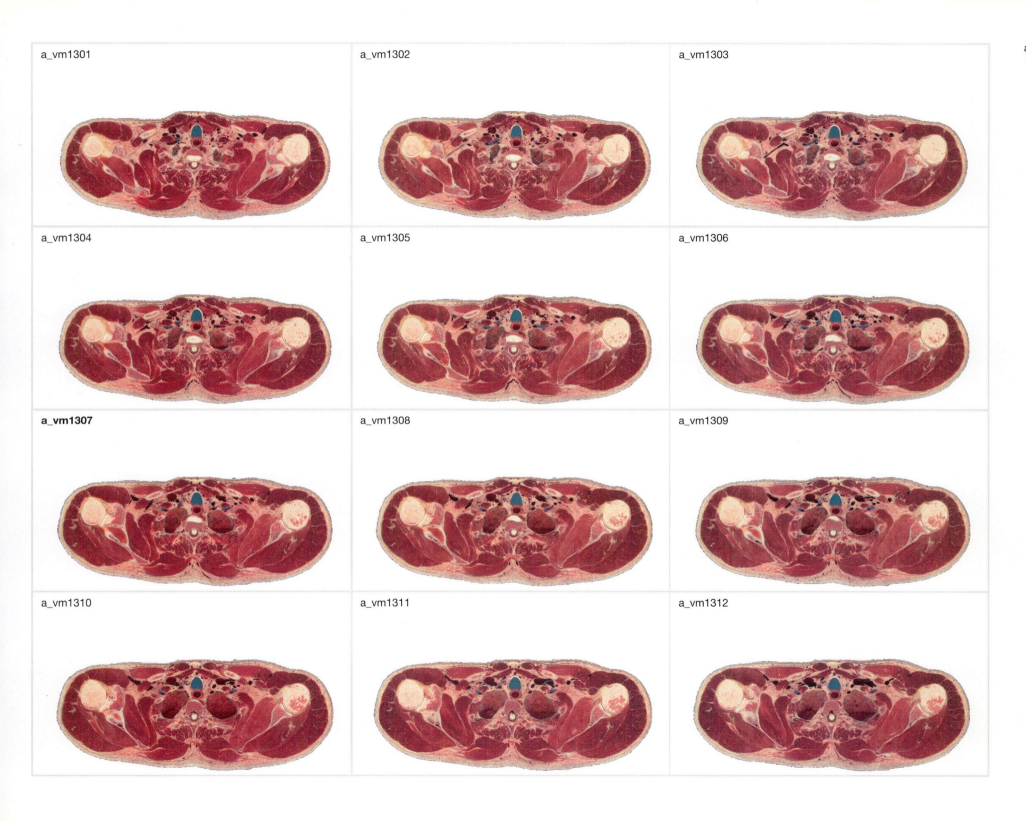

a_vm1301

a_vm1302

a_vm1303

a_vm1304

a_vm1305

a_vm1306

a_vm1307

a_vm1308

a_vm1309

a_vm1310

a_vm1311

a_vm1312

anterior

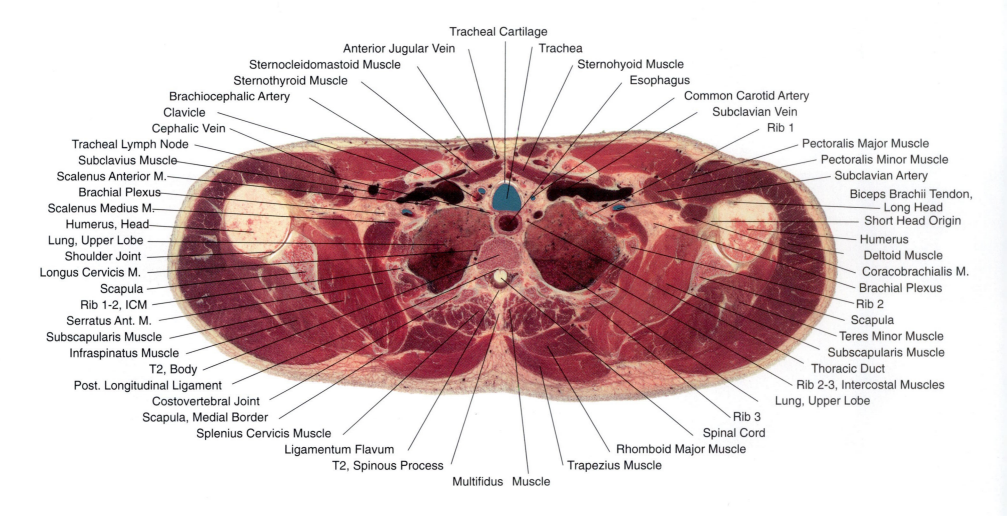

Tracheal Cartilage

Anterior Jugular Vein

Trachea

Sternocleidomastoid Muscle

Sternohyoid Muscle

Sternothyroid Muscle

Esophagus

Brachiocephalic Artery

Common Carotid Artery

Clavicle

Subclavian Vein

Cephalic Vein

Rib 1

Tracheal Lymph Node

Pectoralis Major Muscle

Subclavius Muscle

Pectoralis Minor Muscle

Scalenus Anterior M.

Subclavian Artery

Brachial Plexus

Biceps Brachii Tendon,

Scalenus Medius M.

Long Head

Humerus, Head

Short Head Origin

Lung, Upper Lobe

Humerus

Shoulder Joint

Deltoid Muscle

Longus Cervicis M.

Coracobrachialis M.

Scapula

Brachial Plexus

Rib 1-2, ICM

Rib 2

Serratus Ant. M.

Scapula

Subscapularis Muscle

Teres Minor Muscle

Infraspinatus Muscle

Subscapularis Muscle

T2, Body

Thoracic Duct

Post. Longitudinal Ligament

Rib 2-3, Intercostal Muscles

Costovertebral Joint

Lung, Upper Lobe

Scapula, Medial Border

Splenius Cervicis Muscle

Rib 3

Ligamentum Flavum

Spinal Cord

T2, Spinous Process

Rhomboid Major Muscle

Multifidus Muscle

Trapezius Muscle

right

left

posterior

ICM=Intercostal Muscles

54

a_vm1313

a_vm1314

a_vm1315

a_vm1316

a_vm1317

a_vm1318

a_vm1319

a_vm1320

a_vm1321

a_vm1322

a_vm1323

a_vm1324

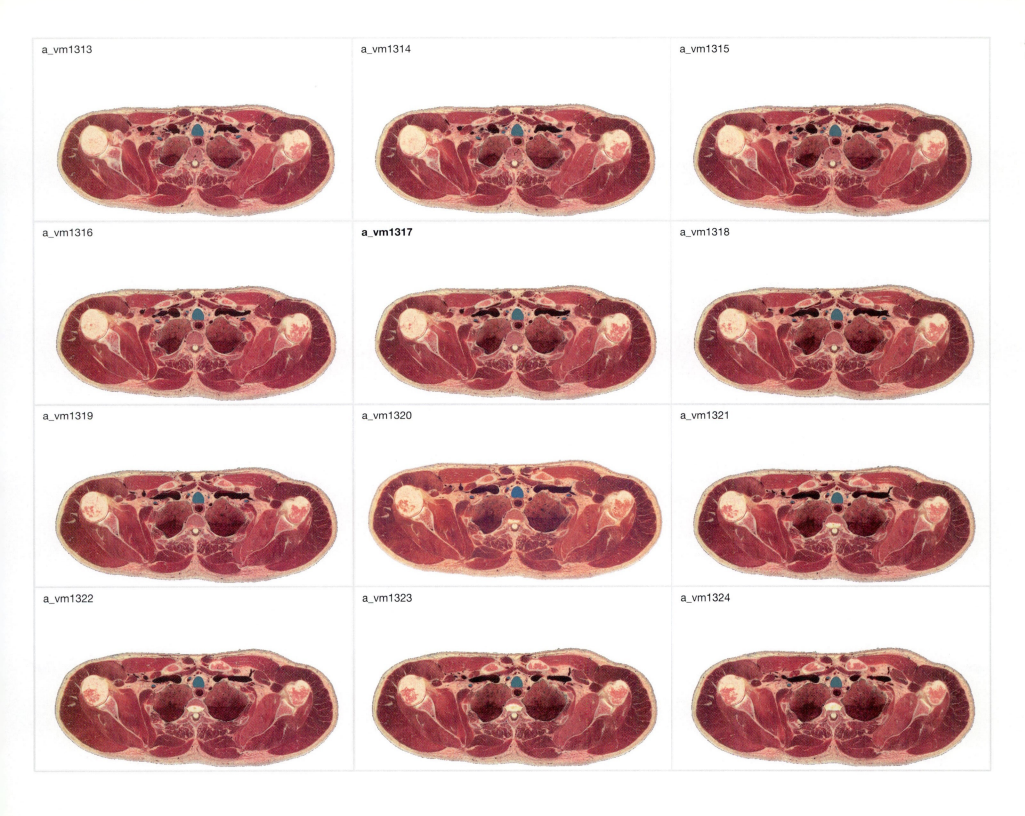

anterior

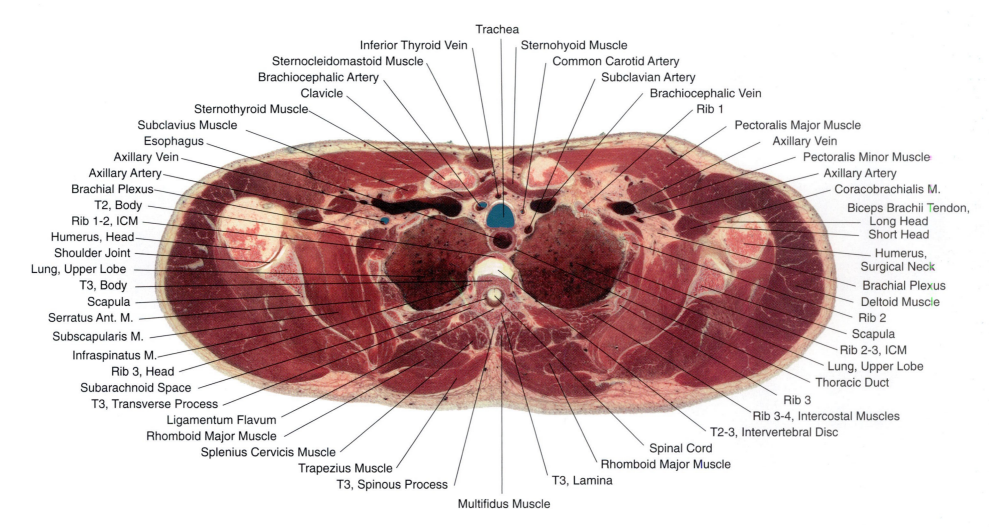

Trachea
Inferior Thyroid Vein
Sternohyoid Muscle
Sternocleidomastoid Muscle
Common Carotid Artery
Brachiocephalic Artery
Subclavian Artery
Clavicle
Brachiocephalic Vein
Sternothyroid Muscle
Rib 1
Subclavius Muscle
Pectoralis Major Muscle
Esophagus
Axillary Vein
Axillary Vein
Pectoralis Minor Muscle
Axillary Artery
Axillary Artery
Brachial Plexus
Coracobrachialis M.
T2, Body
Biceps Brachii Tendon,
Rib 1-2, ICM
Long Head
Humerus, Head
Short Head
Shoulder Joint
Humerus,
Lung, Upper Lobe
Surgical Neck
T3, Body
Brachial Plexus
Scapula
Deltoid Muscle
Serratus Ant. M.
Rib 2
Subscapularis M.
Scapula
Infraspinatus M.
Rib 2-3, ICM
Rib 3, Head
Lung, Upper Lobe
Subarachnoid Space
Thoracic Duct
T3, Transverse Process
Rib 3
Ligamentum Flavum
Rib 3-4, Intercostal Muscles
Rhomboid Major Muscle
T2-3, Intervertebral Disc
Splenius Cervicis Muscle
Spinal Cord
Trapezius Muscle
Rhomboid Major Muscle
T3, Spinous Process
T3, Lamina
Multifidus Muscle

right

left

ICM=Intercostal Muscles

posterior

56

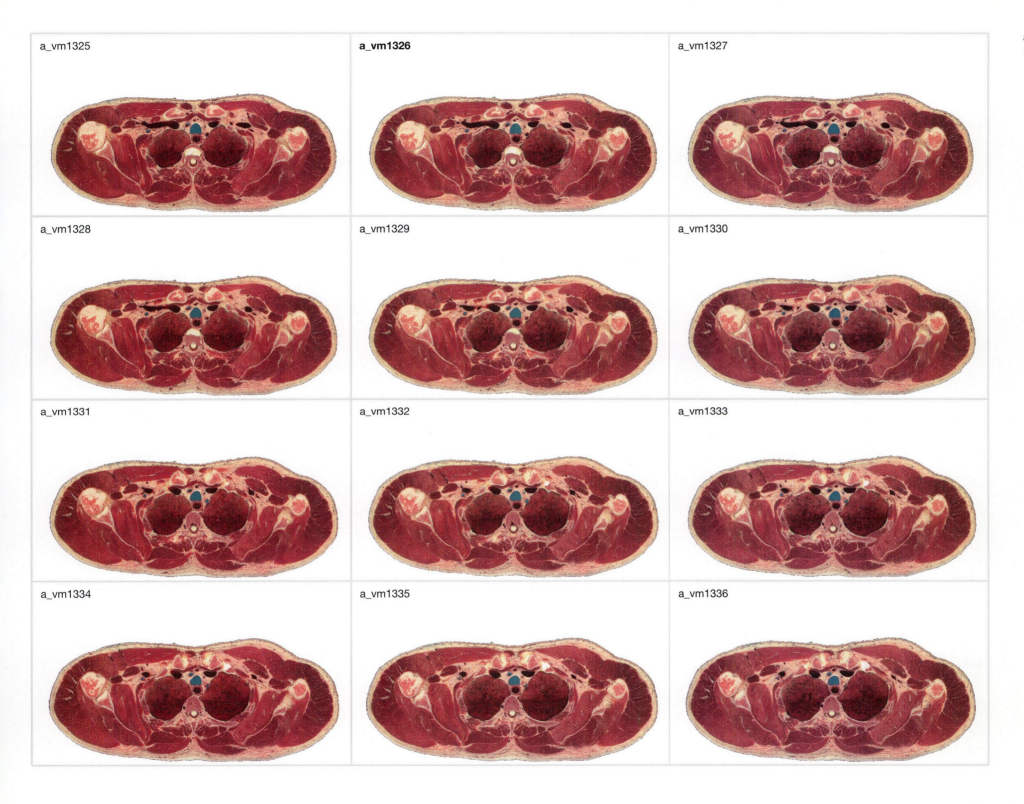

a_vm1325

a_vm1326

a_vm1327

a_vm1328

a_vm1329

a_vm1330

a_vm1331

a_vm1332

a_vm1333

a_vm1334

a_vm1335

a_vm1336

anterior

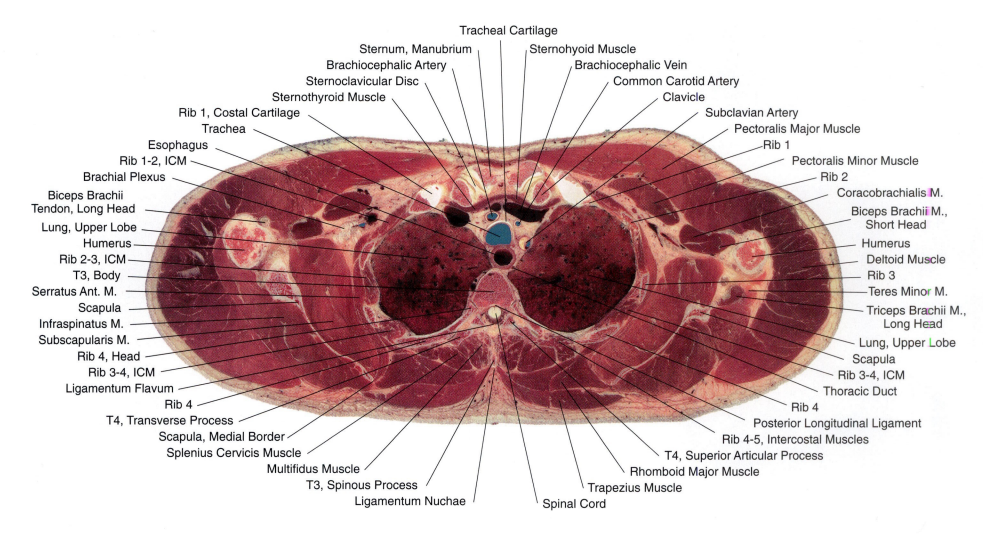

Tracheal Cartilage

Sternum, Manubrium

Sternohyoid Muscle

Brachiocephalic Artery

Brachiocephalic Vein

Sternoclavicular Disc

Common Carotid Artery

Sternothyroid Muscle

Clavicle

Rib 1, Costal Cartilage

Subclavian Artery

Trachea

Pectoralis Major Muscle

Esophagus

Rib 1

Rib 1-2, ICM

Pectoralis Minor Muscle

Brachial Plexus

Rib 2

Biceps Brachii
Tendon, Long Head

Coracobrachialis M.

Biceps Brachii M.,
Short Head

Lung, Upper Lobe

Humerus

Humerus

Rib 2-3, ICM

Deltoid Muscle

T3, Body

Rib 3

Serratus Ant. M.

Teres Minor M.

Scapula

Triceps Brachii M.,
Long Head

Infraspinatus M.

Subscapularis M.

Lung, Upper Lobe

Rib 4, Head

Scapula

Rib 3-4, ICM

Rib 3-4, ICM

Ligamentum Flavum

Thoracic Duct

Rib 4

Rib 4

T4, Transverse Process

Posterior Longitudinal Ligament

Scapula, Medial Border

Rib 4-5, Intercostal Muscles

Splenius Cervicis Muscle

T4, Superior Articular Process

Multifidus Muscle

Rhomboid Major Muscle

T3, Spinous Process

Trapezius Muscle

Ligamentum Nuchae

Spinal Cord

right

left

ICM=Intercostal Muscles

posterior

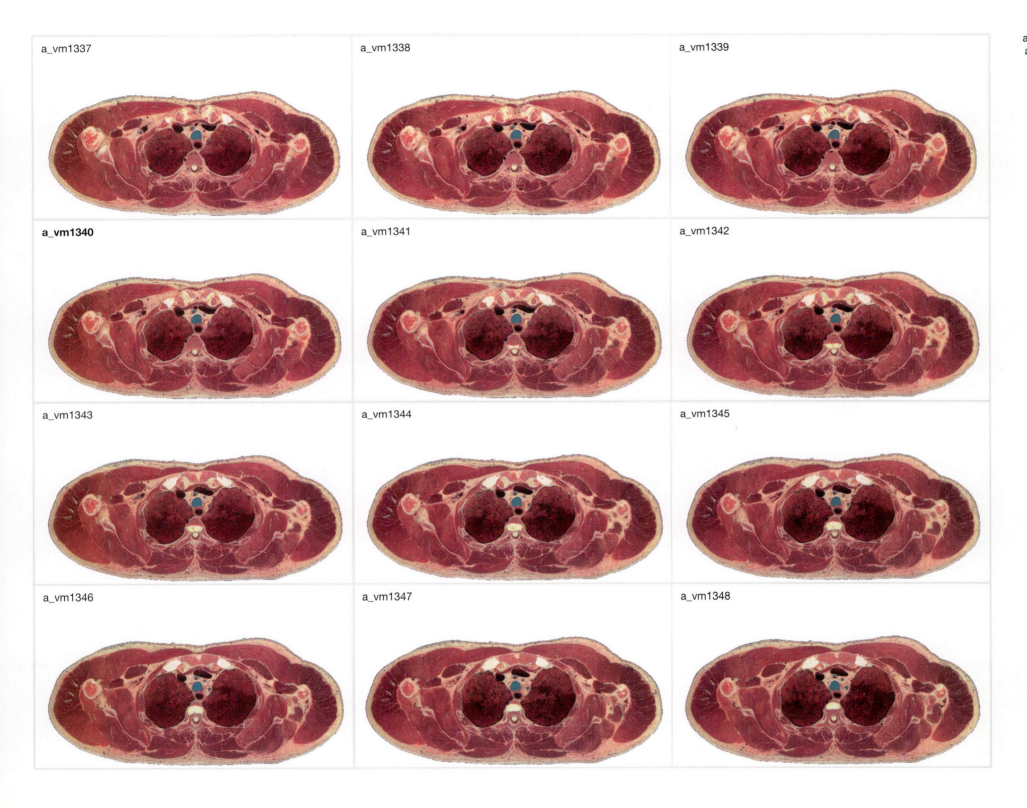

a_vm1337

a_vm1338

a_vm1339

a_vm1340

a_vm1341

a_vm1342

a_vm1343

a_vm1344

a_vm1345

a_vm1346

a_vm1347

a_vm1348

anterior

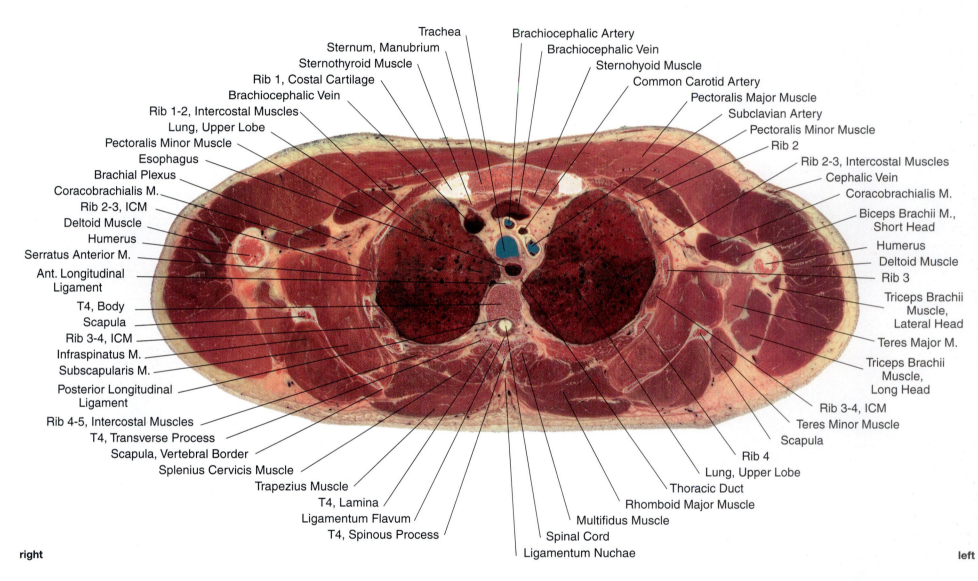

Trachea
Sternum, Manubrium
Sternothyroid Muscle
Rib 1, Costal Cartilage
Brachiocephalic Vein
Rib 1-2, Intercostal Muscles
Lung, Upper Lobe
Pectoralis Minor Muscle
Esophagus
Brachial Plexus
Coracobrachialis M.
Rib 2-3, ICM
Deltoid Muscle
Humerus
Serratus Anterior M.
Ant. Longitudinal Ligament
T4, Body
Scapula
Rib 3-4, ICM
Infraspinatus M.
Subscapularis M.
Posterior Longitudinal Ligament
Rib 4-5, Intercostal Muscles
T4, Transverse Process
Scapula, Vertebral Border
Splenius Cervicis Muscle
Trapezius Muscle
T4, Lamina
Ligamentum Flavum
T4, Spinous Process

Brachiocephalic Artery
Brachiocephalic Vein
Sternohyoid Muscle
Common Carotid Artery
Pectoralis Major Muscle
Subclavian Artery
Pectoralis Minor Muscle
Rib 2
Rib 2-3, Intercostal Muscles
Cephalic Vein
Coracobrachialis M.
Biceps Brachii M., Short Head
Humerus
Deltoid Muscle
Rib 3
Triceps Brachii Muscle, Lateral Head
Teres Major M.
Triceps Brachii Muscle, Long Head
Rib 3-4, ICM
Teres Minor Muscle
Scapula
Rib 4
Lung, Upper Lobe
Thoracic Duct
Rhomboid Major Muscle
Multifidus Muscle
Spinal Cord
Ligamentum Nuchae

right

left

ICM=Intercostal Muscles

posterior

60

a_vm1349

a_vm1350

a_vm1351

a_vm1352

a_vm1353

a_vm1354

a_vm1355

a_vm1356

a_vm1357

a_vm1358

a_vm1359

a_vm1360

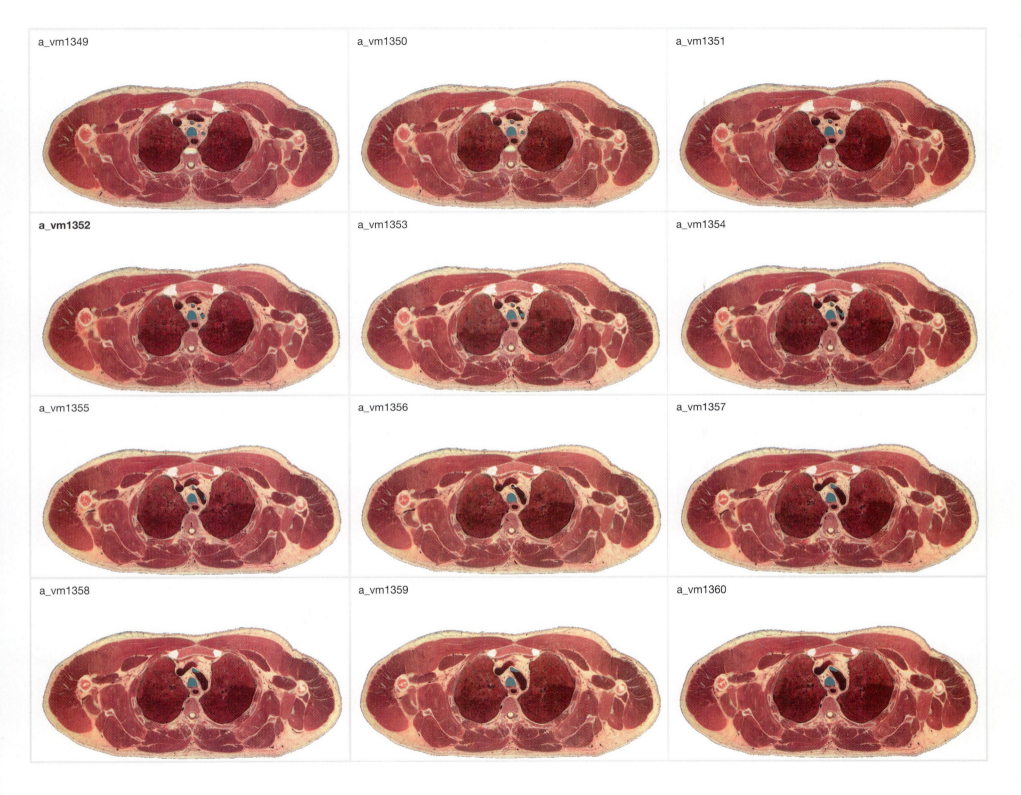

anterior

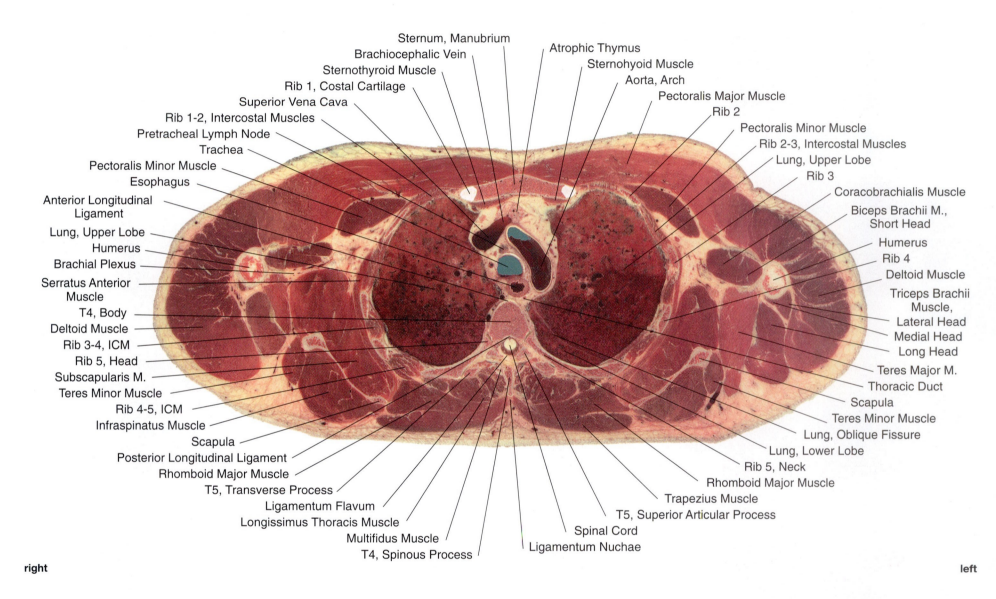

Sternum, Manubrium

Brachiocephalic Vein

Sternothyroid Muscle

Rib 1, Costal Cartilage

Superior Vena Cava

Rib 1-2, Intercostal Muscles

Pretracheal Lymph Node

Trachea

Pectoralis Minor Muscle

Esophagus

Anterior Longitudinal Ligament

Lung, Upper Lobe

Humerus

Brachial Plexus

Serratus Anterior Muscle

T4, Body

Deltoid Muscle

Rib 3-4, ICM

Rib 5, Head

Subscapularis M.

Teres Minor Muscle

Rib 4-5, ICM

Infraspinatus Muscle

Scapula

Posterior Longitudinal Ligament

Rhomboid Major Muscle

T5, Transverse Process

Ligamentum Flavum

Longissimus Thoracis Muscle

Multifidus Muscle

T4, Spinous Process

Atrophic Thymus

Sternohyoid Muscle

Aorta, Arch

Pectoralis Major Muscle

Rib 2

Pectoralis Minor Muscle

Rib 2-3, Intercostal Muscles

Lung, Upper Lobe

Rib 3

Coracobrachialis Muscle

Biceps Brachii M., Short Head

Humerus

Rib 4

Deltoid Muscle

Triceps Brachii Muscle, Lateral Head

Medial Head

Long Head

Teres Major M.

Thoracic Duct

Scapula

Teres Minor Muscle

Lung, Oblique Fissure

Lung, Lower Lobe

Rib 5, Neck

Rhomboid Major Muscle

Trapezius Muscle

T5, Superior Articular Process

Spinal Cord

Ligamentum Nuchae

right

left

ICM=Intercostal Muscles

posterior

62

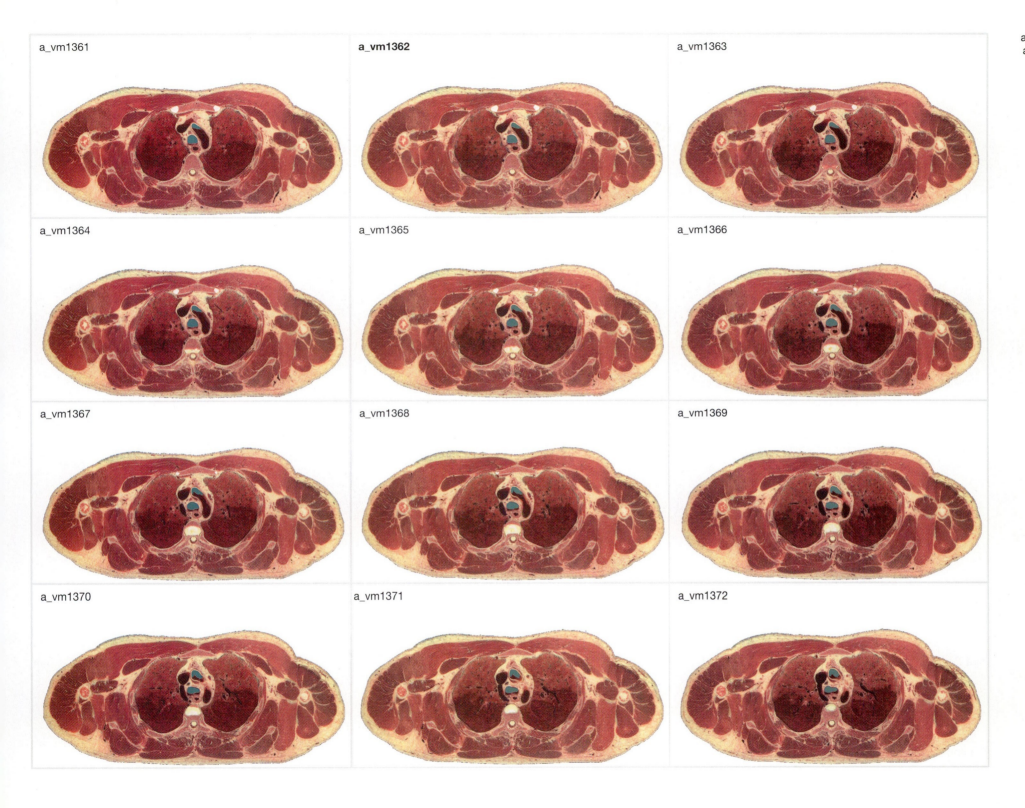

a_vm1361

a_vm1362

a_vm1363

a_vm1364

a_vm1365

a_vm1366

a_vm1367

a_vm1368

a_vm1369

a_vm1370

a_vm1371

a_vm1372

anterior

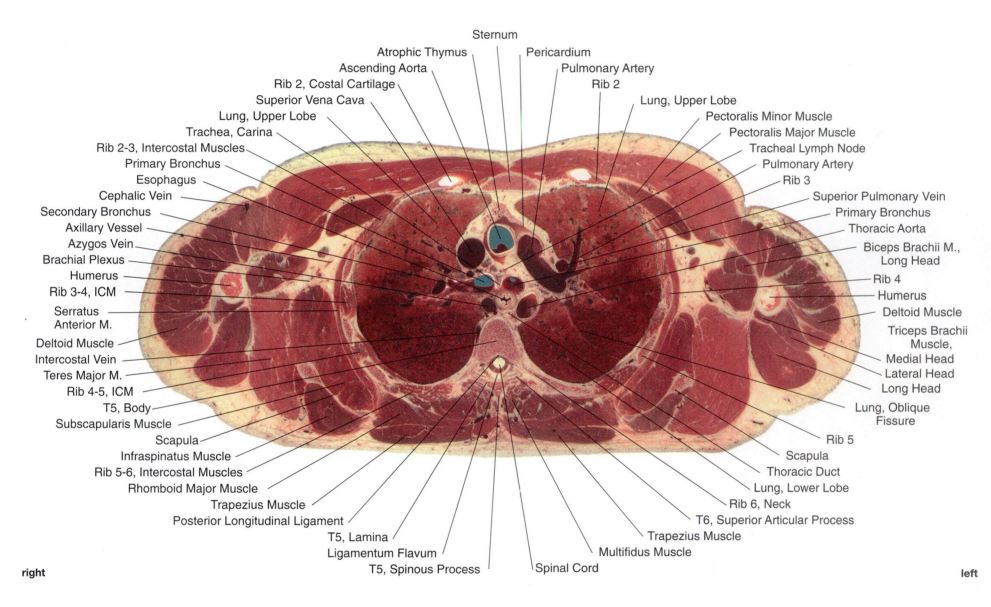

Sternum
Atrophic Thymus
Pericardium
Ascending Aorta
Pulmonary Artery
Rib 2, Costal Cartilage
Rib 2
Superior Vena Cava
Lung, Upper Lobe
Lung, Upper Lobe
Pectoralis Minor Muscle
Trachea, Carina
Pectoralis Major Muscle
Rib 2-3, Intercostal Muscles
Tracheal Lymph Node
Primary Bronchus
Pulmonary Artery
Esophagus
Rib 3
Cephalic Vein
Superior Pulmonary Vein
Secondary Bronchus
Primary Bronchus
Axillary Vessel
Thoracic Aorta
Azygos Vein
Biceps Brachii M.,
Brachial Plexus
Long Head
Humerus
Rib 4
Rib 3-4, ICM
Humerus
Serratus
Deltoid Muscle
Anterior M.
Triceps Brachii
Deltoid Muscle
Muscle,
Intercostal Vein
Medial Head
Teres Major M.
Lateral Head
Rib 4-5, ICM
Long Head
T5, Body
Lung, Oblique
Subscapularis Muscle
Fissure
Scapula
Rib 5
Infraspinatus Muscle
Scapula
Rib 5-6, Intercostal Muscles
Thoracic Duct
Rhomboid Major Muscle
Lung, Lower Lobe
Trapezius Muscle
Rib 6, Neck
Posterior Longitudinal Ligament
T6, Superior Articular Process
T5, Lamina
Trapezius Muscle
Ligamentum Flavum
Multifidus Muscle
T5, Spinous Process
Spinal Cord

right

left

ICM=Intercostal Muscles

posterior

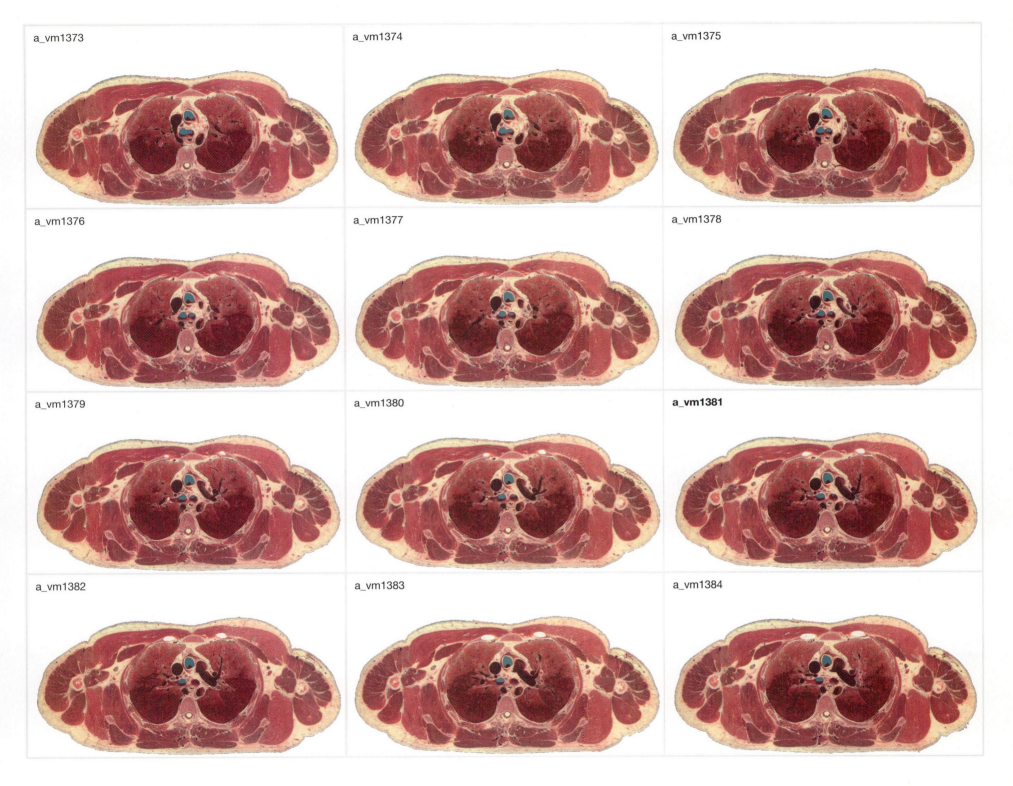

a_vm1373

a_vm1374

a_vm1375

a_vm1376

a_vm1377

a_vm1378

a_vm1379

a_vm1380

a_vm1381

a_vm1382

a_vm1383

a_vm1384

anterior

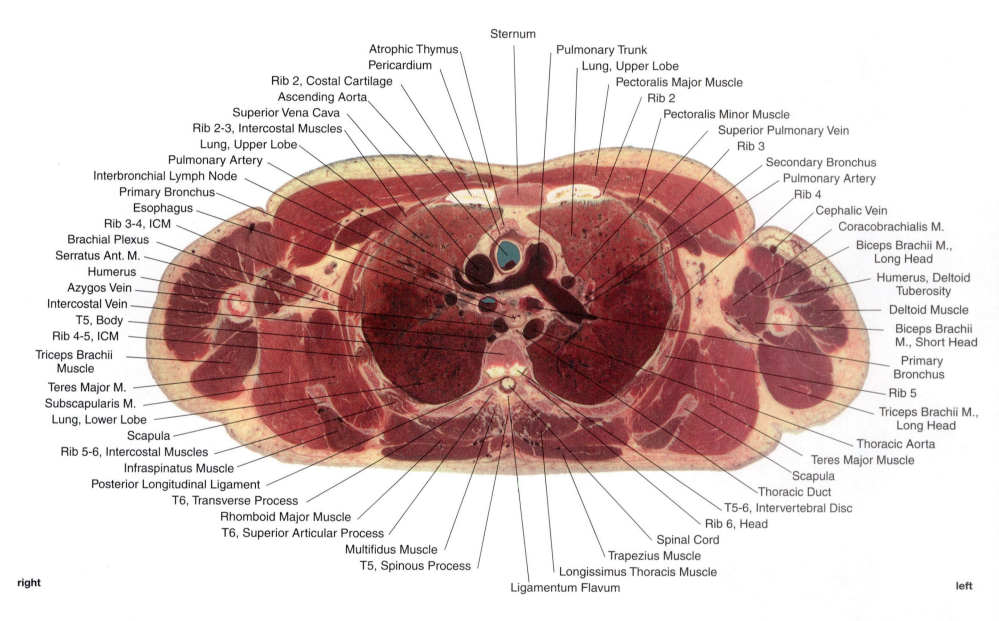

Sternum

Atrophic Thymus
Pericardium
Rib 2, Costal Cartilage
Ascending Aorta
Superior Vena Cava
Rib 2-3, Intercostal Muscles
Lung, Upper Lobe
Pulmonary Artery
Interbronchial Lymph Node
Primary Bronchus
Esophagus
Rib 3-4, ICM
Brachial Plexus
Serratus Ant. M.
Humerus
Azygos Vein
Intercostal Vein
T5, Body
Rib 4-5, ICM
Triceps Brachii
Muscle
Teres Major M.
Subscapularis M.
Lung, Lower Lobe
Scapula
Rib 5-6, Intercostal Muscles
Infraspinatus Muscle
Posterior Longitudinal Ligament
T6, Transverse Process
Rhomboid Major Muscle
T6, Superior Articular Process
Multifidus Muscle
T5, Spinous Process

Pulmonary Trunk
Lung, Upper Lobe
Pectoralis Major Muscle
Rib 2
Pectoralis Minor Muscle
Superior Pulmonary Vein
Rib 3
Secondary Bronchus
Pulmonary Artery
Rib 4
Cephalic Vein
Coracobrachialis M.
Biceps Brachii M.,
Long Head
Humerus, Deltoid
Tuberosity
Deltoid Muscle
Biceps Brachii
M., Short Head
Primary
Bronchus
Rib 5
Triceps Brachii M.,
Long Head
Thoracic Aorta
Teres Major Muscle
Scapula
Thoracic Duct
T5-6, Intervertebral Disc
Rib 6, Head
Spinal Cord
Trapezius Muscle
Longissimus Thoracis Muscle
Ligamentum Flavum

right

left

ICM=Intercostal Muscles

posterior

66

a_vm1385

a_vm1386

a_vm1387

a_vm1388

a_vm1389

a_vm1390

a_vm1391

a_vm1392

a_vm1393

a_vm1394

a_vm1395

a_vm1396

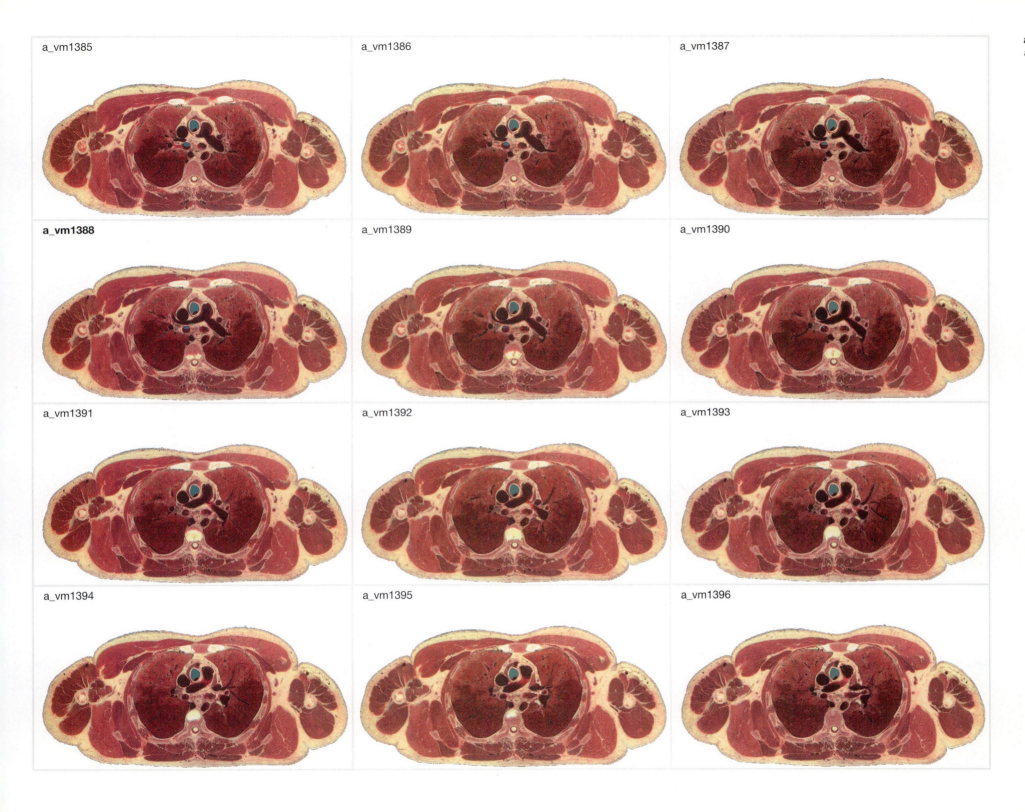

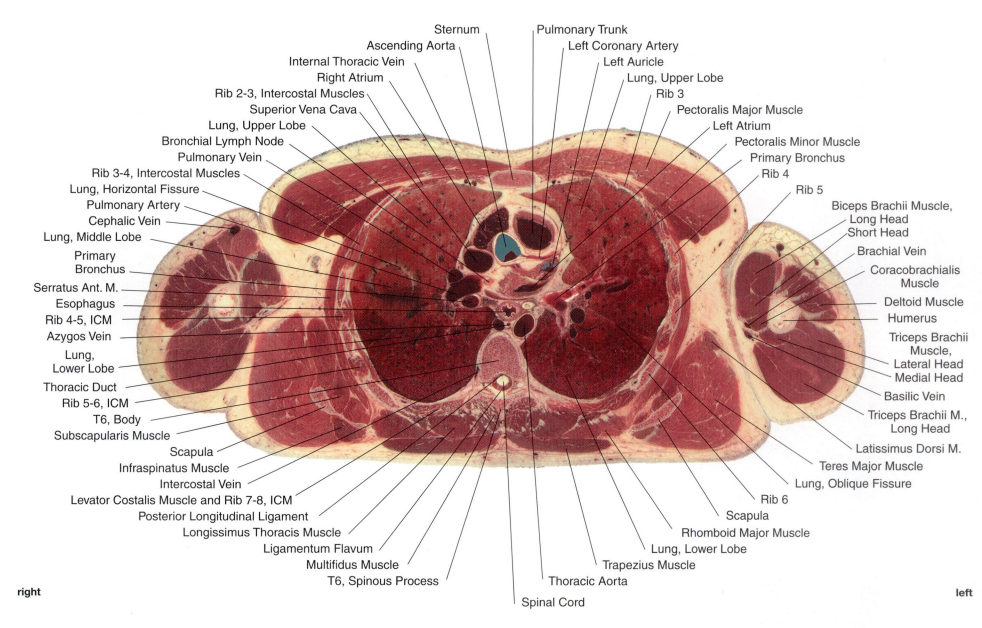

anterior

Sternum
Ascending Aorta
Internal Thoracic Vein
Right Atrium
Rib 2-3, Intercostal Muscles
Superior Vena Cava
Lung, Upper Lobe
Bronchial Lymph Node
Pulmonary Vein
Rib 3-4, Intercostal Muscles
Lung, Horizontal Fissure
Pulmonary Artery
Cephalic Vein
Lung, Middle Lobe
Primary Bronchus
Serratus Ant. M.
Esophagus
Rib 4-5, ICM
Azygos Vein
Lung, Lower Lobe
Thoracic Duct
Rib 5-6, ICM
T6, Body
Subscapularis Muscle
Scapula
Infraspinatus Muscle
Intercostal Vein
Levator Costalis Muscle and Rib 7-8, ICM
Posterior Longitudinal Ligament
Longissimus Thoracis Muscle
Ligamentum Flavum
Multifidus Muscle
T6, Spinous Process

Pulmonary Trunk
Left Coronary Artery
Left Auricle
Lung, Upper Lobe
Rib 3
Pectoralis Major Muscle
Left Atrium
Pectoralis Minor Muscle
Primary Bronchus
Rib 4
Rib 5
Biceps Brachii Muscle,
Long Head
Short Head
Brachial Vein
Coracobrachialis
Muscle
Deltoid Muscle
Humerus
Triceps Brachii
Muscle,
Lateral Head
Medial Head
Basilic Vein
Triceps Brachii M.,
Long Head
Latissimus Dorsi M.
Teres Major Muscle
Lung, Oblique Fissure
Rib 6
Scapula
Rhomboid Major Muscle
Lung, Lower Lobe
Trapezius Muscle
Thoracic Aorta
Spinal Cord

right

left

ICM=Intercostal Muscles

posterior

68

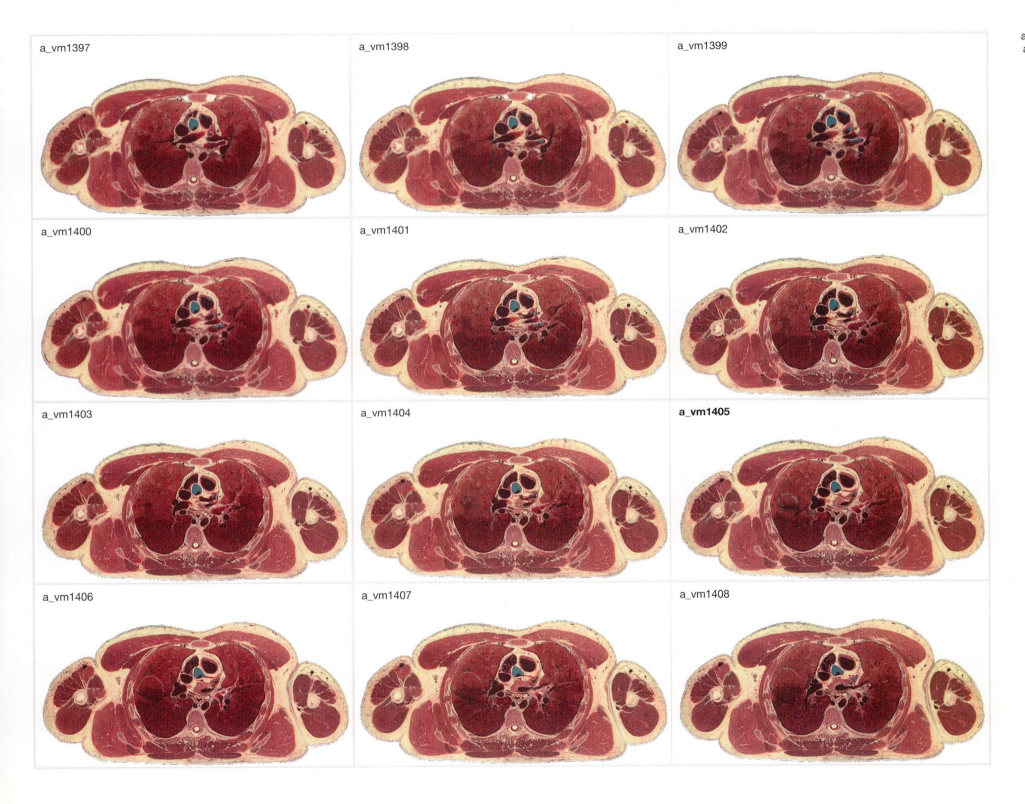

a_vm1397

a_vm1398

a_vm1399

a_vm1400

a_vm1401

a_vm1402

a_vm1403

a_vm1404

a_vm1405

a_vm1406

a_vm1407

a_vm1408

anterior

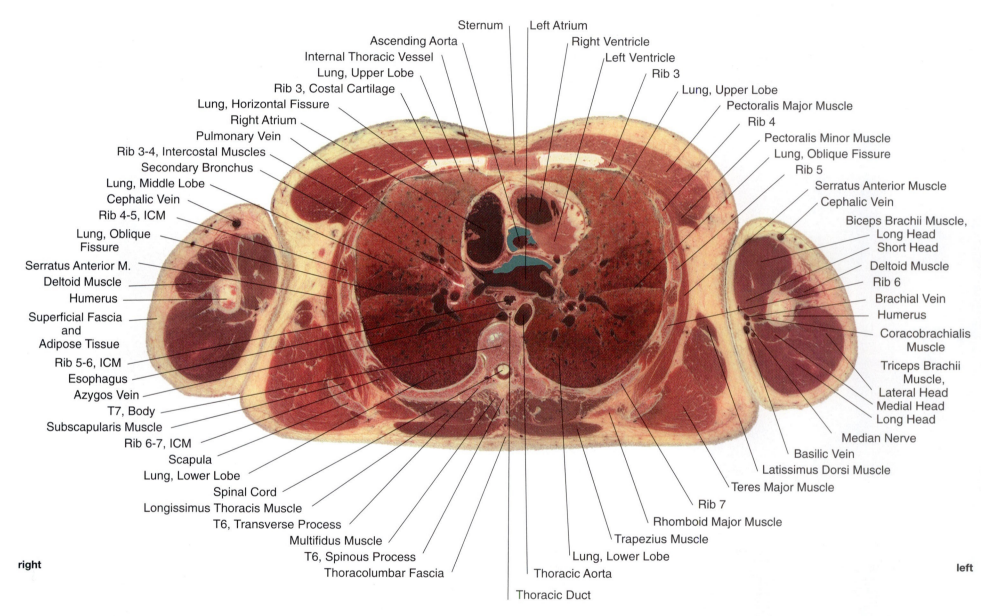

Sternum
Left Atrium
Ascending Aorta
Right Ventricle
Internal Thoracic Vessel
Left Ventricle
Lung, Upper Lobe
Rib 3
Rib 3, Costal Cartilage
Lung, Upper Lobe
Lung, Horizontal Fissure
Pectoralis Major Muscle
Right Atrium
Rib 4
Pulmonary Vein
Pectoralis Minor Muscle
Rib 3-4, Intercostal Muscles
Lung, Oblique Fissure
Secondary Bronchus
Rib 5
Lung, Middle Lobe
Serratus Anterior Muscle
Cephalic Vein
Cephalic Vein
Rib 4-5, ICM
Biceps Brachii Muscle,
Long Head
Lung, Oblique
Short Head
Fissure
Deltoid Muscle
Serratus Anterior M.
Rib 6
Deltoid Muscle
Brachial Vein
Humerus
Humerus
Superficial Fascia
Coracobrachialis
and
Muscle
Adipose Tissue
Triceps Brachii
Rib 5-6, ICM
Muscle,
Esophagus
Lateral Head
Azygos Vein
Medial Head
T7, Body
Long Head
Subscapularis Muscle
Median Nerve
Rib 6-7, ICM
Basilic Vein
Scapula
Latissimus Dorsi Muscle
Lung, Lower Lobe
Teres Major Muscle
Spinal Cord
Longissimus Thoracis Muscle
Rib 7
T6, Transverse Process
Rhomboid Major Muscle
Multifidus Muscle
Trapezius Muscle
T6, Spinous Process
Lung, Lower Lobe
Thoracolumbar Fascia
Thoracic Aorta

Thoracic Duct

right

left

ICM=Intercostal Muscles

posterior

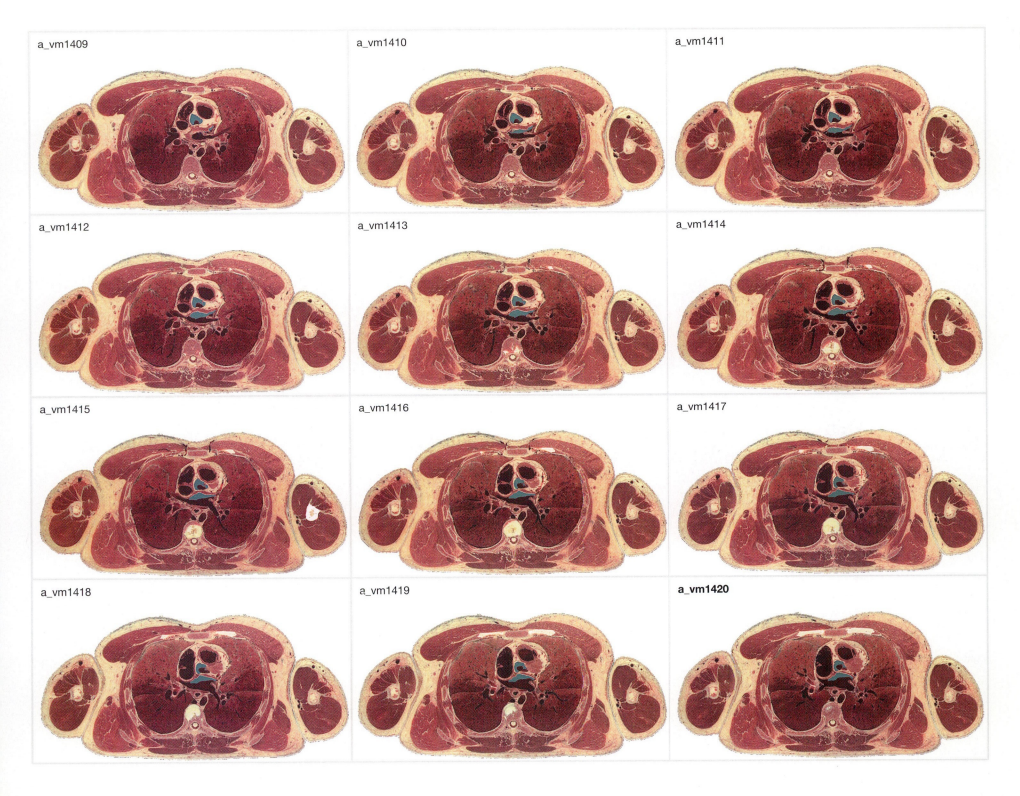

a_vm1409

a_vm1410

a_vm1411

a_vm1412

a_vm1413

a_vm1414

a_vm1415

a_vm1416

a_vm1417

a_vm1418

a_vm1419

a_vm1420

anterior

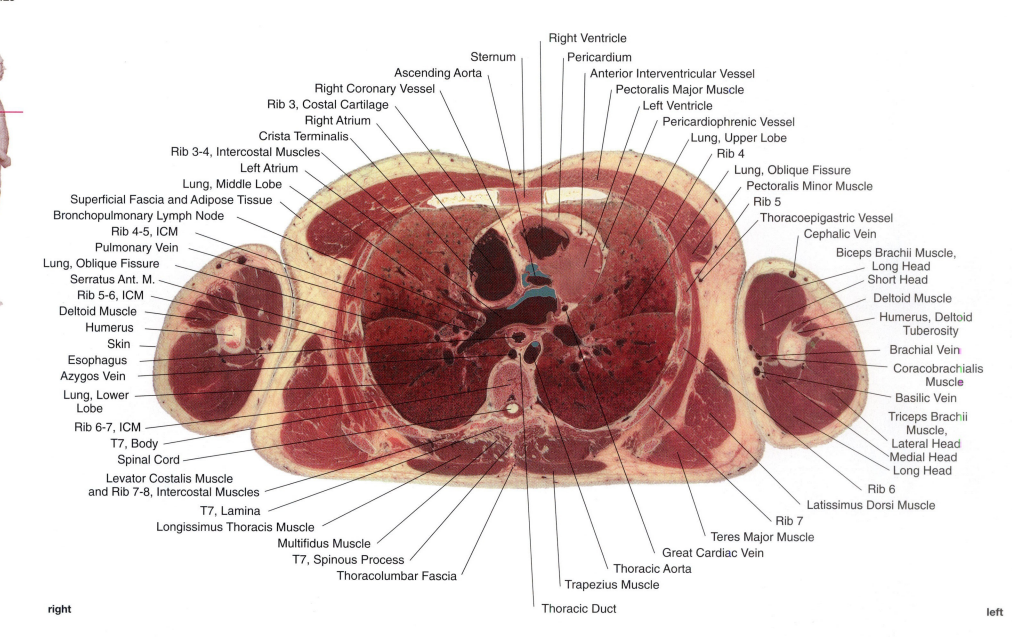

Right Ventricle

Sternum

Pericardium

Ascending Aorta

Anterior Interventricular Vessel

Right Coronary Vessel

Pectoralis Major Muscle

Rib 3, Costal Cartilage

Left Ventricle

Right Atrium

Pericardiophrenic Vessel

Crista Terminalis

Lung, Upper Lobe

Rib 3-4, Intercostal Muscles

Rib 4

Left Atrium

Lung, Oblique Fissure

Lung, Middle Lobe

Pectoralis Minor Muscle

Superficial Fascia and Adipose Tissue

Rib 5

Bronchopulmonary Lymph Node

Thoracoepigastric Vessel

Rib 4-5, ICM

Cephalic Vein

Pulmonary Vein

Biceps Brachii Muscle,
Long Head

Lung, Oblique Fissure

Short Head

Serratus Ant. M.

Deltoid Muscle

Rib 5-6, ICM

Humerus, Deltoid
Tuberosity

Deltoid Muscle

Humerus

Brachial Vein

Skin

Coracobrachialis
Muscle

Esophagus

Azygos Vein

Basilic Vein

Lung, Lower
Lobe

Triceps Brachii
Muscle,
Lateral Head

Rib 6-7, ICM

Medial Head

T7, Body

Long Head

Spinal Cord

Levator Costalis Muscle
and Rib 7-8, Intercostal Muscles

Rib 6

Latissimus Dorsi Muscle

T7, Lamina

Longissimus Thoracis Muscle

Rib 7

Multifidus Muscle

Teres Major Muscle

T7, Spinous Process

Great Cardiac Vein

Thoracolumbar Fascia

Thoracic Aorta

Trapezius Muscle

Thoracic Duct

right

left

ICM=Intercostal Muscles

posterior

72

a_vm1421

a_vm1422

a_vm1423

a_vm1424

a_vm1425

a_vm1426

a_vm1427

a_vm1428

a_vm1429

a_vm1430

a_vm1431

a_vm1432

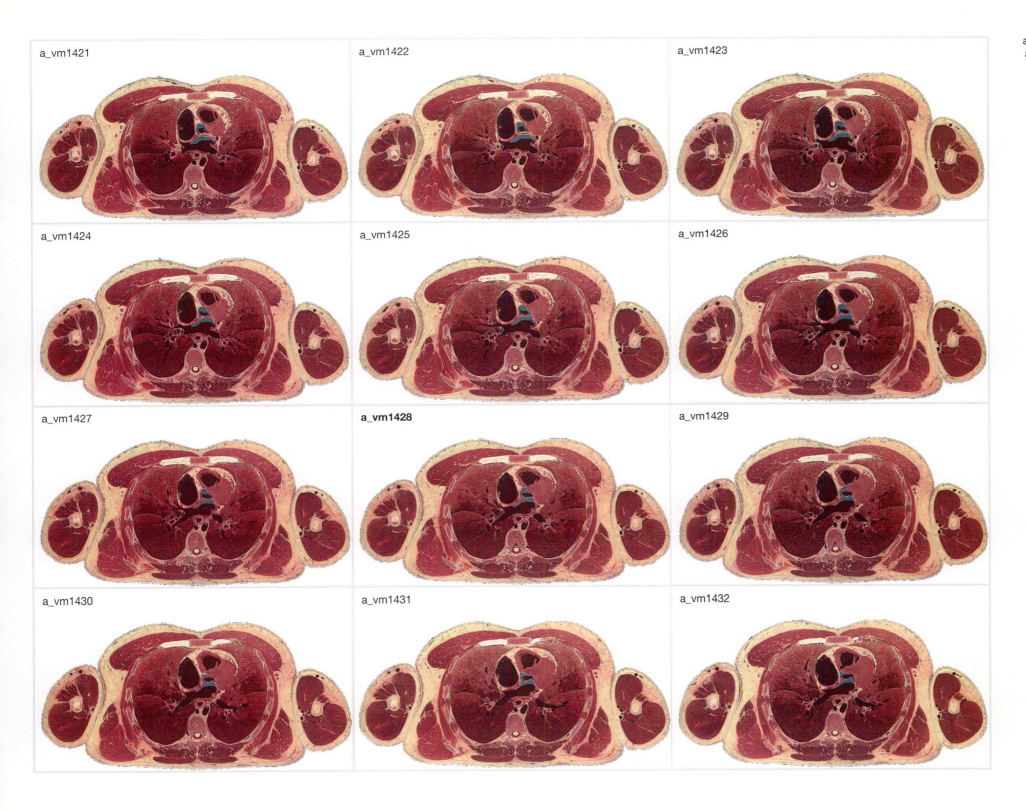

anterior

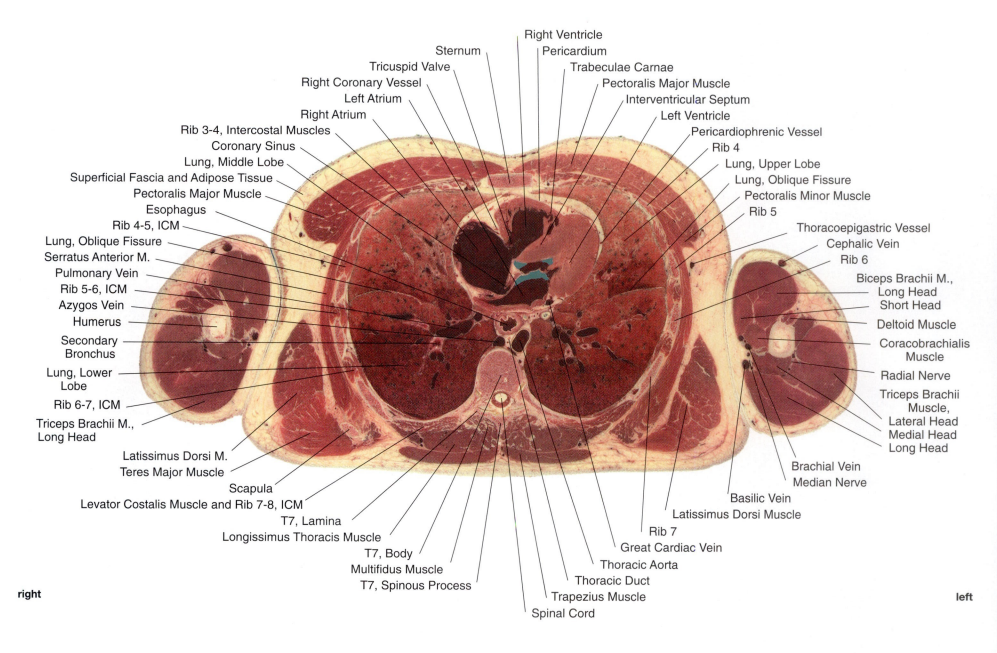

Right Ventricle

Sternum
Pericardium

Tricuspid Valve
Trabeculae Carnae

Right Coronary Vessel
Pectoralis Major Muscle

Left Atrium
Interventricular Septum

Right Atrium
Left Ventricle

Rib 3-4, Intercostal Muscles
Pericardiophrenic Vessel

Coronary Sinus
Rib 4

Lung, Middle Lobe
Lung, Upper Lobe

Superficial Fascia and Adipose Tissue
Lung, Oblique Fissure

Pectoralis Major Muscle
Pectoralis Minor Muscle

Esophagus
Rib 5

Rib 4-5, ICM
Thoracoepigastric Vessel

Lung, Oblique Fissure
Cephalic Vein

Serratus Anterior M.
Rib 6

Pulmonary Vein
Biceps Brachii M.,

Rib 5-6, ICM
Long Head

Azygos Vein
Short Head

Humerus
Deltoid Muscle

Secondary
Coracobrachialis
Bronchus
Muscle

Lung, Lower
Radial Nerve
Lobe

Rib 6-7, ICM
Triceps Brachii
Muscle,

Triceps Brachii M.,
Lateral Head
Long Head
Medial Head
Long Head

Latissimus Dorsi M.
Teres Major Muscle
Brachial Vein

Scapula
Median Nerve

Levator Costalis Muscle and Rib 7-8, ICM

T7, Lamina
Basilic Vein

Longissimus Thoracis Muscle
Latissimus Dorsi Muscle

Rib 7

T7, Body
Great Cardiac Vein

Multifidus Muscle
Thoracic Aorta

T7, Spinous Process
Thoracic Duct

Trapezius Muscle

Spinal Cord

right
left

ICM=Intercostal Muscles

posterior

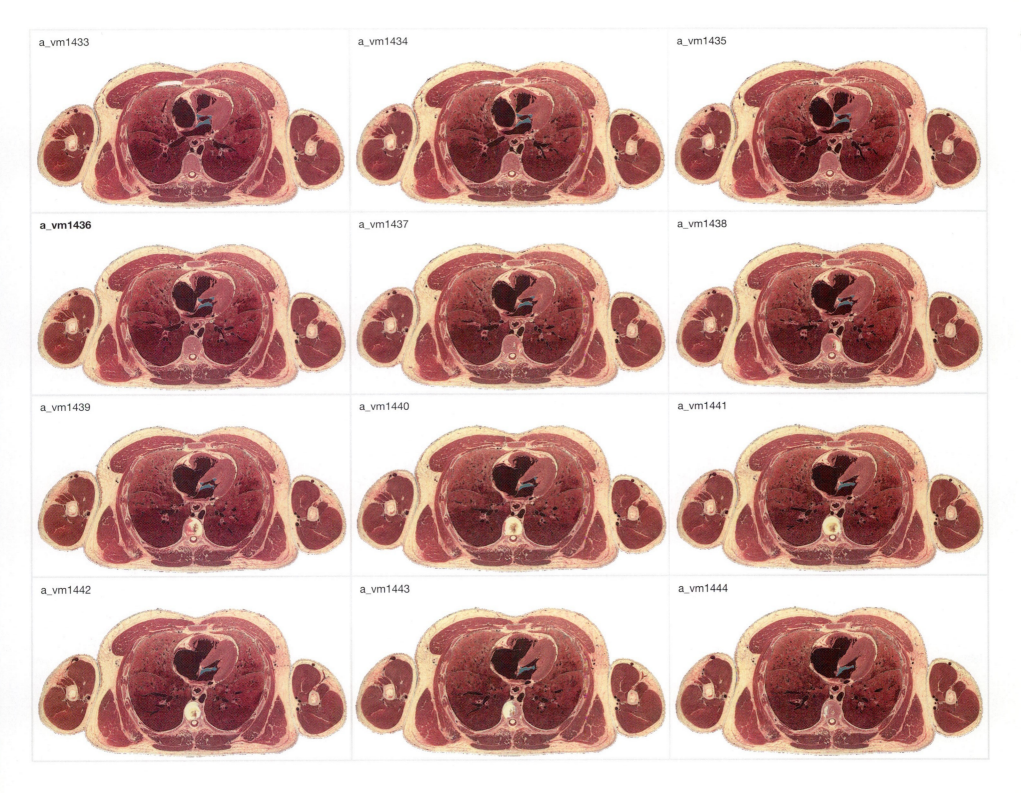

a_vm1433

a_vm1434

a_vm1435

a_vm1436

a_vm1437

a_vm1438

a_vm1439

a_vm1440

a_vm1441

a_vm1442

a_vm1443

a_vm1444

anterior

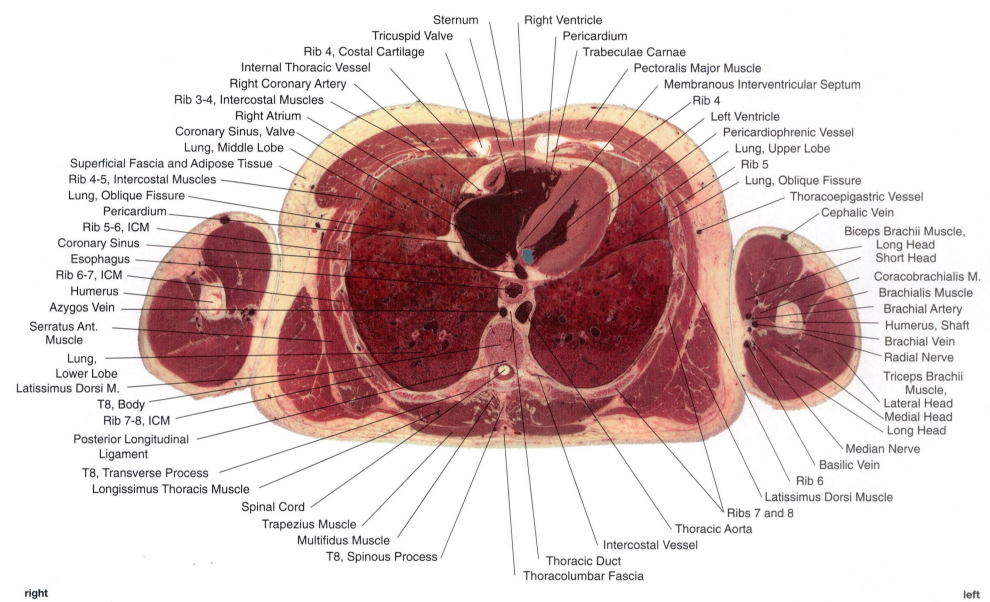

Sternum
Right Ventricle
Tricuspid Valve
Pericardium
Rib 4, Costal Cartilage
Trabeculae Carnae
Internal Thoracic Vessel
Pectoralis Major Muscle
Right Coronary Artery
Membranous Interventricular Septum
Rib 3-4, Intercostal Muscles
Rib 4
Right Atrium
Left Ventricle
Coronary Sinus, Valve
Pericardiophrenic Vessel
Lung, Middle Lobe
Lung, Upper Lobe
Superficial Fascia and Adipose Tissue
Rib 5
Rib 4-5, Intercostal Muscles
Lung, Oblique Fissure
Lung, Oblique Fissure
Thoracoepigastric Vessel
Pericardium
Cephalic Vein
Rib 5-6, ICM
Biceps Brachii Muscle,
Coronary Sinus
Long Head
Esophagus
Short Head
Rib 6-7, ICM
Coracobrachialis M.
Humerus
Brachialis Muscle
Azygos Vein
Brachial Artery
Serratus Ant.
Humerus, Shaft
Muscle
Brachial Vein
Lung,
Radial Nerve
Lower Lobe
Triceps Brachii
Latissimus Dorsi M.
Muscle,
T8, Body
Lateral Head
Rib 7-8, ICM
Medial Head
Posterior Longitudinal
Long Head
Ligament
Median Nerve
T8, Transverse Process
Basilic Vein
Longissimus Thoracis Muscle
Rib 6
Spinal Cord
Latissimus Dorsi Muscle
Trapezius Muscle
Ribs 7 and 8
Multifidus Muscle
Thoracic Aorta
T8, Spinous Process
Intercostal Vessel
Thoracic Duct
Thoracolumbar Fascia

ICM=Intercostal Muscles

posterior

right

left

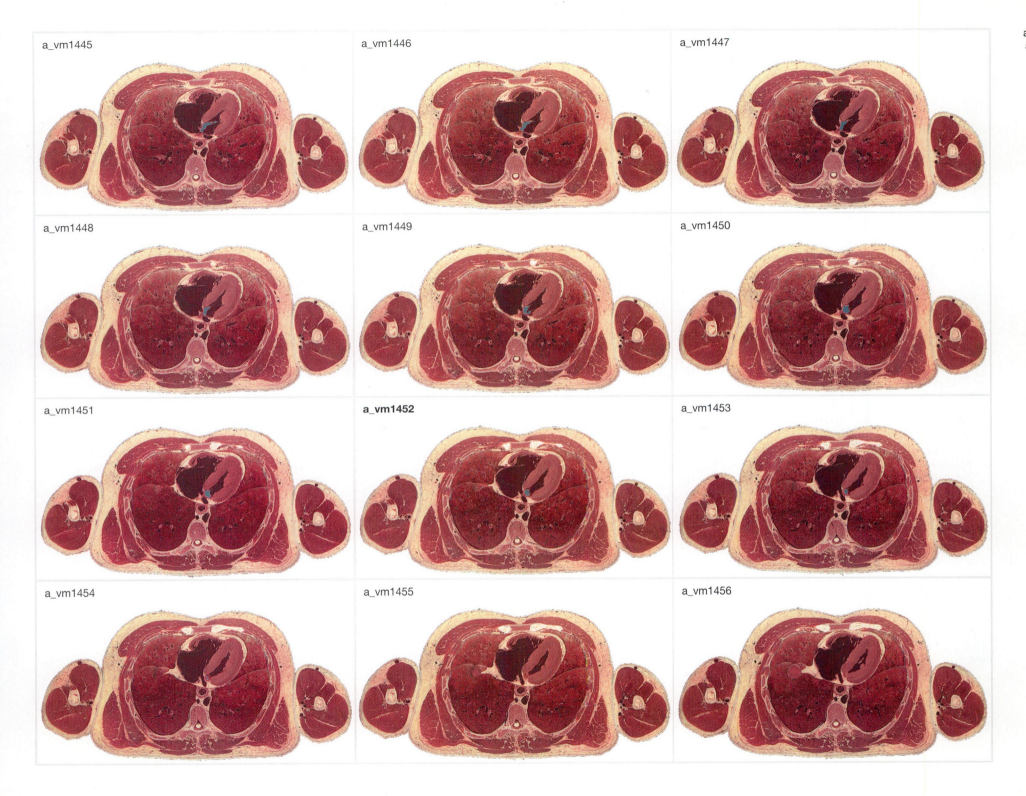

a_vm1445

a_vm1446

a_vm1447

a_vm1448

a_vm1449

a_vm1450

a_vm1451

a_vm1452

a_vm1453

a_vm1454

a_vm1455

a_vm1456

anterior

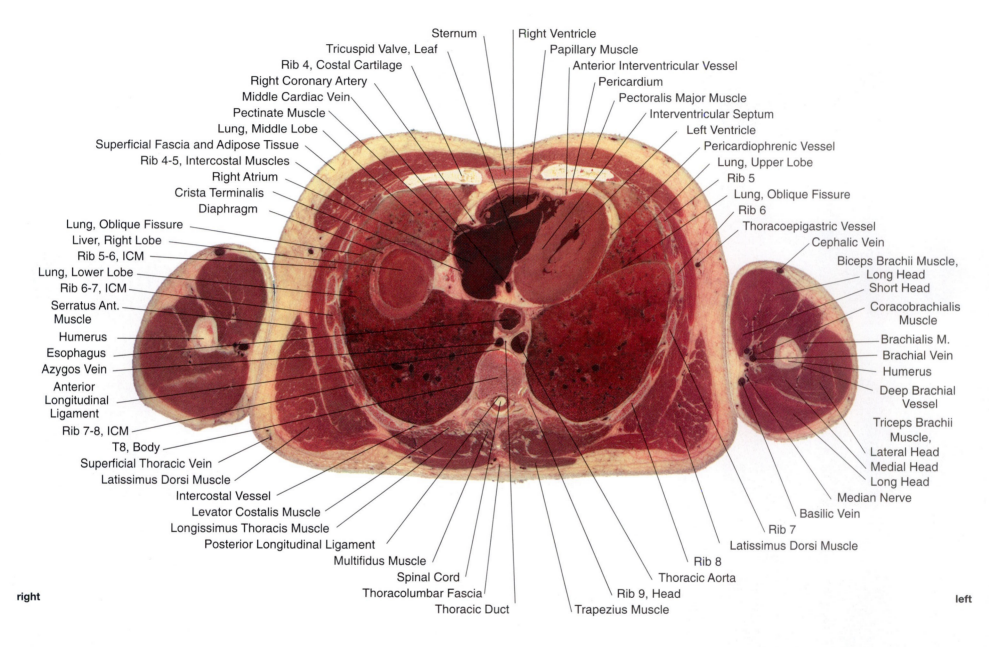

Sternum
Right Ventricle
Tricuspid Valve, Leaf
Papillary Muscle
Rib 4, Costal Cartilage
Anterior Interventricular Vessel
Right Coronary Artery
Pericardium
Middle Cardiac Vein
Pectoralis Major Muscle
Pectinate Muscle
Interventricular Septum
Lung, Middle Lobe
Left Ventricle
Superficial Fascia and Adipose Tissue
Pericardiophrenic Vessel
Rib 4-5, Intercostal Muscles
Lung, Upper Lobe
Right Atrium
Rib 5
Crista Terminalis
Lung, Oblique Fissure
Diaphragm
Rib 6
Lung, Oblique Fissure
Thoracoepigastric Vessel
Liver, Right Lobe
Cephalic Vein
Rib 5-6, ICM
Biceps Brachii Muscle,
Lung, Lower Lobe
Long Head
Rib 6-7, ICM
Short Head
Serratus Ant.
Coracobrachialis
Muscle
Muscle
Humerus
Brachialis M.
Esophagus
Brachial Vein
Azygos Vein
Humerus
Anterior
Deep Brachial
Longitudinal
Vessel
Ligament
Triceps Brachii
Rib 7-8, ICM
Muscle,
T8, Body
Lateral Head
Superficial Thoracic Vein
Medial Head
Latissimus Dorsi Muscle
Long Head
Intercostal Vessel
Median Nerve
Levator Costalis Muscle
Basilic Vein
Longissimus Thoracis Muscle
Rib 7
Posterior Longitudinal Ligament
Latissimus Dorsi Muscle
Multifidus Muscle
Rib 8
Spinal Cord
Thoracic Aorta
Thoracolumbar Fascia
Rib 9, Head
Thoracic Duct
Trapezius Muscle

right

left

ICM=Intercostal Muscles

posterior

a_vm1457

a_vm1458

a_vm1459

a_vm1460

a_vm1461

a_vm1462

a_vm1463

a_vm1464

a_vm1465

a_vm1466

a_vm1467

a_vm1468

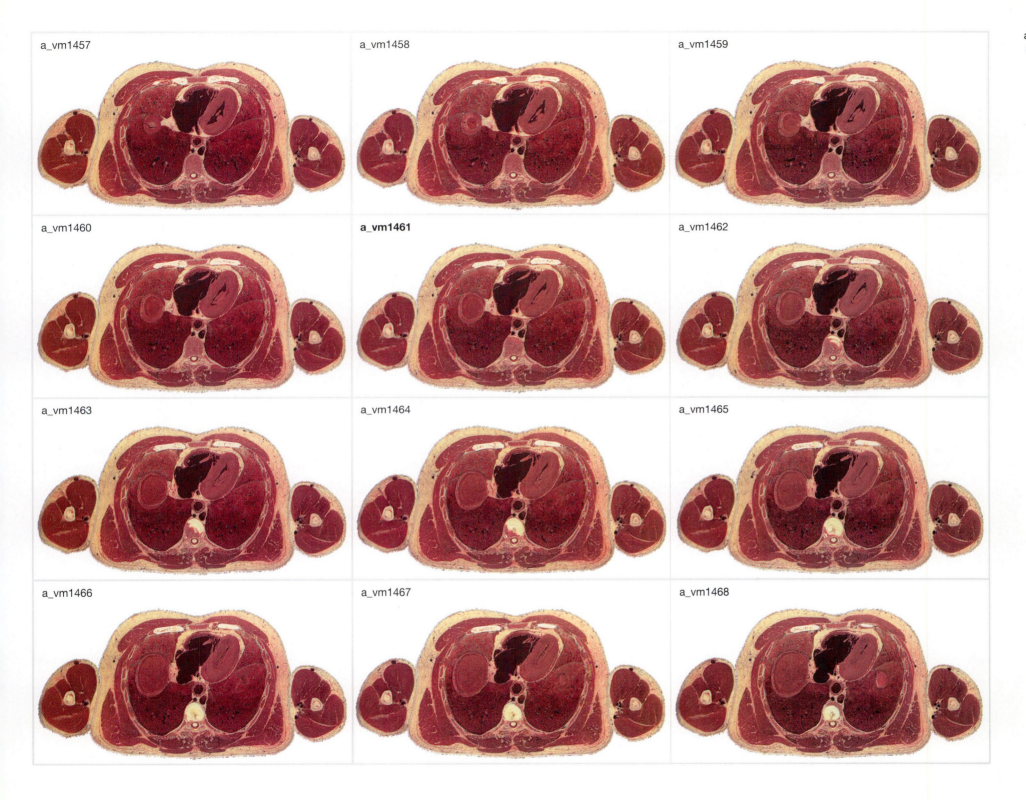

anterior

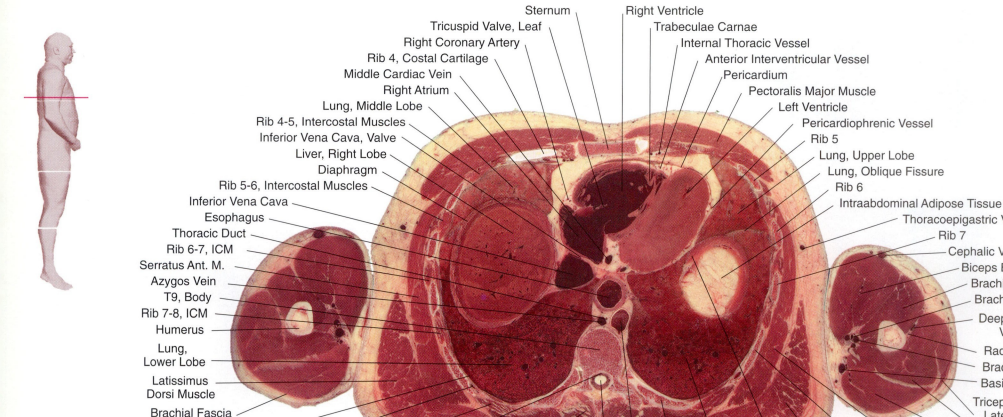

Sternum
Right Ventricle
Tricuspid Valve, Leaf
Trabeculae Carnae
Right Coronary Artery
Internal Thoracic Vessel
Rib 4, Costal Cartilage
Anterior Interventricular Vessel
Middle Cardiac Vein
Pericardium
Right Atrium
Pectoralis Major Muscle
Lung, Middle Lobe
Left Ventricle
Rib 4-5, Intercostal Muscles
Pericardiophrenic Vessel
Inferior Vena Cava, Valve
Rib 5
Liver, Right Lobe
Lung, Upper Lobe
Diaphragm
Lung, Oblique Fissure
Rib 5-6, Intercostal Muscles
Rib 6
Inferior Vena Cava
Intraabdominal Adipose Tissue
Esophagus
Thoracoepigastric Vessel
Thoracic Duct
Rib 7
Rib 6-7, ICM
Cephalic Vein
Serratus Ant. M.
Biceps Brachii M.
Azygos Vein
Brachialis Muscle
T9, Body
Brachial Artery
Rib 7-8, ICM
Deep Brachial
Vessel
Humerus
Lung,
Radial Nerve
Lower Lobe
Brachial Vein
Latissimus
Basilic Vein
Dorsi Muscle
Triceps Brachii M.,
Brachial Fascia
Lateral Head
Pleural Cavity
Medial Head
Intercostal Vessel
Long Head
Rib 8-9, Intercostal Muscles
Serratus Anterior Muscle
Posterior Longitudinal Ligament
Rib 8
Longissimus Thoracis Muscle
Latissimus Dorsi Muscle
Multifidus Muscle
Diaphragm
Trapezius Muscle
Lung, Lower Lobe
T8, Spinous Process
Rib 9
Thoracolumbar Fascia
Thoracic Aorta
Spinal Cord

right

left

ICM=Intercostal Muscles

posterior

80

a_vm1469

a_vm1470

a_vm1471

a_vm1472

a_vm1473

a_vm1474

a_vm1475

a_vm1476

a_vm1477

a_vm1478

a_vm1479

a_vm1480

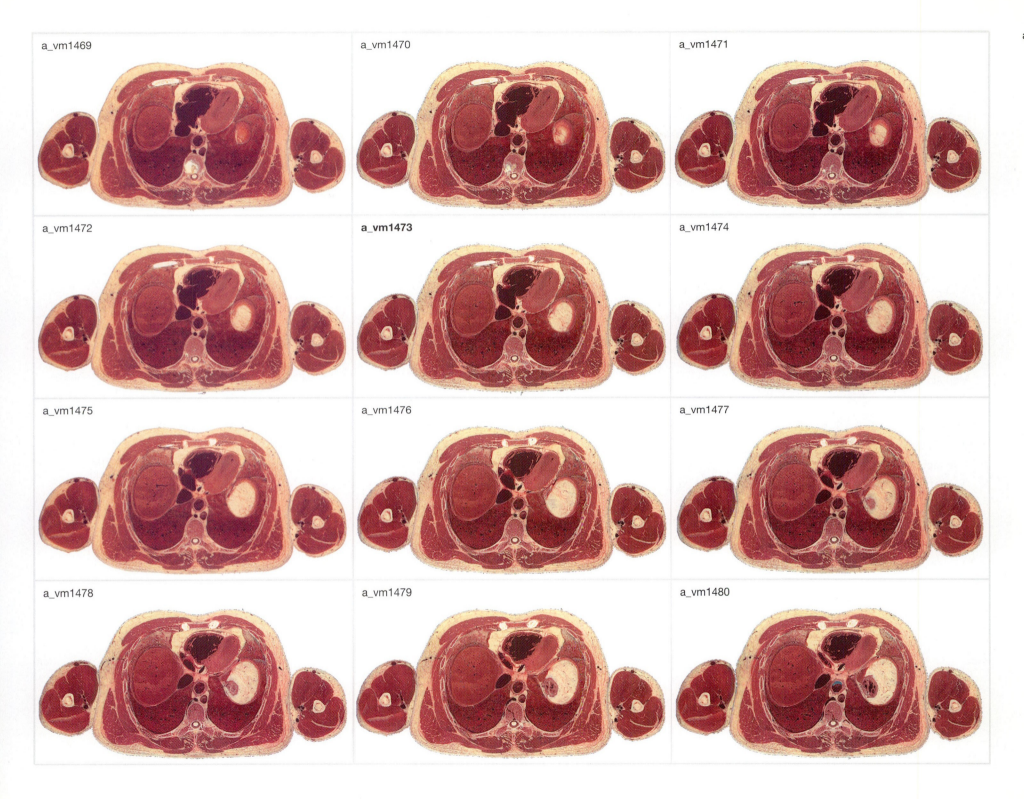

anterior

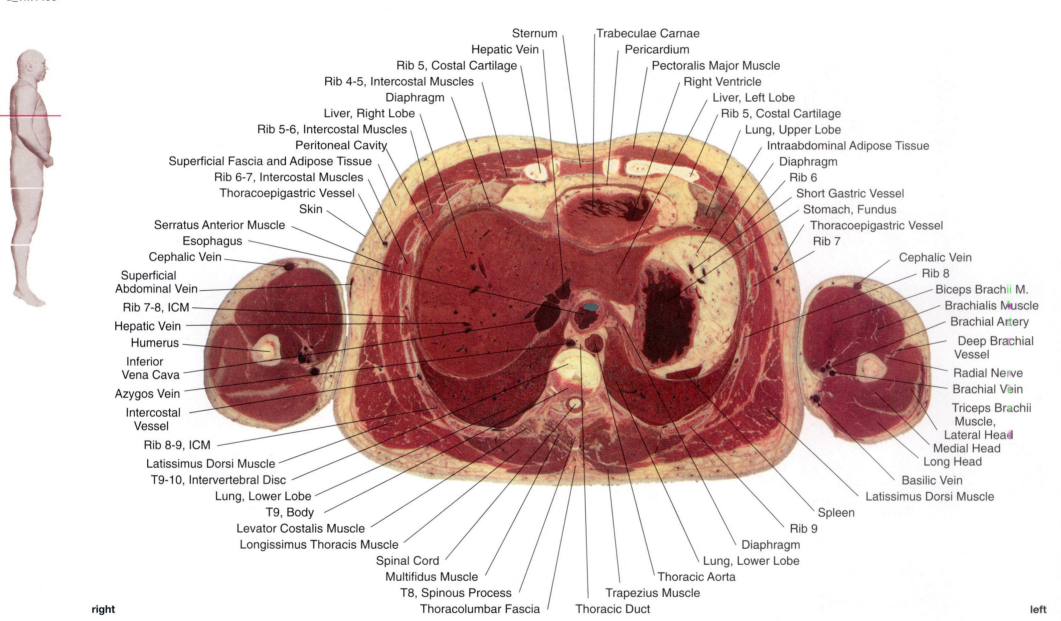

Sternum
Trabeculae Carnae
Hepatic Vein
Pericardium
Rib 5, Costal Cartilage
Pectoralis Major Muscle
Rib 4-5, Intercostal Muscles
Right Ventricle
Diaphragm
Liver, Left Lobe
Liver, Right Lobe
Rib 5, Costal Cartilage
Rib 5-6, Intercostal Muscles
Lung, Upper Lobe
Peritoneal Cavity
Intraabdominal Adipose Tissue
Superficial Fascia and Adipose Tissue
Diaphragm
Rib 6-7, Intercostal Muscles
Rib 6
Thoracoepigastric Vessel
Short Gastric Vessel
Skin
Stomach, Fundus
Serratus Anterior Muscle
Thoracoepigastric Vessel
Esophagus
Rib 7
Cephalic Vein
Cephalic Vein
Superficial
Rib 8
Abdominal Vein
Biceps Brachii M.
Rib 7-8, ICM
Brachialis Muscle
Hepatic Vein
Brachial Artery
Humerus
Deep Brachial Vessel
Inferior
Vena Cava
Radial Nerve
Azygos Vein
Brachial Vein
Intercostal
Triceps Brachii
Vessel
Muscle,
Rib 8-9, ICM
Lateral Head
Latissimus Dorsi Muscle
Medial Head
T9-10, Intervertebral Disc
Long Head
Lung, Lower Lobe
Basilic Vein
T9, Body
Latissimus Dorsi Muscle
Levator Costalis Muscle
Longissimus Thoracis Muscle
Spleen
Spinal Cord
Rib 9
Multifidus Muscle
Diaphragm
T8, Spinous Process
Lung, Lower Lobe
Thoracolumbar Fascia
Thoracic Aorta
Trapezius Muscle
Thoracic Duct

right

left

ICM=Intercostal Muscles

posterior

a_vm1481

a_vm1482

a_vm1483

a_vm1484

a_vm1485

a_vm1486

a_vm1487

a_vm1488

a_vm1489

a_vm1490

a_vm1491

a_vm1492

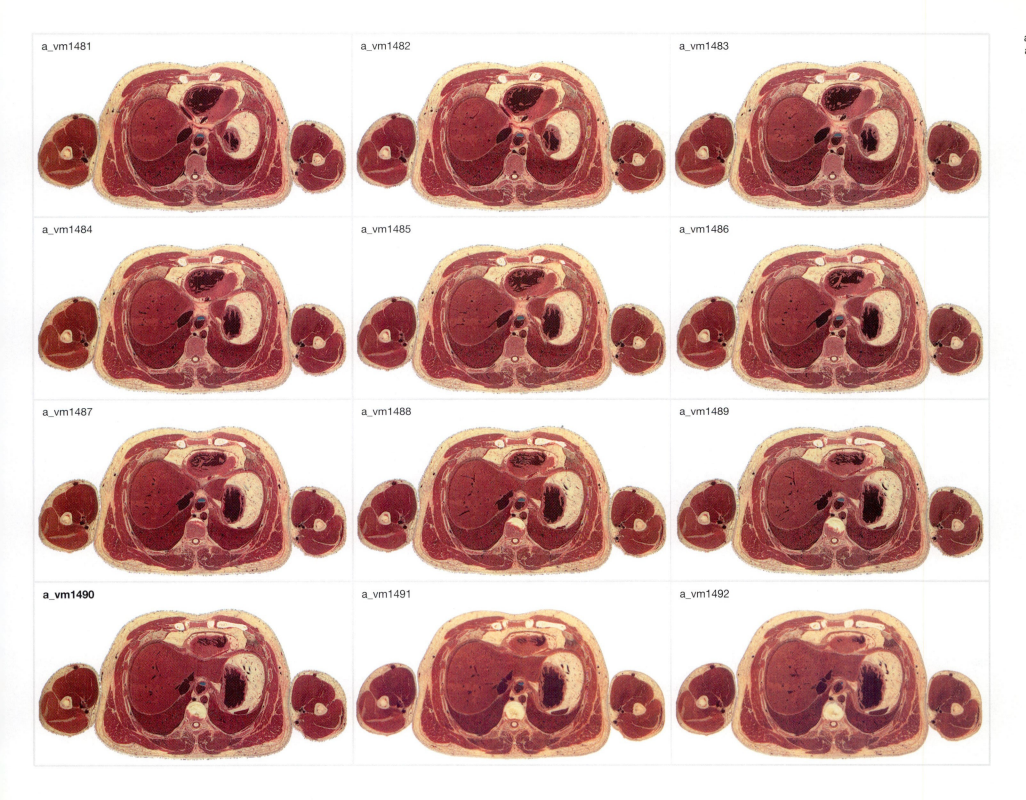

anterior

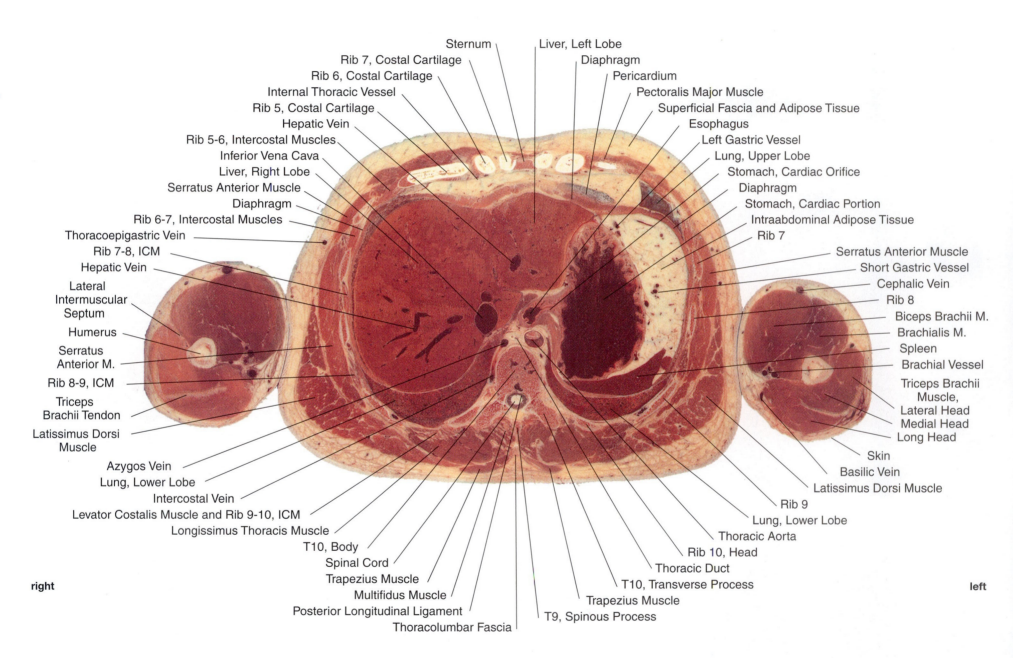

Sternum
Liver, Left Lobe
Rib 7, Costal Cartilage
Diaphragm
Rib 6, Costal Cartilage
Pericardium
Internal Thoracic Vessel
Pectoralis Major Muscle
Rib 5, Costal Cartilage
Superficial Fascia and Adipose Tissue
Hepatic Vein
Esophagus
Rib 5-6, Intercostal Muscles
Left Gastric Vessel
Inferior Vena Cava
Lung, Upper Lobe
Liver, Right Lobe
Stomach, Cardiac Orifice
Serratus Anterior Muscle
Diaphragm
Diaphragm
Stomach, Cardiac Portion
Rib 6-7, Intercostal Muscles
Intraabdominal Adipose Tissue
Thoracoepigastric Vein
Rib 7
Rib 7-8, ICM
Serratus Anterior Muscle
Hepatic Vein
Short Gastric Vessel
Lateral
Cephalic Vein
Intermuscular
Rib 8
Septum
Biceps Brachii M.
Humerus
Brachialis M.
Serratus
Spleen
Anterior M.
Brachial Vessel
Rib 8-9, ICM
Triceps Brachii
Triceps
Muscle,
Brachii Tendon
Lateral Head
Medial Head
Latissimus Dorsi
Long Head
Muscle
Skin
Azygos Vein
Basilic Vein
Lung, Lower Lobe
Latissimus Dorsi Muscle
Intercostal Vein
Levator Costalis Muscle and Rib 9-10, ICM
Rib 9
Longissimus Thoracis Muscle
Lung, Lower Lobe
T10, Body
Thoracic Aorta
Spinal Cord
Rib 10, Head
Trapezius Muscle
Thoracic Duct
Multifidus Muscle
T10, Transverse Process
Posterior Longitudinal Ligament
Trapezius Muscle
Thoracolumbar Fascia
T9, Spinous Process

right

left

ICM=Intercostal Muscles

posterior

84

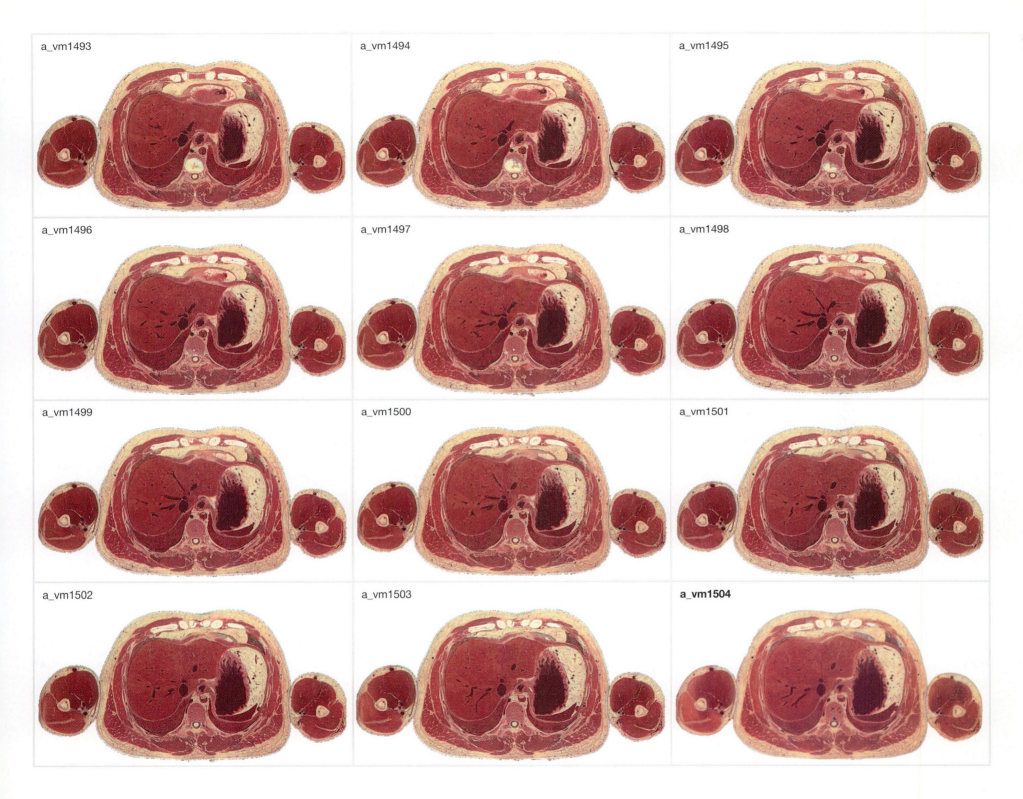

a_vm1493

a_vm1494

a_vm1495

a_vm1496

a_vm1497

a_vm1498

a_vm1499

a_vm1500

a_vm1501

a_vm1502

a_vm1503

a_vm1504

anterior

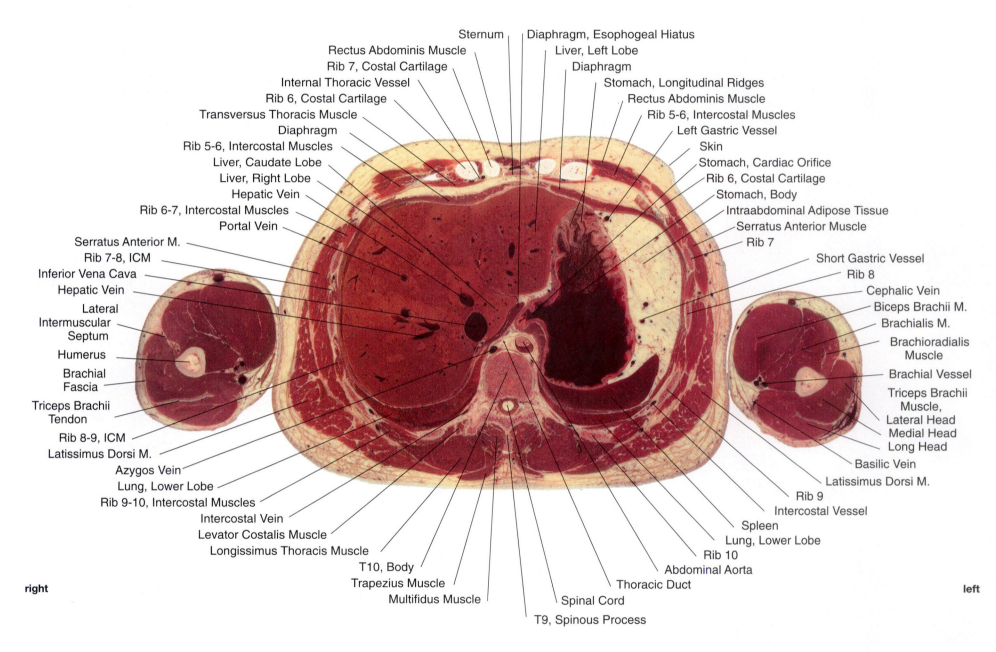

Sternum
Diaphragm, Esophogeal Hiatus
Rectus Abdominis Muscle
Liver, Left Lobe
Rib 7, Costal Cartilage
Diaphragm
Internal Thoracic Vessel
Stomach, Longitudinal Ridges
Rib 6, Costal Cartilage
Rectus Abdominis Muscle
Transversus Thoracis Muscle
Rib 5-6, Intercostal Muscles
Diaphragm
Left Gastric Vessel
Rib 5-6, Intercostal Muscles
Skin
Liver, Caudate Lobe
Stomach, Cardiac Orifice
Liver, Right Lobe
Rib 6, Costal Cartilage
Hepatic Vein
Stomach, Body
Rib 6-7, Intercostal Muscles
Intraabdominal Adipose Tissue
Portal Vein
Serratus Anterior Muscle
Rib 7

Serratus Anterior M.
Rib 7-8, ICM
Short Gastric Vessel
Inferior Vena Cava
Rib 8
Hepatic Vein
Cephalic Vein
Lateral
Biceps Brachii M.
Intermuscular
Brachialis M.
Septum
Brachioradialis
Humerus
Muscle
Brachial
Fascia
Brachial Vessel
Triceps Brachii
Tendon
Triceps Brachii
Muscle,
Rib 8-9, ICM
Lateral Head
Latissimus Dorsi M.
Medial Head
Long Head
Azygos Vein
Basilic Vein
Lung, Lower Lobe
Latissimus Dorsi M.
Rib 9-10, Intercostal Muscles
Rib 9
Intercostal Vein
Intercostal Vessel
Levator Costalis Muscle
Longissimus Thoracis Muscle
Spleen
Lung, Lower Lobe
T10, Body
Rib 10
Trapezius Muscle
Abdominal Aorta
Multifidus Muscle
Thoracic Duct
Spinal Cord
T9, Spinous Process

right

left

ICM=Intercostal Muscles

posterior

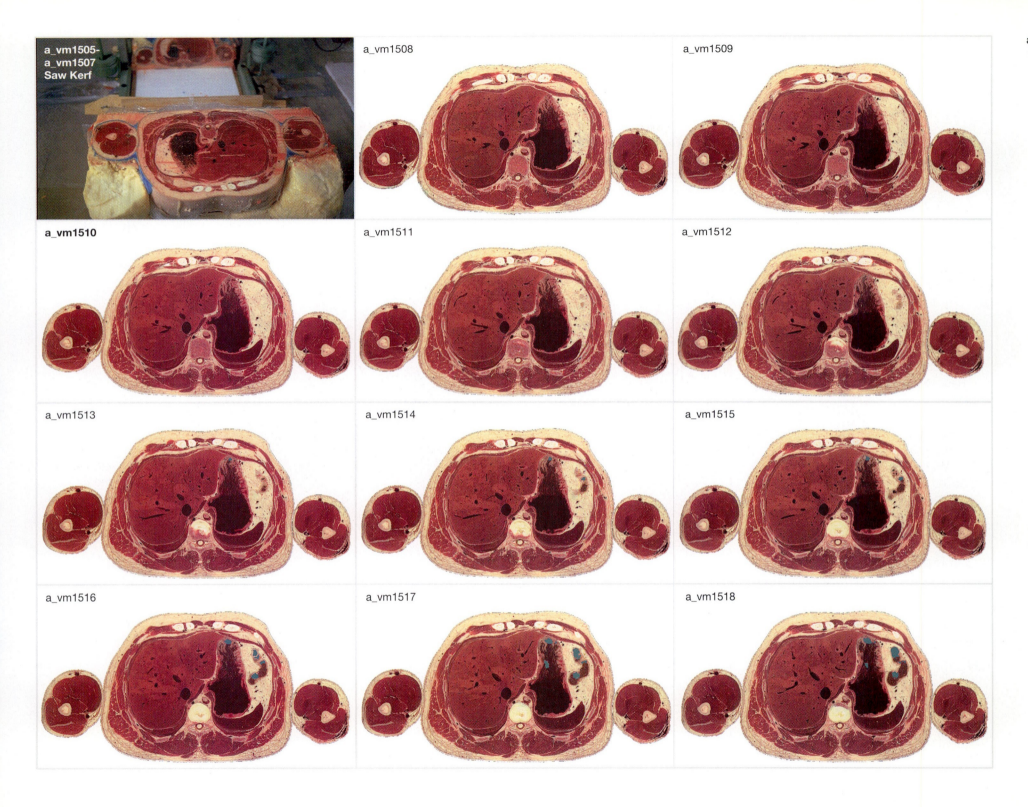

a_vm1505-
a_vm1507
Saw Kerf

a_vm1508

a_vm1509

a_vm1510

a_vm1511

a_vm1512

a_vm1513

a_vm1514

a_vm1515

a_vm1516

a_vm1517

a_vm1518

anterior

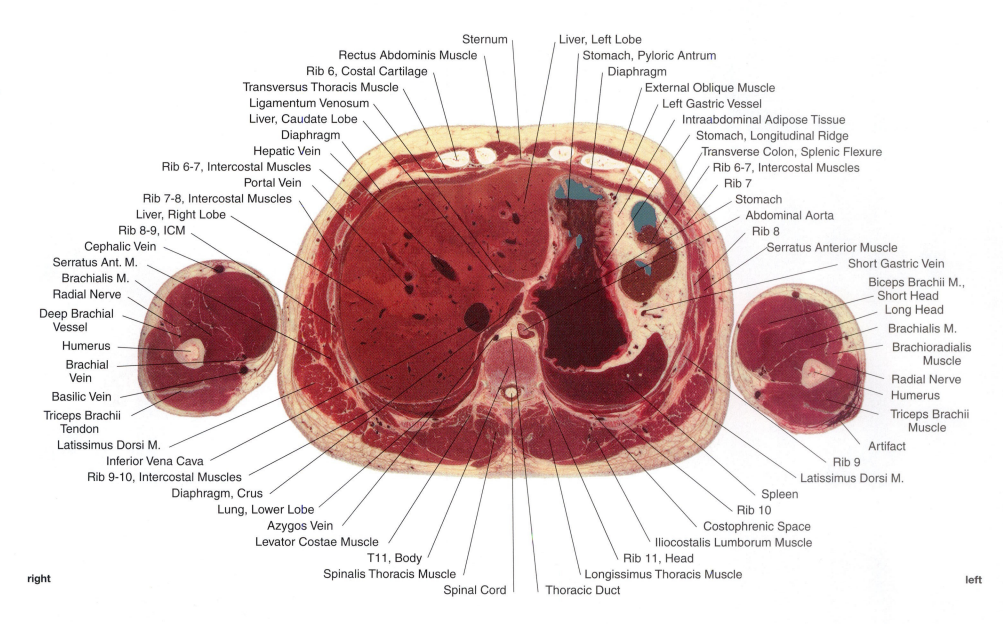

Sternum
Liver, Left Lobe
Rectus Abdominis Muscle
Stomach, Pyloric Antrum
Rib 6, Costal Cartilage
Diaphragm
Transversus Thoracis Muscle
External Oblique Muscle
Ligamentum Venosum
Left Gastric Vessel
Liver, Caudate Lobe
Intraabdominal Adipose Tissue
Diaphragm
Stomach, Longitudinal Ridge
Hepatic Vein
Transverse Colon, Splenic Flexure
Rib 6-7, Intercostal Muscles
Rib 6-7, Intercostal Muscles
Portal Vein
Rib 7
Rib 7-8, Intercostal Muscles
Stomach
Liver, Right Lobe
Abdominal Aorta
Rib 8-9, ICM
Rib 8
Cephalic Vein
Serratus Anterior Muscle
Serratus Ant. M.
Short Gastric Vein
Brachialis M.
Biceps Brachii M.,
Radial Nerve
Short Head
Deep Brachial
Long Head
Vessel
Brachialis M.
Humerus
Brachioradialis
Brachial
Muscle
Vein
Radial Nerve
Basilic Vein
Humerus
Triceps Brachii
Triceps Brachii
Tendon
Muscle
Latissimus Dorsi M.
Artifact
Inferior Vena Cava
Rib 9
Rib 9-10, Intercostal Muscles
Latissimus Dorsi M.
Diaphragm, Crus
Spleen
Lung, Lower Lobe
Rib 10
Azygos Vein
Costophrenic Space
Levator Costae Muscle
Iliocostalis Lumborum Muscle
T11, Body
Rib 11, Head
Spinalis Thoracis Muscle
Longissimus Thoracis Muscle
Spinal Cord
Thoracic Duct

right

left

ICM=Intercostal Muscles

posterior

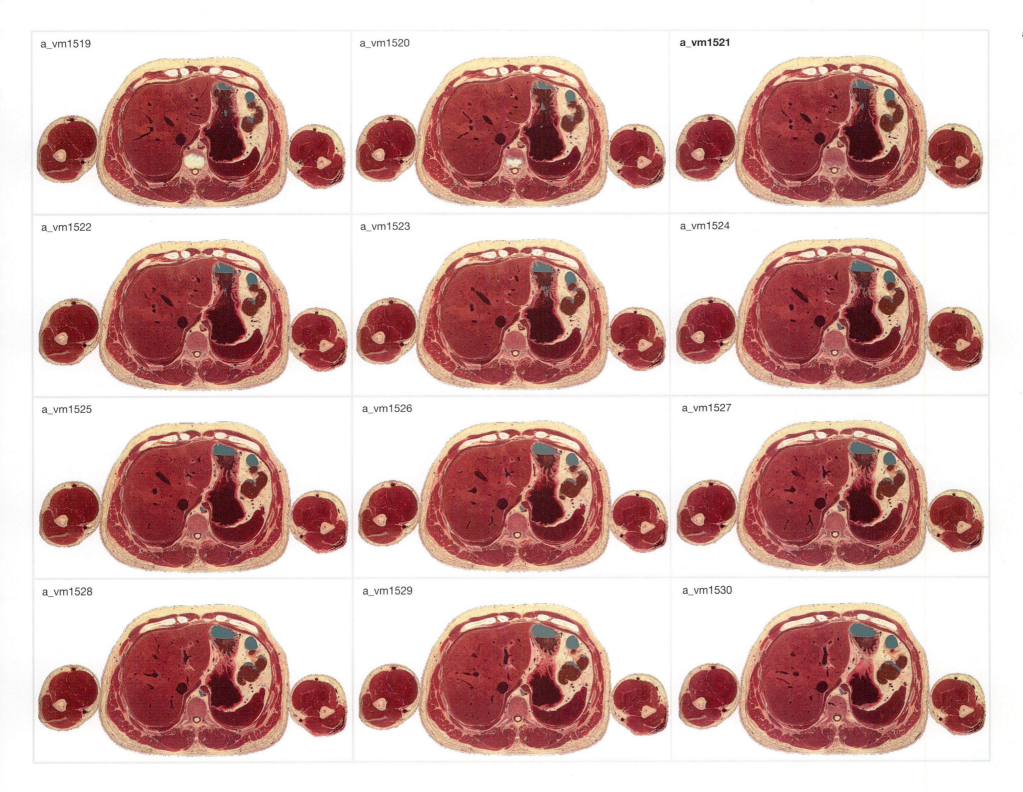

a_vm1519

a_vm1520

a_vm1521

a_vm1522

a_vm1523

a_vm1524

a_vm1525

a_vm1526

a_vm1527

a_vm1528

a_vm1529

a_vm1530

anterior

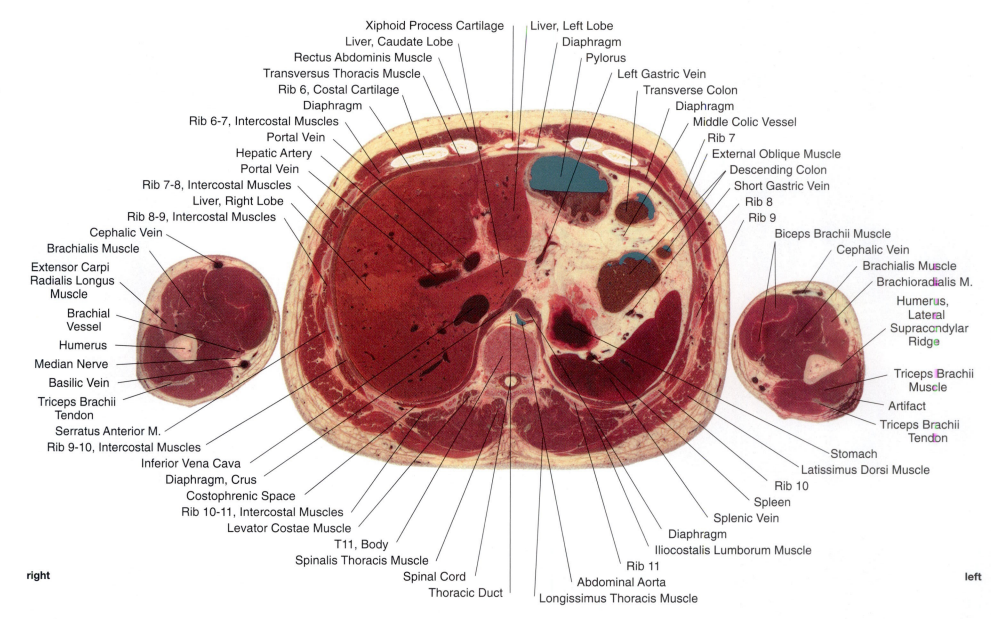

Xiphoid Process Cartilage

Liver, Caudate Lobe

Rectus Abdominis Muscle

Transversus Thoracis Muscle

Rib 6, Costal Cartilage

Diaphragm

Rib 6-7, Intercostal Muscles

Portal Vein

Hepatic Artery

Portal Vein

Rib 7-8, Intercostal Muscles

Liver, Right Lobe

Rib 8-9, Intercostal Muscles

Cephalic Vein

Brachialis Muscle

Extensor Carpi Radialis Longus Muscle

Brachial Vessel

Humerus

Median Nerve

Basilic Vein

Triceps Brachii Tendon

Serratus Anterior M.

Rib 9-10, Intercostal Muscles

Inferior Vena Cava

Diaphragm, Crus

Costophrenic Space

Rib 10-11, Intercostal Muscles

Levator Costae Muscle

T11, Body

Spinalis Thoracis Muscle

Spinal Cord

Thoracic Duct

Liver, Left Lobe

Diaphragm

Pylorus

Left Gastric Vein

Transverse Colon

Diaphragm

Middle Colic Vessel

Rib 7

External Oblique Muscle

Descending Colon

Short Gastric Vein

Rib 8

Rib 9

Biceps Brachii Muscle

Cephalic Vein

Brachialis Muscle

Brachioradialis M.

Humerus, Lateral Supracondylar Ridge

Triceps Brachii Muscle

Artifact

Triceps Brachii Tendon

Stomach

Latissimus Dorsi Muscle

Rib 10

Spleen

Splenic Vein

Diaphragm

Iliocostalis Lumborum Muscle

Rib 11

Abdominal Aorta

Longissimus Thoracis Muscle

right

left

ICM=Intercostal Muscles

posterior

90

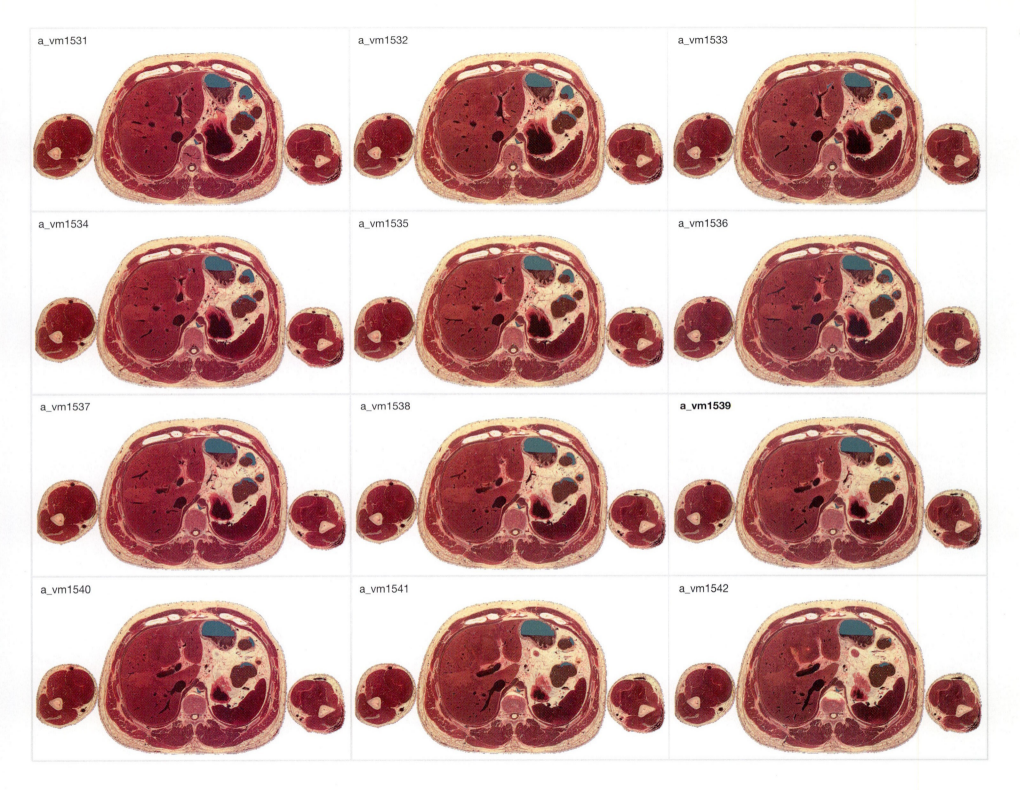

a_vm1531

a_vm1532

a_vm1533

a_vm1534

a_vm1535

a_vm1536

a_vm1537

a_vm1538

a_vm1539

a_vm1540

a_vm1541

a_vm1542

anterior

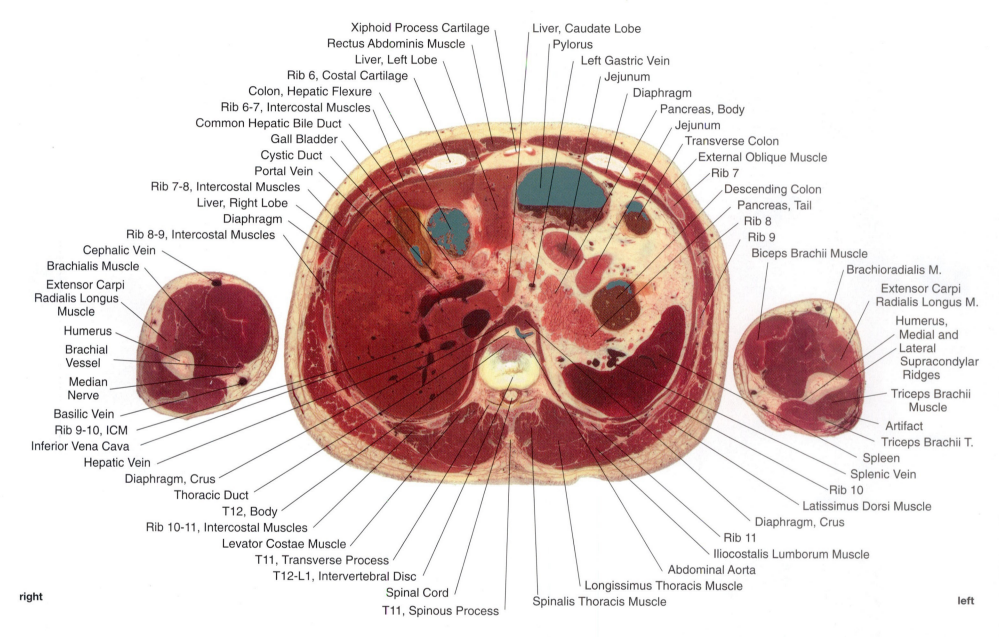

Xiphoid Process Cartilage
Rectus Abdominis Muscle
Liver, Left Lobe
Rib 6, Costal Cartilage
Colon, Hepatic Flexure
Rib 6-7, Intercostal Muscles
Common Hepatic Bile Duct
Gall Bladder
Cystic Duct
Portal Vein
Rib 7-8, Intercostal Muscles
Liver, Right Lobe
Diaphragm
Rib 8-9, Intercostal Muscles
Cephalic Vein
Brachialis Muscle
Extensor Carpi Radialis Longus Muscle
Humerus
Brachial Vessel
Median Nerve
Basilic Vein
Rib 9-10, ICM
Inferior Vena Cava
Hepatic Vein
Diaphragm, Crus
Thoracic Duct
T12, Body
Rib 10-11, Intercostal Muscles
Levator Costae Muscle
T11, Transverse Process
T12-L1, Intervertebral Disc
Spinal Cord
T11, Spinous Process

Liver, Caudate Lobe
Pylorus
Left Gastric Vein
Jejunum
Diaphragm
Pancreas, Body
Jejunum
Transverse Colon
External Oblique Muscle
Rib 7
Descending Colon
Pancreas, Tail
Rib 8
Rib 9
Biceps Brachii Muscle
Brachioradialis M.
Extensor Carpi Radialis Longus M.
Humerus, Medial and Lateral Supracondylar Ridges
Triceps Brachii Muscle
Artifact
Triceps Brachii T.
Spleen
Splenic Vein
Rib 10
Latissimus Dorsi Muscle
Diaphragm, Crus
Rib 11
Iliocostalis Lumborum Muscle
Abdominal Aorta
Longissimus Thoracis Muscle
Spinalis Thoracis Muscle

right

left

ICM=Intercostal Muscles

posterior

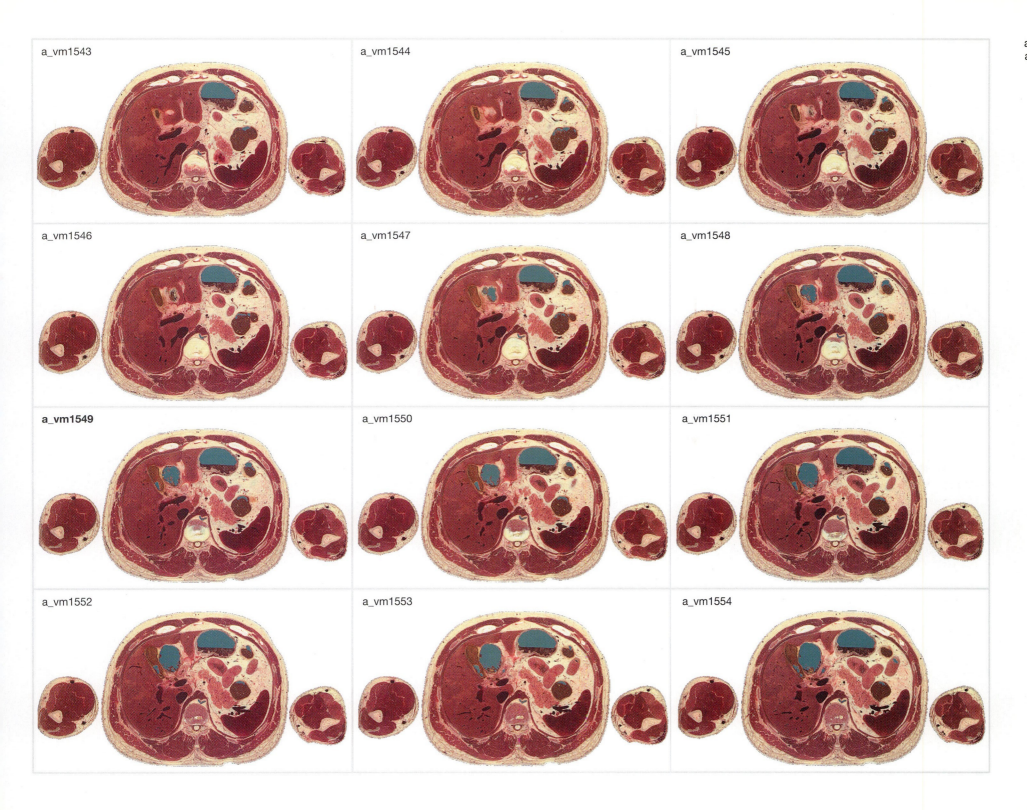

a_vm1543

a_vm1544

a_vm1545

a_vm1546

a_vm1547

a_vm1548

a_vm1549

a_vm1550

a_vm1551

a_vm1552

a_vm1553

a_vm1554

anterior

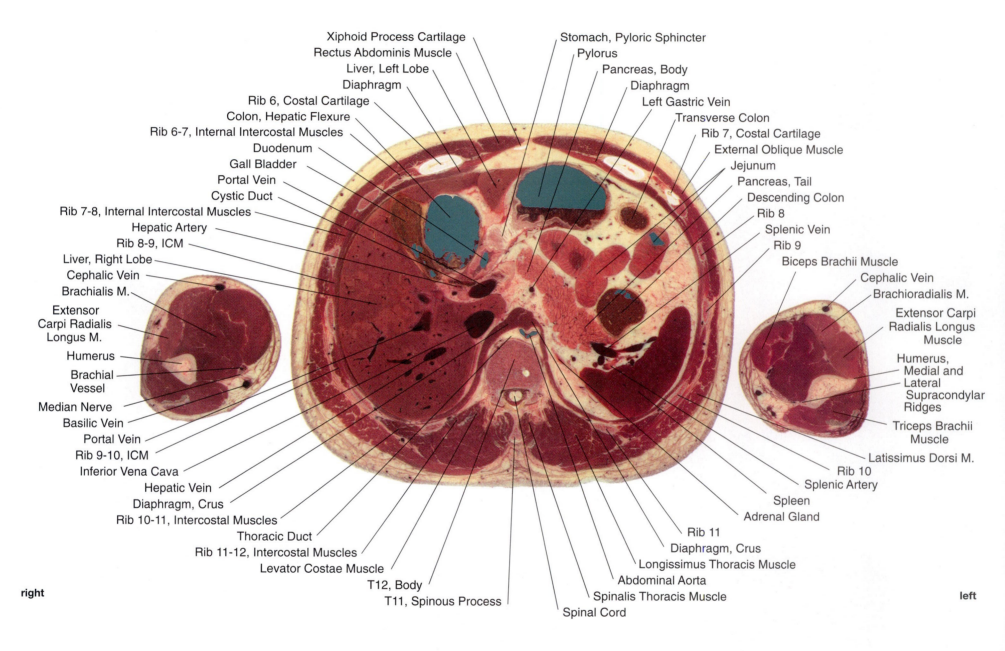

Xiphoid Process Cartilage
Rectus Abdominis Muscle
Liver, Left Lobe
Diaphragm
Rib 6, Costal Cartilage
Colon, Hepatic Flexure
Rib 6-7, Internal Intercostal Muscles
Duodenum
Gall Bladder
Portal Vein
Cystic Duct
Rib 7-8, Internal Intercostal Muscles
Hepatic Artery
Rib 8-9, ICM
Liver, Right Lobe
Cephalic Vein
Brachialis M.
Extensor Carpi Radialis Longus M.
Humerus
Brachial Vessel
Median Nerve
Basilic Vein
Portal Vein
Rib 9-10, ICM
Inferior Vena Cava
Hepatic Vein
Diaphragm, Crus
Rib 10-11, Intercostal Muscles
Thoracic Duct
Rib 11-12, Intercostal Muscles
Levator Costae Muscle
T12, Body
T11, Spinous Process

Stomach, Pyloric Sphincter
Pylorus
Pancreas, Body
Diaphragm
Left Gastric Vein
Transverse Colon
Rib 7, Costal Cartilage
External Oblique Muscle
Jejunum
Pancreas, Tail
Descending Colon
Rib 8
Splenic Vein
Rib 9
Biceps Brachii Muscle
Cephalic Vein
Brachioradialis M.
Extensor Carpi Radialis Longus Muscle
Humerus, Medial and Lateral Supracondylar Ridges
Triceps Brachii Muscle
Latissimus Dorsi M.
Rib 10
Splenic Artery
Spleen
Adrenal Gland
Rib 11
Diaphragm, Crus
Longissimus Thoracis Muscle
Abdominal Aorta
Spinalis Thoracis Muscle
Spinal Cord

right

left

ICM=Intercostal Muscles

posterior

94

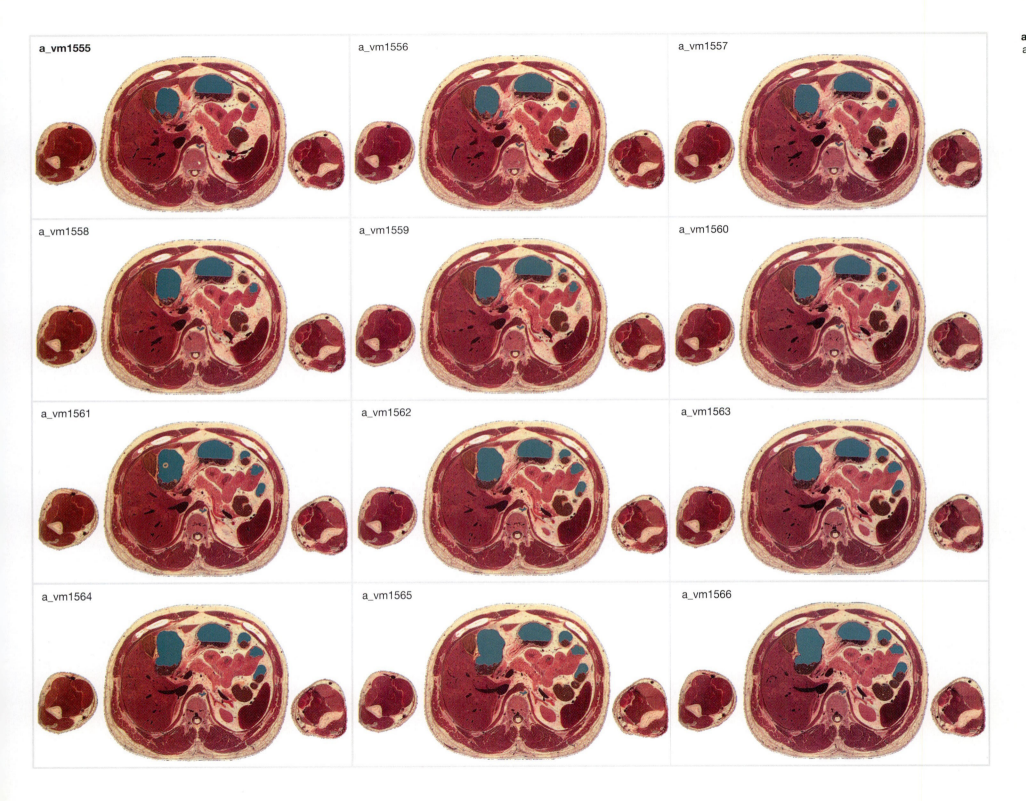

a_vm1555

a_vm1556

a_vm1557

a_vm1558

a_vm1559

a_vm1560

a_vm1561

a_vm1562

a_vm1563

a_vm1564

a_vm1565

a_vm1566

anterior

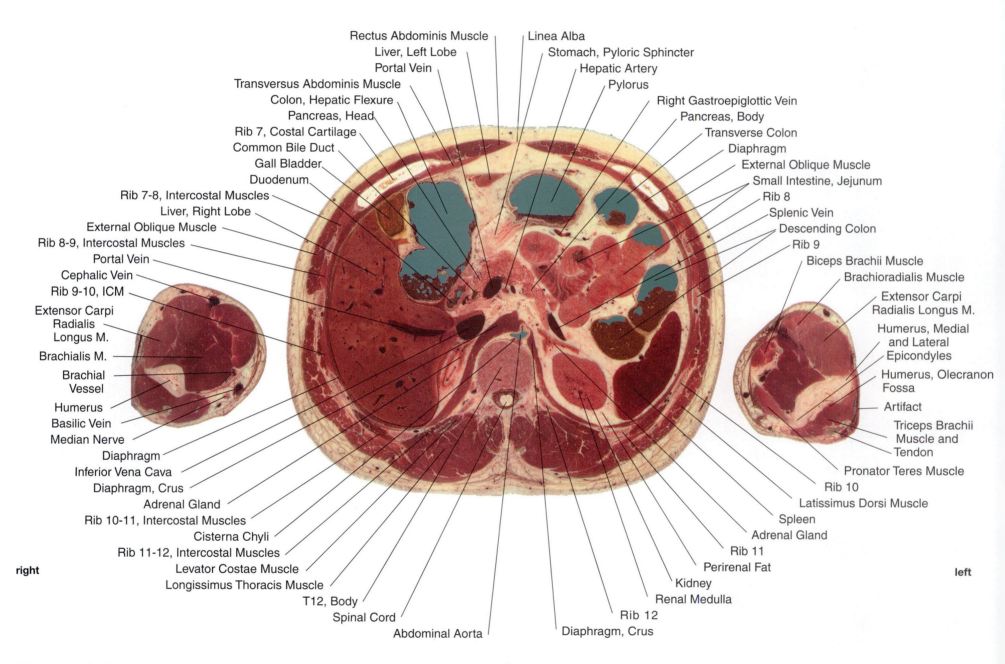

Rectus Abdominis Muscle
Liver, Left Lobe
Portal Vein
Transversus Abdominis Muscle
Colon, Hepatic Flexure
Pancreas, Head
Rib 7, Costal Cartilage
Common Bile Duct
Gall Bladder
Duodenum
Rib 7-8, Intercostal Muscles
Liver, Right Lobe
External Oblique Muscle
Rib 8-9, Intercostal Muscles
Portal Vein
Cephalic Vein
Rib 9-10, ICM
Extensor Carpi Radialis Longus M.
Brachialis M.
Brachial Vessel
Humerus
Basilic Vein
Median Nerve
Diaphragm
Inferior Vena Cava
Diaphragm, Crus
Adrenal Gland
Rib 10-11, Intercostal Muscles
Cisterna Chyli
Rib 11-12, Intercostal Muscles
Levator Costae Muscle
Longissimus Thoracis Muscle
T12, Body
Spinal Cord
Abdominal Aorta

Linea Alba
Stomach, Pyloric Sphincter
Hepatic Artery
Pylorus
Right Gastroepiglottic Vein
Pancreas, Body
Transverse Colon
Diaphragm
External Oblique Muscle
Small Intestine, Jejunum
Rib 8
Splenic Vein
Descending Colon
Rib 9
Biceps Brachii Muscle
Brachioradialis Muscle
Extensor Carpi Radialis Longus M.
Humerus, Medial and Lateral Epicondyles
Humerus, Olecranon Fossa
Artifact
Triceps Brachii Muscle and Tendon
Pronator Teres Muscle
Rib 10
Latissimus Dorsi Muscle
Spleen
Adrenal Gland
Rib 11
Perirenal Fat
Kidney
Renal Medulla
Rib 12
Diaphragm, Crus

right

left

ICM=Intercostal Muscles

posterior

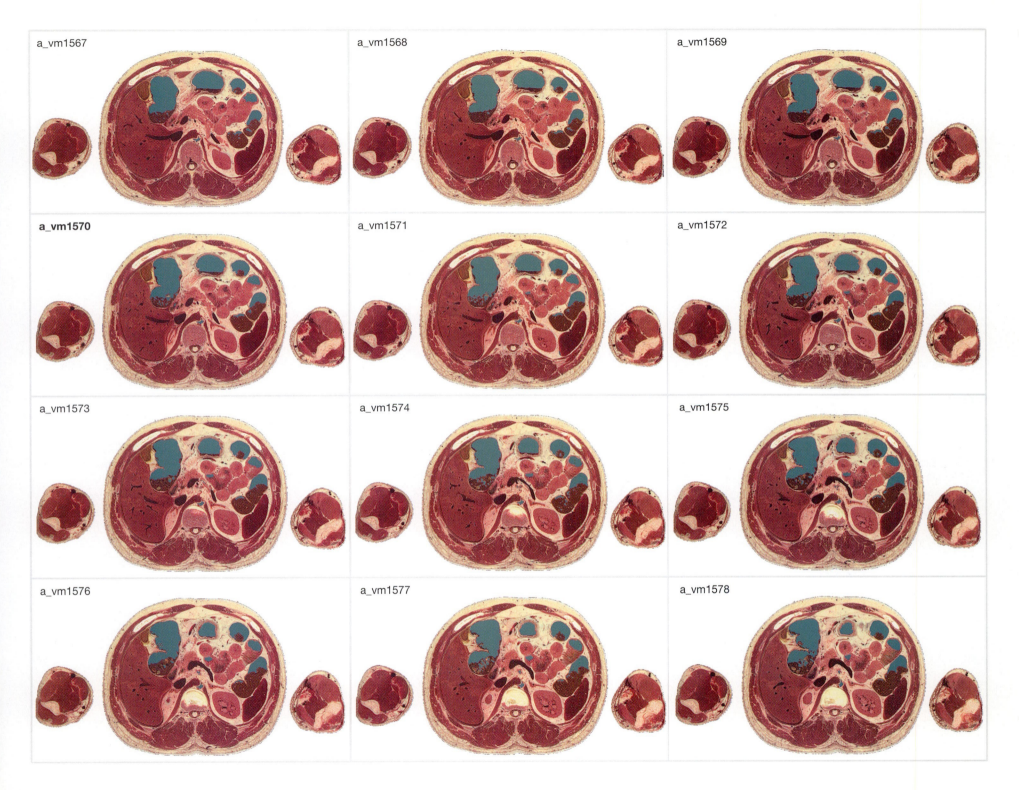

a_vm1567

a_vm1568

a_vm1569

a_vm1570

a_vm1571

a_vm1572

a_vm1573

a_vm1574

a_vm1575

a_vm1576

a_vm1577

a_vm1578

anterior

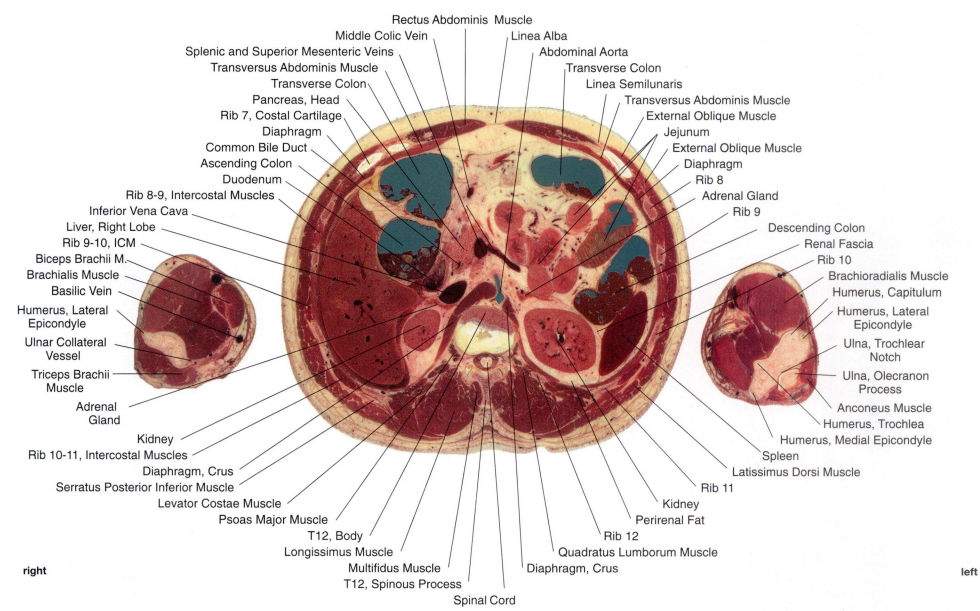

Rectus Abdominis Muscle
Middle Colic Vein
Linea Alba
Splenic and Superior Mesenteric Veins
Abdominal Aorta
Transversus Abdominis Muscle
Transverse Colon
Transverse Colon
Linea Semilunaris
Pancreas, Head
Transversus Abdominis Muscle
Rib 7, Costal Cartilage
External Oblique Muscle
Diaphragm
Jejunum
Common Bile Duct
External Oblique Muscle
Ascending Colon
Diaphragm
Duodenum
Rib 8
Rib 8-9, Intercostal Muscles
Adrenal Gland
Inferior Vena Cava
Rib 9
Liver, Right Lobe
Descending Colon
Rib 9-10, ICM
Renal Fascia
Biceps Brachii M.
Rib 10
Brachialis Muscle
Brachioradialis Muscle
Basilic Vein
Humerus, Capitulum
Humerus, Lateral
Epicondyle
Humerus, Lateral
Epicondyle
Ulnar Collateral
Vessel
Ulna, Trochlear
Notch
Triceps Brachii
Muscle
Ulna, Olecranon
Process
Adrenal
Gland
Anconeus Muscle
Kidney
Humerus, Trochlea
Rib 10-11, Intercostal Muscles
Humerus, Medial Epicondyle
Diaphragm, Crus
Spleen
Serratus Posterior Inferior Muscle
Latissimus Dorsi Muscle
Levator Costae Muscle
Rib 11
Psoas Major Muscle
Kidney
T12, Body
Perirenal Fat
Longissimus Muscle
Rib 12
Multifidus Muscle
Quadratus Lumborum Muscle
T12, Spinous Process
Diaphragm, Crus
Spinal Cord

right

left

ICM=Intercostal Muscles

posterior

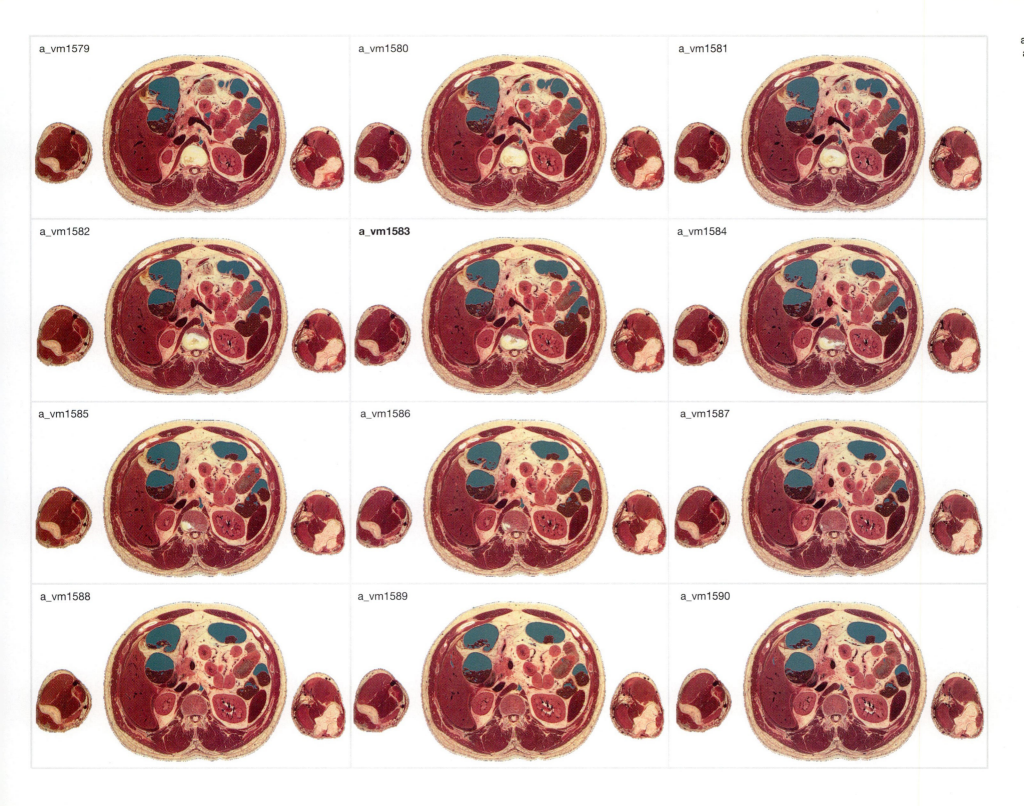

a_vm1579

a_vm1580

a_vm1581

a_vm1582

a_vm1583

a_vm1584

a_vm1585

a_vm1586

a_vm1587

a_vm1588

a_vm1589

a_vm1590

anterior

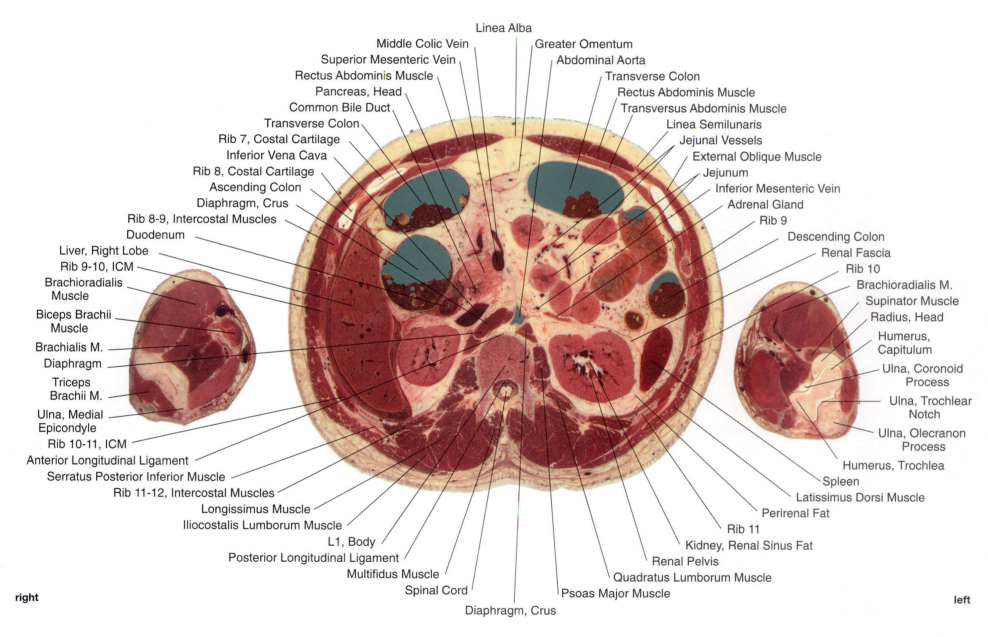

Linea Alba

Middle Colic Vein

Superior Mesenteric Vein

Rectus Abdominis Muscle

Pancreas, Head

Common Bile Duct

Transverse Colon

Rib 7, Costal Cartilage

Inferior Vena Cava

Rib 8, Costal Cartilage

Ascending Colon

Diaphragm, Crus

Rib 8-9, Intercostal Muscles

Duodenum

Liver, Right Lobe

Rib 9-10, ICM

Brachioradialis Muscle

Biceps Brachii Muscle

Brachialis M.

Diaphragm

Triceps Brachii M.

Ulna, Medial Epicondyle

Rib 10-11, ICM

Anterior Longitudinal Ligament

Serratus Posterior Inferior Muscle

Rib 11-12, Intercostal Muscles

Longissimus Muscle

Iliocostalis Lumborum Muscle

L1, Body

Posterior Longitudinal Ligament

Multifidus Muscle

Spinal Cord

Diaphragm, Crus

Psoas Major Muscle

Quadratus Lumborum Muscle

Renal Pelvis

Kidney, Renal Sinus Fat

Rib 11

Perirenal Fat

Latissimus Dorsi Muscle

Spleen

Humerus, Trochlea

Ulna, Olecranon Process

Ulna, Trochlear Notch

Ulna, Coronoid Process

Humerus, Capitulum

Radius, Head

Supinator Muscle

Brachioradialis M.

Rib 10

Rib 9

Renal Fascia

Descending Colon

Adrenal Gland

Inferior Mesenteric Vein

Jejunum

External Oblique Muscle

Jejunal Vessels

Linea Semilunaris

Transversus Abdominis Muscle

Rectus Abdominis Muscle

Transverse Colon

Abdominal Aorta

Greater Omentum

Greater Omentum

right

left

ICM=Intercostal Muscles

posterior

100

a_vm1591

a_vm1592

a_vm1593

a_vm1594

a_vm1595

a_vm1596

a_vm1597

a_vm1598

a_vm1599

a_vm1600

a_vm1601

a_vm1602

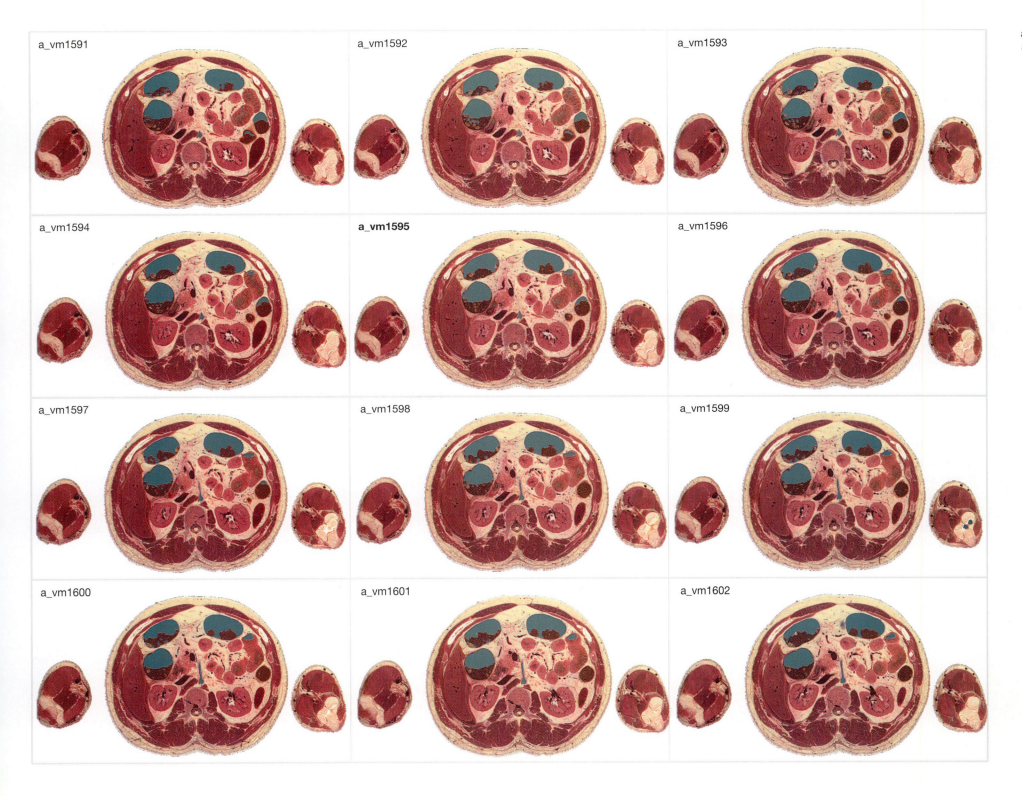

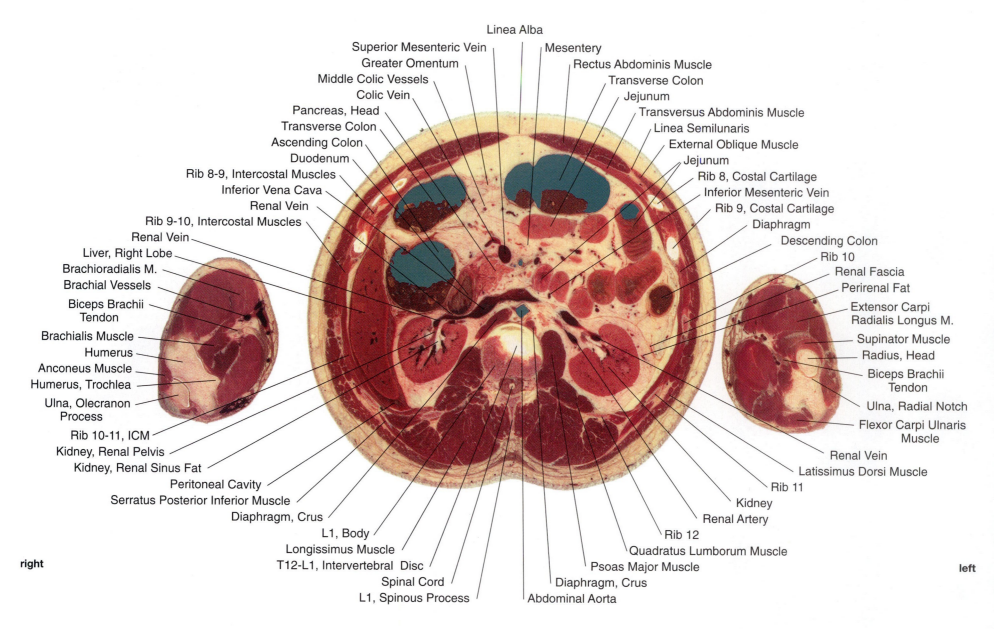

anterior

Linea Alba

Superior Mesenteric Vein
Greater Omentum
Middle Colic Vessels
Colic Vein
Pancreas, Head
Transverse Colon
Ascending Colon
Duodenum
Rib 8-9, Intercostal Muscles
Inferior Vena Cava
Renal Vein
Rib 9-10, Intercostal Muscles
Renal Vein
Liver, Right Lobe
Brachioradialis M.
Brachial Vessels
Biceps Brachii Tendon
Brachialis Muscle
Humerus
Anconeus Muscle
Humerus, Trochlea
Ulna, Olecranon Process
Rib 10-11, ICM
Kidney, Renal Pelvis
Kidney, Renal Sinus Fat
Peritoneal Cavity
Serratus Posterior Inferior Muscle
Diaphragm, Crus
L1, Body
Longissimus Muscle
T12-L1, Intervertebral Disc
Spinal Cord
L1, Spinous Process

Mesentery
Rectus Abdominis Muscle
Transverse Colon
Jejunum
Transversus Abdominis Muscle
Linea Semilunaris
External Oblique Muscle
Jejunum
Rib 8, Costal Cartilage
Inferior Mesenteric Vein
Rib 9, Costal Cartilage
Diaphragm
Descending Colon
Rib 10
Renal Fascia
Perirenal Fat
Extensor Carpi Radialis Longus M.
Supinator Muscle
Radius, Head
Biceps Brachii Tendon
Ulna, Radial Notch
Flexor Carpi Ulnaris Muscle
Renal Vein
Latissimus Dorsi Muscle
Rib 11
Kidney
Renal Artery
Rib 12
Quadratus Lumborum Muscle
Psoas Major Muscle
Diaphragm, Crus
Abdominal Aorta

right

left

ICM=Intercostal Muscles

posterior

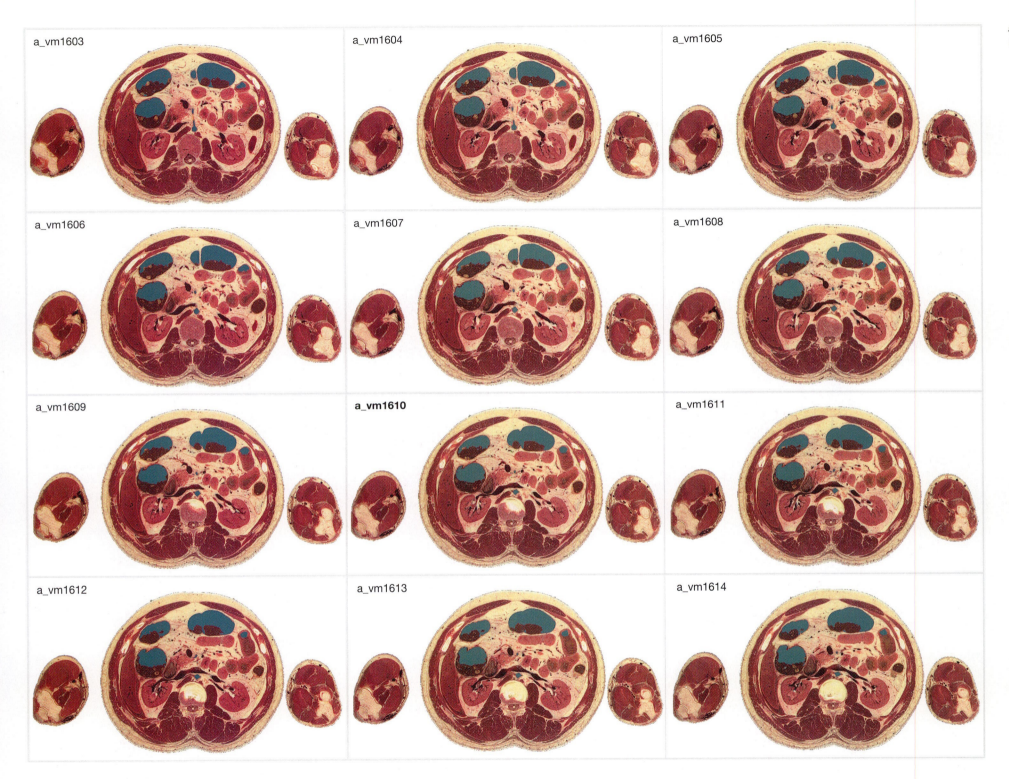

a_vm1603

a_vm1604

a_vm1605

a_vm1606

a_vm1607

a_vm1608

a_vm1609

a_vm1610

a_vm1611

a_vm1612

a_vm1613

a_vm1614

anterior

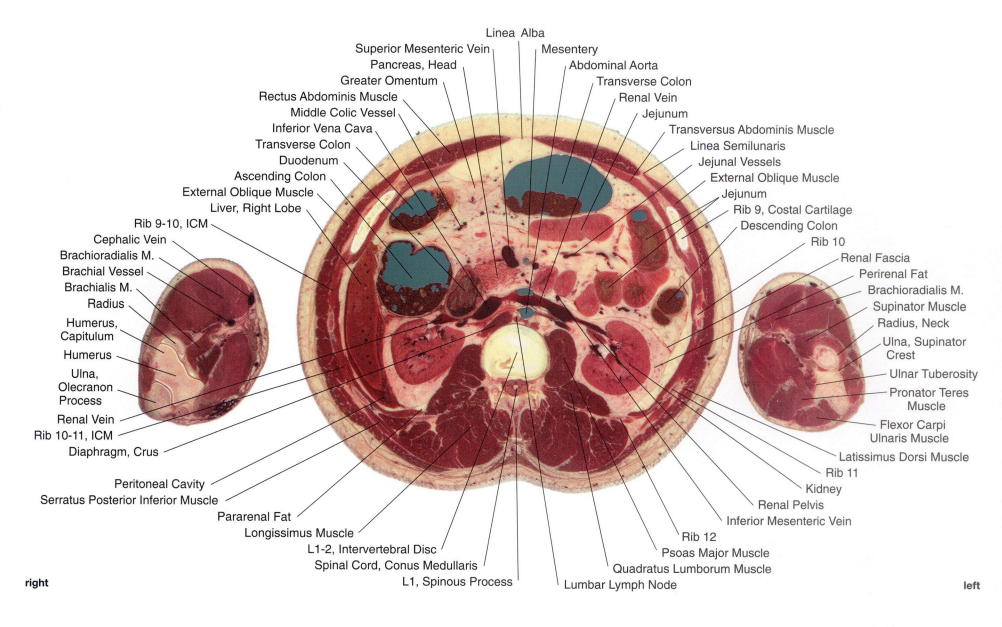

Linea Alba

Superior Mesenteric Vein

Mesentery

Pancreas, Head

Abdominal Aorta

Greater Omentum

Transverse Colon

Rectus Abdominis Muscle

Renal Vein

Middle Colic Vessel

Jejunum

Inferior Vena Cava

Transversus Abdominis Muscle

Transverse Colon

Linea Semilunaris

Duodenum

Jejunal Vessels

Ascending Colon

External Oblique Muscle

External Oblique Muscle

Jejunum

Liver, Right Lobe

Rib 9, Costal Cartilage

Rib 9-10, ICM

Descending Colon

Cephalic Vein

Rib 10

Brachioradialis M.

Renal Fascia

Brachial Vessel

Perirenal Fat

Brachialis M.

Brachioradialis M.

Radius

Supinator Muscle

Humerus,
Capitulum

Radius, Neck

Ulna, Supinator
Crest

Humerus

Ulna,
Olecranon
Process

Ulnar Tuberosity

Pronator Teres
Muscle

Renal Vein

Rib 10-11, ICM

Flexor Carpi
Ulnaris Muscle

Diaphragm, Crus

Latissimus Dorsi Muscle

Peritoneal Cavity

Rib 11

Serratus Posterior Inferior Muscle

Kidney

Pararenal Fat

Renal Pelvis

Longissimus Muscle

Inferior Mesenteric Vein

L1-2, Intervertebral Disc

Rib 12

Spinal Cord, Conus Medullaris

Psoas Major Muscle

L1, Spinous Process

Quadratus Lumborum Muscle

Lumbar Lymph Node

right

left

ICM=Intercostal Muscles

posterior

a_vm1615

a_vm1616

a_vm1617

a_vm1618

a_vm1619

a_vm1620

a_vm1621

a_vm1622

a_vm1623

a_vm1624

a_vm1625

a_vm1626

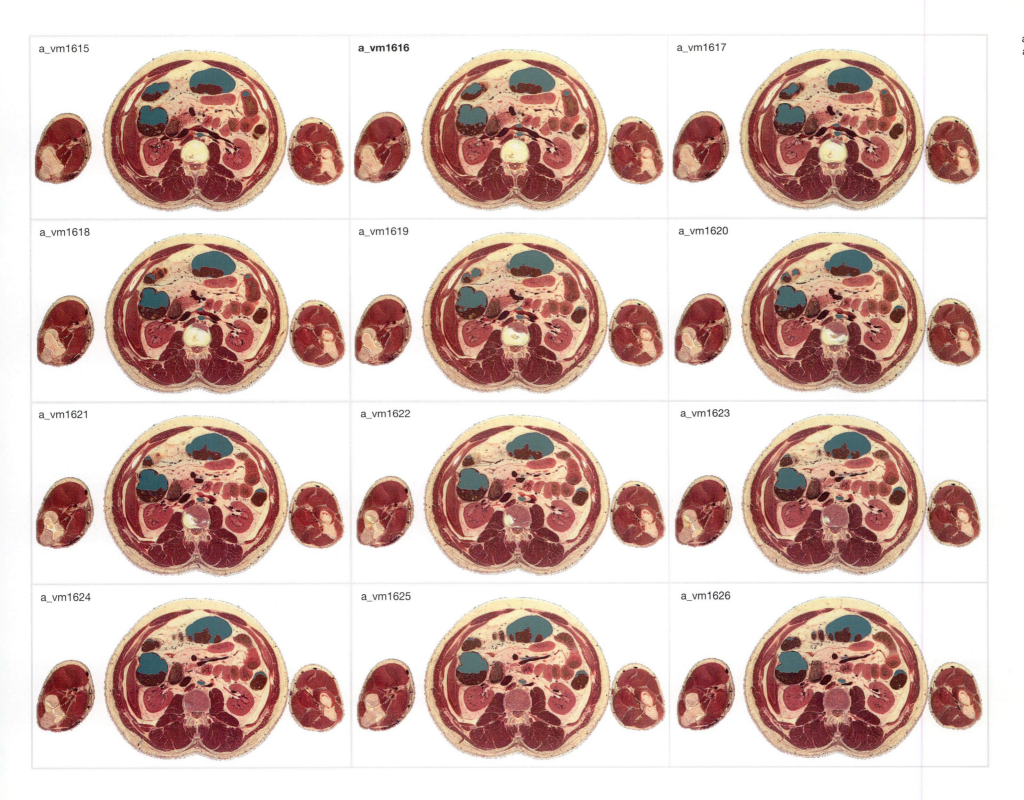

anterior

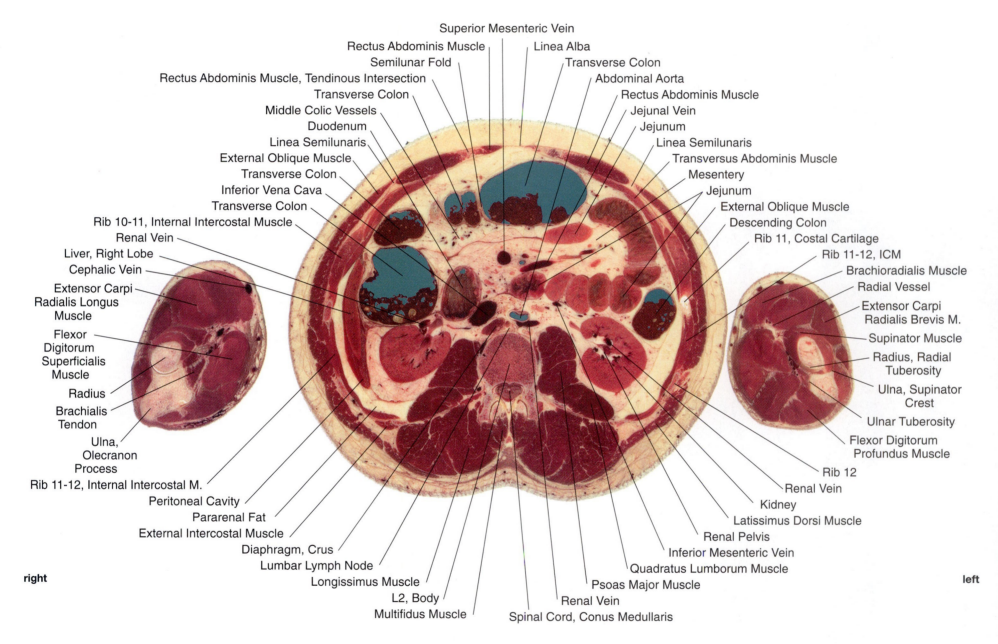

Superior Mesenteric Vein

Rectus Abdominis Muscle

Semilunar Fold

Rectus Abdominis Muscle, Tendinous Intersection

Transverse Colon

Middle Colic Vessels

Duodenum

Linea Semilunaris

External Oblique Muscle

Transverse Colon

Inferior Vena Cava

Transverse Colon

Rib 10-11, Internal Intercostal Muscle

Renal Vein

Liver, Right Lobe

Cephalic Vein

Extensor Carpi Radialis Longus Muscle

Flexor Digitorum Superficialis Muscle

Radius

Brachialis Tendon

Ulna, Olecranon Process

Rib 11-12, Internal Intercostal M.

Peritoneal Cavity

Pararenal Fat

External Intercostal Muscle

Diaphragm, Crus

Lumbar Lymph Node

Longissimus Muscle

L2, Body

Multifidus Muscle

Linea Alba

Transverse Colon

Abdominal Aorta

Rectus Abdominis Muscle

Jejunal Vein

Jejunum

Linea Semilunaris

Transversus Abdominis Muscle

Mesentery

Jejunum

External Oblique Muscle

Descending Colon

Rib 11, Costal Cartilage

Rib 11-12, ICM

Brachioradialis Muscle

Radial Vessel

Extensor Carpi Radialis Brevis M.

Supinator Muscle

Radius, Radial Tuberosity

Ulna, Supinator Crest

Ulnar Tuberosity

Flexor Digitorum Profundus Muscle

Rib 12

Renal Vein

Kidney

Latissimus Dorsi Muscle

Renal Pelvis

Inferior Mesenteric Vein

Quadratus Lumborum Muscle

Psoas Major Muscle

Renal Vein

Spinal Cord, Conus Medullaris

right

left

ICM=Intercostal Muscles

posterior

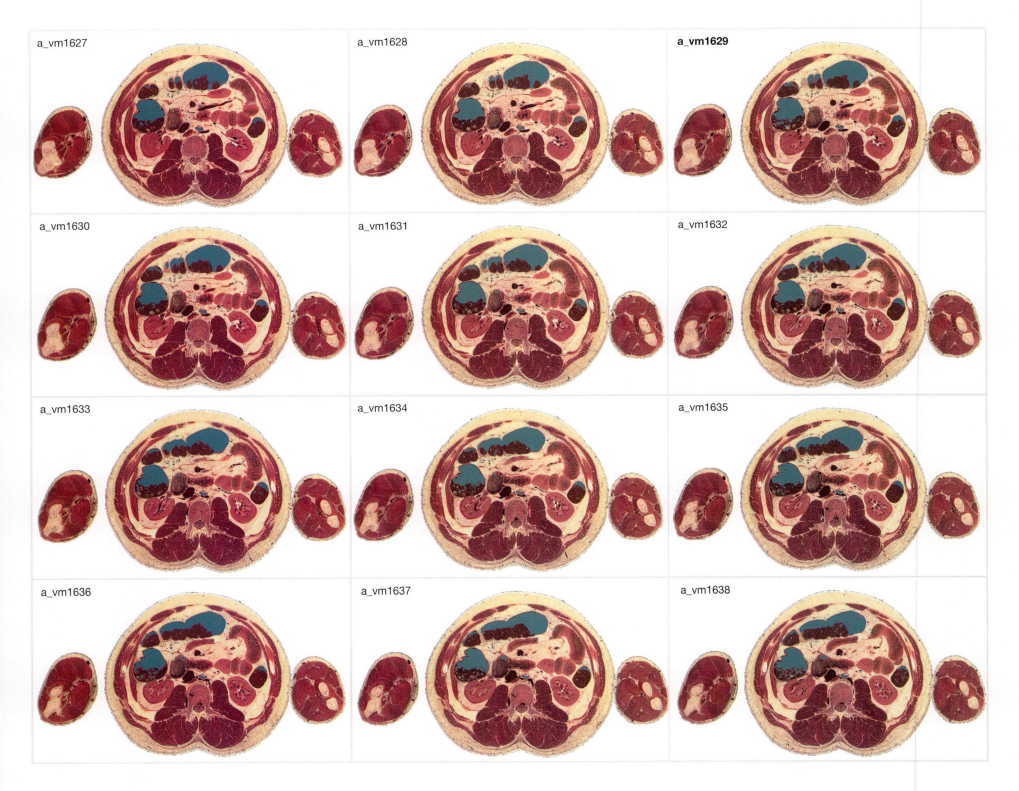

a_vm1627

a_vm1628

a_vm1629

a_vm1630

a_vm1631

a_vm1632

a_vm1633

a_vm1634

a_vm1635

a_vm1636

a_vm1637

a_vm1638

anterior

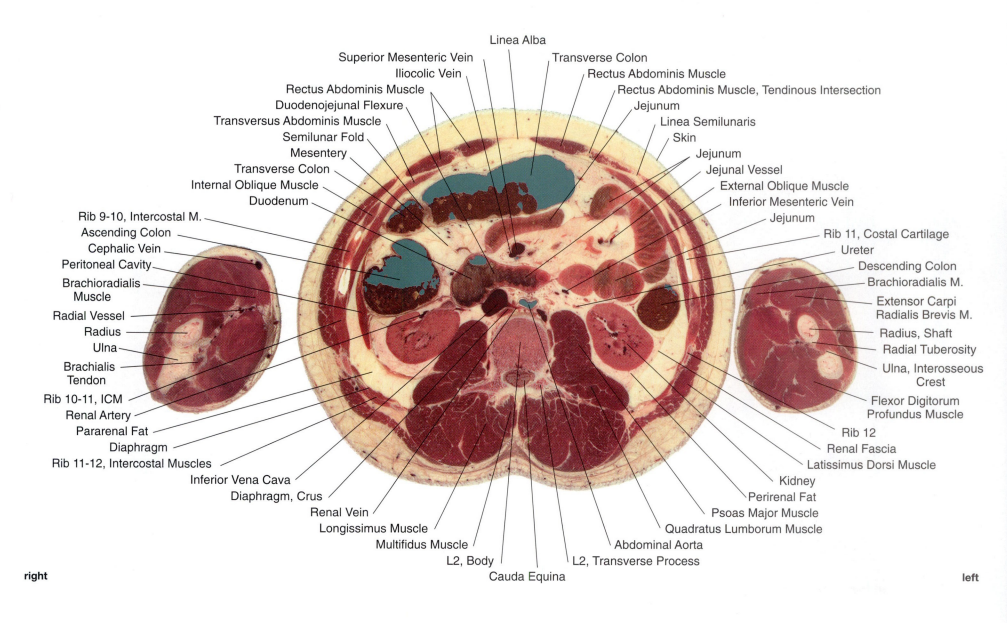

Linea Alba

Superior Mesenteric Vein
Iliocolic Vein
Rectus Abdominis Muscle
Duodenojejunal Flexure
Transversus Abdominis Muscle
Semilunar Fold
Mesentery
Transverse Colon
Internal Oblique Muscle
Duodenum

Transverse Colon
Rectus Abdominis Muscle
Rectus Abdominis Muscle, Tendinous Intersection
Jejunum
Linea Semilunaris
Skin
Jejunum
Jejunal Vessel
External Oblique Muscle
Inferior Mesenteric Vein
Jejunum

Rib 9-10, Intercostal M.
Ascending Colon
Cephalic Vein
Peritoneal Cavity
Brachioradialis Muscle
Radial Vessel
Radius
Ulna
Brachialis Tendon
Rib 10-11, ICM
Renal Artery
Pararenal Fat
Diaphragm
Rib 11-12, Intercostal Muscles
Inferior Vena Cava
Diaphragm, Crus
Renal Vein
Longissimus Muscle
Multifidus Muscle
L2, Body
Cauda Equina

Rib 11, Costal Cartilage
Ureter
Descending Colon
Brachioradialis M.
Extensor Carpi Radialis Brevis M.
Radius, Shaft
Radial Tuberosity
Ulna, Interosseous Crest
Flexor Digitorum Profundus Muscle
Rib 12
Renal Fascia
Latissimus Dorsi Muscle
Kidney
Perirenal Fat
Psoas Major Muscle
Quadratus Lumborum Muscle
Abdominal Aorta
L2, Transverse Process

right

left

ICM=Intercostal Muscles

posterior

108

a_vm1639

a_vm1640

a_vm1641

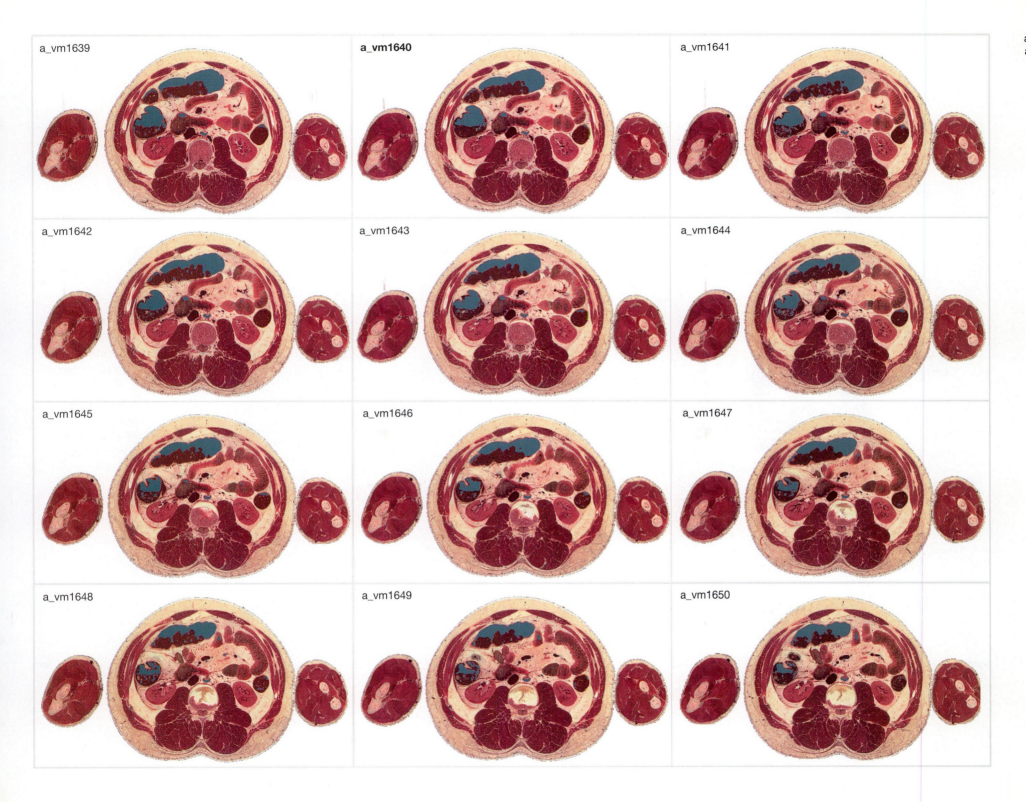

a_vm1642

a_vm1643

a_vm1644

a_vm1645

a_vm1646

a_vm1647

a_vm1648

a_vm1649

a_vm1650

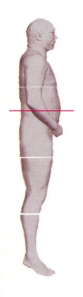

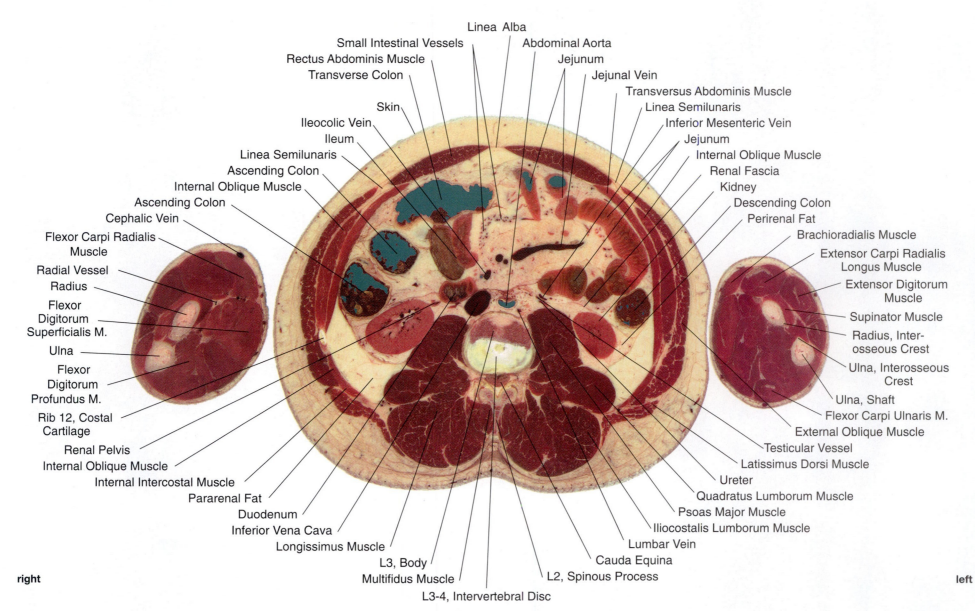

anterior

Linea Alba

Small Intestinal Vessels
Rectus Abdominis Muscle
Transverse Colon

Abdominal Aorta
Jejunum

Jejunal Vein

Transversus Abdominis Muscle

Skin

Linea Semilunaris

Ileocolic Vein

Inferior Mesenteric Vein

Ileum

Jejunum

Linea Semilunaris

Internal Oblique Muscle

Ascending Colon

Renal Fascia

Internal Oblique Muscle

Kidney

Ascending Colon

Descending Colon

Cephalic Vein

Perirenal Fat

Flexor Carpi Radialis
Muscle

Brachioradialis Muscle

Radial Vessel

Extensor Carpi Radialis
Longus Muscle

Radius

Extensor Digitorum
Muscle

Flexor
Digitorum
Superficialis M.

Supinator Muscle

Radius, Inter-
osseous Crest

Ulna

Flexor
Digitorum
Profundus M.

Ulna, Interosseous
Crest

Ulna, Shaft

Rib 12, Costal
Cartilage

Flexor Carpi Ulnaris M.

Renal Pelvis

External Oblique Muscle

Internal Oblique Muscle

Testicular Vessel

Internal Intercostal Muscle

Latissimus Dorsi Muscle

Pararenal Fat

Ureter

Duodenum

Quadratus Lumborum Muscle

Inferior Vena Cava

Psoas Major Muscle

Longissimus Muscle

Iliocostalis Lumborum Muscle

L3, Body

Lumbar Vein

Multifidus Muscle

Cauda Equina

L3-4, Intervertebral Disc

L2, Spinous Process

right

left

posterior

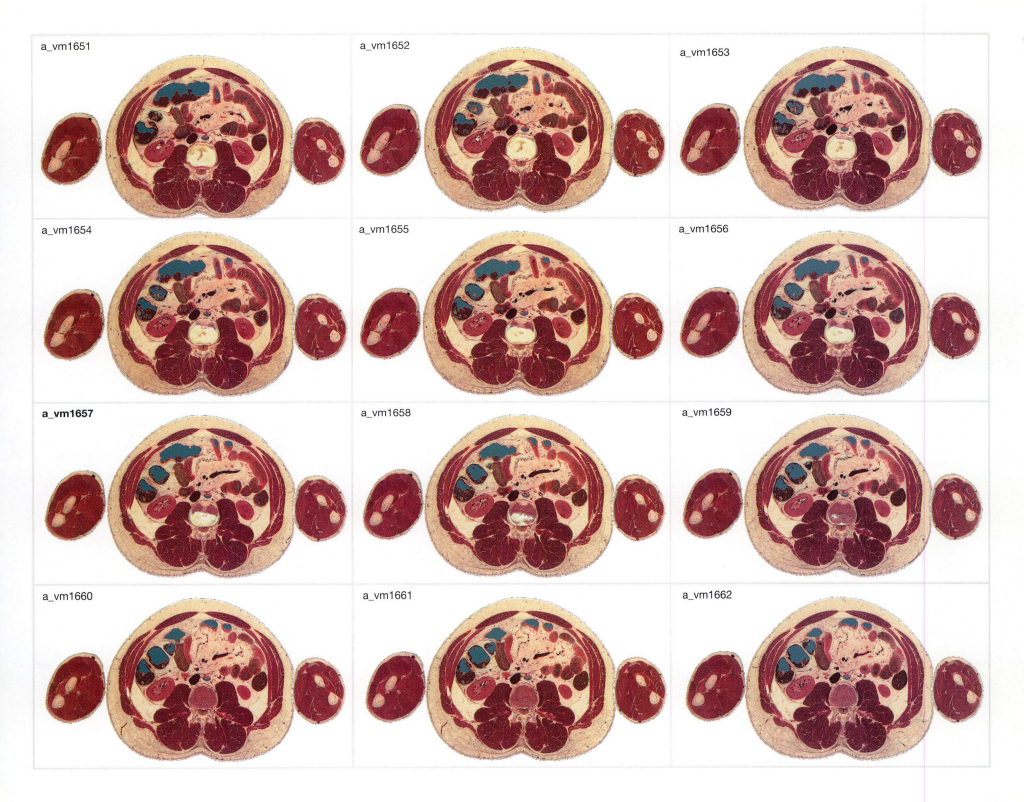

a_vm1651

a_vm1652

a_vm1651-
a_vm1662

a_vm1653

a_vm1654

a_vm1655

a_vm1656

a_vm1657

a_vm1658

a_vm1659

a_vm1660

a_vm1661

a_vm1662

anterior

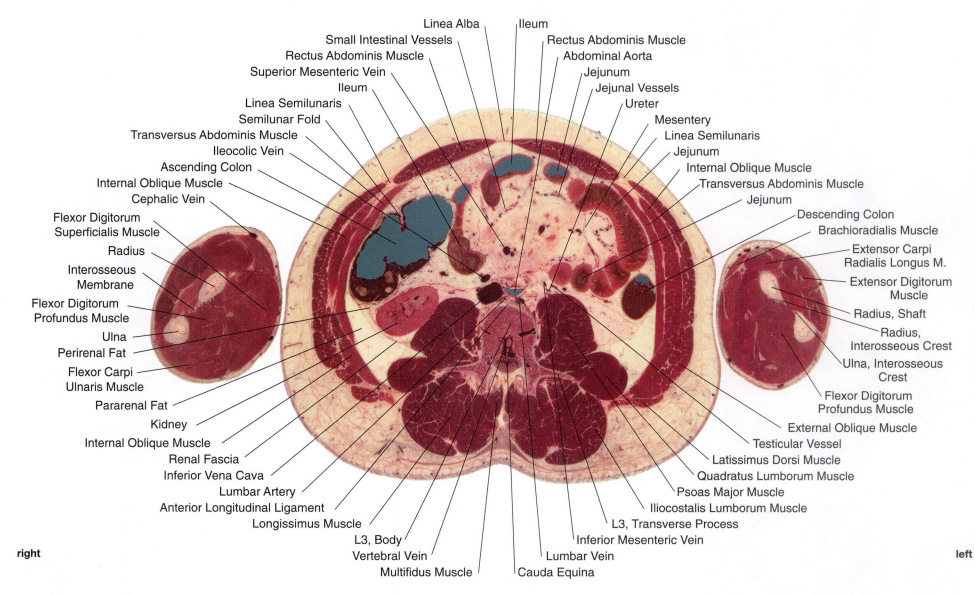

Linea Alba
Small Intestinal Vessels
Rectus Abdominis Muscle
Superior Mesenteric Vein
Ileum
Linea Semilunaris
Semilunar Fold
Transversus Abdominis Muscle
Ileocolic Vein
Ascending Colon
Internal Oblique Muscle
Cephalic Vein
Flexor Digitorum
Superficialis Muscle
Radius
Interosseous
Membrane
Flexor Digitorum
Profundus Muscle
Ulna
Perirenal Fat
Flexor Carpi
Ulnaris Muscle
Pararenal Fat
Kidney
Internal Oblique Muscle
Renal Fascia
Inferior Vena Cava
Lumbar Artery
Anterior Longitudinal Ligament
Longissimus Muscle
L3, Body
Vertebral Vein
Multifidus Muscle

Ileum
Rectus Abdominis Muscle
Abdominal Aorta
Jejunum
Jejunal Vessels
Ureter
Mesentery
Linea Semilunaris
Jejunum
Internal Oblique Muscle
Transversus Abdominis Muscle
Jejunum
Descending Colon
Brachioradialis Muscle
Extensor Carpi
Radialis Longus M.
Extensor Digitorum
Muscle
Radius, Shaft
Radius,
Interosseous Crest
Ulna, Interosseous
Crest
Flexor Digitorum
Profundus Muscle
External Oblique Muscle
Testicular Vessel
Latissimus Dorsi Muscle
Quadratus Lumborum Muscle
Psoas Major Muscle
Iliocostalis Lumborum Muscle
L3, Transverse Process
Inferior Mesenteric Vein
Lumbar Vein
Cauda Equina

right

left

posterior

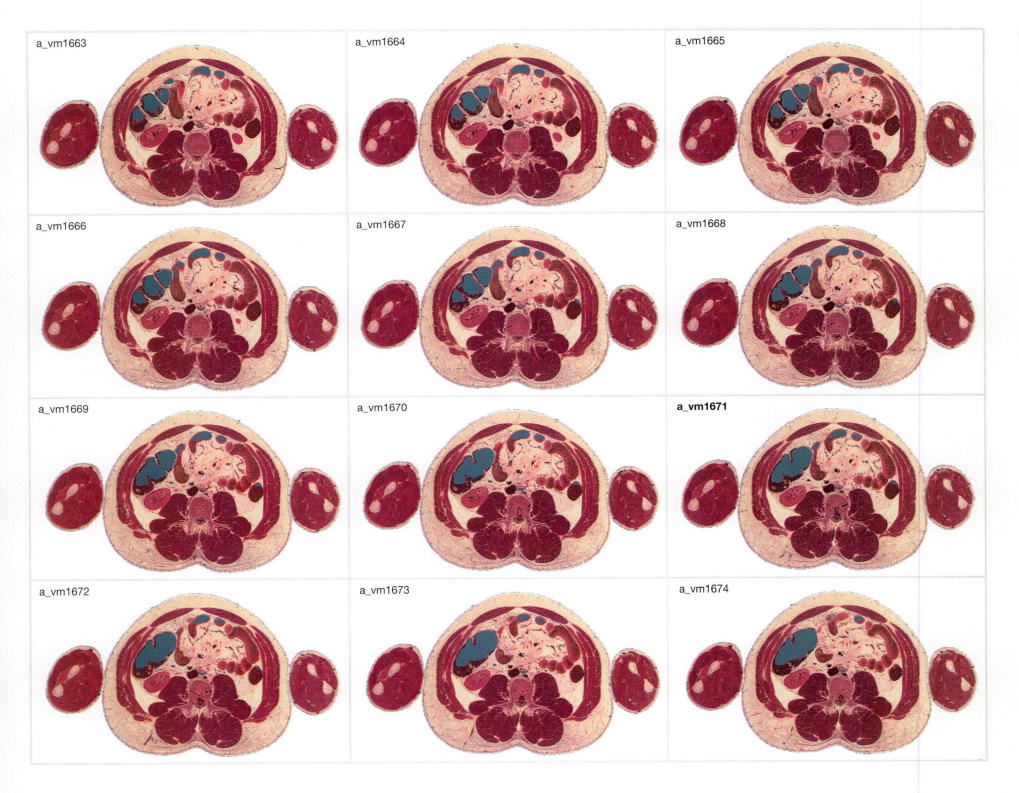

a_vm1663

a_vm1664

a_vm1665

a_vm1666

a_vm1667

a_vm1668

a_vm1669

a_vm1670

a_vm1671

a_vm1672

a_vm1673

a_vm1674

anterior

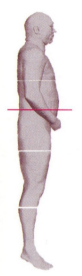

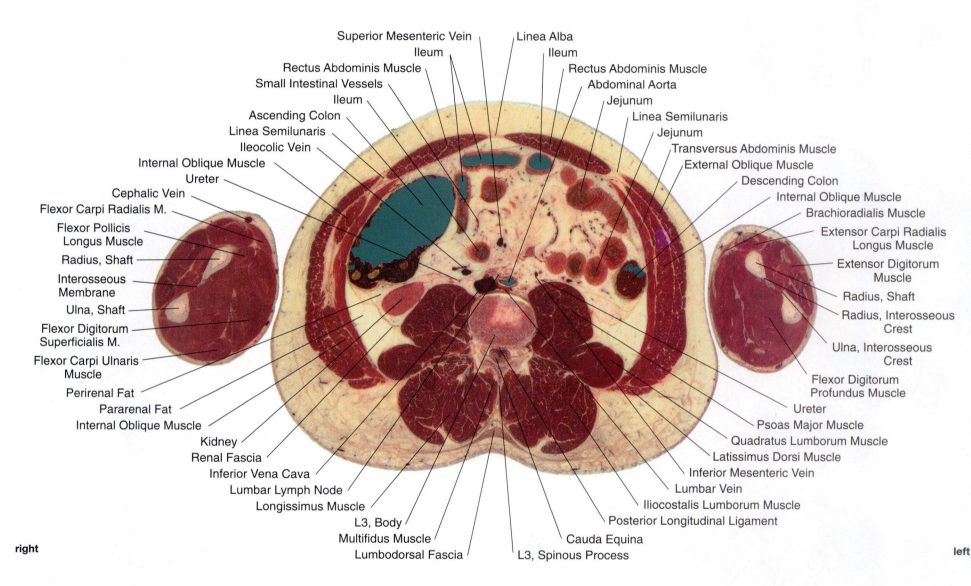

Superior Mesenteric Vein
Ileum
Rectus Abdominis Muscle
Small Intestinal Vessels
Ileum
Ascending Colon
Linea Semilunaris
Ileocolic Vein
Internal Oblique Muscle
Ureter
Cephalic Vein
Flexor Carpi Radialis M.
Flexor Pollicis
Longus Muscle
Radius, Shaft
Interosseous
Membrane
Ulna, Shaft
Flexor Digitorum
Superficialis M.
Flexor Carpi Ulnaris
Muscle
Perirenal Fat
Pararenal Fat
Internal Oblique Muscle
Kidney
Renal Fascia
Inferior Vena Cava
Lumbar Lymph Node
Longissimus Muscle
L3, Body
Multifidus Muscle
Lumbodorsal Fascia

Linea Alba
Ileum
Rectus Abdominis Muscle
Abdominal Aorta
Jejunum
Linea Semilunaris
Jejunum
Transversus Abdominis Muscle
External Oblique Muscle
Descending Colon
Internal Oblique Muscle
Brachioradialis Muscle
Extensor Carpi Radialis
Longus Muscle
Extensor Digitorum
Muscle
Radius, Shaft
Radius, Interosseous
Crest
Ulna, Interosseous
Crest
Flexor Digitorum
Profundus Muscle
Ureter
Psoas Major Muscle
Quadratus Lumborum Muscle
Latissimus Dorsi Muscle
Inferior Mesenteric Vein
Lumbar Vein
Iliocostalis Lumborum Muscle
Posterior Longitudinal Ligament
Cauda Equina
L3, Spinous Process

right

left

posterior

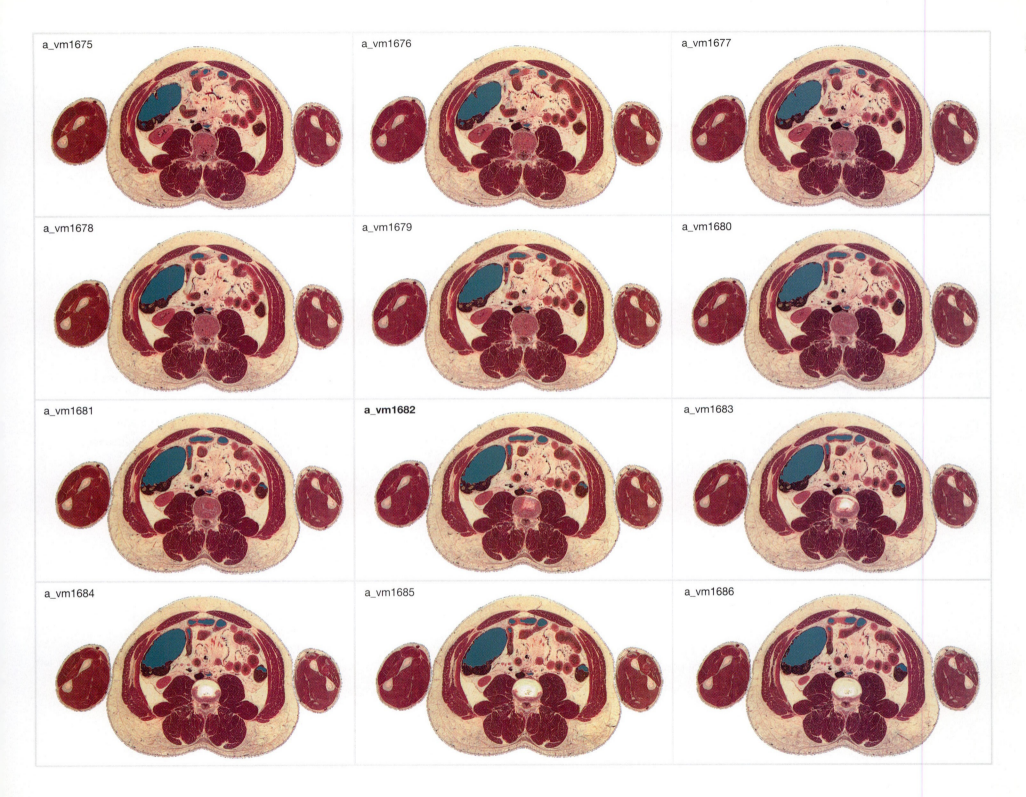

a_vm1675

a_vm1676

a_vm1677

a_vm1678

a_vm1679

a_vm1680

a_vm1681

a_vm1682

a_vm1683

a_vm1684

a_vm1685

a_vm1686

anterior

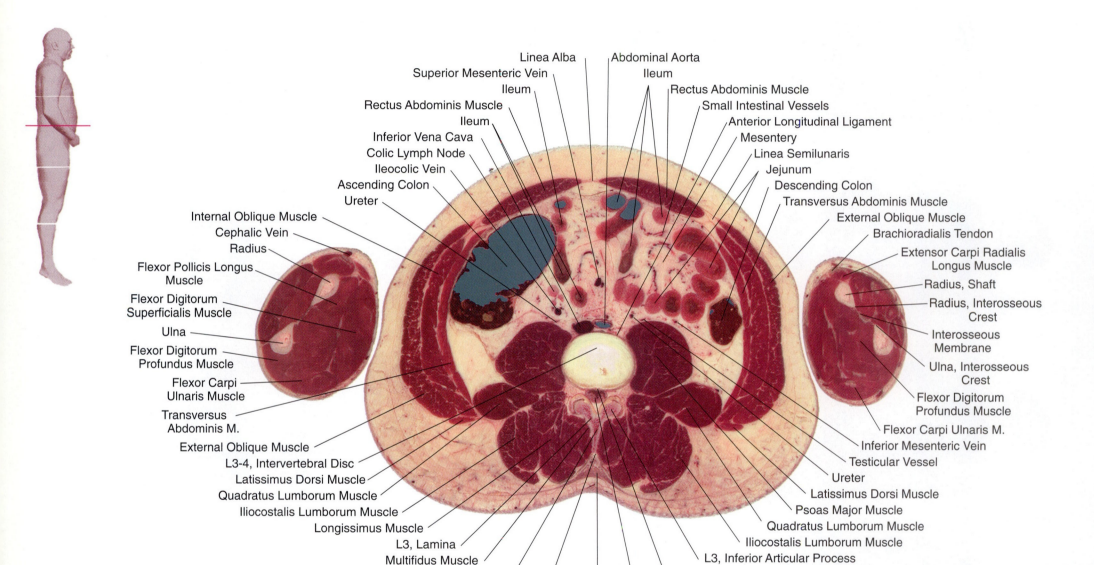

Linea Alba
Abdominal Aorta
Superior Mesenteric Vein
Ileum
Ileum
Rectus Abdominis Muscle
Rectus Abdominis Muscle
Small Intestinal Vessels
Ileum
Anterior Longitudinal Ligament
Inferior Vena Cava
Mesentery
Colic Lymph Node
Linea Semilunaris
Ileocolic Vein
Jejunum
Ascending Colon
Descending Colon
Ureter
Transversus Abdominis Muscle
Internal Oblique Muscle
External Oblique Muscle
Cephalic Vein
Brachioradialis Tendon
Radius
Extensor Carpi Radialis
Flexor Pollicis Longus
Longus Muscle
Muscle
Radius, Shaft
Flexor Digitorum
Radius, Interosseous
Superficialis Muscle
Crest
Ulna
Interosseous
Flexor Digitorum
Membrane
Profundus Muscle
Ulna, Interosseous
Flexor Carpi
Crest
Ulnaris Muscle
Flexor Digitorum
Transversus
Profundus Muscle
Abdominis M.
Flexor Carpi Ulnaris M.
External Oblique Muscle
Inferior Mesenteric Vein
L3-4, Intervertebral Disc
Testicular Vessel
Latissimus Dorsi Muscle
Ureter
Quadratus Lumborum Muscle
Latissimus Dorsi Muscle
Iliocostalis Lumborum Muscle
Psoas Major Muscle
Longissimus Muscle
Quadratus Lumborum Muscle
L3, Lamina
Iliocostalis Lumborum Muscle
Multifidus Muscle
L3, Inferior Articular Process
L3, Spinous Process
Posterior Longitudinal Ligament
Supraspinous Ligament
Cauda Equina
Lumbodorsal Fascia

right

left

posterior

116

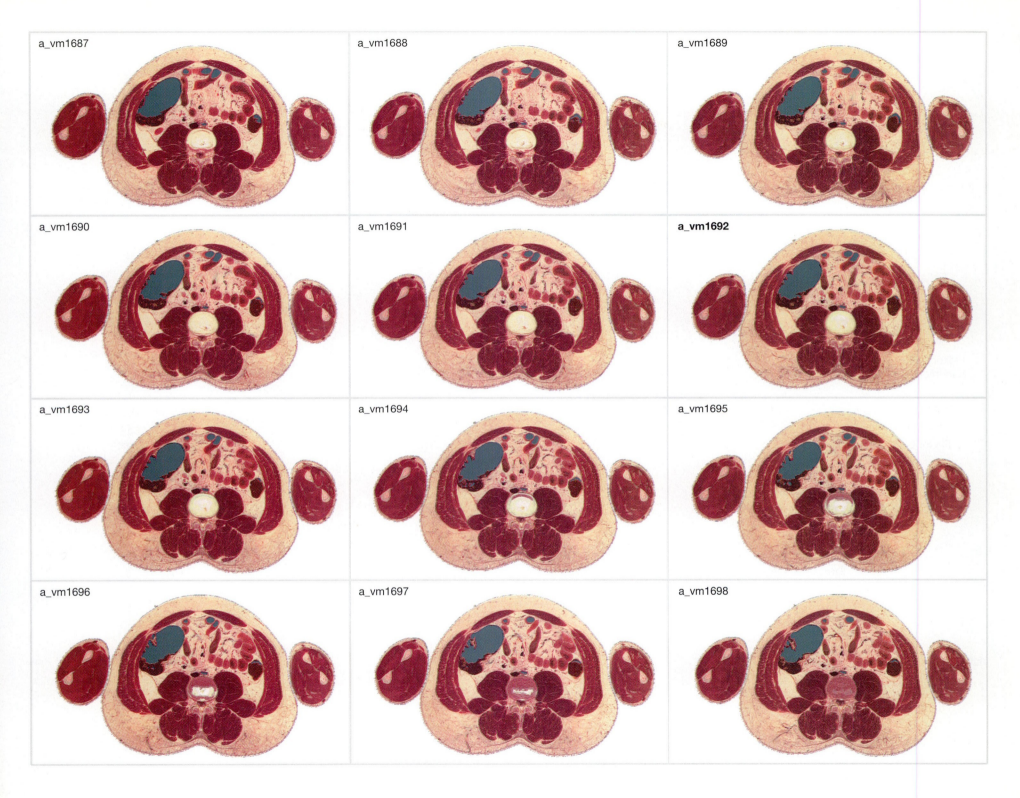

a_vm1687

a_vm1688

a_vm1689

a_vm1690

a_vm1691

a_vm1692

a_vm1693

a_vm1694

a_vm1695

a_vm1696

a_vm1697

a_vm1698

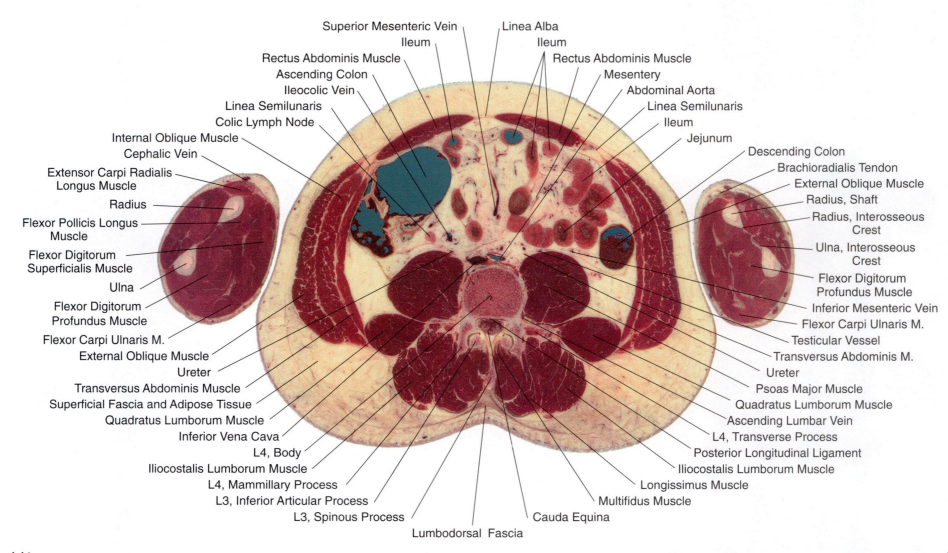

Transverse
a_vm1703

anterior

Superior Mesenteric Vein
Ileum
Rectus Abdominis Muscle
Ascending Colon
Ileocolic Vein
Linea Semilunaris
Colic Lymph Node
Internal Oblique Muscle
Cephalic Vein
Extensor Carpi Radialis
Longus Muscle
Radius
Flexor Pollicis Longus
Muscle
Flexor Digitorum
Superficialis Muscle
Ulna
Flexor Digitorum
Profundus Muscle
Flexor Carpi Ulnaris M.
External Oblique Muscle
Ureter
Transversus Abdominis Muscle
Superficial Fascia and Adipose Tissue
Quadratus Lumborum Muscle
Inferior Vena Cava
L4, Body
Iliocostalis Lumborum Muscle
L4, Mammillary Process
L3, Inferior Articular Process
L3, Spinous Process

Lumbodorsal Fascia

Linea Alba
Ileum
Rectus Abdominis Muscle
Mesentery
Abdominal Aorta
Linea Semilunaris
Ileum
Jejunum
Descending Colon
Brachioradialis Tendon
External Oblique Muscle
Radius, Shaft
Radius, Interosseous
Crest
Ulna, Interosseous
Crest
Flexor Digitorum
Profundus Muscle
Inferior Mesenteric Vein
Flexor Carpi Ulnaris M.
Testicular Vessel
Transversus Abdominis M.
Ureter
Psoas Major Muscle
Quadratus Lumborum Muscle
Ascending Lumbar Vein
L4, Transverse Process
Posterior Longitudinal Ligament
Iliocostalis Lumborum Muscle
Longissimus Muscle
Multifidus Muscle
Cauda Equina

right

left

posterior

118

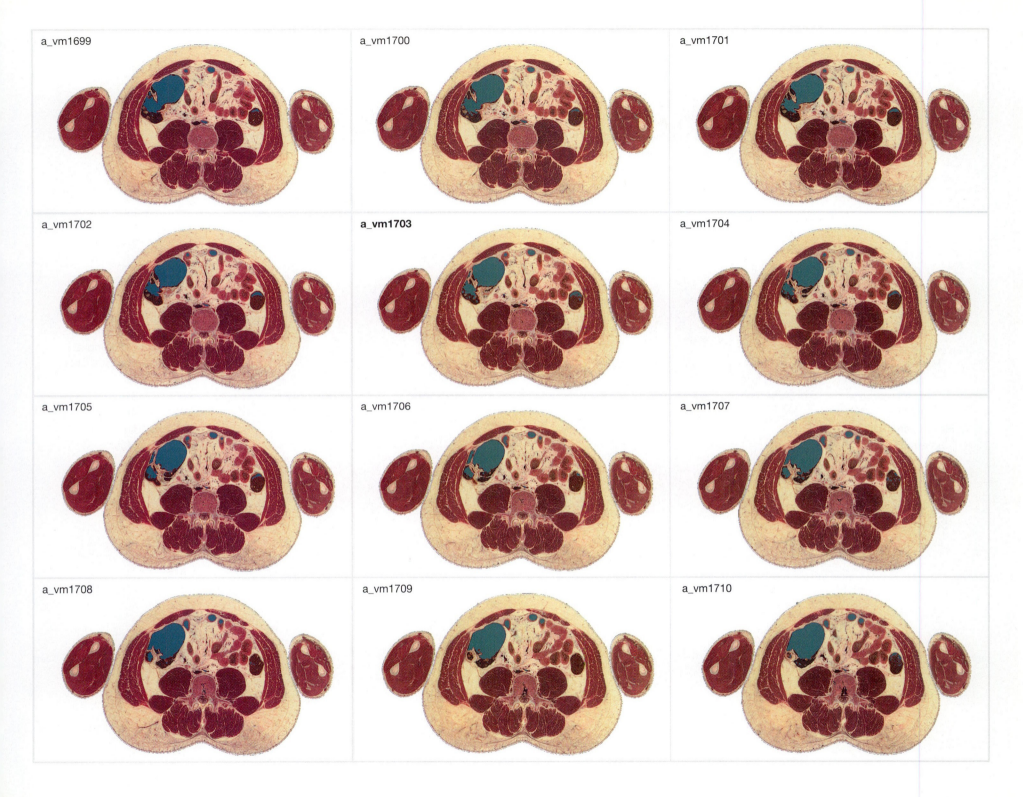

a_vm1699

a_vm1700

a_vm1701

a_vm1702

a_vm1703

a_vm1704

a_vm1705

a_vm1706

a_vm1707

a_vm1708

a_vm1709

a_vm1710

anterior

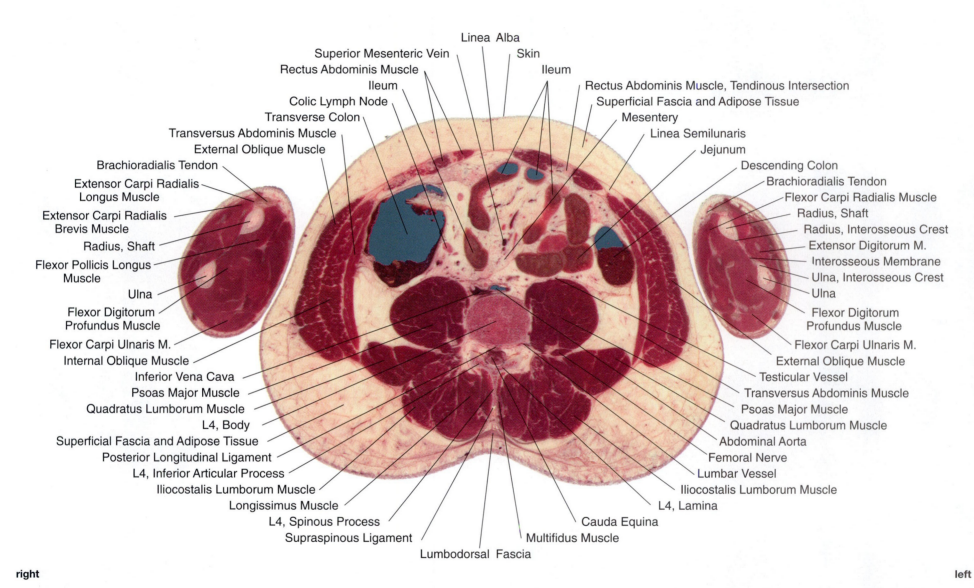

Linea Alba

Superior Mesenteric Vein
Rectus Abdominis Muscle
Ileum
Colic Lymph Node
Transverse Colon
Transversus Abdominis Muscle
External Oblique Muscle
Brachioradialis Tendon
Extensor Carpi Radialis
Longus Muscle
Extensor Carpi Radialis
Brevis Muscle
Radius, Shaft
Flexor Pollicis Longus
Muscle
Ulna
Flexor Digitorum
Profundus Muscle
Flexor Carpi Ulnaris M.
Internal Oblique Muscle
Inferior Vena Cava
Psoas Major Muscle
Quadratus Lumborum Muscle
L4, Body
Superficial Fascia and Adipose Tissue
Posterior Longitudinal Ligament
L4, Inferior Articular Process
Iliocostalis Lumborum Muscle
Longissimus Muscle
L4, Spinous Process
Supraspinous Ligament

Skin
Ileum
Rectus Abdominis Muscle, Tendinous Intersection
Superficial Fascia and Adipose Tissue
Mesentery
Linea Semilunaris
Jejunum
Descending Colon
Brachioradialis Tendon
Flexor Carpi Radialis Muscle
Radius, Shaft
Radius, Interosseous Crest
Extensor Digitorum M.
Interosseous Membrane
Ulna, Interosseous Crest
Ulna
Flexor Digitorum
Profundus Muscle
Flexor Carpi Ulnaris M.
External Oblique Muscle
Testicular Vessel
Transversus Abdominis Muscle
Psoas Major Muscle
Quadratus Lumborum Muscle
Abdominal Aorta
Femoral Nerve
Lumbar Vessel
Iliocostalis Lumborum Muscle
L4, Lamina
Cauda Equina

Multifidus Muscle
Lumbodorsal Fascia

right

left

posterior

120

a_vm1711

a_vm1712

a_vm1713

a_vm1714

a_vm1715

a_vm1716

a_vm1717

a_vm1718

a_vm1719

a_vm1720

a_vm1721

a_vm1722

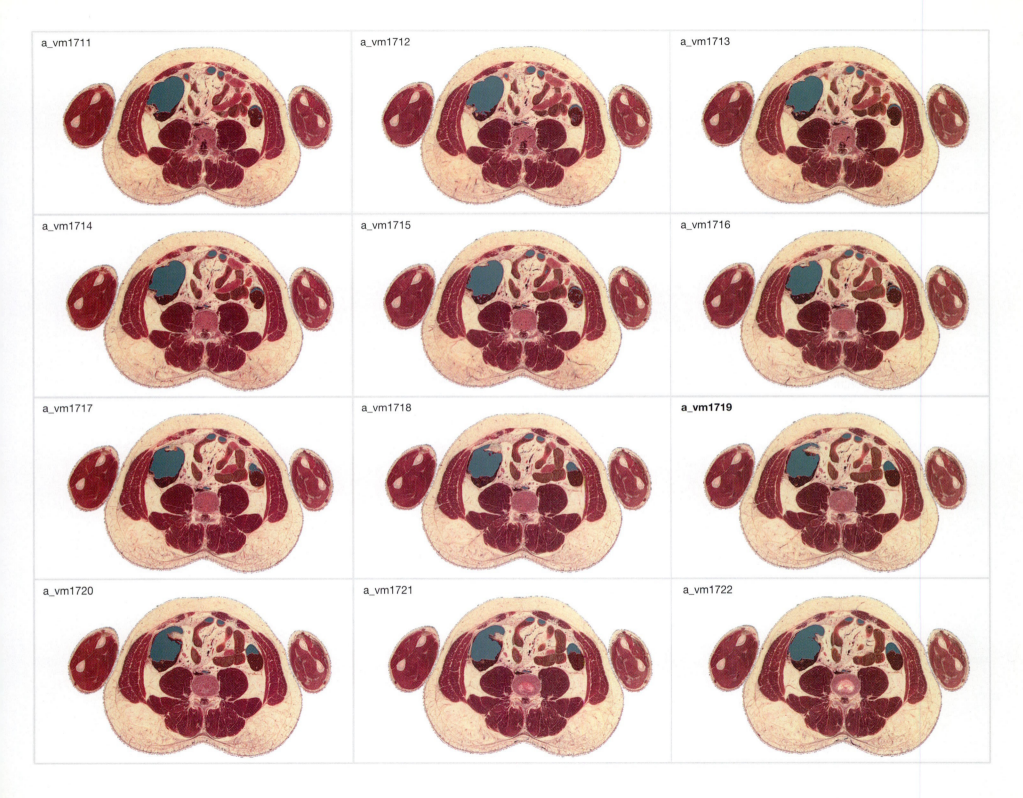

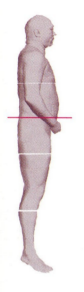

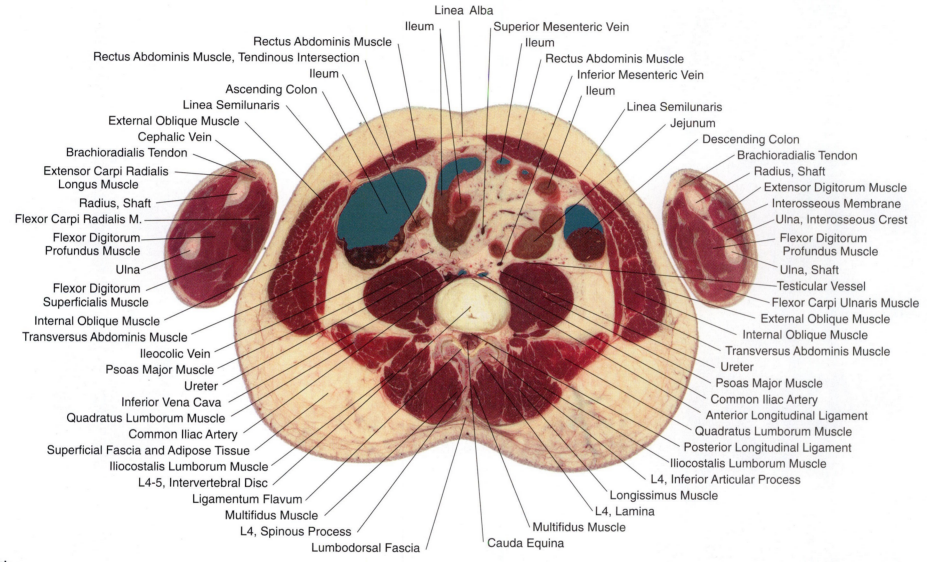

anterior

Linea Alba

Ileum

Superior Mesenteric Vein

Rectus Abdominis Muscle

Ileum

Rectus Abdominis Muscle, Tendinous Intersection

Rectus Abdominis Muscle

Ileum

Inferior Mesenteric Vein

Ascending Colon

Ileum

Linea Semilunaris

Linea Semilunaris

External Oblique Muscle

Jejunum

Cephalic Vein

Descending Colon

Brachioradialis Tendon

Brachioradialis Tendon

Extensor Carpi Radialis
Longus Muscle

Radius, Shaft

Extensor Digitorum Muscle

Radius, Shaft

Interosseous Membrane

Flexor Carpi Radialis M.

Ulna, Interosseous Crest

Flexor Digitorum
Profundus Muscle

Flexor Digitorum
Profundus Muscle

Ulna

Ulna, Shaft

Flexor Digitorum
Superficialis Muscle

Testicular Vessel

Internal Oblique Muscle

Flexor Carpi Ulnaris Muscle

Transversus Abdominis Muscle

External Oblique Muscle

Ileocolic Vein

Internal Oblique Muscle

Psoas Major Muscle

Transversus Abdominis Muscle

Ureter

Ureter

Inferior Vena Cava

Psoas Major Muscle

Quadratus Lumborum Muscle

Common Iliac Artery

Common Iliac Artery

Anterior Longitudinal Ligament

Superficial Fascia and Adipose Tissue

Quadratus Lumborum Muscle

Iliocostalis Lumborum Muscle

Posterior Longitudinal Ligament

L4-5, Intervertebral Disc

Iliocostalis Lumborum Muscle

Ligamentum Flavum

L4, Inferior Articular Process

Multifidus Muscle

Longissimus Muscle

L4, Spinous Process

L4, Lamina

Multifidus Muscle

Lumbodorsal Fascia

Cauda Equina

right

left

posterior

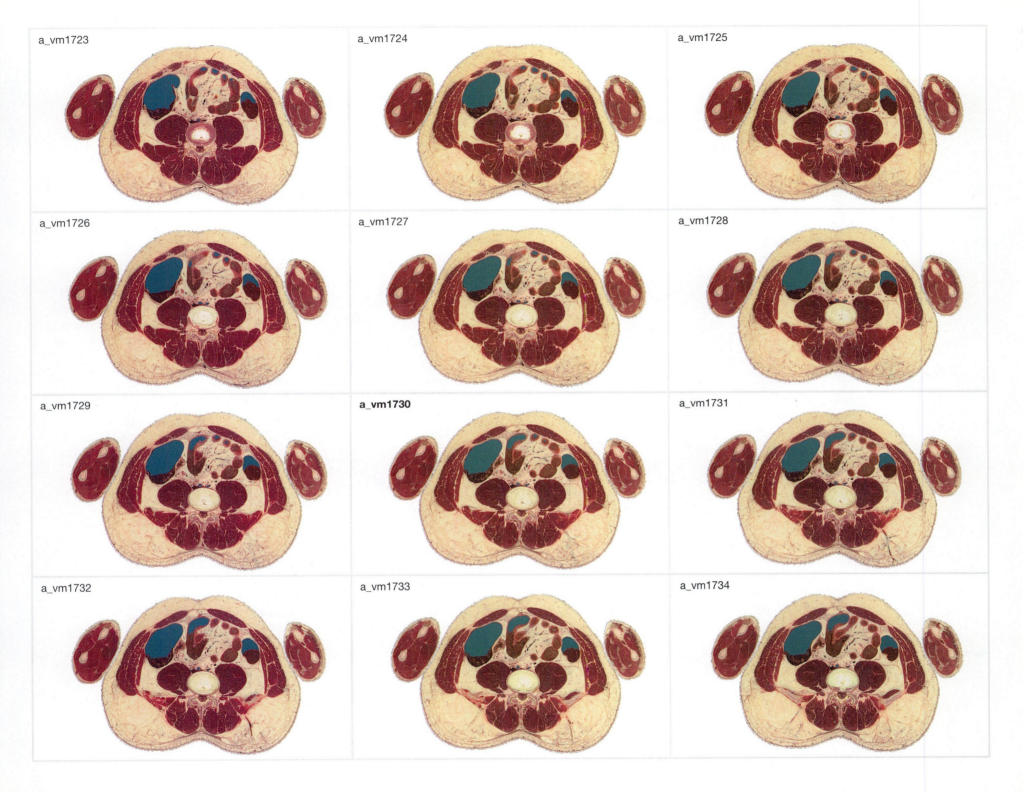

a_vm1723

a_vm1724

a_vm1725

a_vm1726

a_vm1727

a_vm1728

a_vm1729

a_vm1730

a_vm1731

a_vm1732

a_vm1733

a_vm1734

anterior

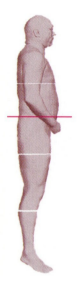

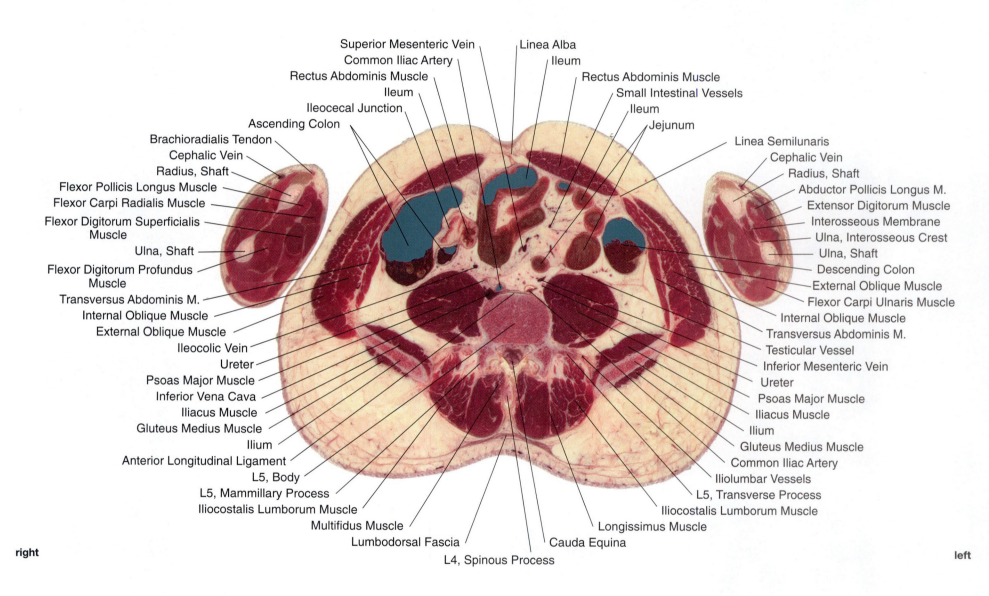

Superior Mesenteric Vein
Common Iliac Artery
Rectus Abdominis Muscle
Ileum
Ileocecal Junction
Ascending Colon
Brachioradialis Tendon
Cephalic Vein
Radius, Shaft
Flexor Pollicis Longus Muscle
Flexor Carpi Radialis Muscle
Flexor Digitorum Superficialis Muscle
Ulna, Shaft
Flexor Digitorum Profundus Muscle
Transversus Abdominis M.
Internal Oblique Muscle
External Oblique Muscle
Ileocolic Vein
Ureter
Psoas Major Muscle
Inferior Vena Cava
Iliacus Muscle
Gluteus Medius Muscle
Ilium
Anterior Longitudinal Ligament
L5, Body
L5, Mammillary Process
Iliocostalis Lumborum Muscle

Linea Alba
Ileum
Rectus Abdominis Muscle
Small Intestinal Vessels
Ileum
Jejunum
Linea Semilunaris
Cephalic Vein
Radius, Shaft
Abductor Pollicis Longus M.
Extensor Digitorum Muscle
Interosseous Membrane
Ulna, Interosseous Crest
Ulna, Shaft
Descending Colon
External Oblique Muscle
Flexor Carpi Ulnaris Muscle
Internal Oblique Muscle
Transversus Abdominis M.
Testicular Vessel
Inferior Mesenteric Vein
Ureter
Psoas Major Muscle
Iliacus Muscle
Ilium
Gluteus Medius Muscle
Common Iliac Artery
Iliolumbar Vessels
L5, Transverse Process
Iliocostalis Lumborum Muscle

Multifidus Muscle
Lumbodorsal Fascia
L4, Spinous Process
Cauda Equina
Longissimus Muscle

right

left

posterior

a_vm1735

a_vm1736

a_vm1737

a_vm1738

a_vm1739

a_vm1740

a_vm1741

a_vm1742

a_vm1743

a_vm1744

a_vm1745

a_vm1746

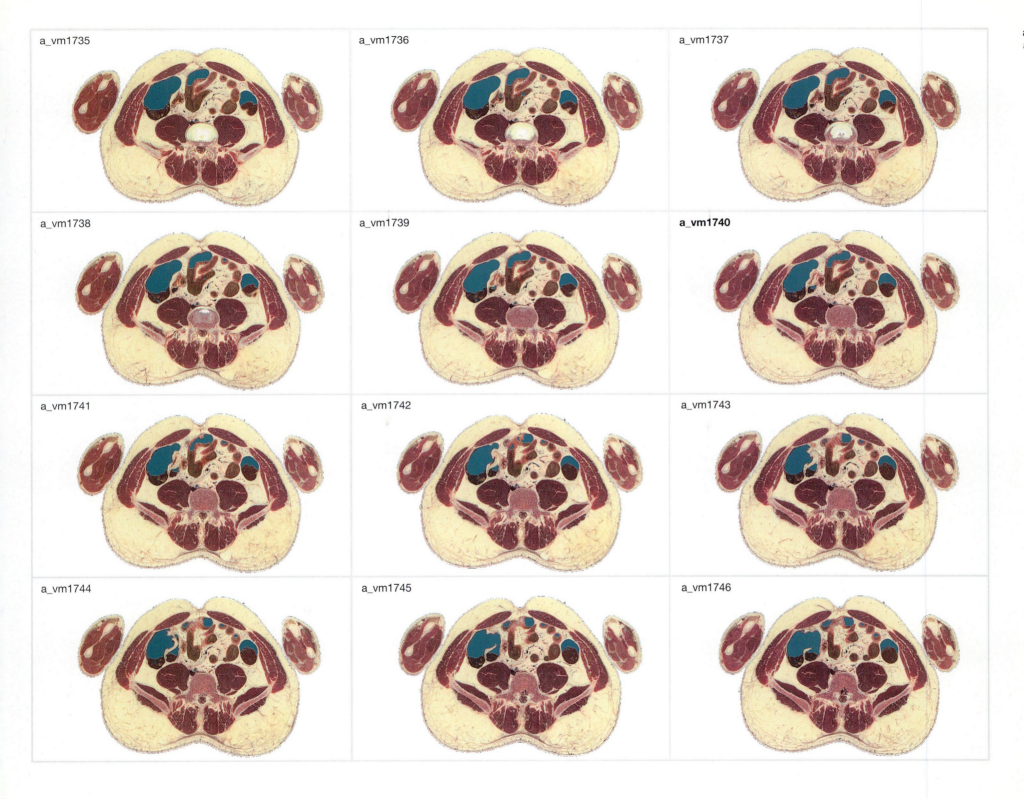

anterior

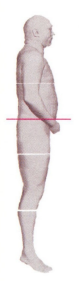

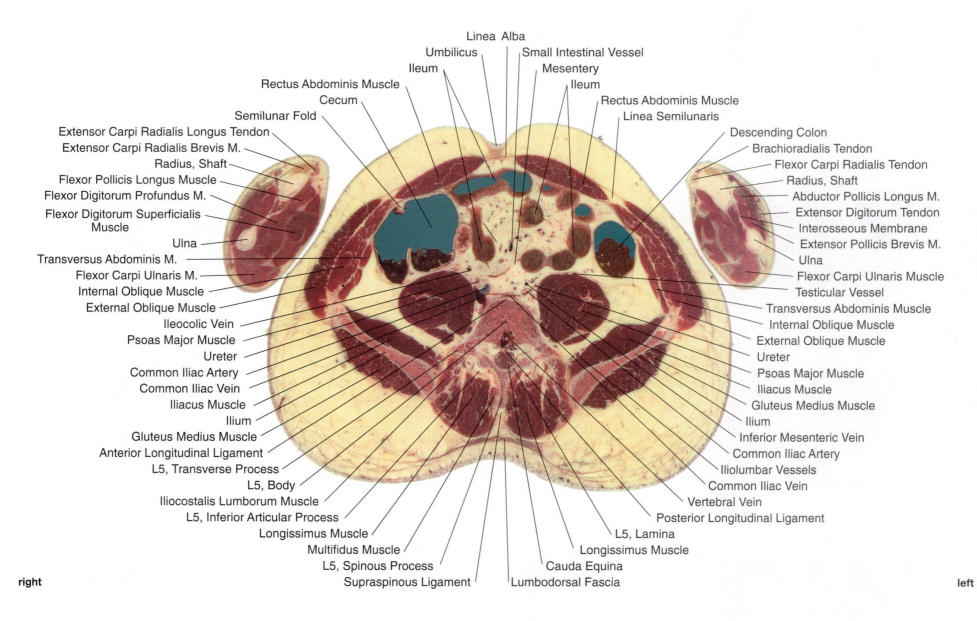

Linea Alba

Umbilicus

Small Intestinal Vessel

Ileum

Mesentery

Ileum

Rectus Abdominis Muscle

Cecum

Rectus Abdominis Muscle

Semilunar Fold

Linea Semilunaris

Extensor Carpi Radialis Longus Tendon

Descending Colon

Extensor Carpi Radialis Brevis M.

Brachioradialis Tendon

Radius, Shaft

Flexor Carpi Radialis Tendon

Flexor Pollicis Longus Muscle

Radius, Shaft

Flexor Digitorum Profundus M.

Abductor Pollicis Longus M.

Flexor Digitorum Superficialis
Muscle

Extensor Digitorum Tendon

Interosseous Membrane

Ulna

Extensor Pollicis Brevis M.

Transversus Abdominis M.

Ulna

Flexor Carpi Ulnaris M.

Flexor Carpi Ulnaris Muscle

Internal Oblique Muscle

Testicular Vessel

External Oblique Muscle

Transversus Abdominis Muscle

Ileocolic Vein

Internal Oblique Muscle

Psoas Major Muscle

External Oblique Muscle

Ureter

Ureter

Common Iliac Artery

Psoas Major Muscle

Common Iliac Vein

Iliacus Muscle

Iliacus Muscle

Gluteus Medius Muscle

Ilium

Ilium

Gluteus Medius Muscle

Inferior Mesenteric Vein

Anterior Longitudinal Ligament

Common Iliac Artery

L5, Transverse Process

Iliolumbar Vessels

L5, Body

Common Iliac Vein

Iliocostalis Lumborum Muscle

Vertebral Vein

L5, Inferior Articular Process

Posterior Longitudinal Ligament

Longissimus Muscle

L5, Lamina

Multifidus Muscle

Longissimus Muscle

L5, Spinous Process

Cauda Equina

Supraspinous Ligament

Lumbodorsal Fascia

right

left

posterior

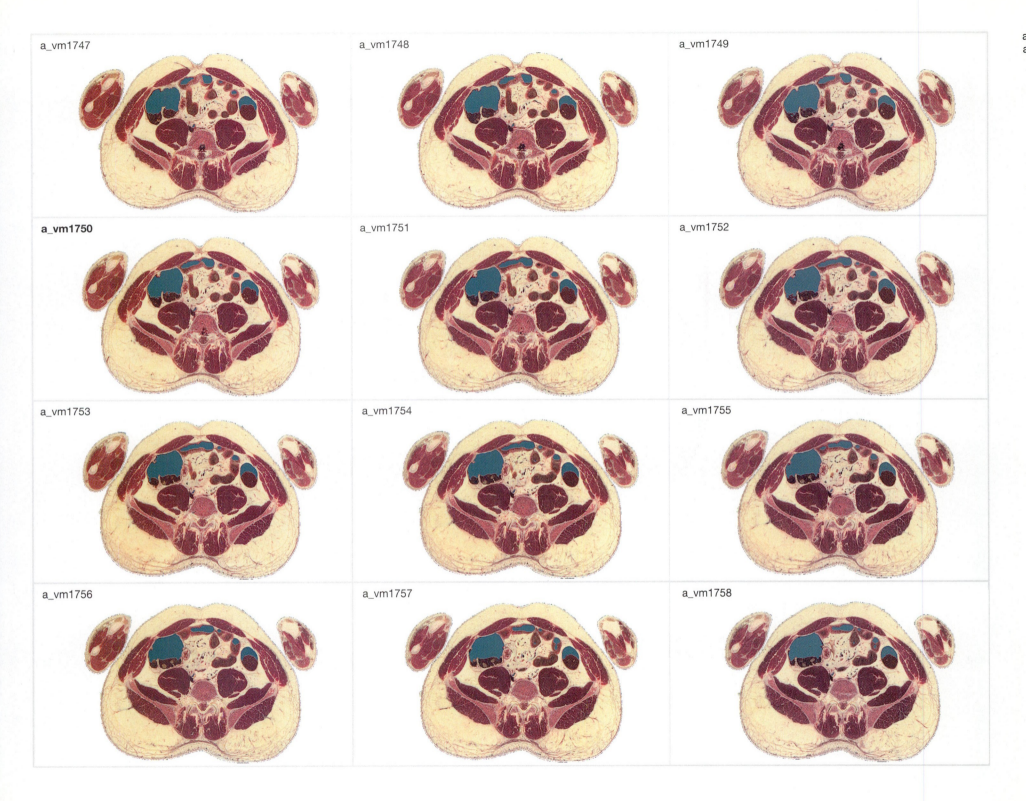

a_vm1747

a_vm1748

a_vm1749

a_vm1750

a_vm1751

a_vm1752

a_vm1753

a_vm1754

a_vm1755

a_vm1756

a_vm1757

a_vm1758

anterior

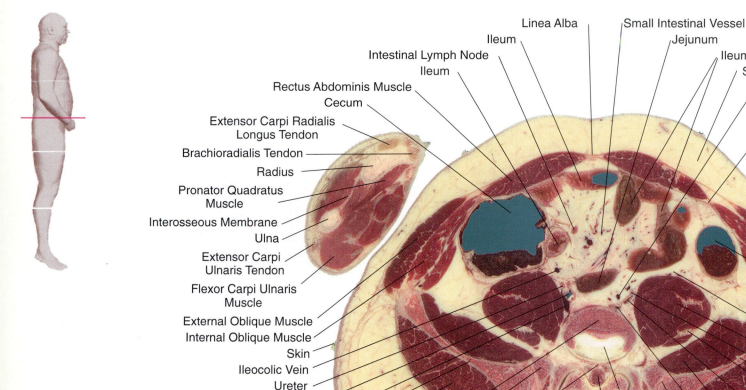
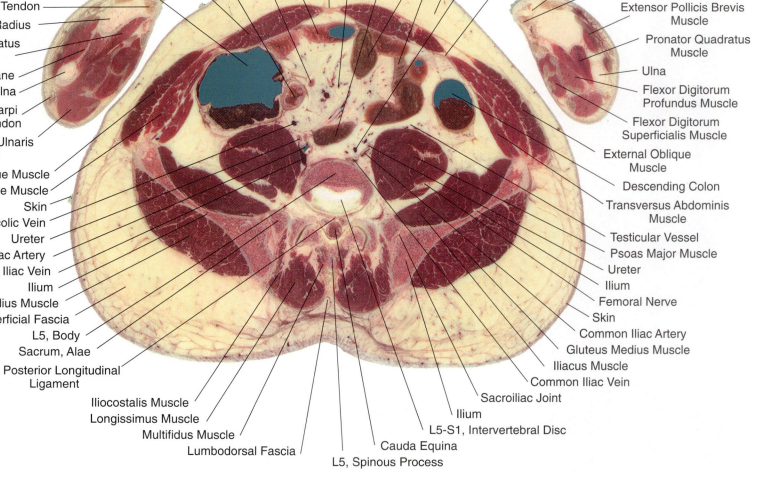

Linea Alba

Small Intestinal Vessel

Ileum

Jejunum

Intestinal Lymph Node

Ileum

Ileum

Superficial Fascia

Rectus Abdominis Muscle

Inferior Mesenteric Vein

Cecum

Linea Semilunaris

Extensor Carpi Radialis
Longus Tendon

Radius, Styloid Process

Flexor Carpi Radialis Tendon

Brachioradialis Tendon

Extensor Pollicis Brevis
Muscle

Radius

Pronator Quadratus
Muscle

Pronator Quadratus
Muscle

Interosseous Membrane

Ulna

Ulna

Extensor Carpi
Ulnaris Tendon

Flexor Digitorum
Profundus Muscle

Flexor Carpi Ulnaris
Muscle

Flexor Digitorum
Superficialis Muscle

External Oblique Muscle

External Oblique
Muscle

Internal Oblique Muscle

Skin

Descending Colon

Ileocolic Vein

Transversus Abdominis
Muscle

Ureter

Common Iliac Artery

Testicular Vessel

Common Iliac Vein

Psoas Major Muscle

Ilium

Ureter

Gluteus Medius Muscle

Ilium

Superficial Fascia

Femoral Nerve

L5, Body

Skin

Sacrum, Alae

Common Iliac Artery

Posterior Longitudinal
Ligament

Gluteus Medius Muscle

Iliacus Muscle

Common Iliac Vein

Iliocostalis Muscle

Sacroiliac Joint

Longissimus Muscle

Ilium

Multifidus Muscle

L5-S1, Intervertebral Disc

Lumbodorsal Fascia

Cauda Equina

L5, Spinous Process

right

left

posterior

128

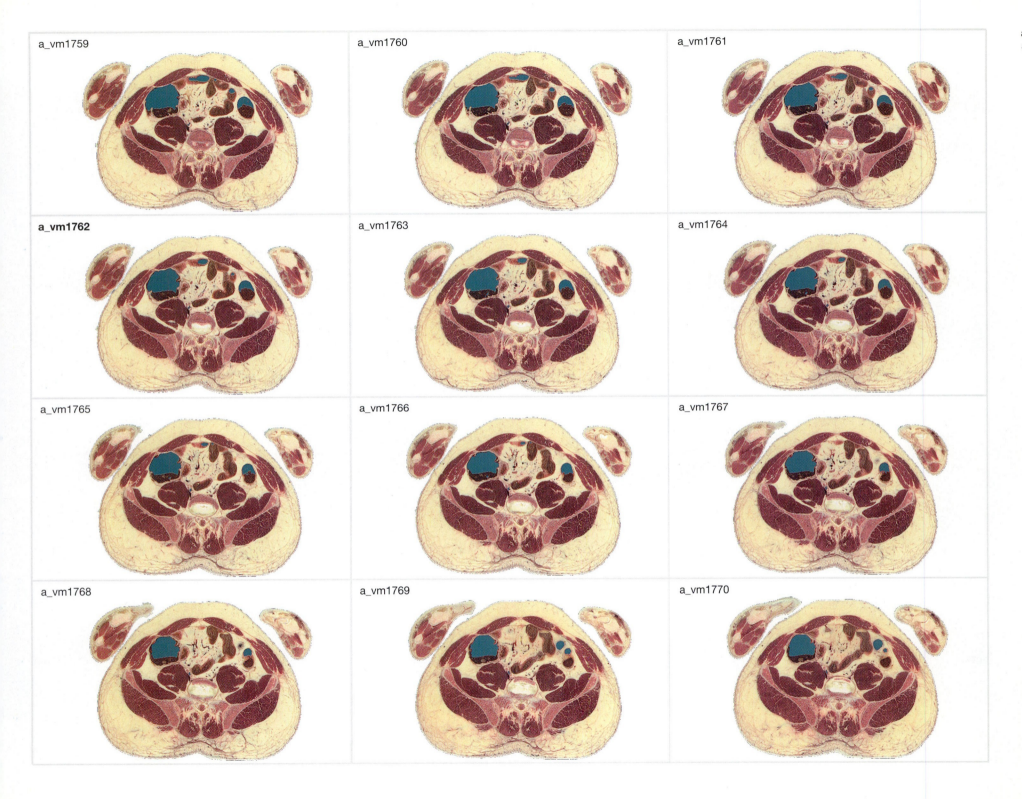

a_vm1759

a_vm1760

a_vm1761

a_vm1762

a_vm1763

a_vm1764

a_vm1765

a_vm1766

a_vm1767

a_vm1768

a_vm1769

a_vm1770

anterior

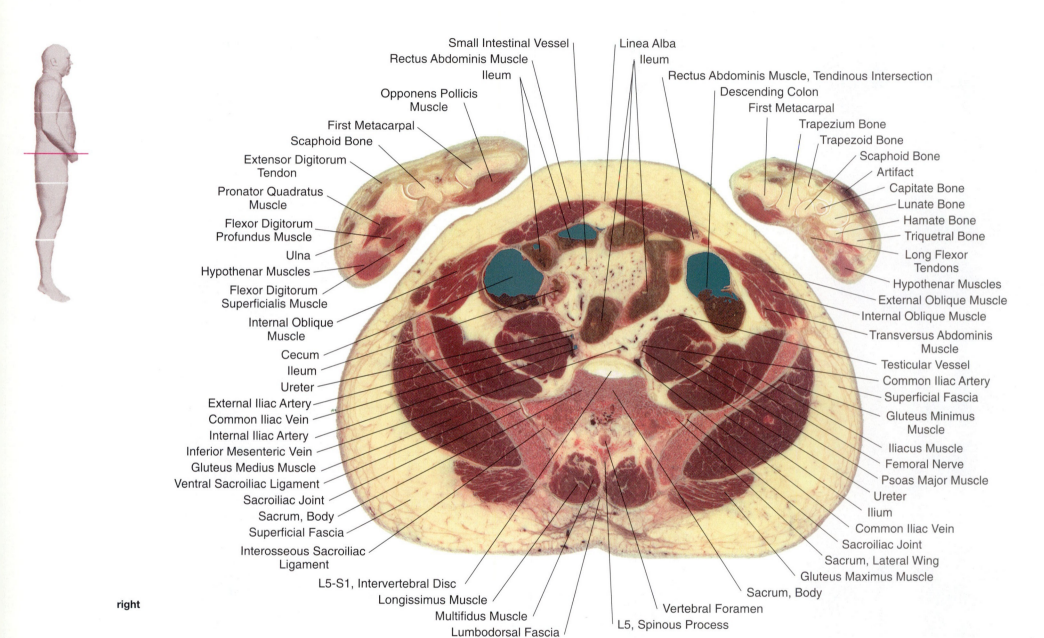

Small Intestinal Vessel
Rectus Abdominis Muscle
Ileum
Opponens Pollicis
Muscle
First Metacarpal
Scaphoid Bone
Extensor Digitorum
Tendon
Pronator Quadratus
Muscle
Flexor Digitorum
Profundus Muscle
Ulna
Hypothenar Muscles
Flexor Digitorum
Superficialis Muscle
Internal Oblique
Muscle
Cecum
Ileum
Ureter
External Iliac Artery
Common Iliac Vein
Internal Iliac Artery
Inferior Mesenteric Vein
Gluteus Medius Muscle
Ventral Sacroiliac Ligament
Sacroiliac Joint
Sacrum, Body
Superficial Fascia
Interosseous Sacroiliac
Ligament

Linea Alba
Ileum
Rectus Abdominis Muscle, Tendinous Intersection
Descending Colon
First Metacarpal
Trapezium Bone
Trapezoid Bone
Scaphoid Bone
Artifact
Capitate Bone
Lunate Bone
Hamate Bone
Triquetral Bone
Long Flexor
Tendons
Hypothenar Muscles
External Oblique Muscle
Internal Oblique Muscle
Transversus Abdominis
Muscle
Testicular Vessel
Common Iliac Artery
Superficial Fascia
Gluteus Minimus
Muscle
Iliacus Muscle
Femoral Nerve
Psoas Major Muscle
Ureter
Ilium
Common Iliac Vein
Sacroiliac Joint
Sacrum, Lateral Wing
Gluteus Maximus Muscle
Sacrum, Body

L5-S1, Intervertebral Disc
Longissimus Muscle
Multifidus Muscle
Lumbodorsal Fascia
Vertebral Foramen
L5, Spinous Process

right

left

posterior

a_vm1771

a_vm1772

a_vm1773

a_vm1774

a_vm1775

a_vm1776

a_vm1777

a_vm1778

a_vm1779

a_vm1780

a_vm1781

a_vm1782

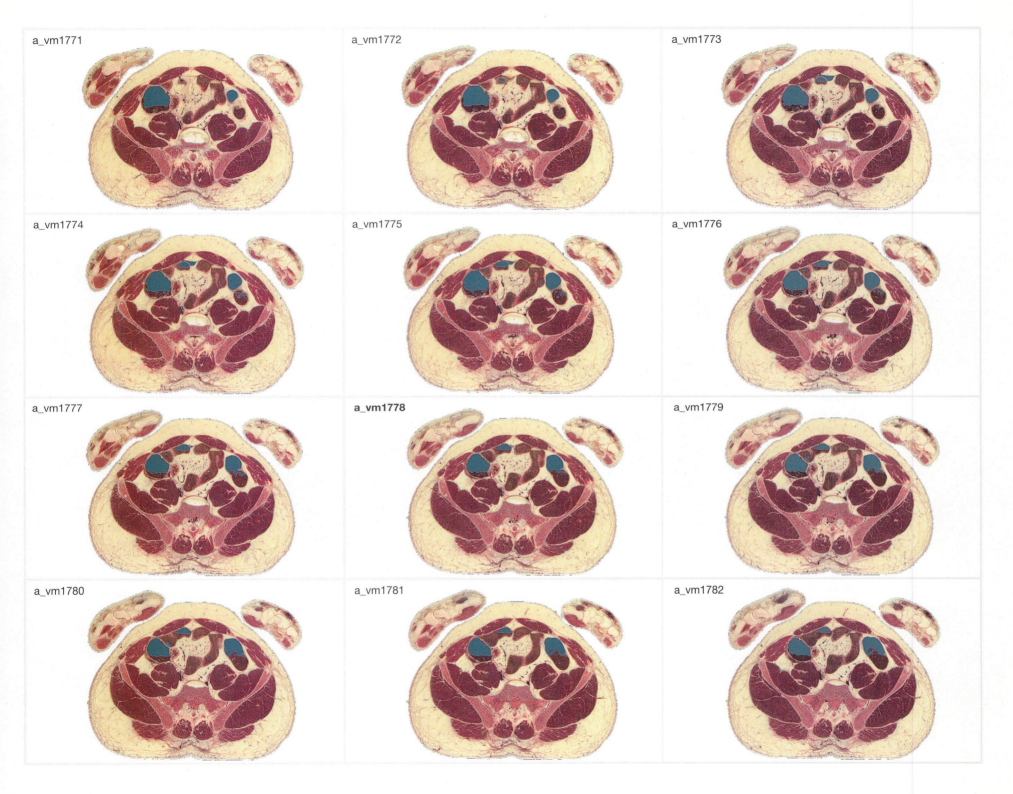

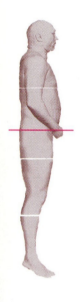

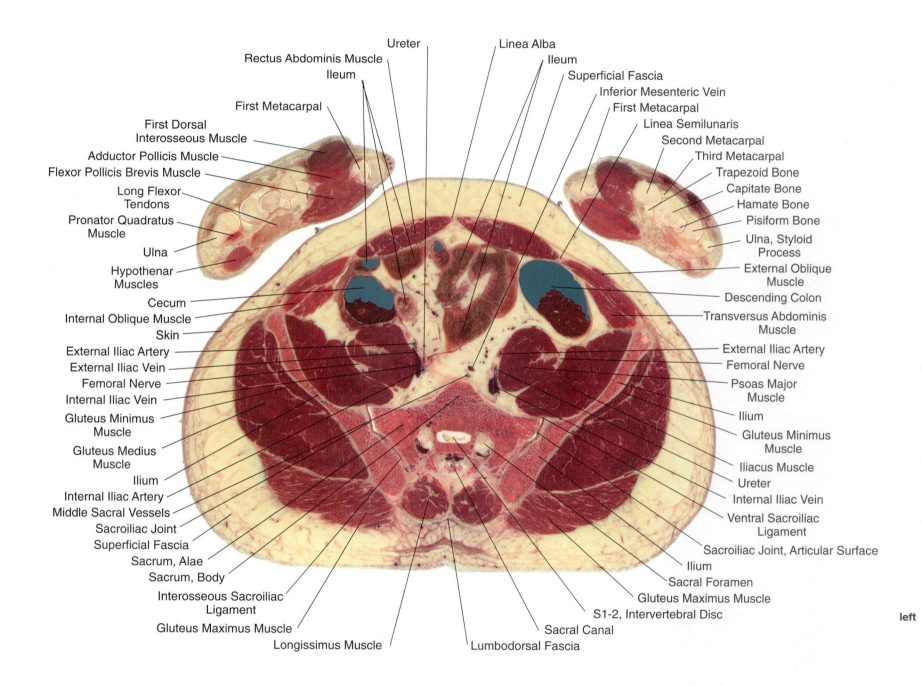

anterior

Ureter

Linea Alba

Rectus Abdominis Muscle

Ileum

Ileum

Superficial Fascia

Inferior Mesenteric Vein

First Metacarpal

First Metacarpal

Linea Semilunaris

First Dorsal
Interosseous Muscle

Second Metacarpal

Adductor Pollicis Muscle

Third Metacarpal

Flexor Pollicis Brevis Muscle

Trapezoid Bone

Long Flexor
Tendons

Capitate Bone

Hamate Bone

Pronator Quadratus
Muscle

Pisiform Bone

Ulna

Ulna, Styloid
Process

Hypothenar
Muscles

External Oblique
Muscle

Cecum

Descending Colon

Internal Oblique Muscle

Transversus Abdominis
Muscle

Skin

External Iliac Artery

External Iliac Artery

External Iliac Vein

Femoral Nerve

Femoral Nerve

Psoas Major
Muscle

Internal Iliac Vein

Gluteus Minimus
Muscle

Ilium

Gluteus Medius
Muscle

Gluteus Minimus
Muscle

Iliacus Muscle

Ilium

Ureter

Internal Iliac Artery

Internal Iliac Vein

Middle Sacral Vessels

Ventral Sacroiliac
Ligament

Sacroiliac Joint

Superficial Fascia

Sacroiliac Joint, Articular Surface

Sacrum, Alae

Ilium

Sacrum, Body

Sacral Foramen

Interosseous Sacroiliac
Ligament

Gluteus Maximus Muscle

Gluteus Maximus Muscle

S1-2, Intervertebral Disc

Longissimus Muscle

Sacral Canal

Lumbodorsal Fascia

right

left

posterior

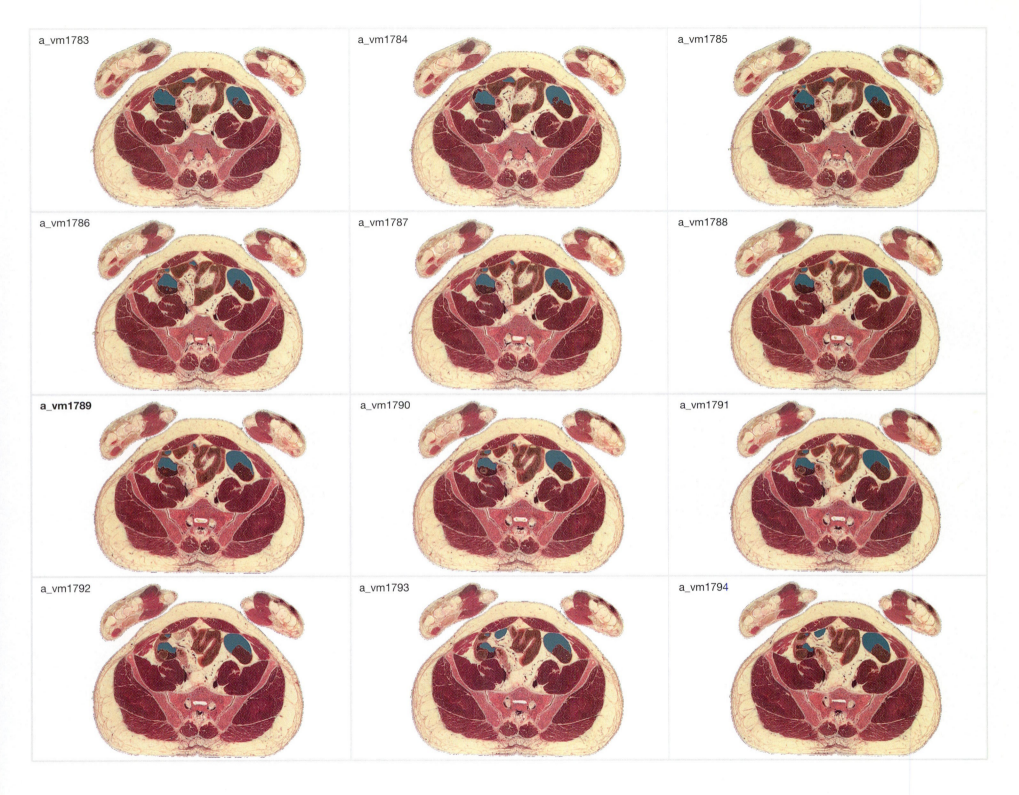

a_vm1783

a_vm1784

a_vm1785

a_vm1786

a_vm1787

a_vm1788

a_vm1789

a_vm1790

a_vm1791

a_vm1792

a_vm1793

a_vm1794

anterior

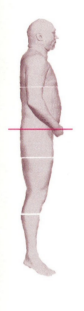

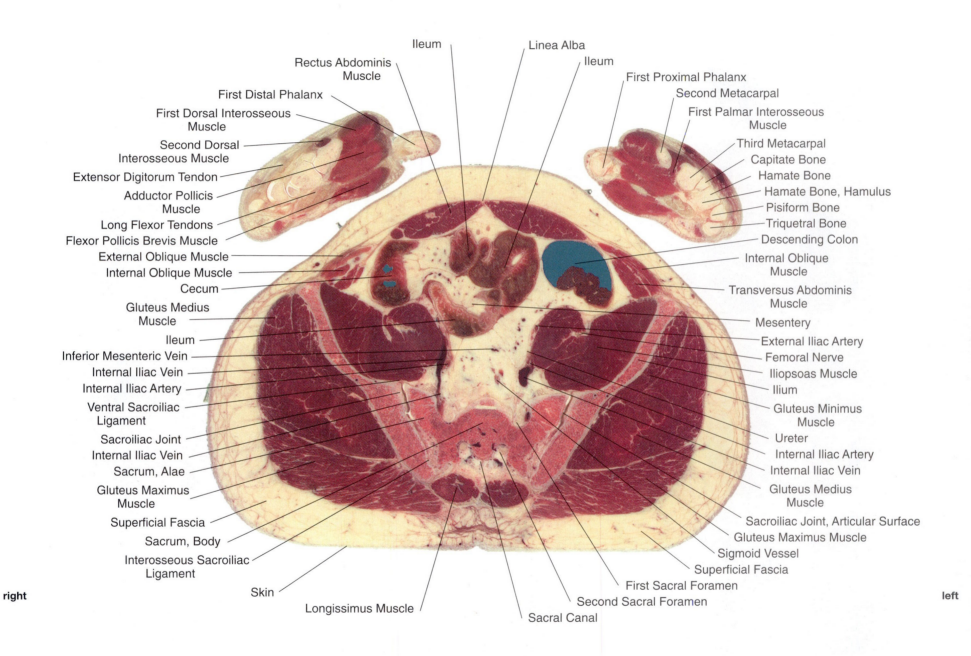

Ileum

Linea Alba

Ileum

Rectus Abdominis
Muscle

First Proximal Phalanx

Second Metacarpal

First Distal Phalanx

First Palmar Interosseous
Muscle

First Dorsal Interosseous
Muscle

Third Metacarpal

Second Dorsal
Interosseous Muscle

Capitate Bone

Hamate Bone

Extensor Digitorum Tendon

Hamate Bone, Hamulus

Adductor Pollicis
Muscle

Pisiform Bone

Long Flexor Tendons

Triquetral Bone

Flexor Pollicis Brevis Muscle

Descending Colon

External Oblique Muscle

Internal Oblique
Muscle

Internal Oblique Muscle

Cecum

Transversus Abdominis
Muscle

Gluteus Medius
Muscle

Mesentery

Ileum

External Iliac Artery

Inferior Mesenteric Vein

Femoral Nerve

Internal Iliac Vein

Iliopsoas Muscle

Internal Iliac Artery

Ilium

Ventral Sacroiliac
Ligament

Gluteus Minimus
Muscle

Sacroiliac Joint

Ureter

Internal Iliac Vein

Internal Iliac Artery

Sacrum, Alae

Internal Iliac Vein

Gluteus Maximus
Muscle

Gluteus Medius
Muscle

Superficial Fascia

Sacroiliac Joint, Articular Surface

Sacrum, Body

Gluteus Maximus Muscle

Interosseous Sacroiliac
Ligament

Sigmoid Vessel

Superficial Fascia

Skin

First Sacral Foramen

Longissimus Muscle

Second Sacral Foramen

Sacral Canal

right

left

posterior

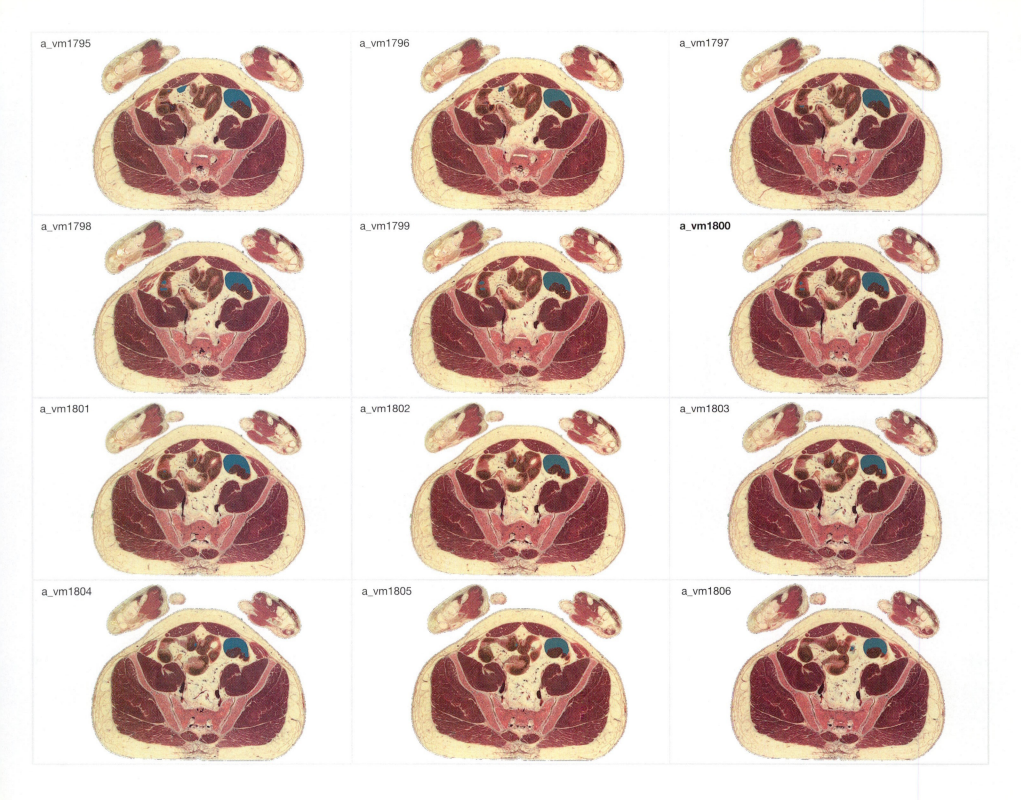

a_vm1795

a_vm1796

a_vm1797

a_vm1798

a_vm1799

a_vm1800

a_vm1801

a_vm1802

a_vm1803

a_vm1804

a_vm1805

a_vm1806

anterior

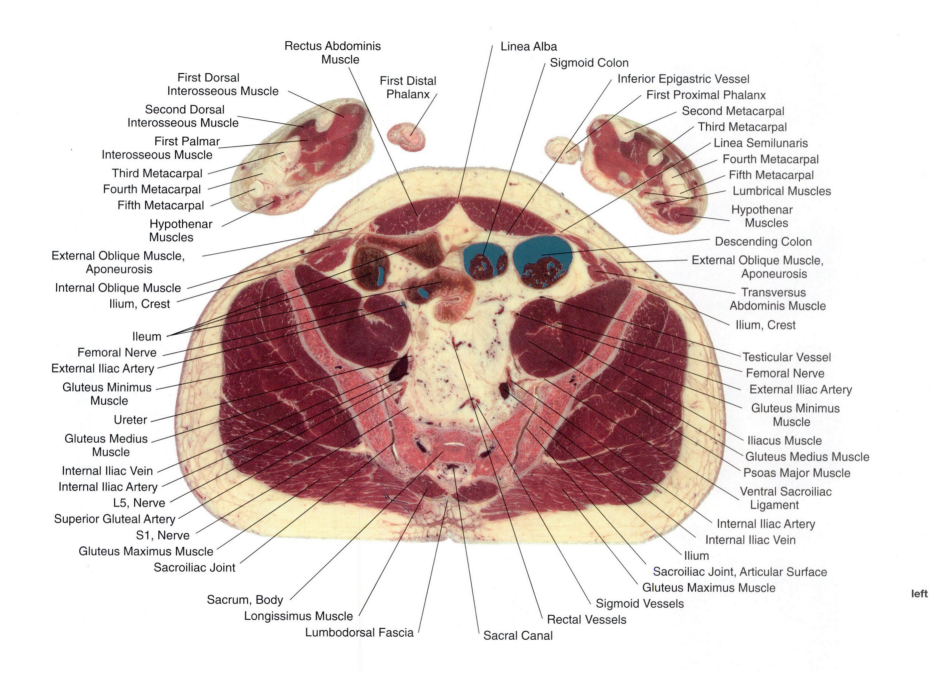

Rectus Abdominis Muscle

Linea Alba

Sigmoid Colon

First Dorsal Interosseous Muscle

First Distal Phalanx

Inferior Epigastric Vessel

First Proximal Phalanx

Second Dorsal Interosseous Muscle

Second Metacarpal

Third Metacarpal

First Palmar Interosseous Muscle

Linea Semilunaris

Third Metacarpal

Fourth Metacarpal

Fourth Metacarpal

Fifth Metacarpal

Fifth Metacarpal

Lumbrical Muscles

Hypothenar Muscles

Hypothenar Muscles

Descending Colon

External Oblique Muscle, Aponeurosis

External Oblique Muscle, Aponeurosis

Internal Oblique Muscle

Transversus Abdominis Muscle

Ilium, Crest

Ilium, Crest

Ileum

Femoral Nerve

Testicular Vessel

External Iliac Artery

Femoral Nerve

Gluteus Minimus Muscle

External Iliac Artery

Ureter

Gluteus Minimus Muscle

Gluteus Medius Muscle

Iliacus Muscle

Internal Iliac Vein

Gluteus Medius Muscle

Internal Iliac Artery

Psoas Major Muscle

L5, Nerve

Ventral Sacroiliac Ligament

Superior Gluteal Artery

Internal Iliac Artery

S1, Nerve

Internal Iliac Vein

Gluteus Maximus Muscle

Ilium

Sacroiliac Joint

Sacroiliac Joint, Articular Surface

Gluteus Maximus Muscle

Sacrum, Body

Sigmoid Vessels

Longissimus Muscle

Rectal Vessels

Lumbodorsal Fascia

Sacral Canal

right

left

posterior

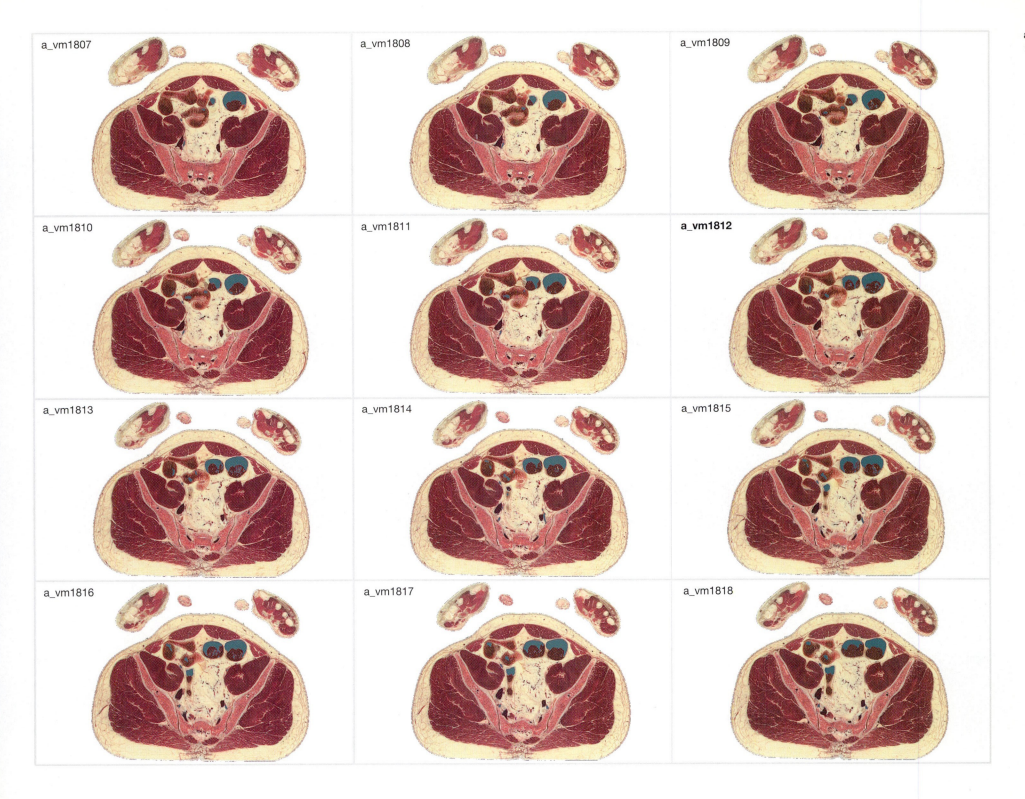

a_vm1807

a_vm1808

a_vm1809

a_vm1810

a_vm1811

a_vm1812

a_vm1813

a_vm1814

a_vm1815

a_vm1816

a_vm1817

a_vm1818

anterior

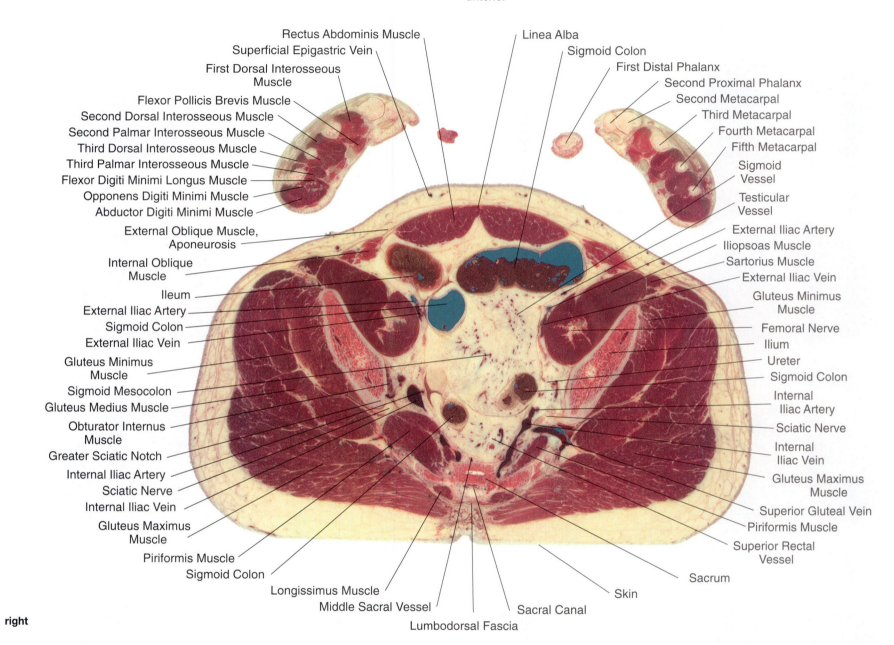

Rectus Abdominis Muscle

Superficial Epigastric Vein

First Dorsal Interosseous Muscle

Flexor Pollicis Brevis Muscle

Second Dorsal Interosseous Muscle

Second Palmar Interosseous Muscle

Third Dorsal Interosseous Muscle

Third Palmar Interosseous Muscle

Flexor Digiti Minimi Longus Muscle

Opponens Digiti Minimi Muscle

Abductor Digiti Minimi Muscle

External Oblique Muscle, Aponeurosis

Internal Oblique Muscle

Ileum

External Iliac Artery

Sigmoid Colon

External Iliac Vein

Gluteus Minimus Muscle

Sigmoid Mesocolon

Gluteus Medius Muscle

Obturator Internus Muscle

Greater Sciatic Notch

Internal Iliac Artery

Sciatic Nerve

Internal Iliac Vein

Gluteus Maximus Muscle

Piriformis Muscle

Sigmoid Colon

Longissimus Muscle

Middle Sacral Vessel

Lumbodorsal Fascia

Sacral Canal

Skin

Sacrum

Linea Alba

Sigmoid Colon

First Distal Phalanx

Second Proximal Phalanx

Second Metacarpal

Third Metacarpal

Fourth Metacarpal

Fifth Metacarpal

Sigmoid Vessel

Testicular Vessel

External Iliac Artery

Iliopsoas Muscle

Sartorius Muscle

External Iliac Vein

Gluteus Minimus Muscle

Femoral Nerve

Ilium

Ureter

Sigmoid Colon

Internal Iliac Artery

Sciatic Nerve

Internal Iliac Vein

Gluteus Maximus Muscle

Superior Gluteal Vein

Piriformis Muscle

Superior Rectal Vessel

right

left

posterior

138

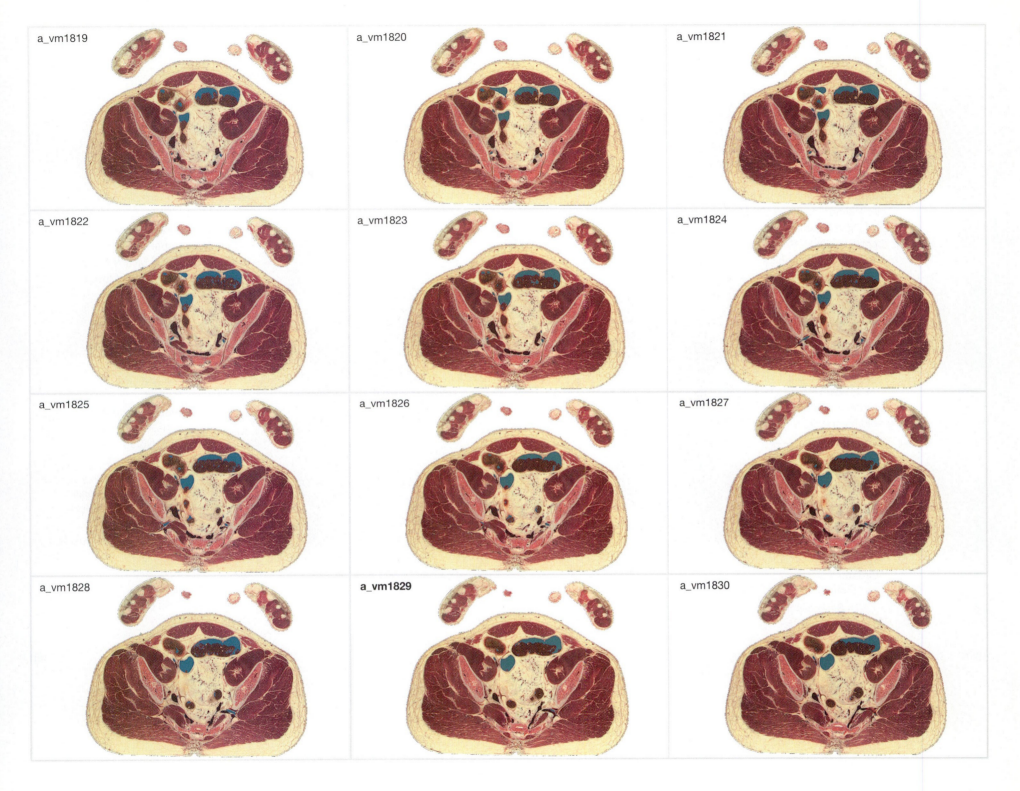

a_vm1819

a_vm1820

a_vm1821

a_vm1822

a_vm1823

a_vm1824

a_vm1825

a_vm1826

a_vm1827

a_vm1828

a_vm1829

a_vm1830

anterior

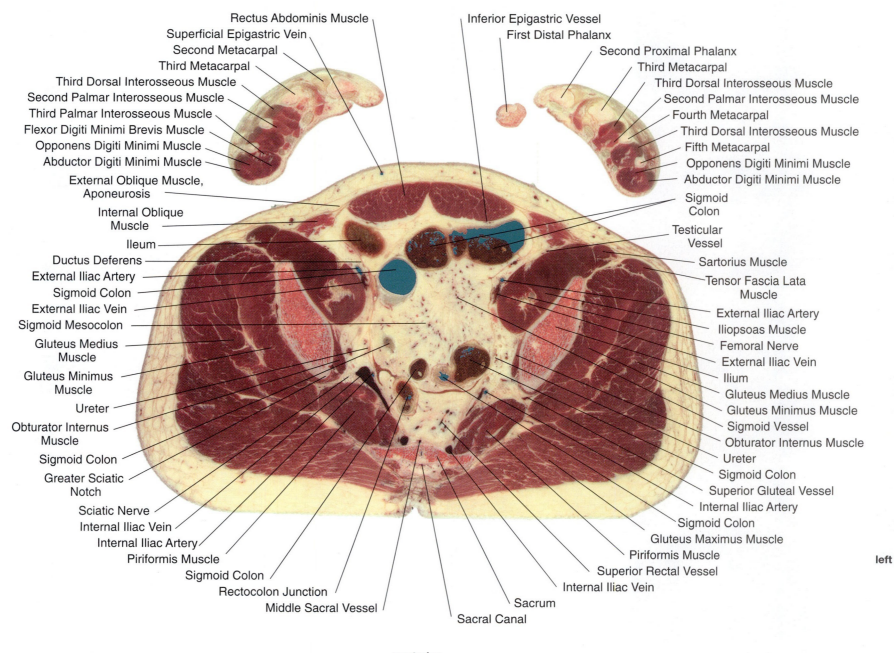

Rectus Abdominis Muscle

Superficial Epigastric Vein

Second Metacarpal

Third Metacarpal

Third Dorsal Interosseous Muscle

Second Palmar Interosseous Muscle

Third Palmar Interosseous Muscle

Flexor Digiti Minimi Brevis Muscle

Opponens Digiti Minimi Muscle

Abductor Digiti Minimi Muscle

External Oblique Muscle, Aponeurosis

Internal Oblique Muscle

Ileum

Ductus Deferens

External Iliac Artery

Sigmoid Colon

External Iliac Vein

Sigmoid Mesocolon

Gluteus Medius Muscle

Gluteus Minimus Muscle

Ureter

Obturator Internus Muscle

Sigmoid Colon

Greater Sciatic Notch

Sciatic Nerve

Internal Iliac Vein

Internal Iliac Artery

Piriformis Muscle

Sigmoid Colon

Rectocolon Junction

Middle Sacral Vessel

Inferior Epigastric Vessel

First Distal Phalanx

Second Proximal Phalanx

Third Metacarpal

Third Dorsal Interosseous Muscle

Second Palmar Interosseous Muscle

Fourth Metacarpal

Third Dorsal Interosseous Muscle

Fifth Metacarpal

Opponens Digiti Minimi Muscle

Abductor Digiti Minimi Muscle

Sigmoid Colon

Testicular Vessel

Sartorius Muscle

Tensor Fascia Lata Muscle

External Iliac Artery

Iliopsoas Muscle

Femoral Nerve

External Iliac Vein

Ilium

Gluteus Medius Muscle

Gluteus Minimus Muscle

Sigmoid Vessel

Obturator Internus Muscle

Ureter

Sigmoid Colon

Superior Gluteal Vessel

Internal Iliac Artery

Sigmoid Colon

Gluteus Maximus Muscle

Piriformis Muscle

Superior Rectal Vessel

Internal Iliac Vein

Sacrum

Sacral Canal

right

left

posterior

a_vm1831

a_vm1832

a_vm1833

a_vm1834

a_vm1835

a_vm1836

a_vm1837

a_vm1838

a_vm1839

a_vm1840

a_vm1841

a_vm1842

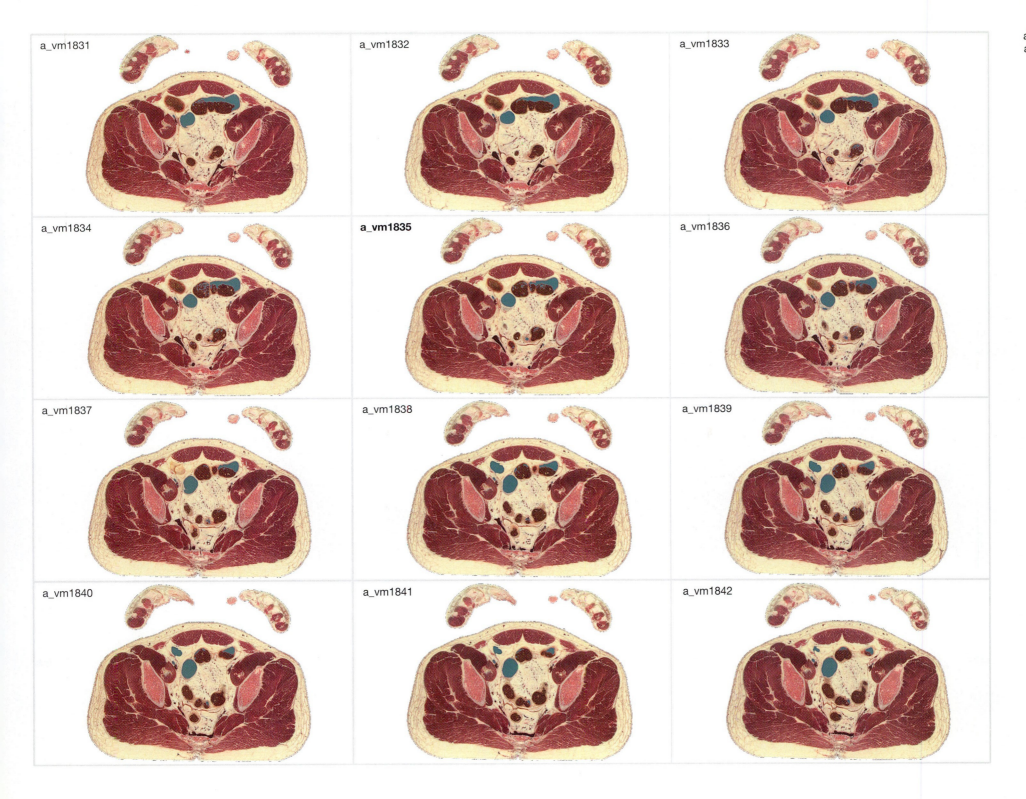

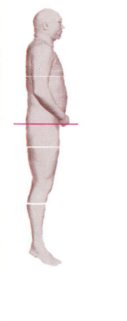

anterior

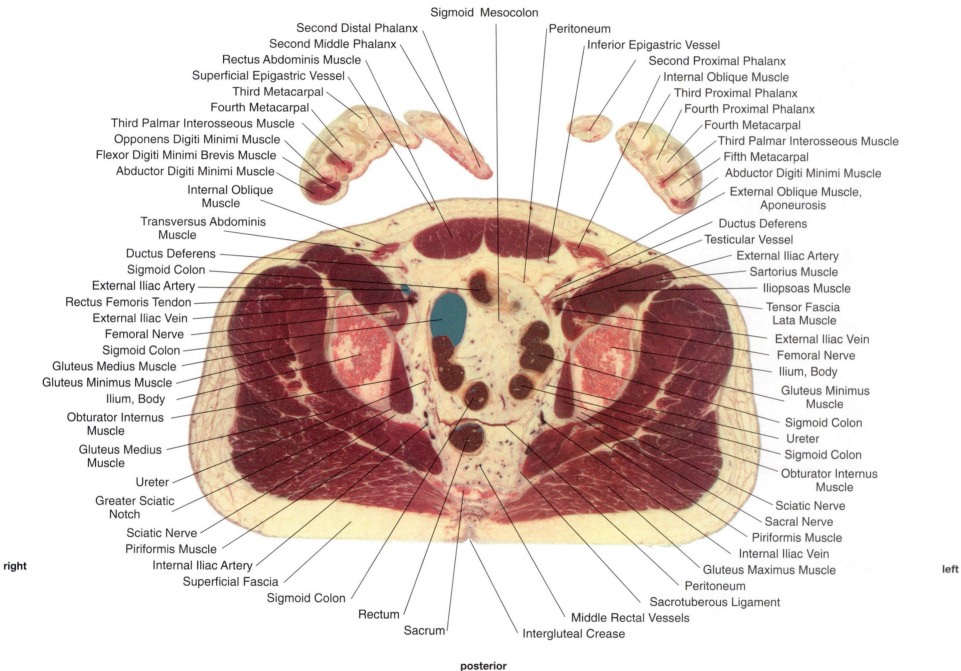

Sigmoid Mesocolon

Second Distal Phalanx

Peritoneum

Second Middle Phalanx

Inferior Epigastric Vessel

Rectus Abdominis Muscle

Second Proximal Phalanx

Superficial Epigastric Vessel

Internal Oblique Muscle

Third Metacarpal

Third Proximal Phalanx

Fourth Metacarpal

Fourth Proximal Phalanx

Third Palmar Interosseous Muscle

Fourth Metacarpal

Opponens Digiti Minimi Muscle

Third Palmar Interosseous Muscle

Flexor Digiti Minimi Brevis Muscle

Fifth Metacarpal

Abductor Digiti Minimi Muscle

Abductor Digiti Minimi Muscle

Internal Oblique
Muscle

External Oblique Muscle,
Aponeurosis

Transversus Abdominis
Muscle

Ductus Deferens

Ductus Deferens

Testicular Vessel

Sigmoid Colon

External Iliac Artery

External Iliac Artery

Sartorius Muscle

Rectus Femoris Tendon

Iliopsoas Muscle

External Iliac Vein

Tensor Fascia
Lata Muscle

Femoral Nerve

Sigmoid Colon

External Iliac Vein

Gluteus Medius Muscle

Femoral Nerve

Gluteus Minimus Muscle

Ilium, Body

Ilium, Body

Gluteus Minimus
Muscle

Obturator Internus
Muscle

Sigmoid Colon

Gluteus Medius
Muscle

Ureter

Ureter

Sigmoid Colon

Greater Sciatic
Notch

Obturator Internus
Muscle

Sciatic Nerve

Sciatic Nerve

Piriformis Muscle

Sacral Nerve

Internal Iliac Artery

Piriformis Muscle

Superficial Fascia

Internal Iliac Vein

Sigmoid Colon

Gluteus Maximus Muscle

Rectum

Peritoneum

Sacrum

Sacrotuberous Ligament

Middle Rectal Vessels

Intergluteal Crease

right

left

posterior

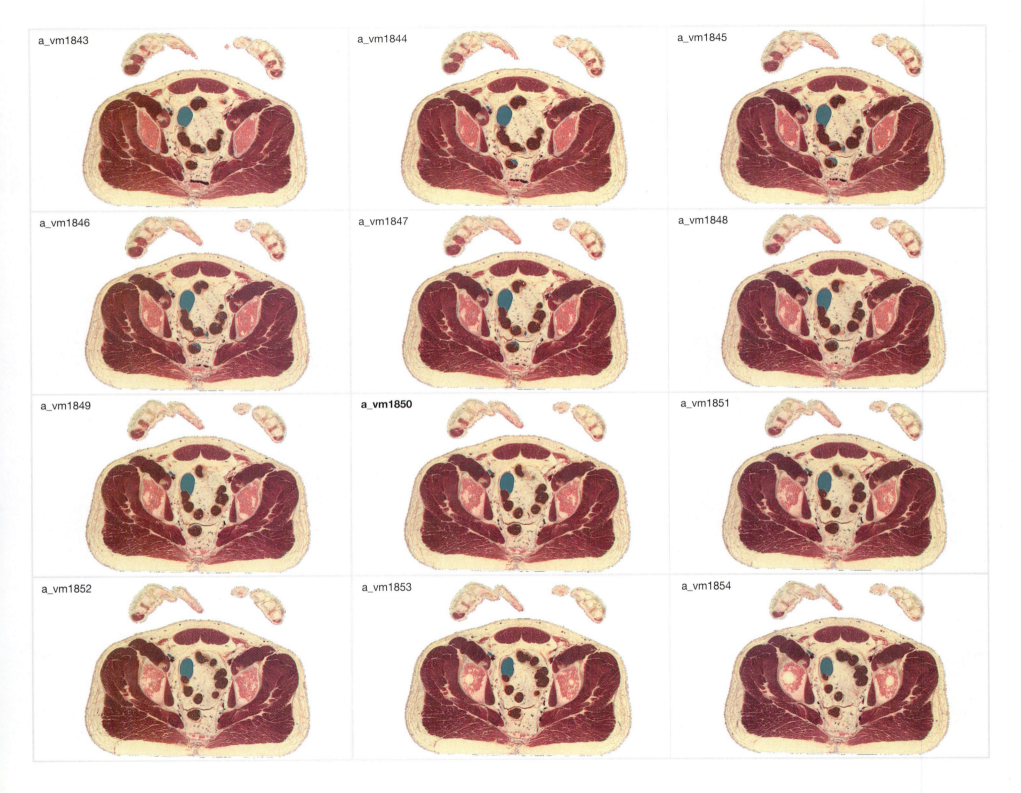

a_vm1843

a_vm1844

a_vm1845

a_vm1846

a_vm1847

a_vm1848

a_vm1849

a_vm1850

a_vm1851

a_vm1852

a_vm1853

a_vm1854

anterior

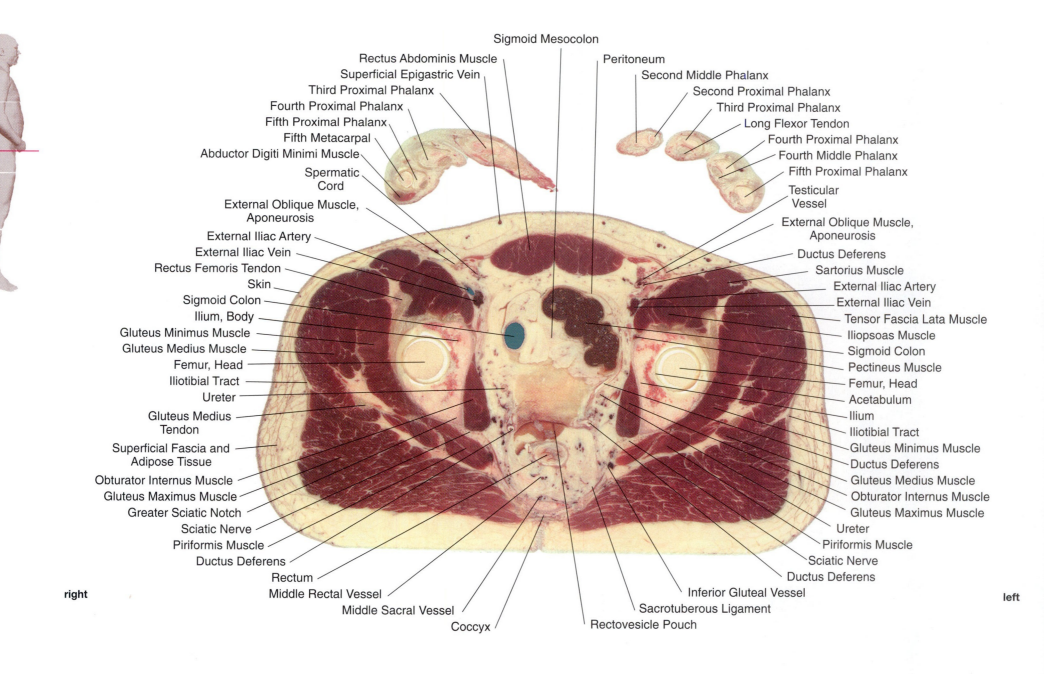

Sigmoid Mesocolon

Rectus Abdominis Muscle

Peritoneum

Superficial Epigastric Vein

Second Middle Phalanx

Third Proximal Phalanx

Second Proximal Phalanx

Fourth Proximal Phalanx

Third Proximal Phalanx

Fifth Proximal Phalanx

Long Flexor Tendon

Fifth Metacarpal

Fourth Proximal Phalanx

Abductor Digiti Minimi Muscle

Fourth Middle Phalanx

Spermatic Cord

Fifth Proximal Phalanx

External Oblique Muscle, Aponeurosis

Testicular Vessel

External Iliac Artery

External Oblique Muscle, Aponeurosis

External Iliac Vein

Ductus Deferens

Rectus Femoris Tendon

Sartorius Muscle

Skin

External Iliac Artery

Sigmoid Colon

External Iliac Vein

Ilium, Body

Tensor Fascia Lata Muscle

Gluteus Minimus Muscle

Iliopsoas Muscle

Gluteus Medius Muscle

Sigmoid Colon

Femur, Head

Pectineus Muscle

Iliotibial Tract

Femur, Head

Ureter

Acetabulum

Gluteus Medius Tendon

Ilium

Iliotibial Tract

Superficial Fascia and Adipose Tissue

Gluteus Minimus Muscle

Obturator Internus Muscle

Ductus Deferens

Gluteus Maximus Muscle

Gluteus Medius Muscle

Greater Sciatic Notch

Obturator Internus Muscle

Sciatic Nerve

Gluteus Maximus Muscle

Piriformis Muscle

Ureter

Ductus Deferens

Piriformis Muscle

Rectum

Sciatic Nerve

Middle Rectal Vessel

Ductus Deferens

Middle Sacral Vessel

Inferior Gluteal Vessel

Coccyx

Sacrotuberous Ligament

Rectovesicle Pouch

right

left

posterior

144

a_vm1855

a_vm1856

a_vm1857

a_vm1858

a_vm1859

a_vm1860

a_vm1861

a_vm1862

a_vm1863

a_vm1864

a_vm1865

a_vm1866

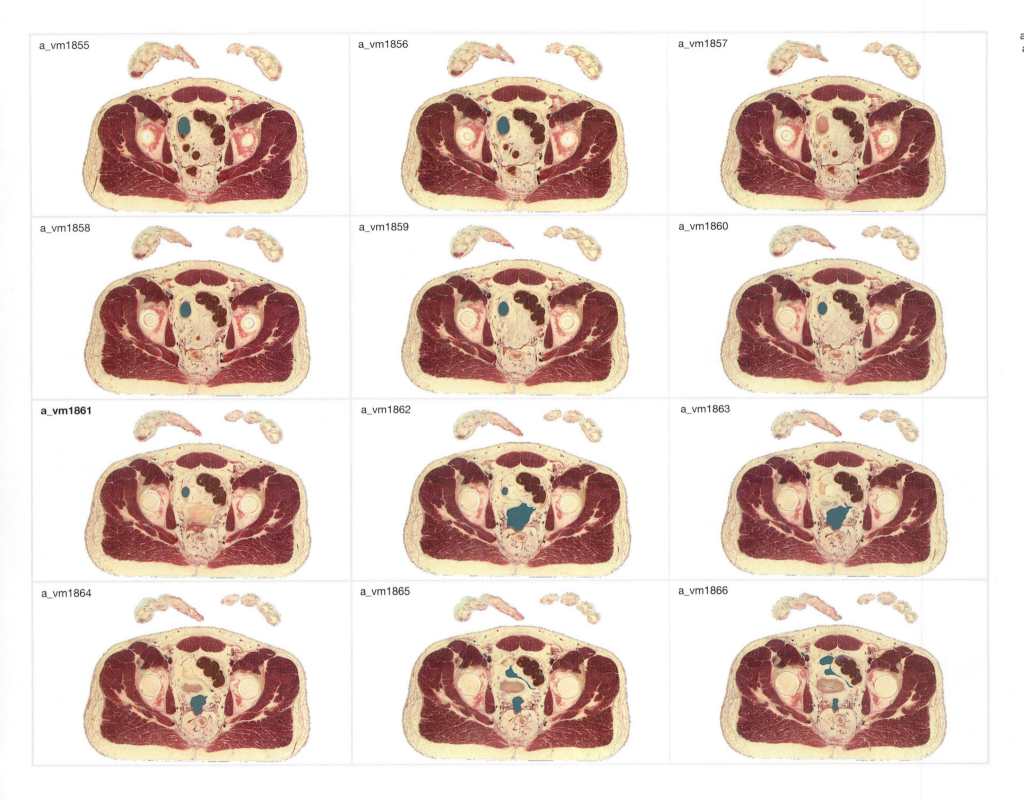

anterior

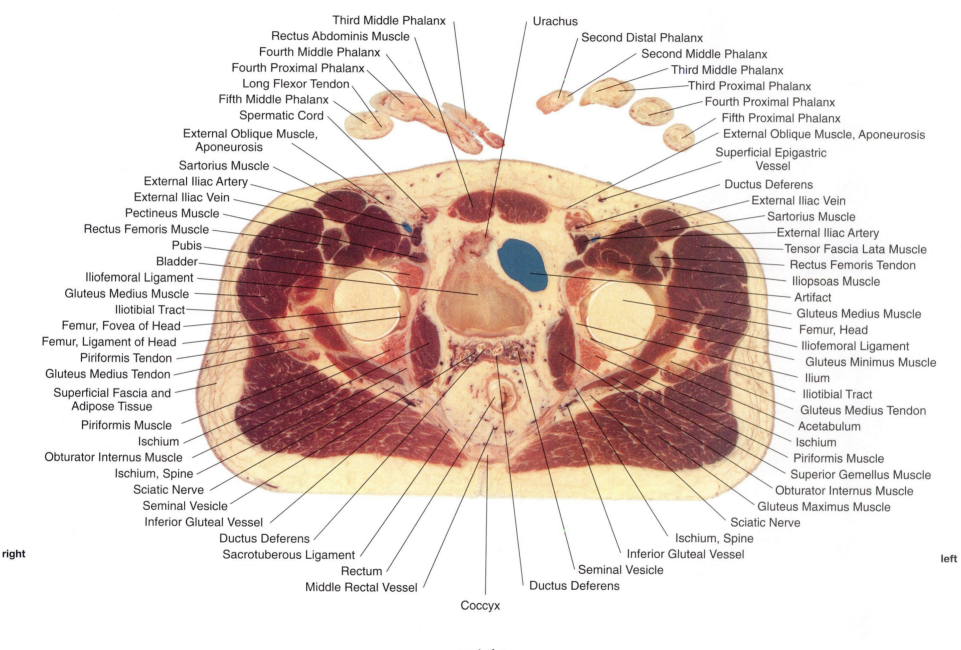

Third Middle Phalanx

Rectus Abdominis Muscle

Fourth Middle Phalanx

Fourth Proximal Phalanx

Long Flexor Tendon

Fifth Middle Phalanx

Spermatic Cord

External Oblique Muscle,
Aponeurosis

Sartorius Muscle

External Iliac Artery

External Iliac Vein

Pectineus Muscle

Rectus Femoris Muscle

Pubis

Bladder

Iliofemoral Ligament

Gluteus Medius Muscle

Iliotibial Tract

Femur, Fovea of Head

Femur, Ligament of Head

Piriformis Tendon

Gluteus Medius Tendon

Superficial Fascia and
Adipose Tissue

Piriformis Muscle

Ischium

Obturator Internus Muscle

Ischium, Spine

Sciatic Nerve

Seminal Vesicle

Inferior Gluteal Vessel

Ductus Deferens

Sacrotuberous Ligament

Rectum

Middle Rectal Vessel

Urachus

Second Distal Phalanx

Second Middle Phalanx

Third Middle Phalanx

Third Proximal Phalanx

Fourth Proximal Phalanx

Fifth Proximal Phalanx

External Oblique Muscle, Aponeurosis

Superficial Epigastric
Vessel

Ductus Deferens

External Iliac Vein

Sartorius Muscle

External Iliac Artery

Tensor Fascia Lata Muscle

Rectus Femoris Tendon

Iliopsoas Muscle

Artifact

Gluteus Medius Muscle

Femur, Head

Iliofemoral Ligament

Gluteus Minimus Muscle

Ilium

Iliotibial Tract

Gluteus Medius Tendon

Acetabulum

Ischium

Piriformis Muscle

Superior Gemellus Muscle

Obturator Internus Muscle

Gluteus Maximus Muscle

Sciatic Nerve

Ischium, Spine

Inferior Gluteal Vessel

Seminal Vesicle

Ductus Deferens

Coccyx

right

left

posterior

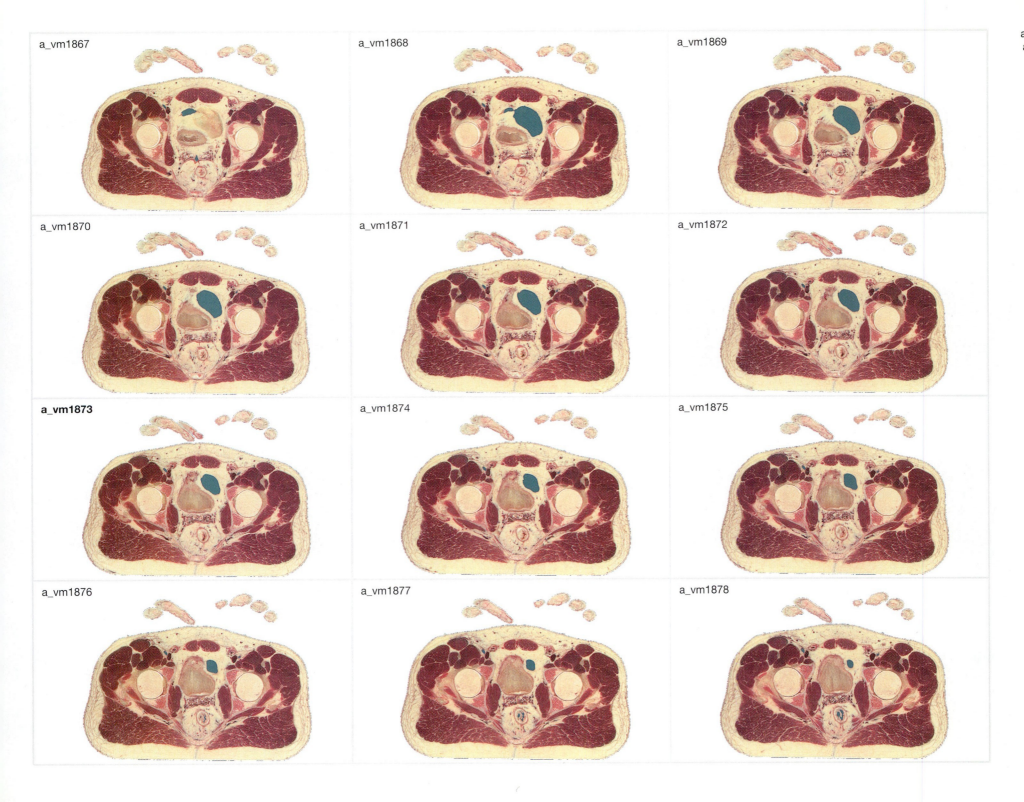

a_vm1867

a_vm1868

a_vm1869

a_vm1870

a_vm1871

a_vm1872

a_vm1873

a_vm1874

a_vm1875

a_vm1876

a_vm1877

a_vm1878

anterior

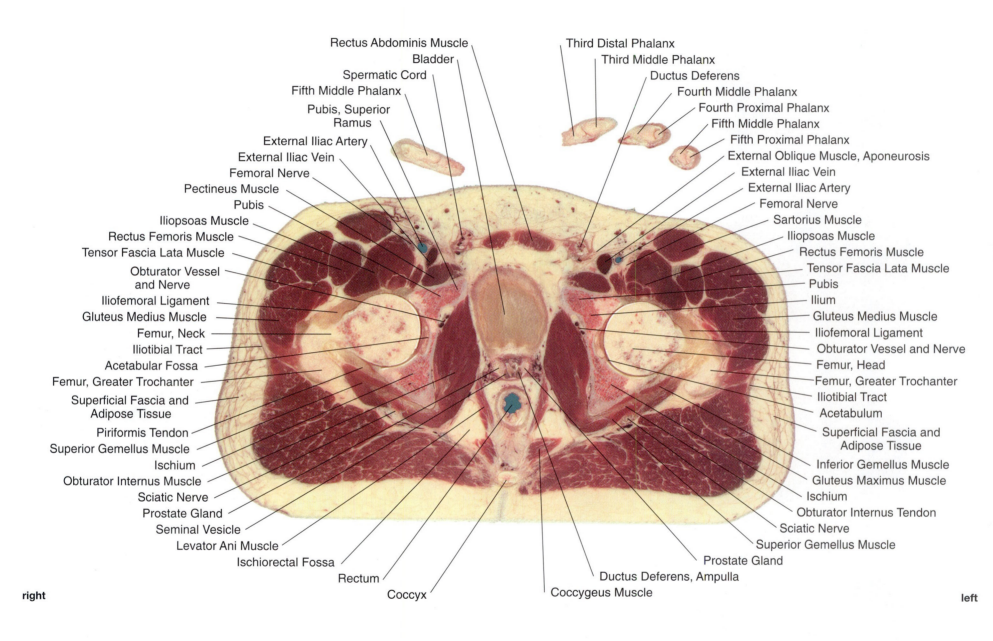

Rectus Abdominis Muscle

Bladder

Spermatic Cord

Fifth Middle Phalanx

Pubis, Superior
Ramus

External Iliac Artery

External Iliac Vein

Femoral Nerve

Pectineus Muscle

Pubis

Iliopsoas Muscle

Rectus Femoris Muscle

Tensor Fascia Lata Muscle

Obturator Vessel
and Nerve

Iliofemoral Ligament

Gluteus Medius Muscle

Femur, Neck

Iliotibial Tract

Acetabular Fossa

Femur, Greater Trochanter

Superficial Fascia and
Adipose Tissue

Piriformis Tendon

Superior Gemellus Muscle

Ischium

Obturator Internus Muscle

Sciatic Nerve

Prostate Gland

Seminal Vesicle

Levator Ani Muscle

Ischiorectal Fossa

Rectum

Coccyx

Third Distal Phalanx

Third Middle Phalanx

Ductus Deferens

Fourth Middle Phalanx

Fourth Proximal Phalanx

Fifth Middle Phalanx

Fifth Proximal Phalanx

External Oblique Muscle, Aponeurosis

External Iliac Vein

External Iliac Artery

Femoral Nerve

Sartorius Muscle

Iliopsoas Muscle

Rectus Femoris Muscle

Tensor Fascia Lata Muscle

Pubis

Ilium

Gluteus Medius Muscle

Iliofemoral Ligament

Obturator Vessel and Nerve

Femur, Head

Femur, Greater Trochanter

Iliotibial Tract

Acetabulum

Superficial Fascia and
Adipose Tissue

Inferior Gemellus Muscle

Gluteus Maximus Muscle

Ischium

Obturator Internus Tendon

Sciatic Nerve

Superior Gemellus Muscle

Prostate Gland

Ductus Deferens, Ampulla

Coccygeus Muscle

right

left

posterior

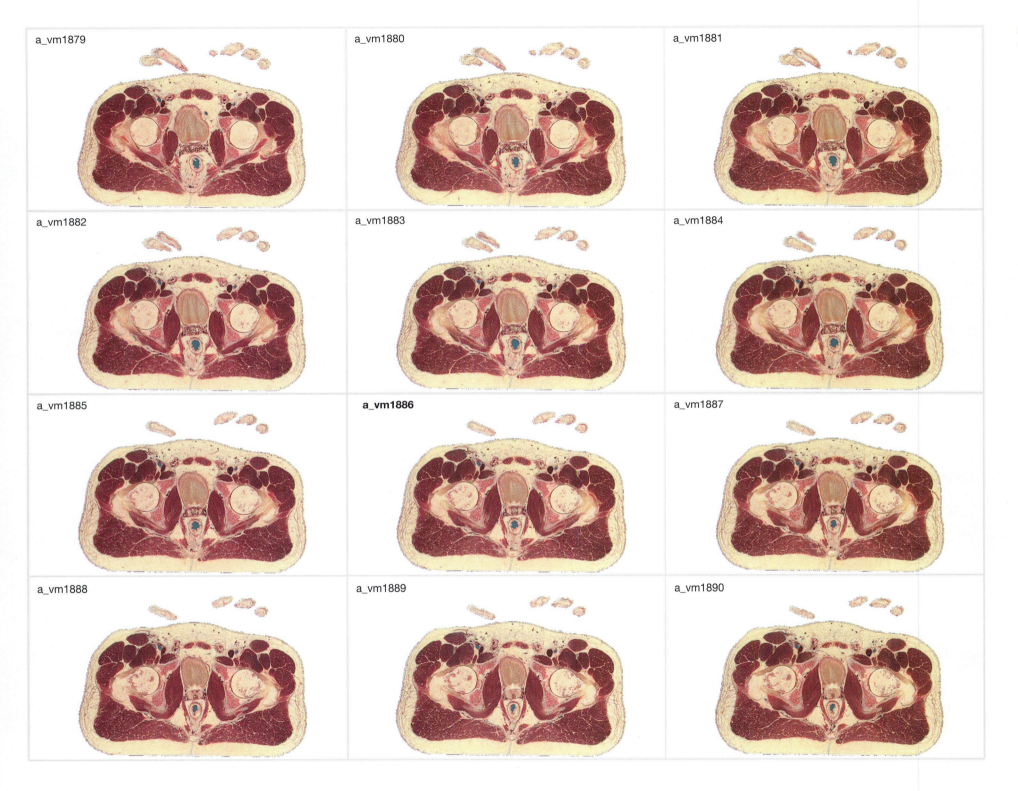

a_vm1879

a_vm1880

a_vm1881

a_vm1882

a_vm1883

a_vm1884

a_vm1885

a_vm1886

a_vm1887

a_vm1888

a_vm1889

a_vm1890

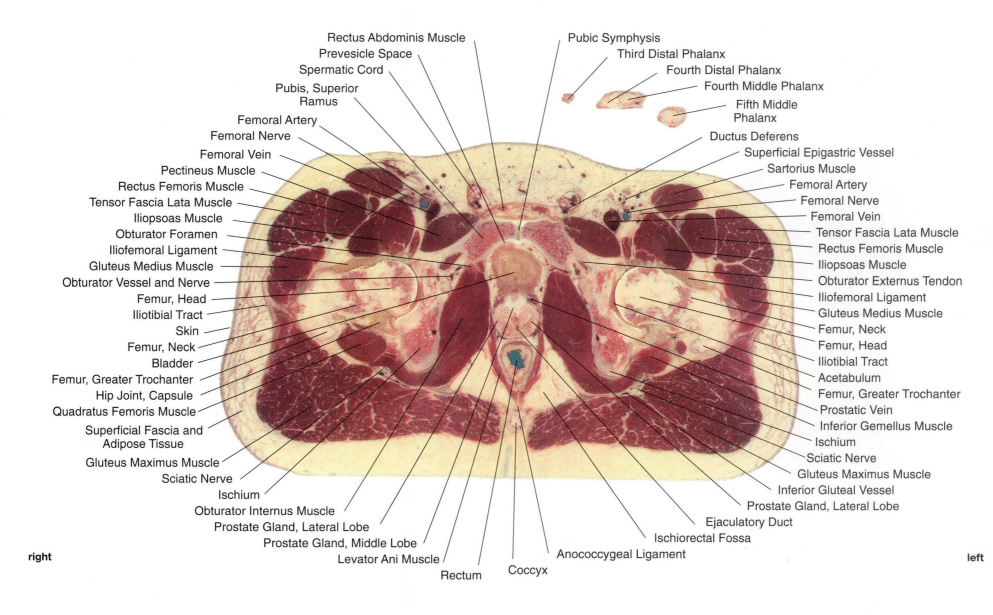

Transverse
a_vm1897

anterior

Rectus Abdominis Muscle
Prevesicle Space
Spermatic Cord
Pubis, Superior Ramus
Femoral Artery
Femoral Nerve
Femoral Vein
Pectineus Muscle
Rectus Femoris Muscle
Tensor Fascia Lata Muscle
Iliopsoas Muscle
Obturator Foramen
Iliofemoral Ligament
Gluteus Medius Muscle
Obturator Vessel and Nerve
Femur, Head
Iliotibial Tract
Skin
Femur, Neck
Bladder
Femur, Greater Trochanter
Hip Joint, Capsule
Quadratus Femoris Muscle
Superficial Fascia and Adipose Tissue
Gluteus Maximus Muscle
Sciatic Nerve
Ischium
Obturator Internus Muscle
Prostate Gland, Lateral Lobe
Prostate Gland, Middle Lobe
Levator Ani Muscle
Rectum
Coccyx
Anococcygeal Ligament

Pubic Symphysis
Third Distal Phalanx
Fourth Distal Phalanx
Fourth Middle Phalanx
Fifth Middle Phalanx
Ductus Deferens
Superficial Epigastric Vessel
Sartorius Muscle
Femoral Artery
Femoral Nerve
Femoral Vein
Tensor Fascia Lata Muscle
Rectus Femoris Muscle
Iliopsoas Muscle
Obturator Externus Tendon
Iliofemoral Ligament
Gluteus Medius Muscle
Femur, Neck
Femur, Head
Iliotibial Tract
Acetabulum
Femur, Greater Trochanter
Prostatic Vein
Inferior Gemellus Muscle
Ischium
Sciatic Nerve
Gluteus Maximus Muscle
Inferior Gluteal Vessel
Prostate Gland, Lateral Lobe
Ejaculatory Duct
Ischiorectal Fossa

right

left

posterior

150

a_vm1891

a_vm1892

a_vm1893

a_vm1894

a_vm1895

a_vm1896

a_vm1897

a_vm1898

a_vm1899

a_vm1900

a_vm1901

a_vm1902

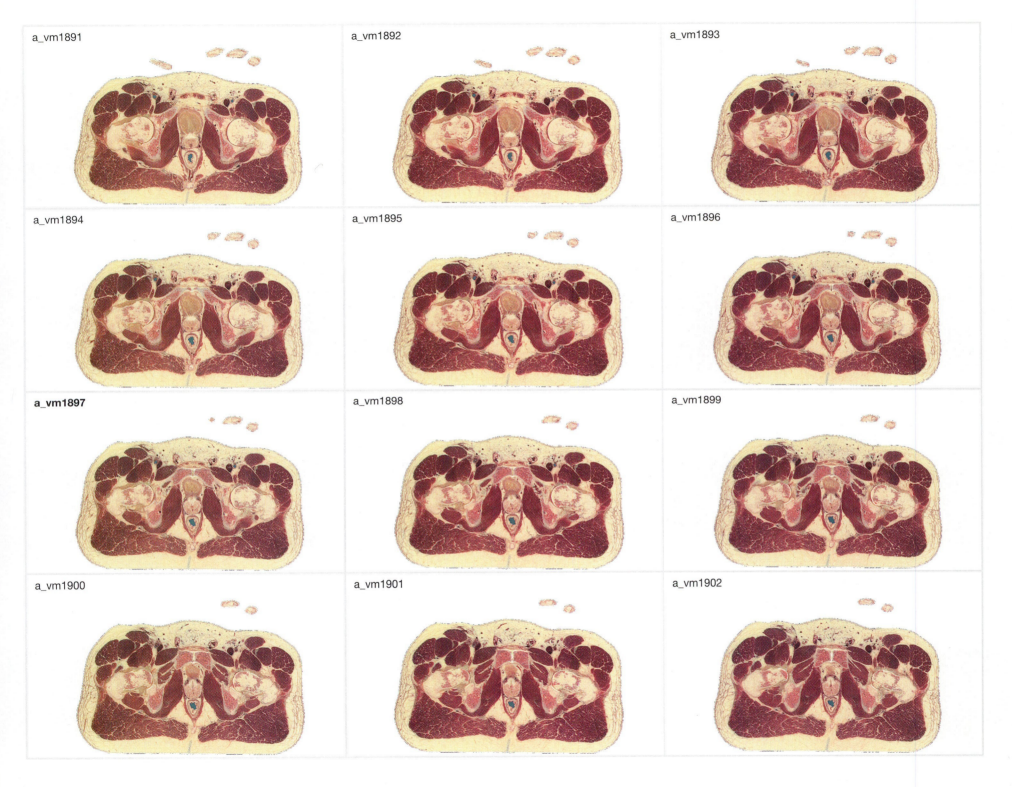

anterior

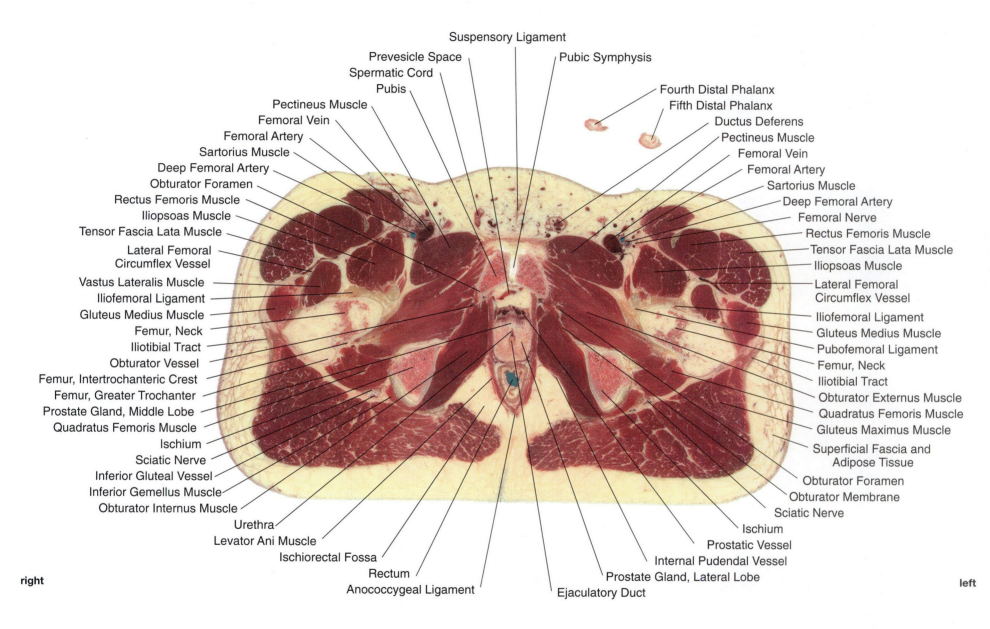

Suspensory Ligament

Prevesicle Space

Spermatic Cord

Pubis

Pectineus Muscle

Femoral Vein

Femoral Artery

Sartorius Muscle

Deep Femoral Artery

Obturator Foramen

Rectus Femoris Muscle

Iliopsoas Muscle

Tensor Fascia Lata Muscle

Lateral Femoral
Circumflex Vessel

Vastus Lateralis Muscle

Iliofemoral Ligament

Gluteus Medius Muscle

Femur, Neck

Iliotibial Tract

Obturator Vessel

Femur, Intertrochanteric Crest

Femur, Greater Trochanter

Prostate Gland, Middle Lobe

Quadratus Femoris Muscle

Ischium

Sciatic Nerve

Inferior Gluteal Vessel

Inferior Gemellus Muscle

Obturator Internus Muscle

Urethra

Levator Ani Muscle

Ischiorectal Fossa

Rectum

Anococcygeal Ligament

Pubic Symphysis

Fourth Distal Phalanx

Fifth Distal Phalanx

Ductus Deferens

Pectineus Muscle

Femoral Vein

Femoral Artery

Sartorius Muscle

Deep Femoral Artery

Femoral Nerve

Rectus Femoris Muscle

Tensor Fascia Lata Muscle

Iliopsoas Muscle

Lateral Femoral
Circumflex Vessel

Iliofemoral Ligament

Gluteus Medius Muscle

Pubofemoral Ligament

Femur, Neck

Iliotibial Tract

Obturator Externus Muscle

Quadratus Femoris Muscle

Gluteus Maximus Muscle

Superficial Fascia and
Adipose Tissue

Obturator Foramen

Obturator Membrane

Sciatic Nerve

Ischium

Prostatic Vessel

Internal Pudendal Vessel

Prostate Gland, Lateral Lobe

Ejaculatory Duct

right

left

posterior

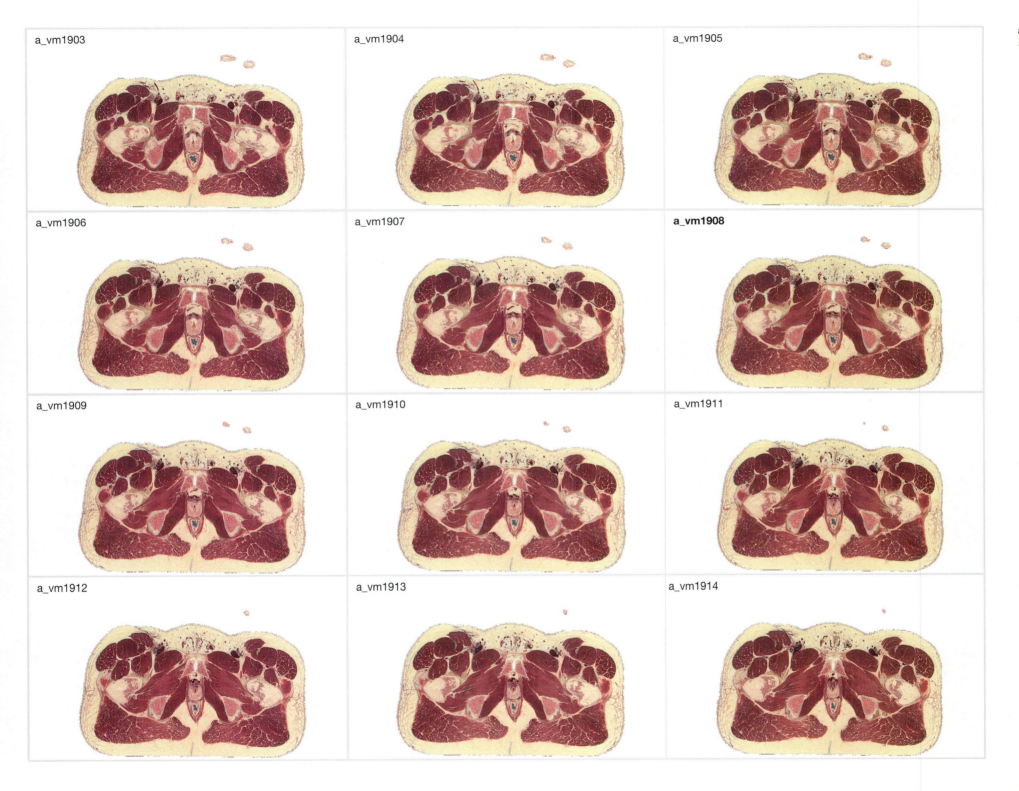

a_vm1903

a_vm1904

a_vm1905

a_vm1906

a_vm1907

a_vm1908

a_vm1909

a_vm1910

a_vm1911

a_vm1912

a_vm1913

a_vm1914

anterior

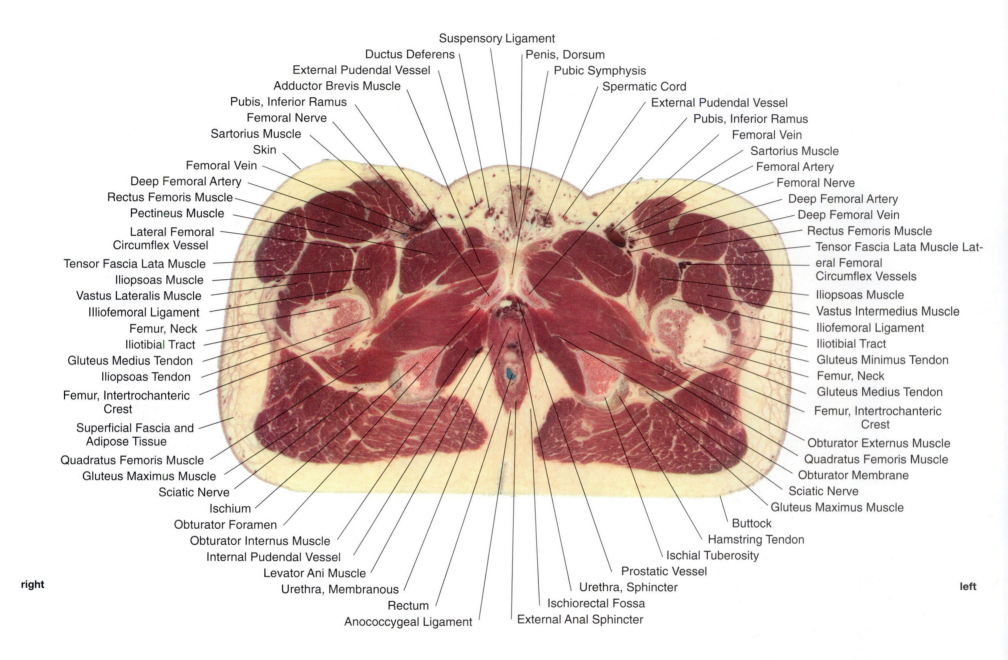

Suspensory Ligament
Ductus Deferens
External Pudendal Vessel
Adductor Brevis Muscle
Pubis, Inferior Ramus
Femoral Nerve
Sartorius Muscle
Skin
Femoral Vein
Deep Femoral Artery
Rectus Femoris Muscle
Pectineus Muscle
Lateral Femoral
Circumflex Vessel
Tensor Fascia Lata Muscle
Iliopsoas Muscle
Vastus Lateralis Muscle
Illiofemoral Ligament
Femur, Neck
Iliotibial Tract
Gluteus Medius Tendon
Iliopsoas Tendon
Femur, Intertrochanteric
Crest
Superficial Fascia and
Adipose Tissue
Quadratus Femoris Muscle
Gluteus Maximus Muscle
Sciatic Nerve
Ischium
Obturator Foramen
Obturator Internus Muscle
Internal Pudendal Vessel
Levator Ani Muscle
Urethra, Membranous
Rectum
Anococcygeal Ligament

Penis, Dorsum
Pubic Symphysis
Spermatic Cord
External Pudendal Vessel
Pubis, Inferior Ramus
Femoral Vein
Sartorius Muscle
Femoral Artery
Femoral Nerve
Deep Femoral Artery
Deep Femoral Vein
Rectus Femoris Muscle
Tensor Fascia Lata Muscle Lat-
eral Femoral
Circumflex Vessels
Iliopsoas Muscle
Vastus Intermedius Muscle
Iliofemoral Ligament
Iliotibial Tract
Gluteus Minimus Tendon
Femur, Neck
Gluteus Medius Tendon
Femur, Intertrochanteric
Crest
Obturator Externus Muscle
Quadratus Femoris Muscle
Obturator Membrane
Sciatic Nerve
Gluteus Maximus Muscle
Buttock
Hamstring Tendon
Ischial Tuberosity
Prostatic Vessel
Urethra, Sphincter
Ischiorectal Fossa
External Anal Sphincter

right

left

posterior

154

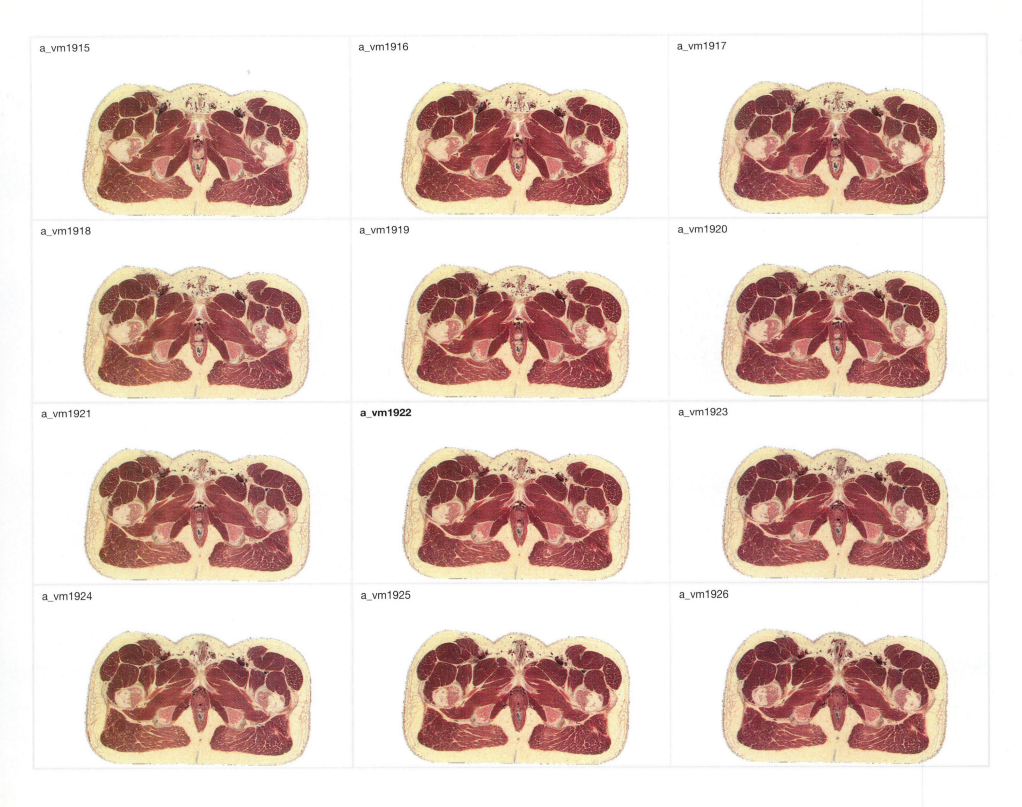

a_vm1915

a_vm1916

a_vm1917

a_vm1918

a_vm1919

a_vm1920

a_vm1921

a_vm1922

a_vm1923

a_vm1924

a_vm1925

a_vm1926

anterior

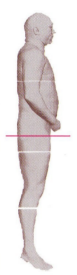

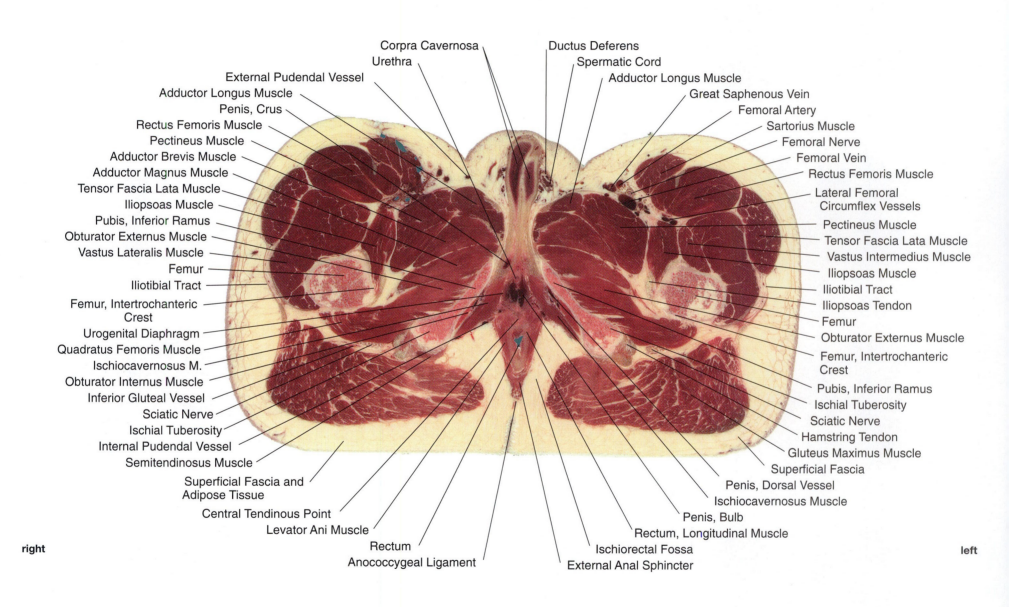

Corpra Cavernosa

Ductus Deferens

Urethra

Spermatic Cord

External Pudendal Vessel

Adductor Longus Muscle

Adductor Longus Muscle

Great Saphenous Vein

Penis, Crus

Femoral Artery

Rectus Femoris Muscle

Sartorius Muscle

Pectineus Muscle

Femoral Nerve

Adductor Brevis Muscle

Femoral Vein

Adductor Magnus Muscle

Rectus Femoris Muscle

Tensor Fascia Lata Muscle

Lateral Femoral
Circumflex Vessels

Iliopsoas Muscle

Pubis, Inferior Ramus

Pectineus Muscle

Obturator Externus Muscle

Tensor Fascia Lata Muscle

Vastus Lateralis Muscle

Vastus Intermedius Muscle

Femur

Iliopsoas Muscle

Iliotibial Tract

Iliotibial Tract

Femur, Intertrochanteric
Crest

Iliopsoas Tendon

Femur

Urogenital Diaphragm

Obturator Externus Muscle

Quadratus Femoris Muscle

Femur, Intertrochanteric
Crest

Ischiocavernosus M.

Obturator Internus Muscle

Pubis, Inferior Ramus

Inferior Gluteal Vessel

Ischial Tuberosity

Sciatic Nerve

Sciatic Nerve

Ischial Tuberosity

Hamstring Tendon

Internal Pudendal Vessel

Gluteus Maximus Muscle

Semitendinosus Muscle

Superficial Fascia

Superficial Fascia and
Adipose Tissue

Penis, Dorsal Vessel

Central Tendinous Point

Ischiocavernosus Muscle

Levator Ani Muscle

Penis, Bulb

Rectum

Rectum, Longitudinal Muscle

Anococcygeal Ligament

Ischiorectal Fossa

External Anal Sphincter

right

left

posterior

156

a_vm1927

a_vm1928

a_vm1929

a_vm1930

a_vm1931

a_vm1932

a_vm1933

a_vm1934

a_vm1935

a_vm1936

a_vm1937

a_vm1938

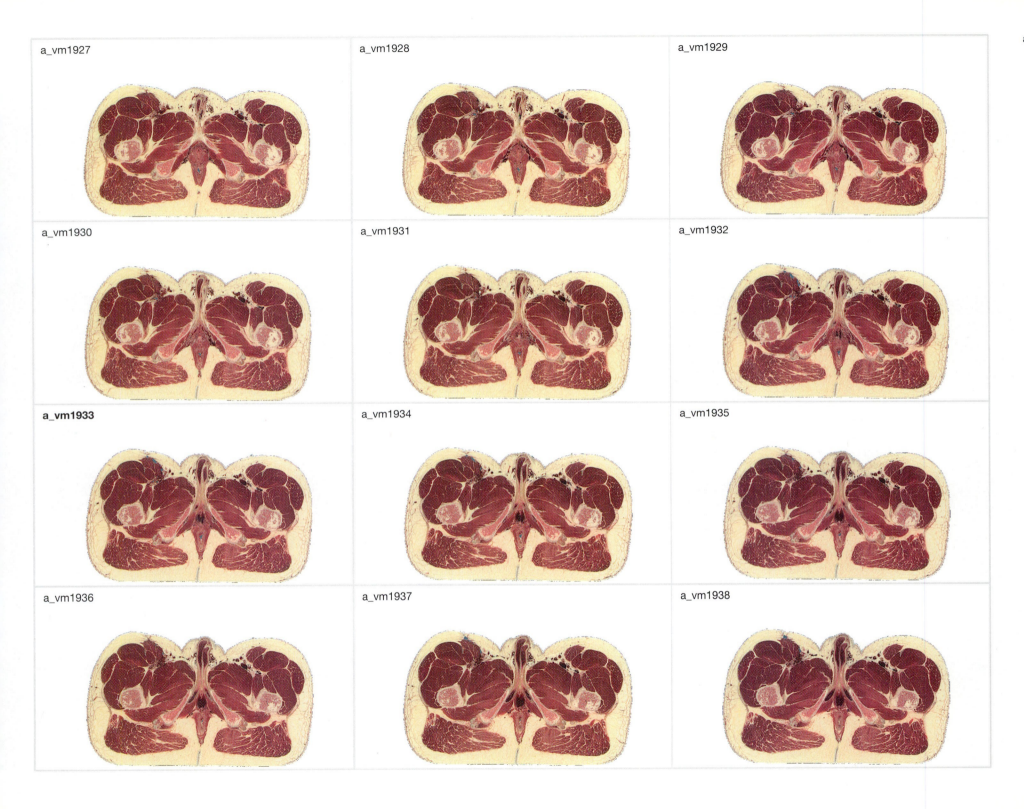

anterior

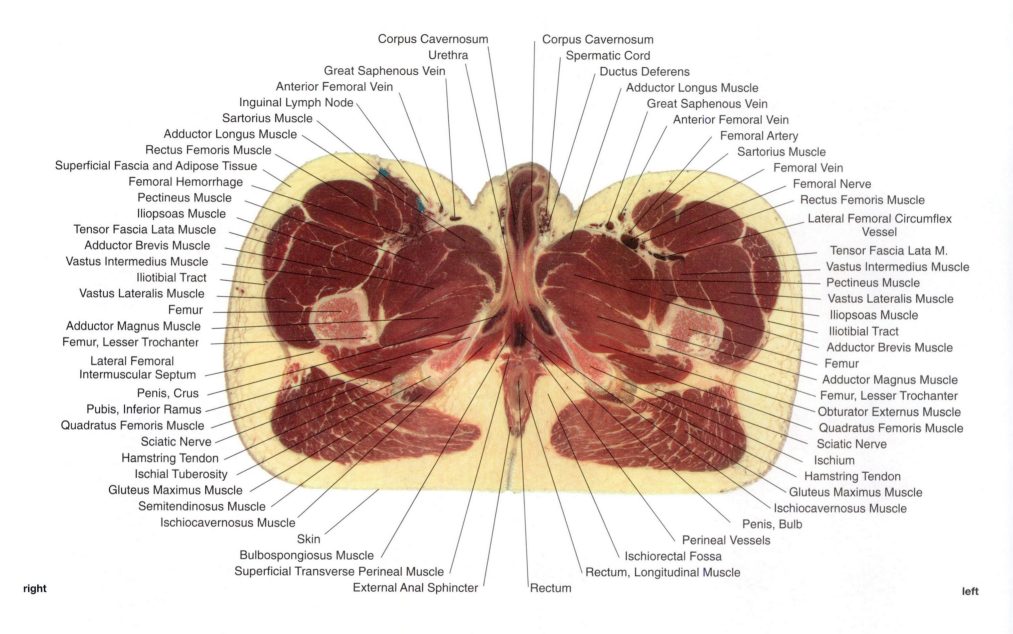

Corpus Cavernosum
Urethra
Great Saphenous Vein
Anterior Femoral Vein
Inguinal Lymph Node
Sartorius Muscle
Adductor Longus Muscle
Rectus Femoris Muscle
Superficial Fascia and Adipose Tissue
Femoral Hemorrhage
Pectineus Muscle
Iliopsoas Muscle
Tensor Fascia Lata Muscle
Adductor Brevis Muscle
Vastus Intermedius Muscle
Iliotibial Tract
Vastus Lateralis Muscle
Femur
Adductor Magnus Muscle
Femur, Lesser Trochanter
Lateral Femoral
Intermuscular Septum
Penis, Crus
Pubis, Inferior Ramus
Quadratus Femoris Muscle
Sciatic Nerve
Hamstring Tendon
Ischial Tuberosity
Gluteus Maximus Muscle
Semitendinosus Muscle
Ischiocavernosus Muscle
Skin
Bulbospongiosus Muscle
Superficial Transverse Perineal Muscle
External Anal Sphincter
Rectum

Corpus Cavernosum
Spermatic Cord
Ductus Deferens
Adductor Longus Muscle
Great Saphenous Vein
Anterior Femoral Vein
Femoral Artery
Sartorius Muscle
Femoral Vein
Femoral Nerve
Rectus Femoris Muscle
Lateral Femoral Circumflex
Vessel
Tensor Fascia Lata M.
Vastus Intermedius Muscle
Pectineus Muscle
Vastus Lateralis Muscle
Iliopsoas Muscle
Iliotibial Tract
Adductor Brevis Muscle
Femur
Adductor Magnus Muscle
Femur, Lesser Trochanter
Obturator Externus Muscle
Quadratus Femoris Muscle
Sciatic Nerve
Ischium
Hamstring Tendon
Gluteus Maximus Muscle
Ischiocavernosus Muscle
Penis, Bulb
Perineal Vessels
Ischiorectal Fossa
Rectum, Longitudinal Muscle

right

left

posterior

158

a_vm1939

a_vm1940

a_vm1941

a_vm1942

a_vm1943

a_vm1944

a_vm1945

a_vm1946

a_vm1947

a_vm1948

a_vm1949

a_vm1950

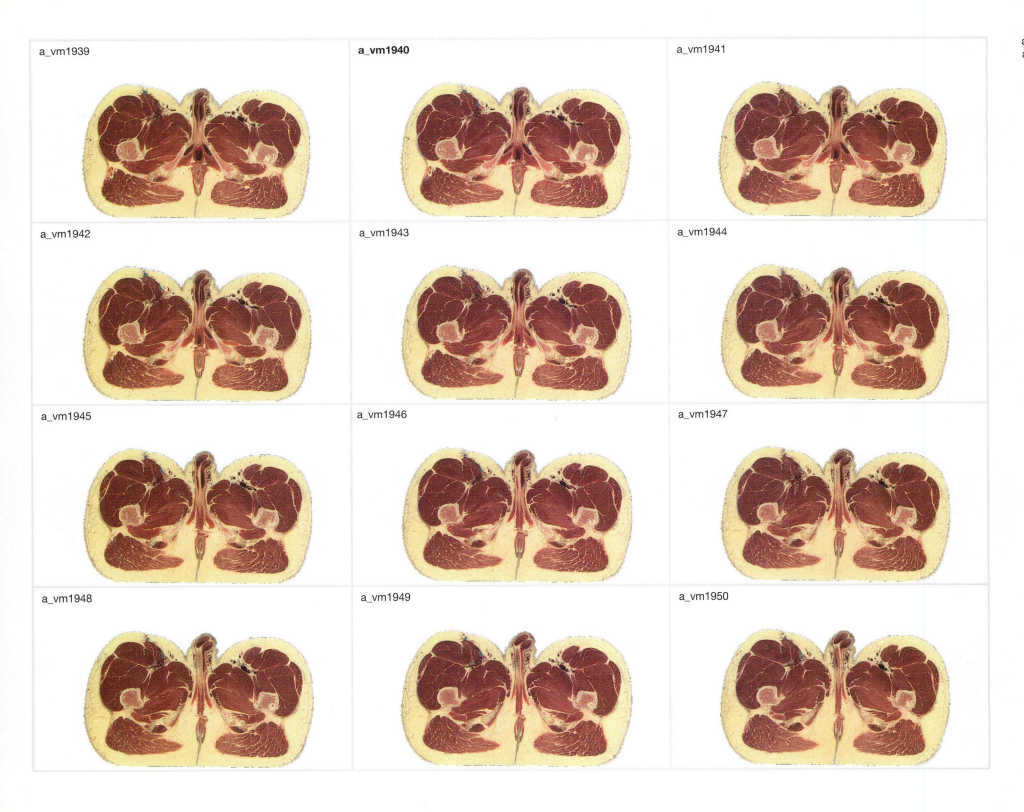

anterior

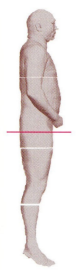

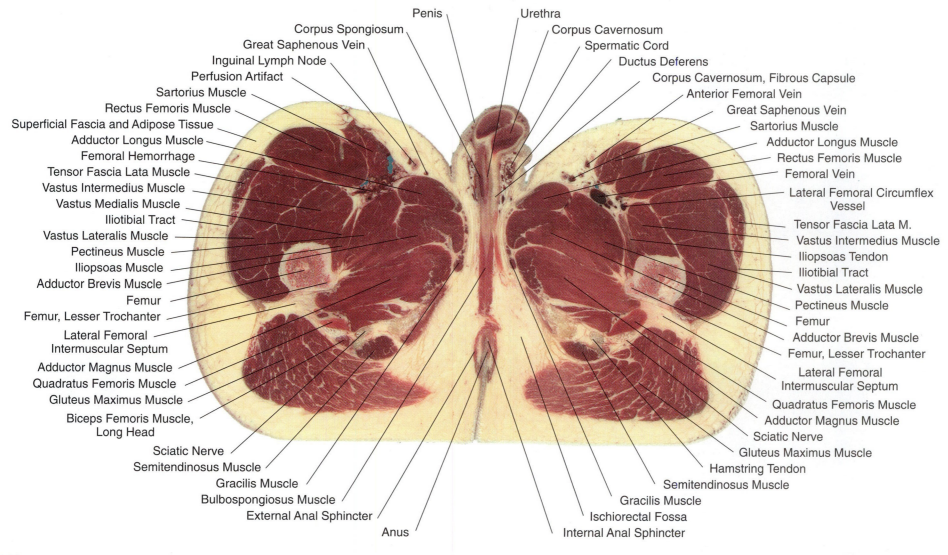

Penis

Urethra

Corpus Spongiosum

Corpus Cavernosum

Great Saphenous Vein

Spermatic Cord

Inguinal Lymph Node

Ductus Deferens

Perfusion Artifact

Corpus Cavernosum, Fibrous Capsule

Sartorius Muscle

Anterior Femoral Vein

Rectus Femoris Muscle

Great Saphenous Vein

Superficial Fascia and Adipose Tissue

Sartorius Muscle

Adductor Longus Muscle

Adductor Longus Muscle

Femoral Hemorrhage

Rectus Femoris Muscle

Tensor Fascia Lata Muscle

Femoral Vein

Vastus Intermedius Muscle

Lateral Femoral Circumflex Vessel

Vastus Medialis Muscle

Iliotibial Tract

Tensor Fascia Lata M.

Vastus Lateralis Muscle

Vastus Intermedius Muscle

Pectineus Muscle

Iliopsoas Tendon

Iliopsoas Muscle

Iliotibial Tract

Adductor Brevis Muscle

Vastus Lateralis Muscle

Femur

Pectineus Muscle

Femur, Lesser Trochanter

Femur

Lateral Femoral
Intermuscular Septum

Adductor Brevis Muscle

Femur, Lesser Trochanter

Adductor Magnus Muscle

Lateral Femoral
Intermuscular Septum

Quadratus Femoris Muscle

Gluteus Maximus Muscle

Quadratus Femoris Muscle

Biceps Femoris Muscle,
Long Head

Adductor Magnus Muscle

Sciatic Nerve

Sciatic Nerve

Semitendinosus Muscle

Gluteus Maximus Muscle

Gracilis Muscle

Hamstring Tendon

Bulbospongiosus Muscle

Semitendinosus Muscle

External Anal Sphincter

Gracilis Muscle

Anus

Ischiorectal Fossa

Internal Anal Sphincter

right

left

posterior

160

a_vm1951

a_vm1952

a_vm1953

a_vm1954

a_vm1955

a_vm1956

a_vm1957

a_vm1958

a_vm1959

a_vm1960

a_vm1961

a_vm1962

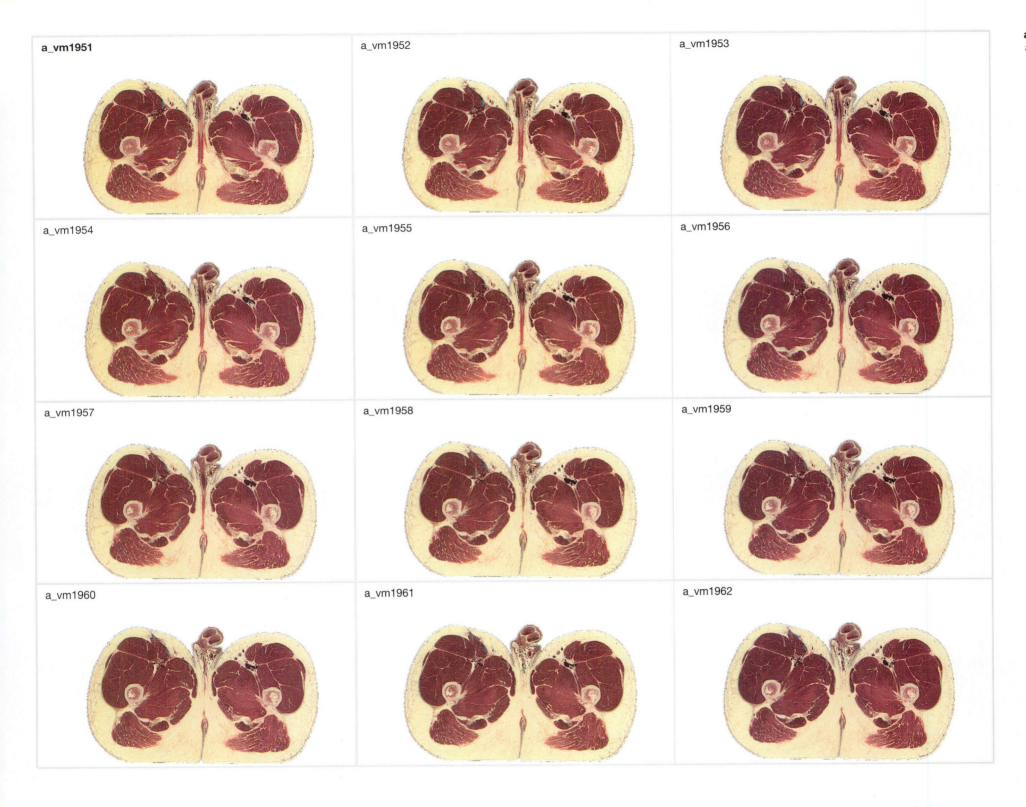

anterior

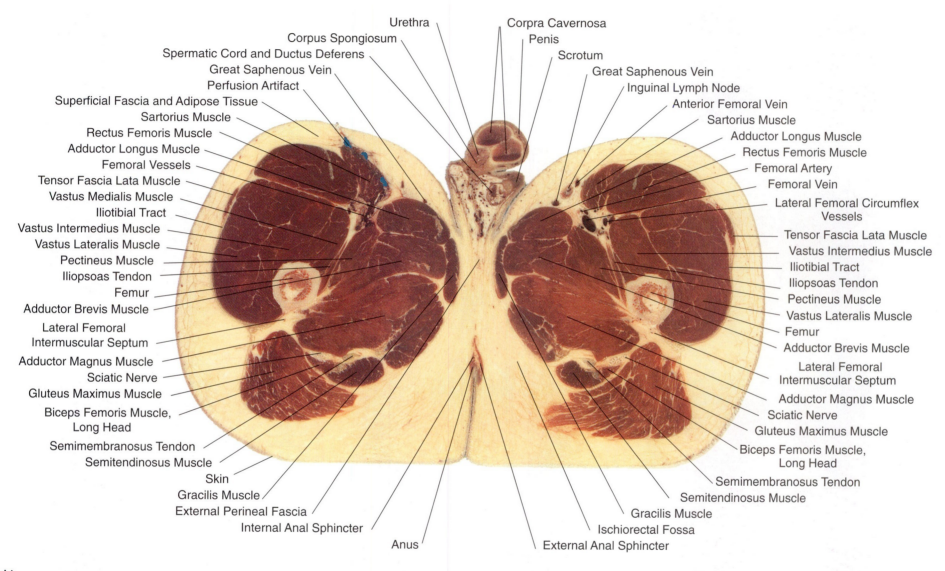

Urethra

Corpus Spongiosum

Spermatic Cord and Ductus Deferens

Great Saphenous Vein

Perfusion Artifact

Superficial Fascia and Adipose Tissue

Sartorius Muscle

Rectus Femoris Muscle

Adductor Longus Muscle

Femoral Vessels

Tensor Fascia Lata Muscle

Vastus Medialis Muscle

Iliotibial Tract

Vastus Intermedius Muscle

Vastus Lateralis Muscle

Pectineus Muscle

Iliopsoas Tendon

Femur

Adductor Brevis Muscle

Lateral Femoral
Intermuscular Septum

Adductor Magnus Muscle

Sciatic Nerve

Gluteus Maximus Muscle

Biceps Femoris Muscle,
Long Head

Semimembranosus Tendon

Semitendinosus Muscle

Skin

Gracilis Muscle

External Perineal Fascia

Internal Anal Sphincter

Anus

Corpra Cavernosa

Penis

Scrotum

Great Saphenous Vein

Inguinal Lymph Node

Anterior Femoral Vein

Sartorius Muscle

Adductor Longus Muscle

Rectus Femoris Muscle

Femoral Artery

Femoral Vein

Lateral Femoral Circumflex
Vessels

Tensor Fascia Lata Muscle

Vastus Intermedius Muscle

Iliotibial Tract

Iliopsoas Tendon

Pectineus Muscle

Vastus Lateralis Muscle

Femur

Adductor Brevis Muscle

Lateral Femoral
Intermuscular Septum

Adductor Magnus Muscle

Sciatic Nerve

Gluteus Maximus Muscle

Biceps Femoris Muscle,
Long Head

Semimembranosus Tendon

Semitendinosus Muscle

Gracilis Muscle

Ischiorectal Fossa

External Anal Sphincter

right

left

posterior

a_vm1963

a_vm1964

a_vm1965

a_vm1966

a_vm1967

a_vm1968

a_vm1969

a_vm1970

a_vm1971

a_vm1972

a_vm1973

a_vm1974

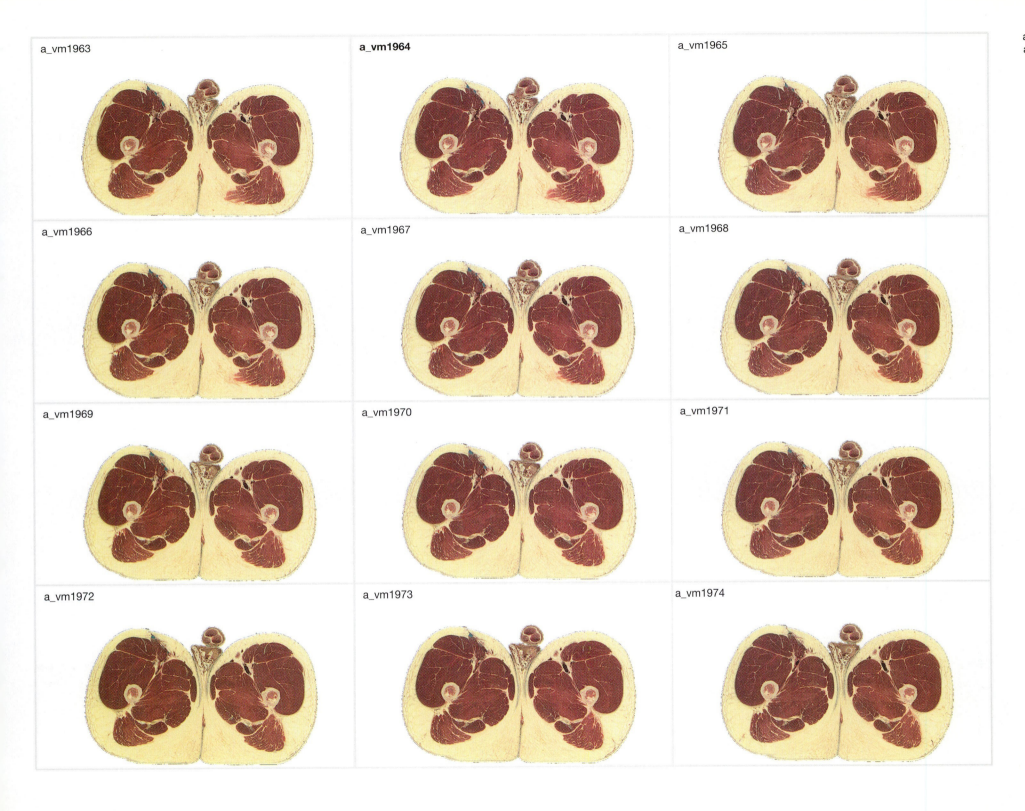

anterior

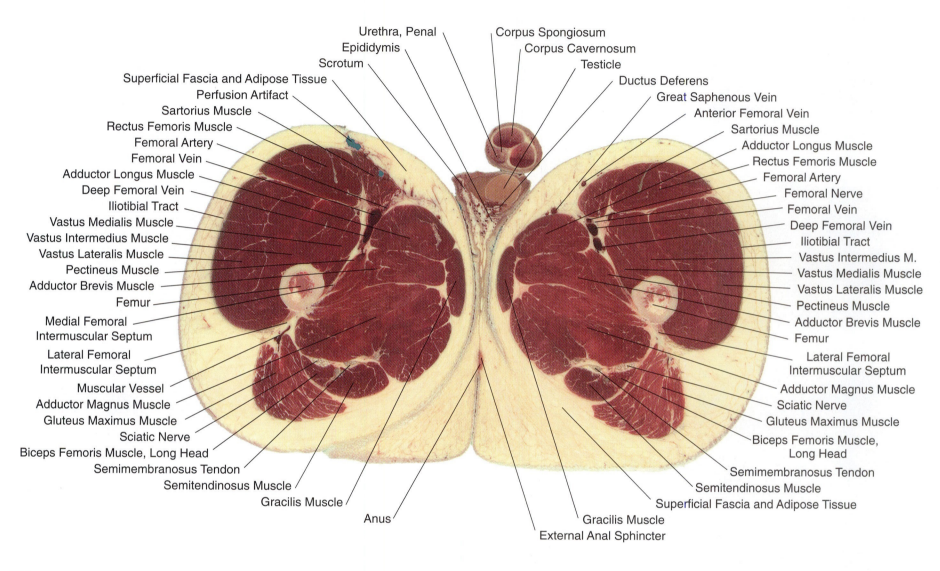

Urethra, Penal

Epididymis

Scrotum

Superficial Fascia and Adipose Tissue

Perfusion Artifact

Sartorius Muscle

Rectus Femoris Muscle

Femoral Artery

Femoral Vein

Adductor Longus Muscle

Deep Femoral Vein

Iliotibial Tract

Vastus Medialis Muscle

Vastus Intermedius Muscle

Vastus Lateralis Muscle

Pectineus Muscle

Adductor Brevis Muscle

Femur

Medial Femoral
Intermuscular Septum

Lateral Femoral
Intermuscular Septum

Muscular Vessel

Adductor Magnus Muscle

Gluteus Maximus Muscle

Sciatic Nerve

Biceps Femoris Muscle, Long Head

Semimembranosus Tendon

Semitendinosus Muscle

Gracilis Muscle

Anus

External Anal Sphincter

Gracilis Muscle

Corpus Spongiosum

Corpus Cavernosum

Testicle

Ductus Deferens

Great Saphenous Vein

Anterior Femoral Vein

Sartorius Muscle

Adductor Longus Muscle

Rectus Femoris Muscle

Femoral Artery

Femoral Nerve

Femoral Vein

Deep Femoral Vein

Iliotibial Tract

Vastus Intermedius M.

Vastus Medialis Muscle

Vastus Lateralis Muscle

Pectineus Muscle

Adductor Brevis Muscle

Femur

Lateral Femoral
Intermuscular Septum

Adductor Magnus Muscle

Sciatic Nerve

Gluteus Maximus Muscle

Biceps Femoris Muscle,
Long Head

Semimembranosus Tendon

Semitendinosus Muscle

Superficial Fascia and Adipose Tissue

right

left

posterior

a_vm1975

a_vm1976

a_vm1977

a_vm1978

a_vm1979

a_vm1980

a_vm1981

a_vm1982

a_vm1983

a_vm1984

a_vm1985

a_vm1986

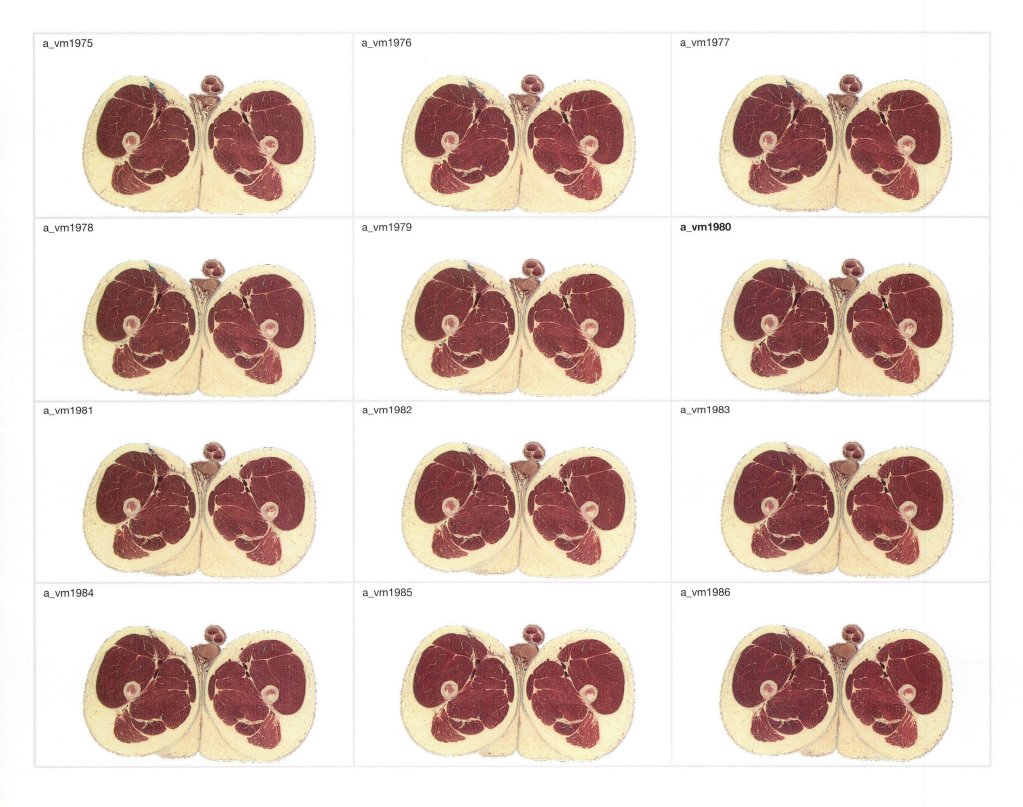

anterior

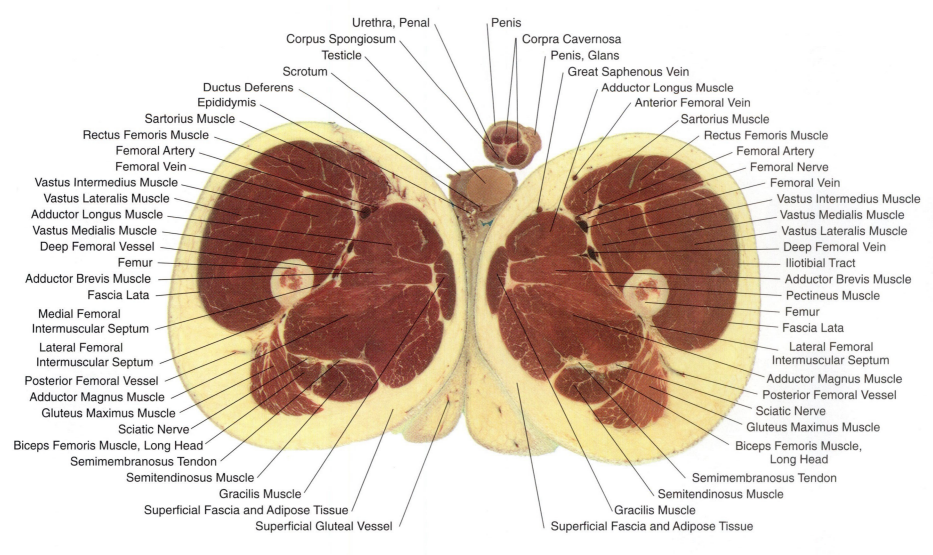

Urethra, Penal
Corpus Spongiosum
Testicle
Scrotum
Ductus Deferens
Epididymis
Sartorius Muscle
Rectus Femoris Muscle
Femoral Artery
Femoral Vein
Vastus Intermedius Muscle
Vastus Lateralis Muscle
Adductor Longus Muscle
Vastus Medialis Muscle
Deep Femoral Vessel
Femur
Adductor Brevis Muscle
Fascia Lata
Medial Femoral
Intermuscular Septum
Lateral Femoral
Intermuscular Septum
Posterior Femoral Vessel
Adductor Magnus Muscle
Gluteus Maximus Muscle
Sciatic Nerve
Biceps Femoris Muscle, Long Head
Semimembranosus Tendon
Semitendinosus Muscle
Gracilis Muscle
Superficial Fascia and Adipose Tissue
Superficial Gluteal Vessel

Penis
Corpra Cavernosa
Penis, Glans
Great Saphenous Vein
Adductor Longus Muscle
Anterior Femoral Vein
Sartorius Muscle
Rectus Femoris Muscle
Femoral Artery
Femoral Nerve
Femoral Vein
Vastus Intermedius Muscle
Vastus Medialis Muscle
Vastus Lateralis Muscle
Deep Femoral Vein
Iliotibial Tract
Adductor Brevis Muscle
Pectineus Muscle
Femur
Fascia Lata
Lateral Femoral
Intermuscular Septum
Adductor Magnus Muscle
Posterior Femoral Vessel
Sciatic Nerve
Gluteus Maximus Muscle
Biceps Femoris Muscle,
Long Head
Semimembranosus Tendon
Semitendinosus Muscle
Gracilis Muscle
Superficial Fascia and Adipose Tissue

right

left

posterior

166

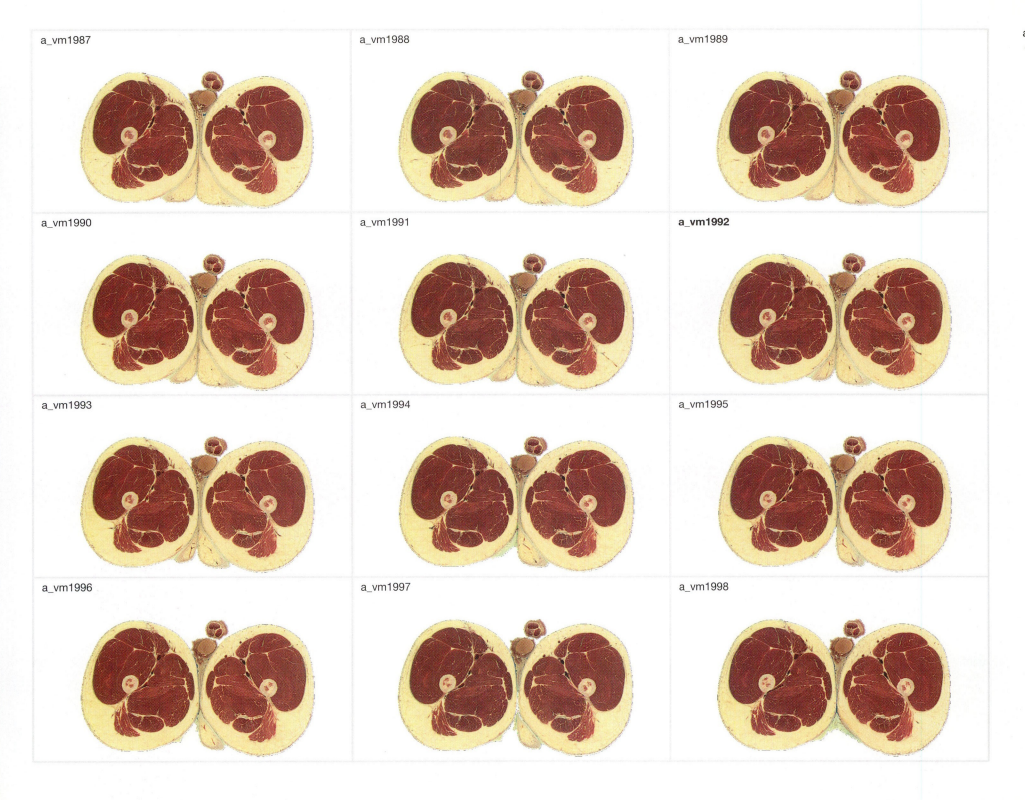

a_vm1987

a_vm1988

a_vm1989

a_vm1990

a_vm1991

a_vm1992

a_vm1993

a_vm1994

a_vm1995

a_vm1996

a_vm1997

a_vm1998

anterior

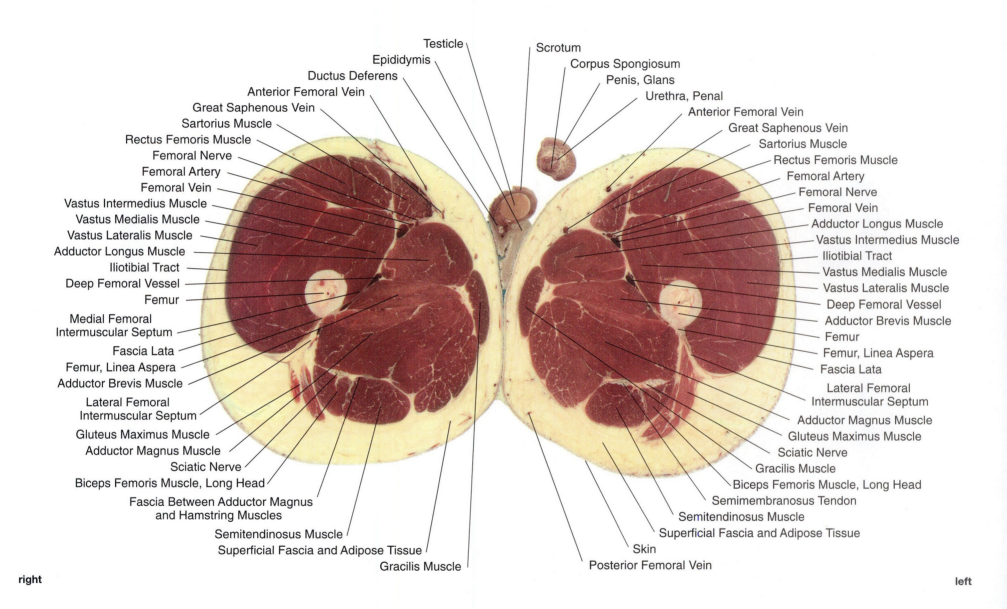

Testicle

Scrotum

Epididymis

Corpus Spongiosum

Ductus Deferens

Penis, Glans

Anterior Femoral Vein

Urethra, Penal

Great Saphenous Vein

Anterior Femoral Vein

Sartorius Muscle

Great Saphenous Vein

Rectus Femoris Muscle

Sartorius Muscle

Femoral Nerve

Rectus Femoris Muscle

Femoral Artery

Femoral Artery

Femoral Vein

Femoral Nerve

Vastus Intermedius Muscle

Femoral Vein

Vastus Medialis Muscle

Adductor Longus Muscle

Vastus Lateralis Muscle

Vastus Intermedius Muscle

Adductor Longus Muscle

Iliotibial Tract

Iliotibial Tract

Vastus Medialis Muscle

Deep Femoral Vessel

Vastus Lateralis Muscle

Femur

Deep Femoral Vessel

Medial Femoral
Intermuscular Septum

Adductor Brevis Muscle

Femur

Fascia Lata

Femur, Linea Aspera

Femur, Linea Aspera

Fascia Lata

Adductor Brevis Muscle

Lateral Femoral
Intermuscular Septum

Lateral Femoral
Intermuscular Septum

Adductor Magnus Muscle

Gluteus Maximus Muscle

Gluteus Maximus Muscle

Adductor Magnus Muscle

Sciatic Nerve

Sciatic Nerve

Gracilis Muscle

Biceps Femoris Muscle, Long Head

Biceps Femoris Muscle, Long Head

Fascia Between Adductor Magnus
and Hamstring Muscles

Semimembranosus Tendon

Semitendinosus Muscle

Semitendinosus Muscle

Superficial Fascia and Adipose Tissue

Superficial Fascia and Adipose Tissue

Skin

Gracilis Muscle

Posterior Femoral Vein

right

left

posterior

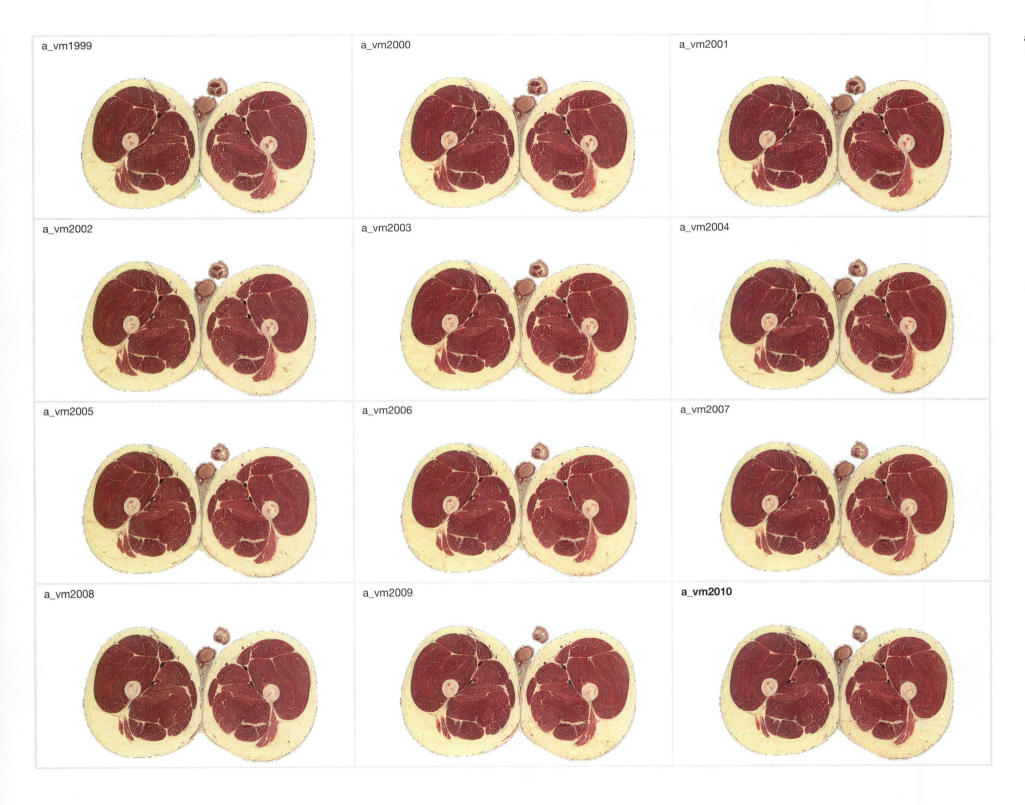

a_vm1999

a_vm2000

a_vm2001

a_vm2002

a_vm2003

a_vm2004

a_vm2005

a_vm2006

a_vm2007

a_vm2008

a_vm2009

a_vm2010

anterior

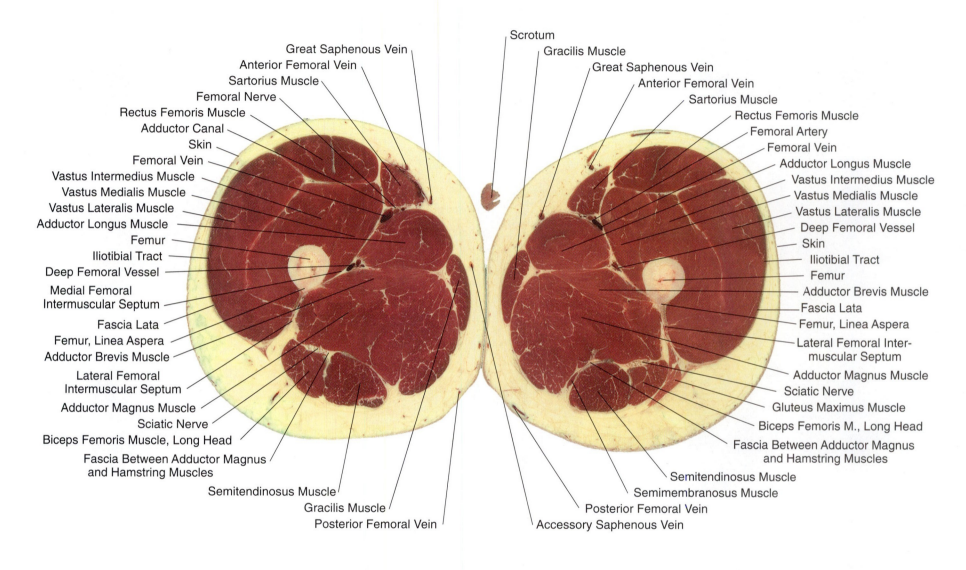

Great Saphenous Vein

Anterior Femoral Vein

Sartorius Muscle

Femoral Nerve

Rectus Femoris Muscle

Adductor Canal

Skin

Femoral Vein

Vastus Intermedius Muscle

Vastus Medialis Muscle

Vastus Lateralis Muscle

Adductor Longus Muscle

Femur

Iliotibial Tract

Deep Femoral Vessel

Medial Femoral
Intermuscular Septum

Fascia Lata

Femur, Linea Aspera

Adductor Brevis Muscle

Lateral Femoral
Intermuscular Septum

Adductor Magnus Muscle

Sciatic Nerve

Biceps Femoris Muscle, Long Head

Fascia Between Adductor Magnus
and Hamstring Muscles

Semitendinosus Muscle

Gracilis Muscle

Posterior Femoral Vein

Scrotum

Gracilis Muscle

Great Saphenous Vein

Anterior Femoral Vein

Sartorius Muscle

Rectus Femoris Muscle

Femoral Artery

Femoral Vein

Adductor Longus Muscle

Vastus Intermedius Muscle

Vastus Medialis Muscle

Vastus Lateralis Muscle

Deep Femoral Vessel

Skin

Iliotibial Tract

Femur

Adductor Brevis Muscle

Fascia Lata

Femur, Linea Aspera

Lateral Femoral Inter-
muscular Septum

Adductor Magnus Muscle

Sciatic Nerve

Gluteus Maximus Muscle

Biceps Femoris M., Long Head

Fascia Between Adductor Magnus
and Hamstring Muscles

Semitendinosus Muscle

Semimembranosus Muscle

Posterior Femoral Vein

Accessory Saphenous Vein

right

left

posterior

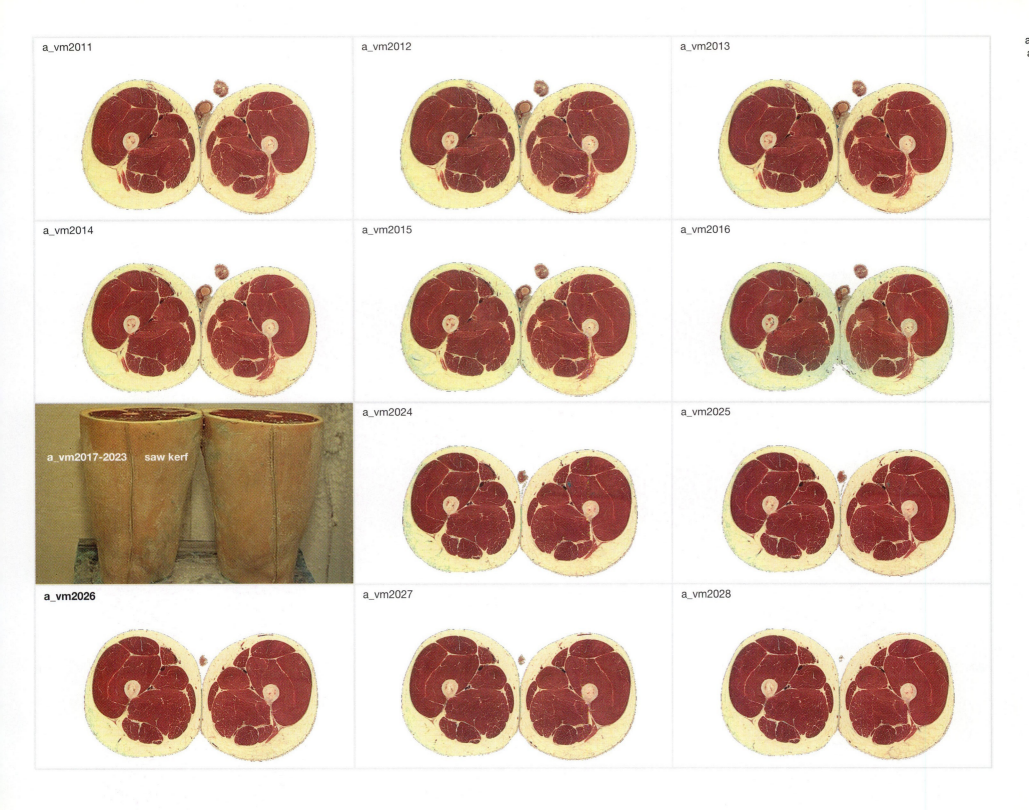

a_vm2011

a_vm2012

a_vm2013

a_vm2014

a_vm2015

a_vm2016

a_vm2017-2023   saw kerf

a_vm2024

a_vm2025

a_vm2026

a_vm2027

a_vm2028

anterior

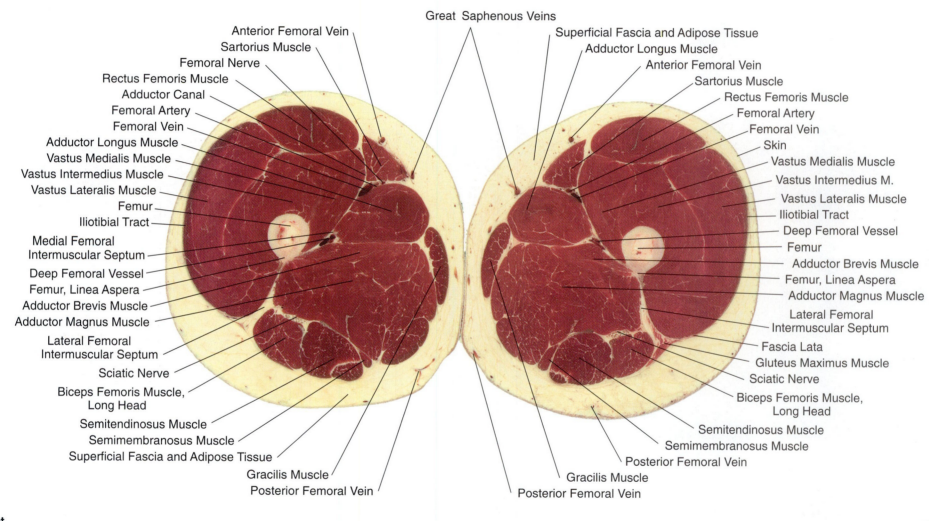

Great Saphenous Veins

Anterior Femoral Vein
Sartorius Muscle
Femoral Nerve
Rectus Femoris Muscle
Adductor Canal
Femoral Artery
Femoral Vein
Adductor Longus Muscle
Vastus Medialis Muscle
Vastus Intermedius Muscle
Vastus Lateralis Muscle
Femur
Iliotibial Tract
Medial Femoral
Intermuscular Septum
Deep Femoral Vessel
Femur, Linea Aspera
Adductor Brevis Muscle
Adductor Magnus Muscle
Lateral Femoral
Intermuscular Septum
Sciatic Nerve
Biceps Femoris Muscle,
Long Head
Semitendinosus Muscle
Semimembranosus Muscle
Superficial Fascia and Adipose Tissue
Gracilis Muscle
Posterior Femoral Vein

Superficial Fascia and Adipose Tissue
Adductor Longus Muscle
Anterior Femoral Vein
Sartorius Muscle
Rectus Femoris Muscle
Femoral Artery
Femoral Vein
Skin
Vastus Medialis Muscle
Vastus Intermedius M.
Vastus Lateralis Muscle
Iliotibial Tract
Deep Femoral Vessel
Femur
Adductor Brevis Muscle
Femur, Linea Aspera
Adductor Magnus Muscle
Lateral Femoral
Intermuscular Septum
Fascia Lata
Gluteus Maximus Muscle
Sciatic Nerve
Biceps Femoris Muscle,
Long Head
Semitendinosus Muscle
Semimembranosus Muscle
Posterior Femoral Vein
Gracilis Muscle
Posterior Femoral Vein

right

left

posterior

172

a_vm2029

a_vm2030

a_vm2031

a_vm2032

a_vm2033

a_vm2034

a_vm2035

a_vm2036

a_vm2037

a_vm2038

a_vm2039

a_vm2040

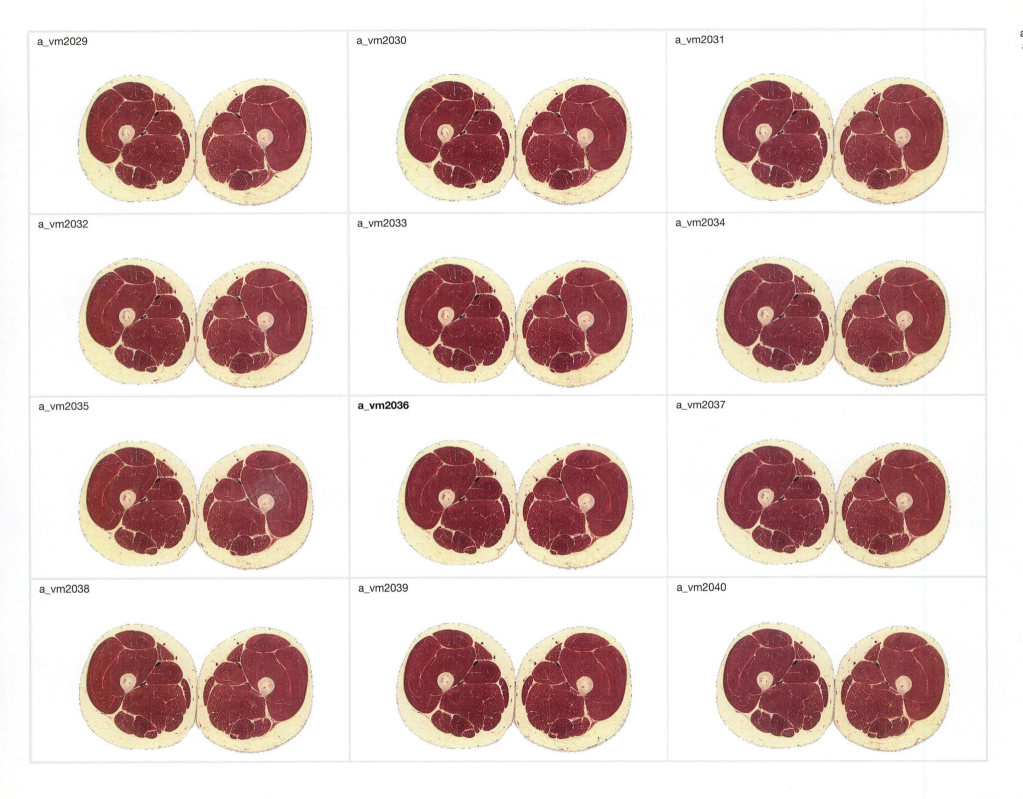

anterior

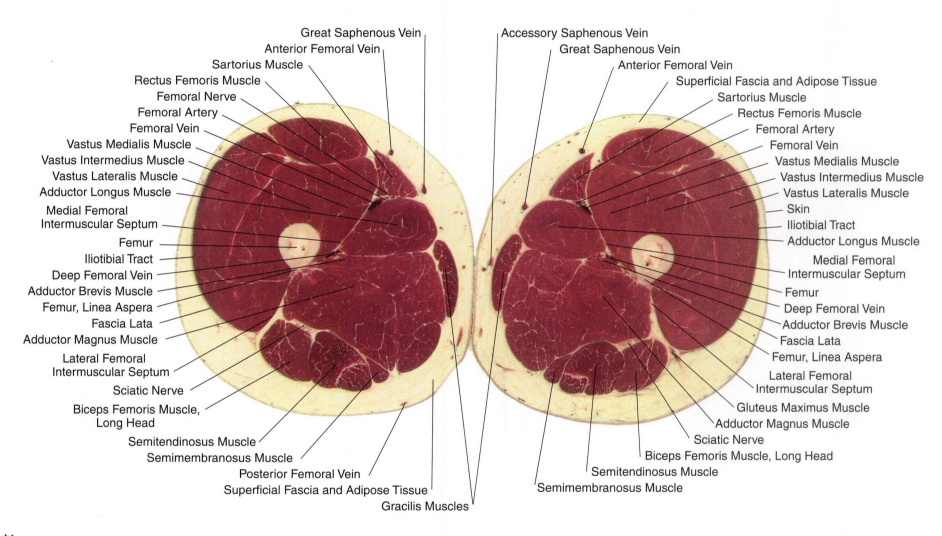

Great Saphenous Vein

Anterior Femoral Vein

Sartorius Muscle

Rectus Femoris Muscle

Femoral Nerve

Femoral Artery

Femoral Vein

Vastus Medialis Muscle

Vastus Intermedius Muscle

Vastus Lateralis Muscle

Adductor Longus Muscle

Medial Femoral
Intermuscular Septum

Femur

Iliotibial Tract

Deep Femoral Vein

Adductor Brevis Muscle

Femur, Linea Aspera

Fascia Lata

Adductor Magnus Muscle

Lateral Femoral
Intermuscular Septum

Sciatic Nerve

Biceps Femoris Muscle,
Long Head

Semitendinosus Muscle

Semimembranosus Muscle

Posterior Femoral Vein

Superficial Fascia and Adipose Tissue

Gracilis Muscles

Accessory Saphenous Vein

Great Saphenous Vein

Anterior Femoral Vein

Superficial Fascia and Adipose Tissue

Sartorius Muscle

Rectus Femoris Muscle

Femoral Artery

Femoral Vein

Vastus Medialis Muscle

Vastus Intermedius Muscle

Vastus Lateralis Muscle

Skin

Iliotibial Tract

Adductor Longus Muscle

Medial Femoral
Intermuscular Septum

Femur

Deep Femoral Vein

Adductor Brevis Muscle

Fascia Lata

Femur, Linea Aspera

Lateral Femoral
Intermuscular Septum

Gluteus Maximus Muscle

Adductor Magnus Muscle

Sciatic Nerve

Biceps Femoris Muscle, Long Head

Semitendinosus Muscle

Semimembranosus Muscle

right

left

posterior

a_vm2041

a_vm2042

a_vm2043

a_vm2044

a_vm2045

a_vm2046

a_vm2047

a_vm2048

a_vm2049

a_vm2050

a_vm2051

a_vm2052

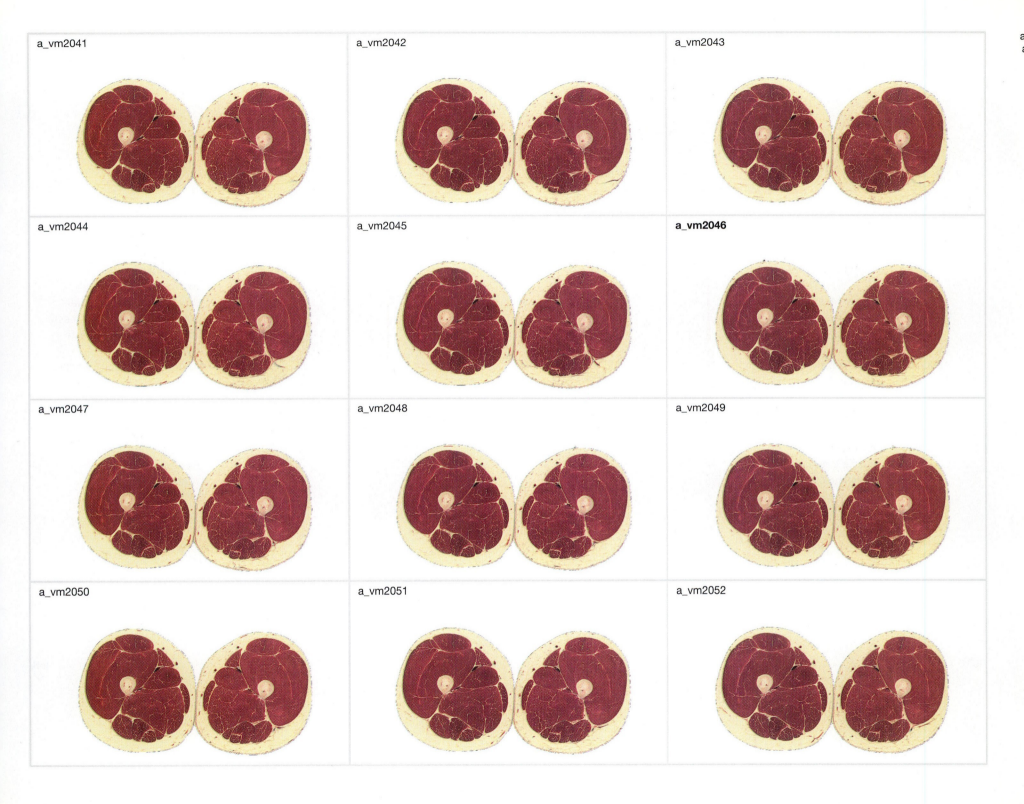

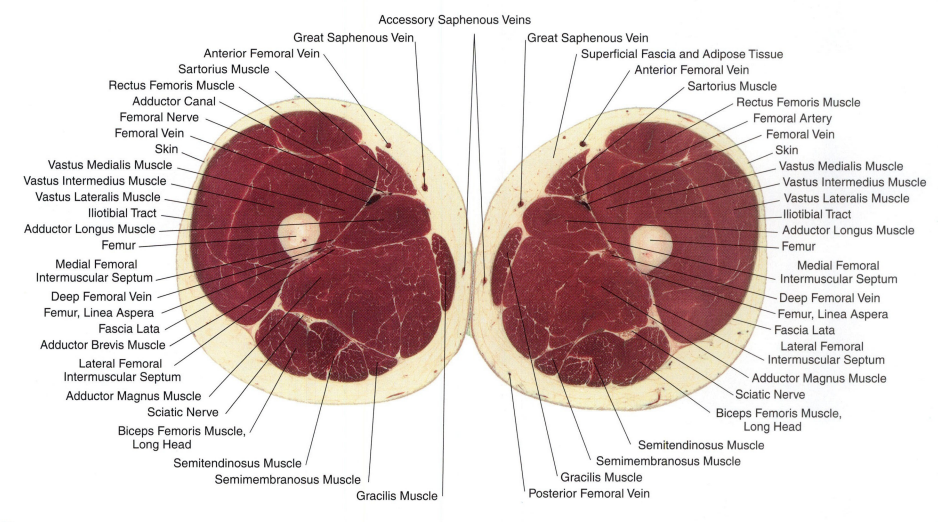

anterior

Accessory Saphenous Veins

Great Saphenous Vein

Anterior Femoral Vein

Sartorius Muscle

Rectus Femoris Muscle

Adductor Canal

Femoral Nerve

Femoral Vein

Skin

Vastus Medialis Muscle

Vastus Intermedius Muscle

Vastus Lateralis Muscle

Iliotibial Tract

Adductor Longus Muscle

Femur

Medial Femoral
Intermuscular Septum

Deep Femoral Vein

Femur, Linea Aspera

Fascia Lata

Adductor Brevis Muscle

Lateral Femoral
Intermuscular Septum

Adductor Magnus Muscle

Sciatic Nerve

Biceps Femoris Muscle,
Long Head

Semitendinosus Muscle

Semimembranosus Muscle

Gracilis Muscle

Great Saphenous Vein

Superficial Fascia and Adipose Tissue

Anterior Femoral Vein

Sartorius Muscle

Rectus Femoris Muscle

Femoral Artery

Femoral Vein

Skin

Vastus Medialis Muscle

Vastus Intermedius Muscle

Vastus Lateralis Muscle

Iliotibial Tract

Adductor Longus Muscle

Femur

Medial Femoral
Intermuscular Septum

Deep Femoral Vein

Femur, Linea Aspera

Fascia Lata

Lateral Femoral
Intermuscular Septum

Adductor Magnus Muscle

Sciatic Nerve

Biceps Femoris Muscle,
Long Head

Semitendinosus Muscle

Semimembranosus Muscle

Gracilis Muscle

Posterior Femoral Vein

right

left

posterior

176

a_vm2053

a_vm2054

a_vm2055

a_vm2056

a_vm2057

a_vm2058

a_vm2069

a_vm2060

a_vm2061

a_vm2062

a_vm2063

a_vm2064

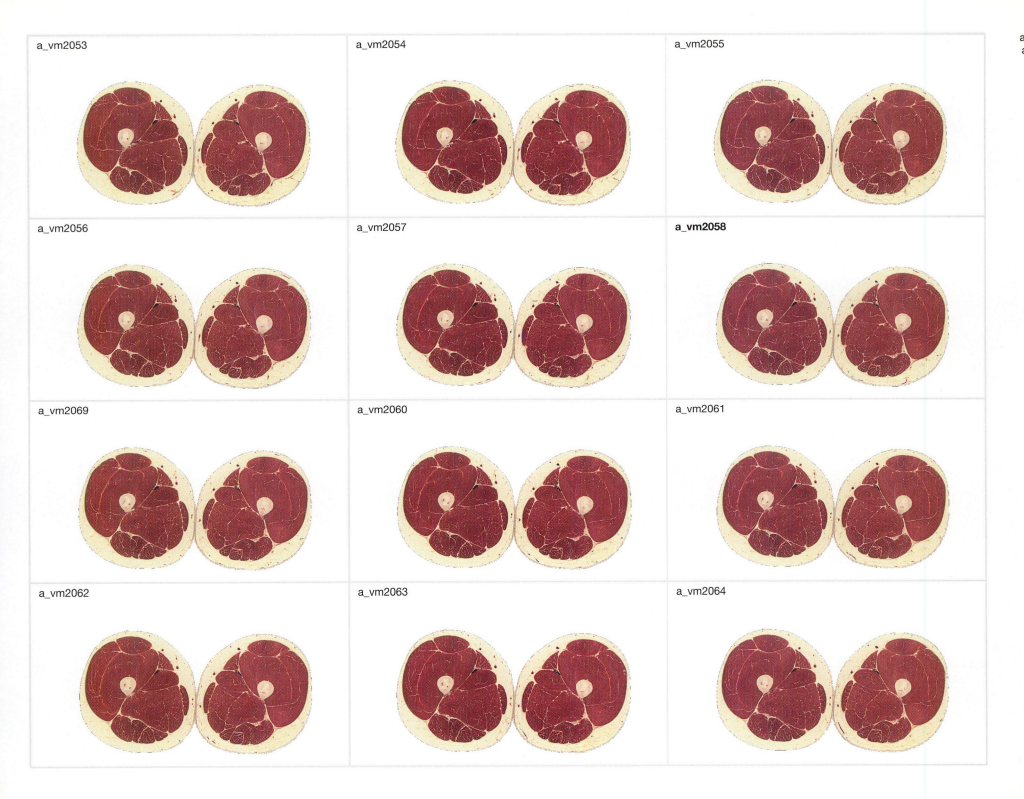

anterior

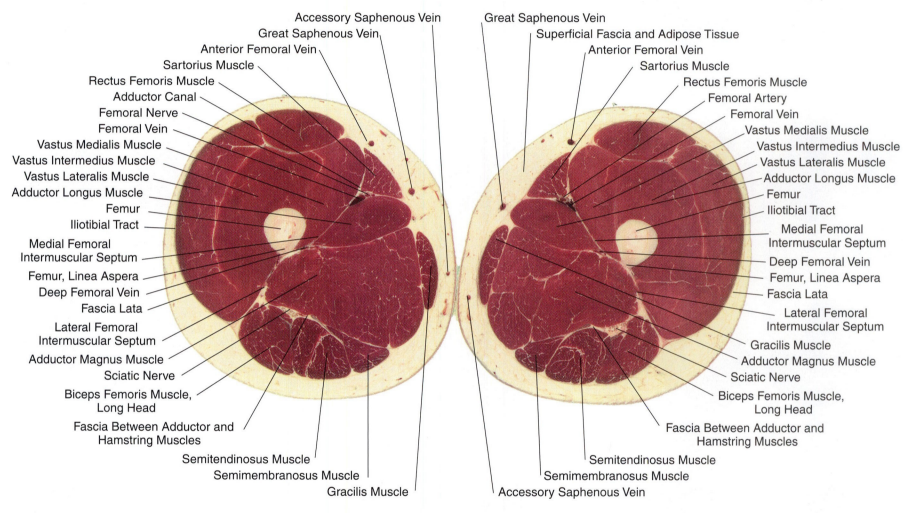

Accessory Saphenous Vein

Great Saphenous Vein

Anterior Femoral Vein

Sartorius Muscle

Rectus Femoris Muscle

Adductor Canal

Femoral Nerve

Femoral Vein

Vastus Medialis Muscle

Vastus Intermedius Muscle

Vastus Lateralis Muscle

Adductor Longus Muscle

Femur

Iliotibial Tract

Medial Femoral
Intermuscular Septum

Femur, Linea Aspera

Deep Femoral Vein

Fascia Lata

Lateral Femoral
Intermuscular Septum

Adductor Magnus Muscle

Sciatic Nerve

Biceps Femoris Muscle,
Long Head

Fascia Between Adductor and
Hamstring Muscles

Semitendinosus Muscle

Semimembranosus Muscle

Gracilis Muscle

Great Saphenous Vein

Superficial Fascia and Adipose Tissue

Anterior Femoral Vein

Sartorius Muscle

Rectus Femoris Muscle

Femoral Artery

Femoral Vein

Vastus Medialis Muscle

Vastus Intermedius Muscle

Vastus Lateralis Muscle

Adductor Longus Muscle

Femur

Iliotibial Tract

Medial Femoral
Intermuscular Septum

Deep Femoral Vein

Femur, Linea Aspera

Fascia Lata

Lateral Femoral
Intermuscular Septum

Gracilis Muscle

Adductor Magnus Muscle

Sciatic Nerve

Biceps Femoris Muscle,
Long Head

Fascia Between Adductor and
Hamstring Muscles

Semitendinosus Muscle

Semimembranosus Muscle

Accessory Saphenous Vein

right

left

posterior

a_vm2065

a_vm2066

a_vm2067

a_vm2068

a_vm2069

a_vm2070

a_vm2071

a_vm2072

a_vm2073

a_vm2074

a_vm2075

a_vm2076

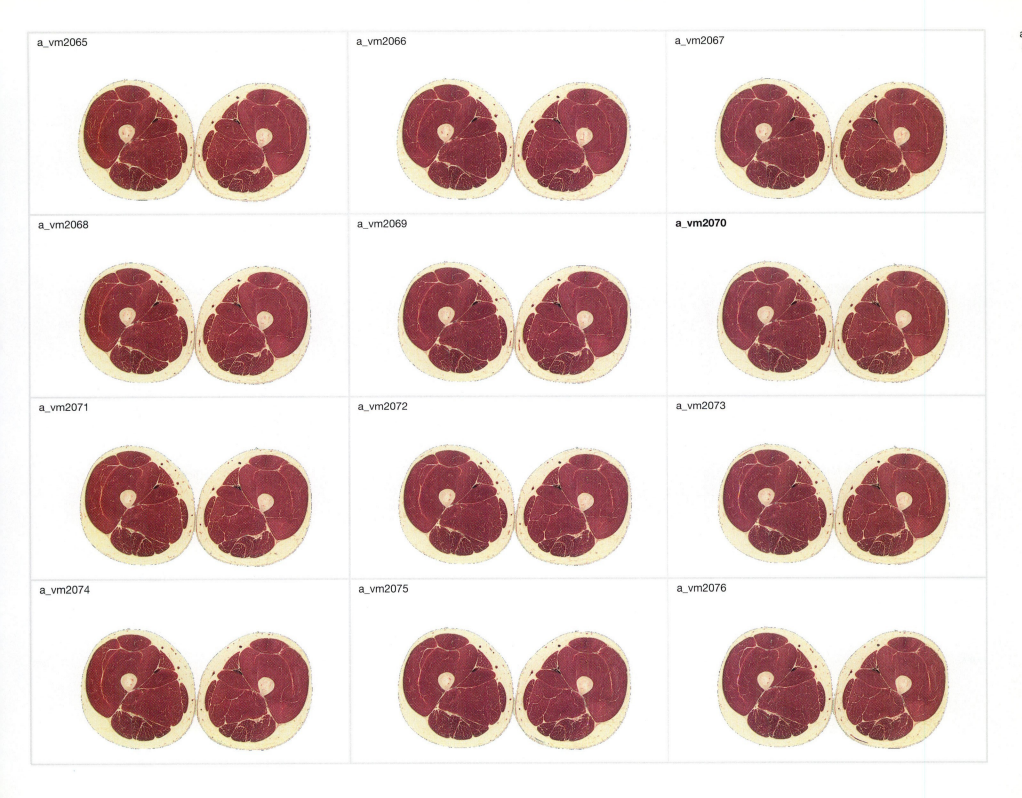

Transverse
a_vm2082

anterior

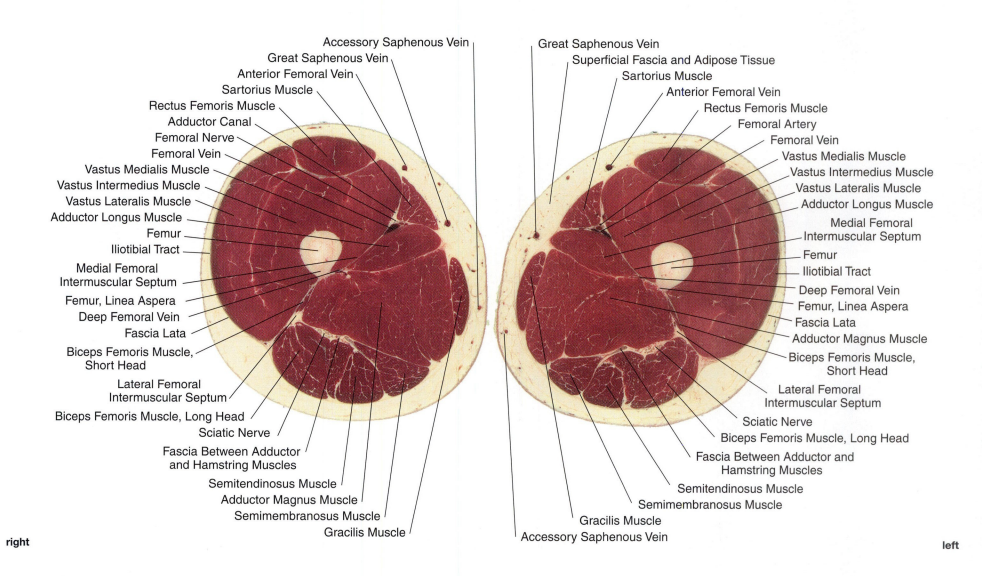

Accessory Saphenous Vein

Great Saphenous Vein

Anterior Femoral Vein

Sartorius Muscle

Rectus Femoris Muscle

Adductor Canal

Femoral Nerve

Femoral Vein

Vastus Medialis Muscle

Vastus Intermedius Muscle

Vastus Lateralis Muscle

Adductor Longus Muscle

Femur

Iliotibial Tract

Medial Femoral
Intermuscular Septum

Femur, Linea Aspera

Deep Femoral Vein

Fascia Lata

Biceps Femoris Muscle,
Short Head

Lateral Femoral
Intermuscular Septum

Biceps Femoris Muscle, Long Head

Sciatic Nerve

Fascia Between Adductor
and Hamstring Muscles

Semitendinosus Muscle

Adductor Magnus Muscle

Semimembranosus Muscle

Gracilis Muscle

Great Saphenous Vein

Superficial Fascia and Adipose Tissue

Sartorius Muscle

Anterior Femoral Vein

Rectus Femoris Muscle

Femoral Artery

Femoral Vein

Vastus Medialis Muscle

Vastus Intermedius Muscle

Vastus Lateralis Muscle

Adductor Longus Muscle

Medial Femoral
Intermuscular Septum

Femur

Iliotibial Tract

Deep Femoral Vein

Femur, Linea Aspera

Fascia Lata

Adductor Magnus Muscle

Biceps Femoris Muscle,
Short Head

Lateral Femoral
Intermuscular Septum

Sciatic Nerve

Biceps Femoris Muscle, Long Head

Fascia Between Adductor and
Hamstring Muscles

Semitendinosus Muscle

Semimembranosus Muscle

Gracilis Muscle

Accessory Saphenous Vein

right

left

posterior

180

a_vm2077

a_vm2078

a_vm2079

a_vm2080

a_vm2081

a_vm2082

a_vm2083

a_vm2084

a_vm2085

a_vm2086

a_vm2087

a_vm2088

anterior

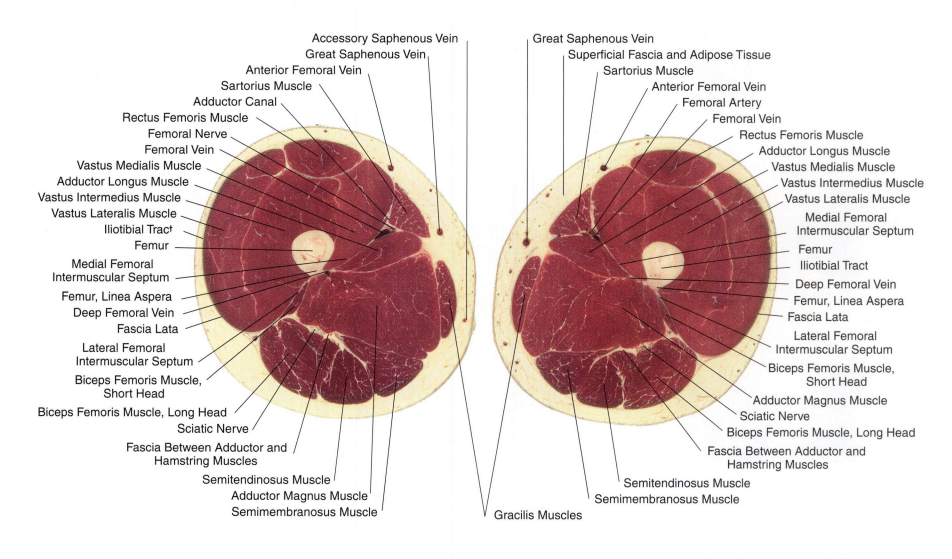

Accessory Saphenous Vein
Great Saphenous Vein
Anterior Femoral Vein
Sartorius Muscle
Adductor Canal
Rectus Femoris Muscle
Femoral Nerve
Femoral Vein
Vastus Medialis Muscle
Adductor Longus Muscle
Vastus Intermedius Muscle
Vastus Lateralis Muscle
Iliotibial Tract
Femur
Medial Femoral
Intermuscular Septum
Femur, Linea Aspera
Deep Femoral Vein
Fascia Lata
Lateral Femoral
Intermuscular Septum
Biceps Femoris Muscle,
Short Head
Biceps Femoris Muscle, Long Head
Sciatic Nerve
Fascia Between Adductor and
Hamstring Muscles
Semitendinosus Muscle
Adductor Magnus Muscle
Semimembranosus Muscle

Great Saphenous Vein
Superficial Fascia and Adipose Tissue
Sartorius Muscle
Anterior Femoral Vein
Femoral Artery
Femoral Vein
Rectus Femoris Muscle
Adductor Longus Muscle
Vastus Medialis Muscle
Vastus Intermedius Muscle
Vastus Lateralis Muscle
Medial Femoral
Intermuscular Septum
Femur
Iliotibial Tract
Deep Femoral Vein
Femur, Linea Aspera
Fascia Lata
Lateral Femoral
Intermuscular Septum
Biceps Femoris Muscle,
Short Head
Adductor Magnus Muscle
Sciatic Nerve
Biceps Femoris Muscle, Long Head
Fascia Between Adductor and
Hamstring Muscles
Semitendinosus Muscle
Semimembranosus Muscle

Gracilis Muscles

right

left

posterior

a_vm2089

a_vm2090

a_vm2091

a_vm2092

a_vm2093

a_vm2094

a_vm2095

a_vm2096

a_vm2097

a_vm2098

a_vm2099

a_vm2100

anterior

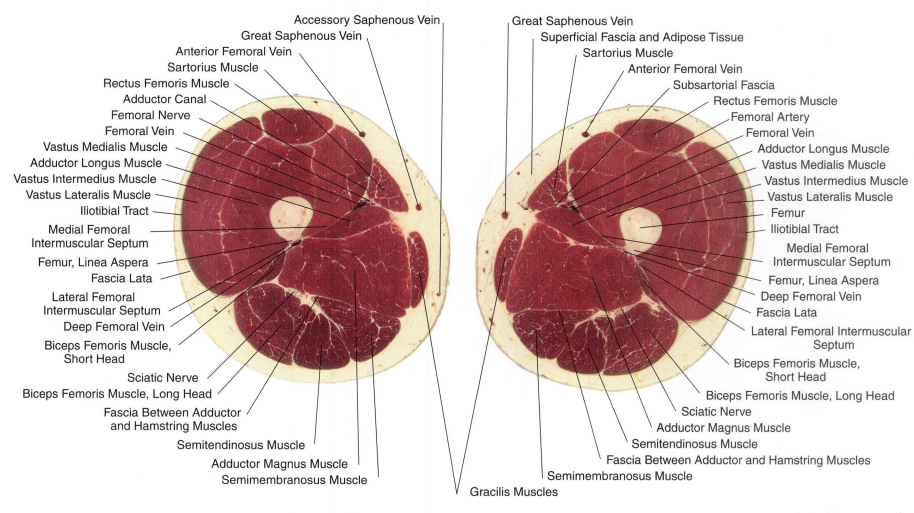

Accessory Saphenous Vein

Great Saphenous Vein

Anterior Femoral Vein

Sartorius Muscle

Rectus Femoris Muscle

Adductor Canal

Femoral Nerve

Femoral Vein

Vastus Medialis Muscle

Adductor Longus Muscle

Vastus Intermedius Muscle

Vastus Lateralis Muscle

Iliotibial Tract

Medial Femoral
Intermuscular Septum

Femur, Linea Aspera

Fascia Lata

Lateral Femoral
Intermuscular Septum

Deep Femoral Vein

Biceps Femoris Muscle,
Short Head

Sciatic Nerve

Biceps Femoris Muscle, Long Head

Fascia Between Adductor
and Hamstring Muscles

Semitendinosus Muscle

Adductor Magnus Muscle

Semimembranosus Muscle

Great Saphenous Vein

Superficial Fascia and Adipose Tissue

Sartorius Muscle

Anterior Femoral Vein

Subsartorial Fascia

Rectus Femoris Muscle

Femoral Artery

Femoral Vein

Adductor Longus Muscle

Vastus Medialis Muscle

Vastus Intermedius Muscle

Vastus Lateralis Muscle

Femur

Iliotibial Tract

Medial Femoral
Intermuscular Septum

Femur, Linea Aspera

Deep Femoral Vein

Fascia Lata

Lateral Femoral Intermuscular
Septum

Biceps Femoris Muscle,
Short Head

Biceps Femoris Muscle, Long Head

Sciatic Nerve

Adductor Magnus Muscle

Semitendinosus Muscle

Fascia Between Adductor and Hamstring Muscles

Semimembranosus Muscle

Gracilis Muscles

right

left

posterior

a_vm2101

a_vm2102

a_vm2103

a_vm2104

a_vm2105

a_vm2106

a_vm2107

a_vm2108

a_vm2109

a_vm2110

a_vm2111

a_vm2112

anterior

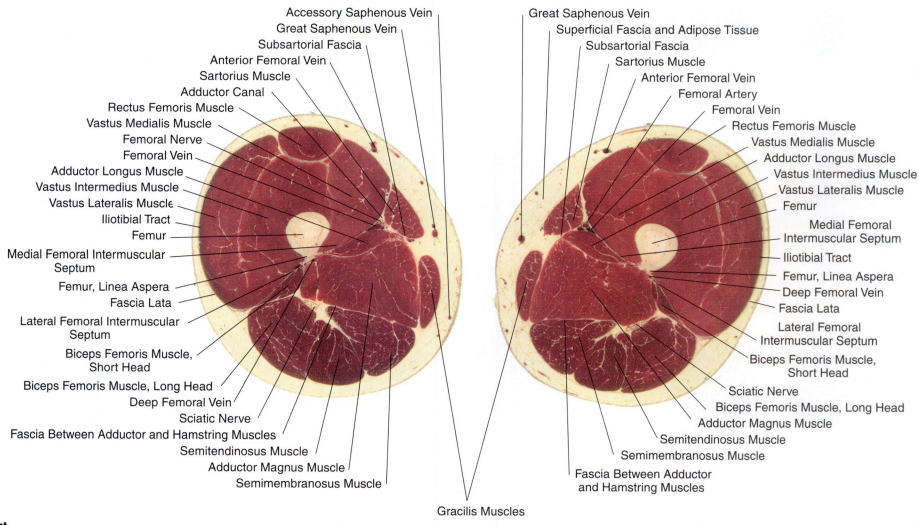

Accessory Saphenous Vein

Great Saphenous Vein

Subsartorial Fascia

Anterior Femoral Vein

Sartorius Muscle

Adductor Canal

Rectus Femoris Muscle

Vastus Medialis Muscle

Femoral Nerve

Femoral Vein

Adductor Longus Muscle

Vastus Intermedius Muscle

Vastus Lateralis Muscle

Iliotibial Tract

Femur

Medial Femoral Intermuscular
Septum

Femur, Linea Aspera

Fascia Lata

Lateral Femoral Intermuscular
Septum

Biceps Femoris Muscle,
Short Head

Biceps Femoris Muscle, Long Head

Deep Femoral Vein

Sciatic Nerve

Fascia Between Adductor and Hamstring Muscles

Semitendinosus Muscle

Adductor Magnus Muscle

Semimembranosus Muscle

Great Saphenous Vein

Superficial Fascia and Adipose Tissue

Subsartorial Fascia

Sartorius Muscle

Anterior Femoral Vein

Femoral Artery

Femoral Vein

Rectus Femoris Muscle

Vastus Medialis Muscle

Adductor Longus Muscle

Vastus Intermedius Muscle

Vastus Lateralis Muscle

Femur

Medial Femoral
Intermuscular Septum

Iliotibial Tract

Femur, Linea Aspera

Deep Femoral Vein

Fascia Lata

Lateral Femoral
Intermuscular Septum

Biceps Femoris Muscle,
Short Head

Sciatic Nerve

Biceps Femoris Muscle, Long Head

Adductor Magnus Muscle

Semitendinosus Muscle

Semimembranosus Muscle

Fascia Between Adductor
and Hamstring Muscles

Gracilis Muscles

right

left

posterior

a_vm2113

a_vm2114

a_vm2115

a_vm2116

a_vm2117

a_vm2118

a_vm2119

a_vm2120

a_vm2121

a_vm2122

a_vm2123

a_vm2124

anterior

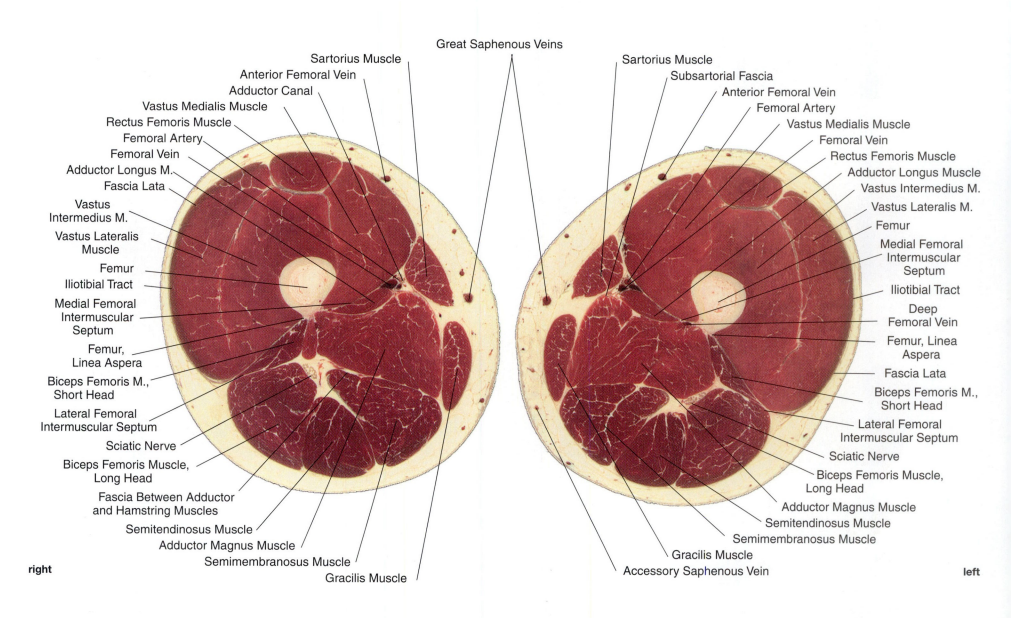

Great Saphenous Veins

Sartorius Muscle

Anterior Femoral Vein

Adductor Canal

Vastus Medialis Muscle

Rectus Femoris Muscle

Femoral Artery

Femoral Vein

Adductor Longus M.

Fascia Lata

Vastus Intermedius M.

Vastus Lateralis Muscle

Femur

Iliotibial Tract

Medial Femoral Intermuscular Septum

Femur, Linea Aspera

Biceps Femoris M., Short Head

Lateral Femoral Intermuscular Septum

Sciatic Nerve

Biceps Femoris Muscle, Long Head

Fascia Between Adductor and Hamstring Muscles

Semitendinosus Muscle

Adductor Magnus Muscle

Semimembranosus Muscle

Gracilis Muscle

Sartorius Muscle

Subsartorial Fascia

Anterior Femoral Vein

Femoral Artery

Vastus Medialis Muscle

Femoral Vein

Rectus Femoris Muscle

Adductor Longus Muscle

Vastus Intermedius M.

Vastus Lateralis M.

Femur

Medial Femoral Intermuscular Septum

Iliotibial Tract

Deep Femoral Vein

Femur, Linea Aspera

Fascia Lata

Biceps Femoris M., Short Head

Lateral Femoral Intermuscular Septum

Sciatic Nerve

Biceps Femoris Muscle, Long Head

Adductor Magnus Muscle

Semitendinosus Muscle

Semimembranosus Muscle

Gracilis Muscle

Accessory Saphenous Vein

right

left

posterior

188

a_vm2125

a_vm2126

a_vm2127

a_vm2128

a_vm2129

a_vm2130

a_vm2131

a_vm2132

a_vm2133

a_vm2134

a_vm2135

a_vm2136

anterior

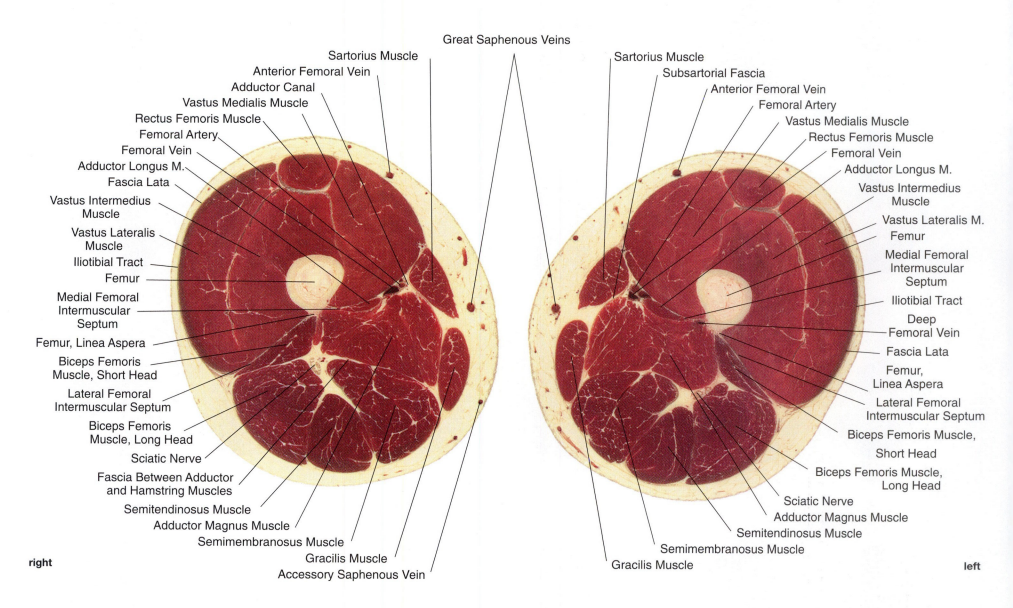

Great Saphenous Veins

Sartorius Muscle
Anterior Femoral Vein
Adductor Canal
Vastus Medialis Muscle
Rectus Femoris Muscle
Femoral Artery
Femoral Vein
Adductor Longus M.
Fascia Lata
Vastus Intermedius Muscle
Vastus Lateralis Muscle
Iliotibial Tract
Femur
Medial Femoral Intermuscular Septum
Femur, Linea Aspera
Biceps Femoris Muscle, Short Head
Lateral Femoral Intermuscular Septum
Biceps Femoris Muscle, Long Head
Sciatic Nerve
Fascia Between Adductor and Hamstring Muscles
Semitendinosus Muscle
Adductor Magnus Muscle
Semimembranosus Muscle
Gracilis Muscle
Accessory Saphenous Vein

Sartorius Muscle
Subsartorial Fascia
Anterior Femoral Vein
Femoral Artery
Vastus Medialis Muscle
Rectus Femoris Muscle
Femoral Vein
Adductor Longus M.
Vastus Intermedius Muscle
Vastus Lateralis M.
Femur
Medial Femoral Intermuscular Septum
Iliotibial Tract
Deep Femoral Vein
Fascia Lata
Femur, Linea Aspera
Lateral Femoral Intermuscular Septum
Biceps Femoris Muscle, Short Head
Biceps Femoris Muscle, Long Head
Sciatic Nerve
Adductor Magnus Muscle
Semitendinosus Muscle
Semimembranosus Muscle
Gracilis Muscle

right

left

posterior

190

a_vm2137

a_vm2138

a_vm2139

a_vm2140

a_vm2141

a_vm2142

a_vm2143

a_vm2144

a_vm2145

a_vm2146

a_vm2147

a_vm2148

anterior

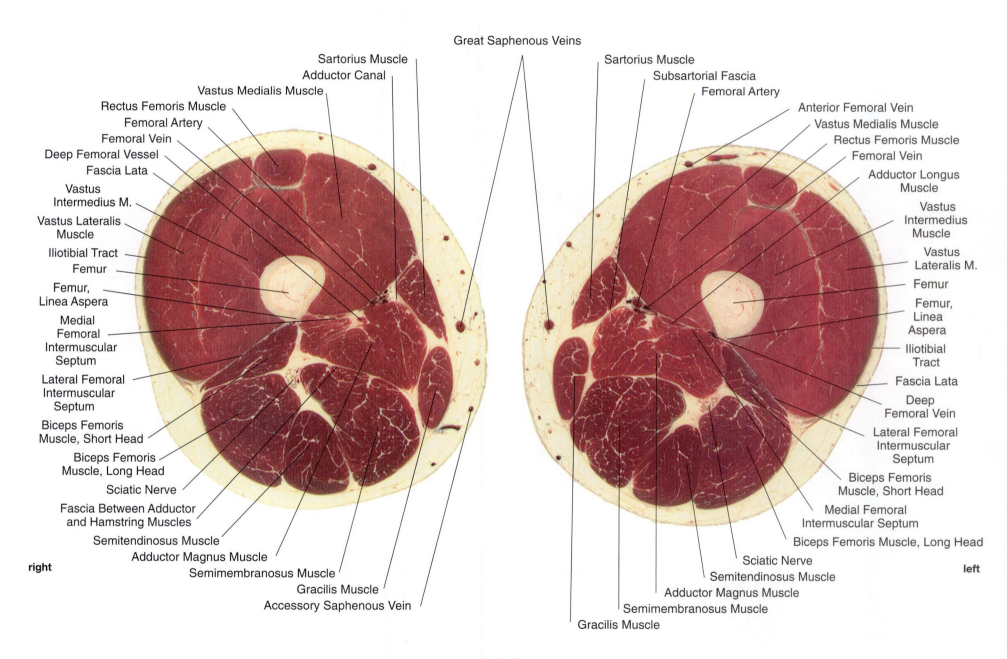

Great Saphenous Veins

Sartorius Muscle
Adductor Canal
Vastus Medialis Muscle
Rectus Femoris Muscle
Femoral Artery
Femoral Vein
Deep Femoral Vessel
Fascia Lata
Vastus Intermedius M.
Vastus Lateralis Muscle
Iliotibial Tract
Femur
Femur, Linea Aspera
Medial Femoral Intermuscular Septum
Lateral Femoral Intermuscular Septum
Biceps Femoris Muscle, Short Head
Biceps Femoris Muscle, Long Head
Sciatic Nerve
Fascia Between Adductor and Hamstring Muscles
Semitendinosus Muscle
Adductor Magnus Muscle
Semimembranosus Muscle
Gracilis Muscle
Accessory Saphenous Vein

Sartorius Muscle
Subsartorial Fascia
Femoral Artery
Anterior Femoral Vein
Vastus Medialis Muscle
Rectus Femoris Muscle
Femoral Vein
Adductor Longus Muscle
Vastus Intermedius Muscle
Vastus Lateralis M.
Femur
Femur, Linea Aspera
Iliotibial Tract
Fascia Lata
Deep Femoral Vein
Lateral Femoral Intermuscular Septum
Biceps Femoris Muscle, Short Head
Medial Femoral Intermuscular Septum
Biceps Femoris Muscle, Long Head
Sciatic Nerve
Semitendinosus Muscle
Adductor Magnus Muscle
Semimembranosus Muscle
Gracilis Muscle

right

left

posterior

192

a_vm2149

a_vm2150

a_vm2151

a_vm2152

a_vm2153

a_vm2154

a_vm2155

a_vm2156

a_vm2157

a_vm2158

a_vm2159

a_vm2160

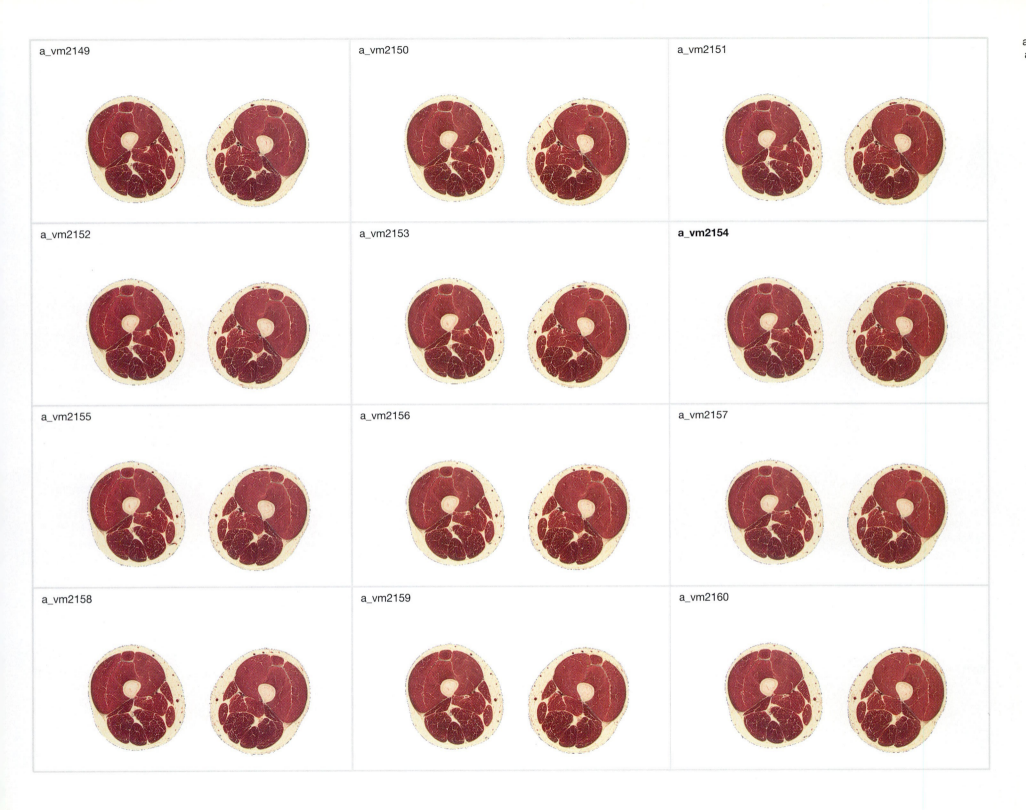

anterior

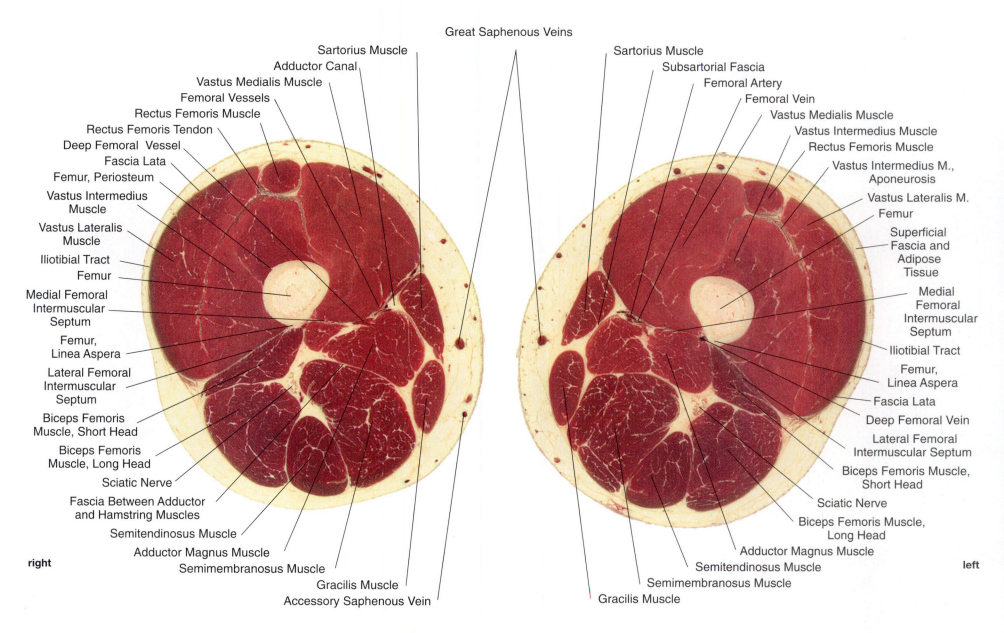

Great Saphenous Veins

Sartorius Muscle
Adductor Canal
Vastus Medialis Muscle
Femoral Vessels
Rectus Femoris Muscle
Rectus Femoris Tendon
Deep Femoral Vessel
Fascia Lata
Femur, Periosteum
Vastus Intermedius
Muscle
Vastus Lateralis
Muscle
Iliotibial Tract
Femur
Medial Femoral
Intermuscular
Septum
Femur,
Linea Aspera
Lateral Femoral
Intermuscular
Septum
Biceps Femoris
Muscle, Short Head
Biceps Femoris
Muscle, Long Head
Sciatic Nerve
Fascia Between Adductor
and Hamstring Muscles
Semitendinosus Muscle
Adductor Magnus Muscle
Semimembranosus Muscle

Gracilis Muscle
Accessory Saphenous Vein

Sartorius Muscle
Subsartorial Fascia
Femoral Artery
Femoral Vein
Vastus Medialis Muscle
Vastus Intermedius Muscle
Rectus Femoris Muscle
Vastus Intermedius M.,
Aponeurosis
Vastus Lateralis M.
Femur
Superficial
Fascia and
Adipose
Tissue
Medial
Femoral
Intermuscular
Septum
Iliotibial Tract
Femur,
Linea Aspera
Fascia Lata
Deep Femoral Vein
Lateral Femoral
Intermuscular Septum
Biceps Femoris Muscle,
Short Head
Sciatic Nerve
Biceps Femoris Muscle,
Long Head
Adductor Magnus Muscle
Semitendinosus Muscle
Semimembranosus Muscle
Gracilis Muscle

right

left

posterior

a_vm2161

a_vm2162

a_vm2163

a_vm2164

a_vm2165

a_vm2166

a_vm2167

a_vm2168

a_vm2169

a_vm2170

a_vm2171

a_vm2172

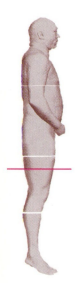

anterior

Great Saphenous Veins

Sartorius Muscle
Adductor Canal
Vastus Medialis Muscle
Femoral Vessels
Rectus Femoris Muscle
Rectus Femoris Tendon
Femur, Periosteum
Fascia Lata
Vastus Intermedius M.
Vastus Lateralis M.
Superficial Fascia
and Adipose Tissue
Iliotibial Tract
Femur
Medial Femoral
Intermuscular
Septum
Femur,
Linea Aspera
Lateral Femoral
Intermuscular
Septum
Biceps Femoris M.,
Short Head
Long Head
Common Peroneal N.
Tibial Nerve
Fascia Between Adductor
and Hamstring Muscles
Semitendinosus Muscle
Adductor Magnus Muscle
Semimembranosus Muscle
Gracilis Muscle
Skin

Sartorius Muscle
Subsartorial Fascia
Femoral Vessels
Vastus Medialis Muscle
Vastus Intermedius Muscle
Rectus Femoris Muscle and Tendon
Vastus Intermedius
Muscle, Aponeurosis
Vastus Lateralis M.
Superficial Fascia
and Adipose Tissue
Femur
Medial Femoral
Intermuscular
Septum
Deep Femoral
Vein
Iliotibial Tract
Fascia Lata
Femur,
Linea Aspera
Lateral Femoral
Intermuscular
Septum
Biceps Femoris M.,
Short Head
Adductor Magnus M.
Sciatic Nerve
Biceps Femoris Muscle,
Long Head
Semitendinosus Muscle
Semimembranosus Muscle
Gracilis Muscle

Accessory Saphenous Veins

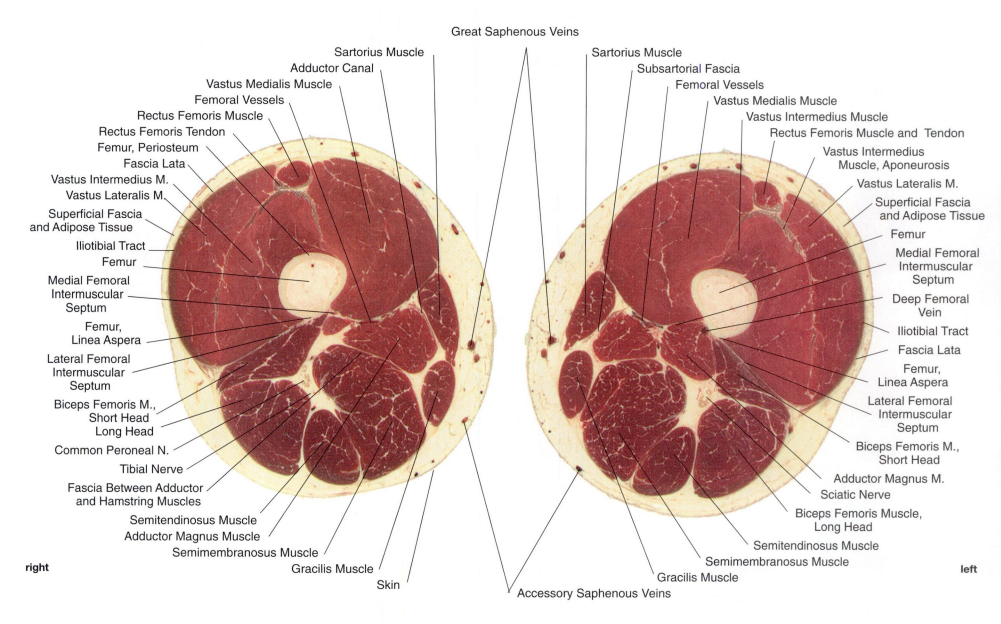

right

left

posterior

a_vm2173

a_vm2174

a_vm2175

a_vm2176

a_vm2177

a_vm2178

a_vm2179

a_vm2180

a_vm2181

a_vm2182

a_vm2183

a_vm2184

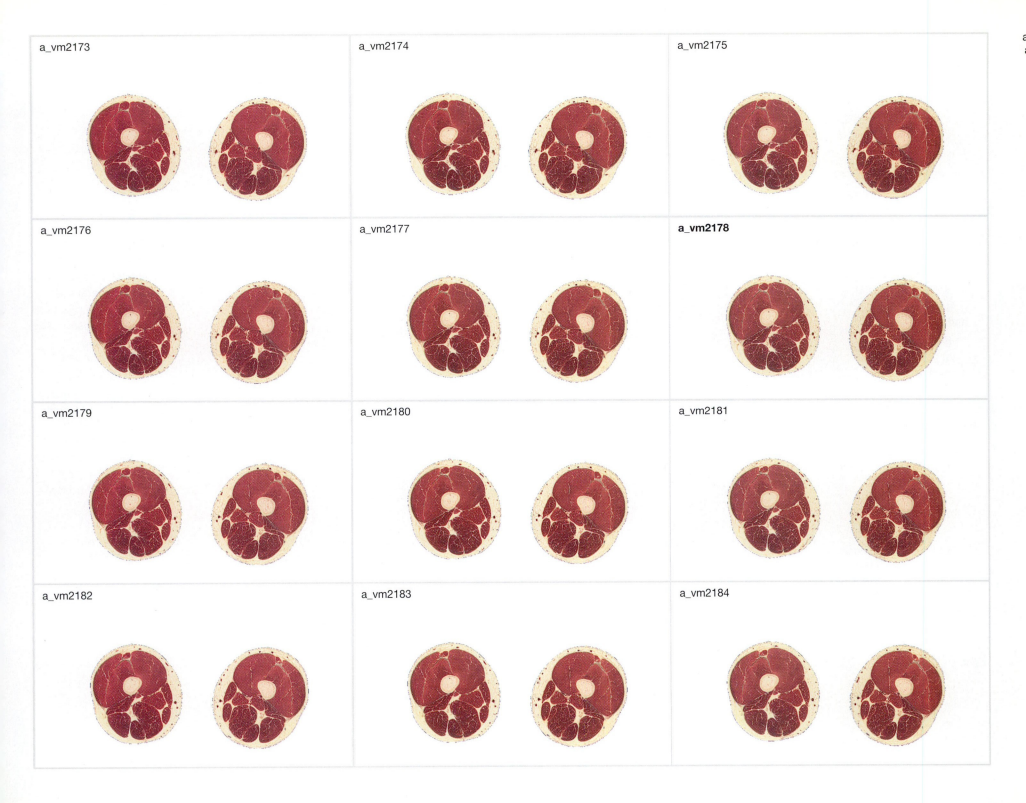

anterior

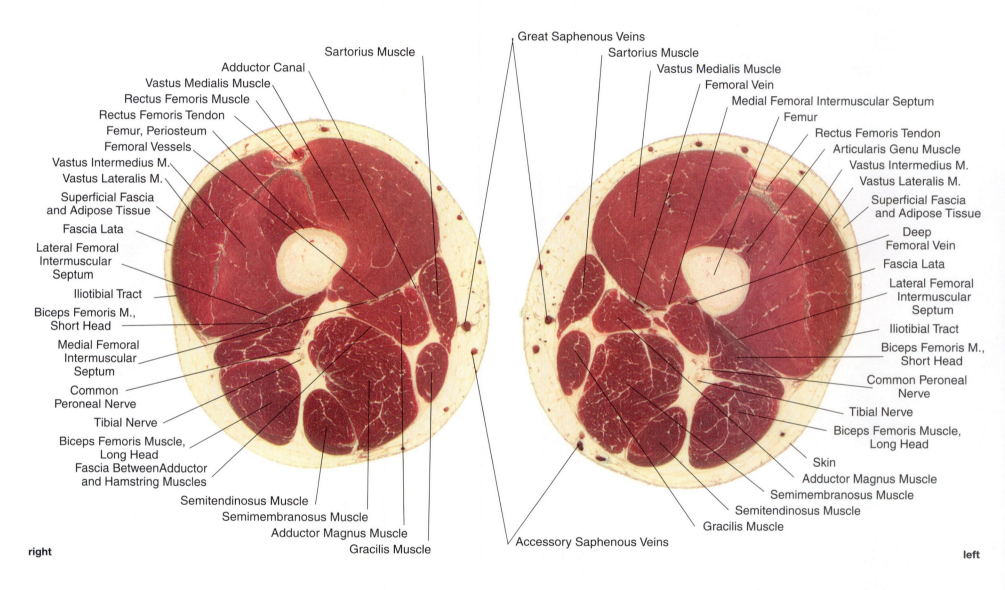

Great Saphenous Veins

Sartorius Muscle

Sartorius Muscle

Adductor Canal

Vastus Medialis Muscle

Vastus Medialis Muscle

Rectus Femoris Muscle

Femoral Vein

Rectus Femoris Tendon

Medial Femoral Intermuscular Septum

Femur, Periosteum

Femur

Femoral Vessels

Rectus Femoris Tendon

Vastus Intermedius M.

Articularis Genu Muscle

Vastus Lateralis M.

Vastus Intermedius M.

Superficial Fascia
and Adipose Tissue

Vastus Lateralis M.

Superficial Fascia
and Adipose Tissue

Fascia Lata

Deep
Femoral Vein

Lateral Femoral
Intermuscular
Septum

Fascia Lata

Iliotibial Tract

Lateral Femoral
Intermuscular
Septum

Biceps Femoris M.,
Short Head

Iliotibial Tract

Medial Femoral
Intermuscular
Septum

Biceps Femoris M.,
Short Head

Common Peroneal
Nerve

Common
Peroneal Nerve

Tibial Nerve

Tibial Nerve

Biceps Femoris Muscle,
Long Head

Biceps Femoris Muscle,
Long Head

Fascia BetweenAdductor
and Hamstring Muscles

Skin

Adductor Magnus Muscle

Semitendinosus Muscle

Semimembranosus Muscle

Semimembranosus Muscle

Semitendinosus Muscle

Adductor Magnus Muscle

Gracilis Muscle

Gracilis Muscle

Accessory Saphenous Veins

right

left

posterior

198

a_vm2185

a_vm2186

a_vm2187

a_vm2188

a_vm2189

a_vm2190

a_vm2191

a_vm2192

a_vm2193

a_vm2194

a_vm2195

a_vm2196

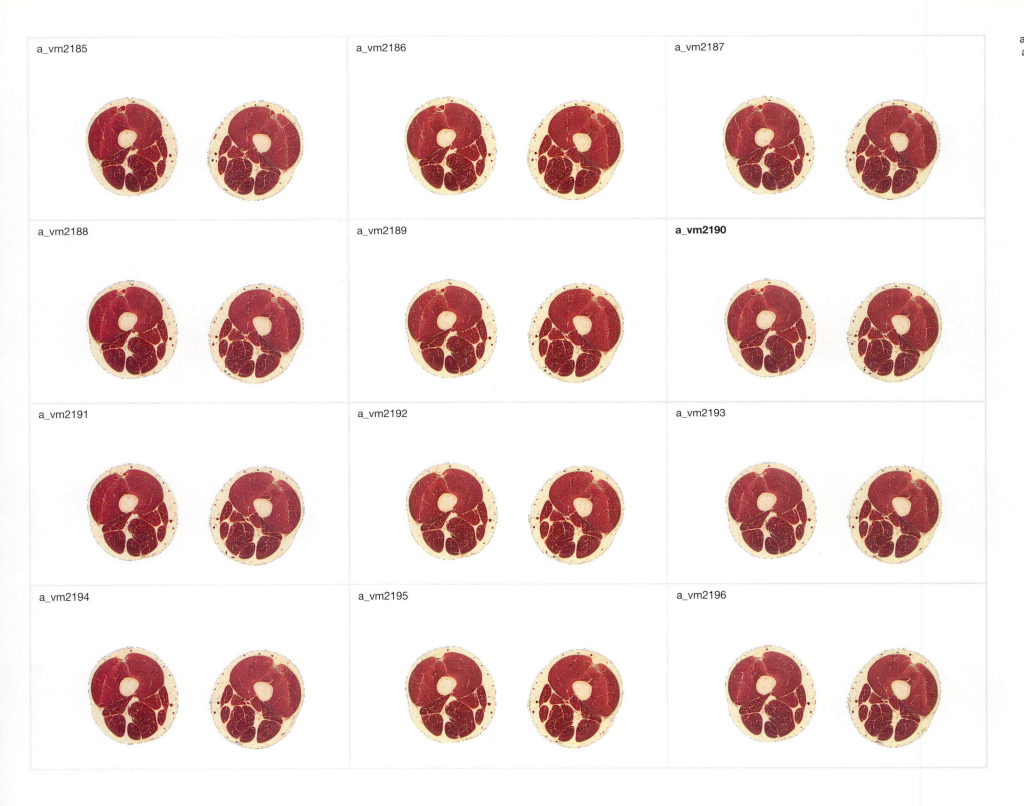

anterior

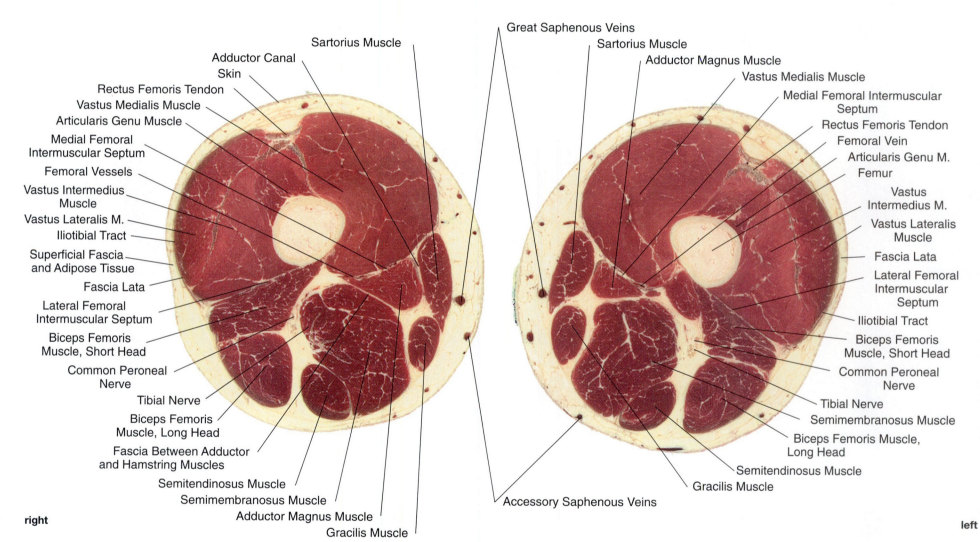

Great Saphenous Veins

Sartorius Muscle

Sartorius Muscle

Adductor Magnus Muscle

Adductor Canal

Vastus Medialis Muscle

Skin

Medial Femoral Intermuscular
Septum

Rectus Femoris Tendon

Vastus Medialis Muscle

Rectus Femoris Tendon

Articularis Genu Muscle

Femoral Vein

Medial Femoral
Intermuscular Septum

Articularis Genu M.

Femur

Femoral Vessels

Vastus
Intermedius M.

Vastus Intermedius
Muscle

Vastus Lateralis
Muscle

Vastus Lateralis M.

Fascia Lata

Iliotibial Tract

Lateral Femoral
Intermuscular
Septum

Superficial Fascia
and Adipose Tissue

Iliotibial Tract

Fascia Lata

Biceps Femoris
Muscle, Short Head

Lateral Femoral
Intermuscular Septum

Common Peroneal
Nerve

Biceps Femoris
Muscle, Short Head

Tibial Nerve

Common Peroneal
Nerve

Biceps Femoris
Muscle, Long Head

Tibial Nerve

Fascia Between Adductor
and Hamstring Muscles

Semimembranosus Muscle

Biceps Femoris Muscle,
Long Head

Semitendinosus Muscle

Semitendinosus Muscle

Semimembranosus Muscle

Gracilis Muscle

Adductor Magnus Muscle

Accessory Saphenous Veins

Gracilis Muscle

right

left

posterior

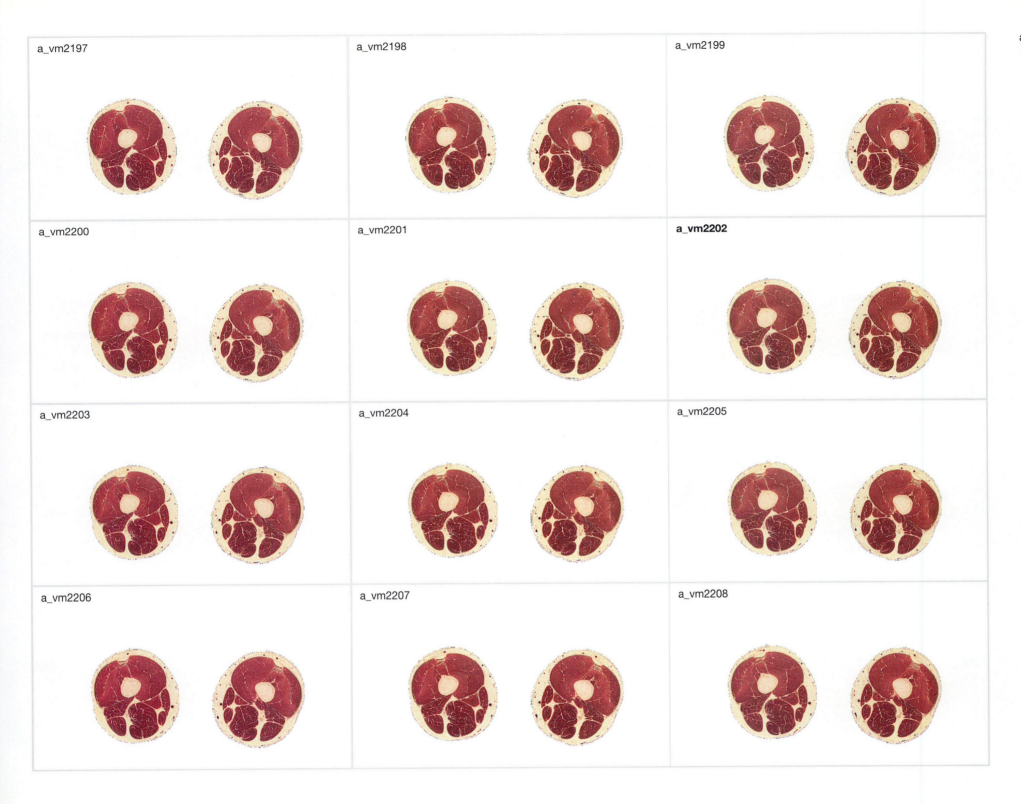

a_vm2197

a_vm2198

a_vm2199

a_vm2200

a_vm2201

a_vm2202

a_vm2203

a_vm2204

a_vm2205

a_vm2206

a_vm2207

a_vm2208

anterior

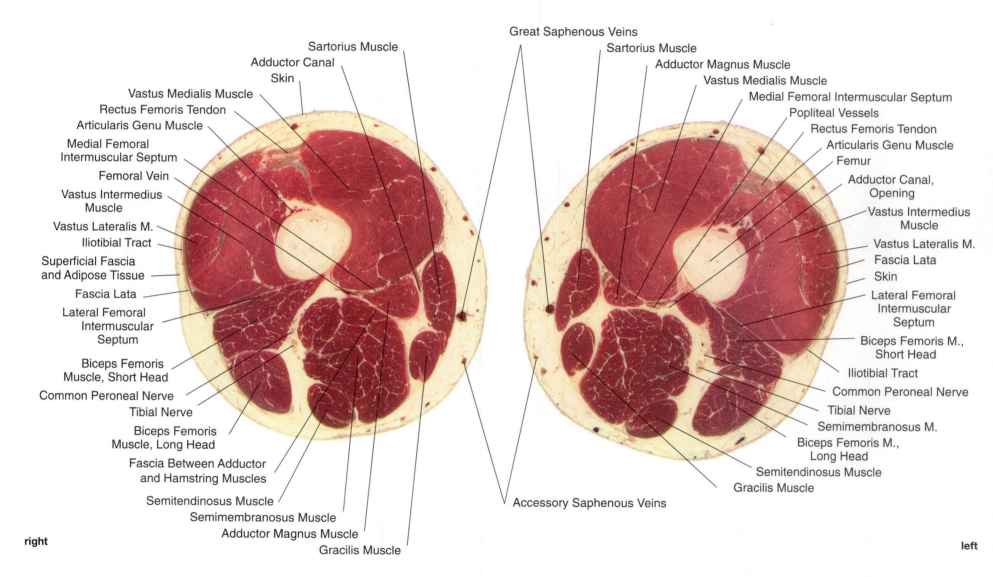

Great Saphenous Veins

Sartorius Muscle

Adductor Canal

Skin

Sartorius Muscle

Adductor Magnus Muscle

Vastus Medialis Muscle

Vastus Medialis Muscle

Rectus Femoris Tendon

Medial Femoral Intermuscular Septum

Articularis Genu Muscle

Popliteal Vessels

Medial Femoral
Intermuscular Septum

Rectus Femoris Tendon

Femoral Vein

Articularis Genu Muscle

Vastus Intermedius
Muscle

Femur

Vastus Lateralis M.

Adductor Canal,
Opening

Iliotibial Tract

Vastus Intermedius
Muscle

Superficial Fascia
and Adipose Tissue

Vastus Lateralis M.

Fascia Lata

Fascia Lata

Skin

Lateral Femoral
Intermuscular
Septum

Lateral Femoral
Intermuscular
Septum

Biceps Femoris
Muscle, Short Head

Biceps Femoris M.,
Short Head

Common Peroneal Nerve

Iliotibial Tract

Tibial Nerve

Common Peroneal Nerve

Biceps Femoris
Muscle, Long Head

Tibial Nerve

Fascia Between Adductor
and Hamstring Muscles

Semimembranosus M.

Semitendinosus Muscle

Biceps Femoris M.,
Long Head

Semimembranosus Muscle

Semitendinosus Muscle

Adductor Magnus Muscle

Gracilis Muscle

Gracilis Muscle

Accessory Saphenous Veins

right

left

posterior

a_vm2209

a_vm2210

a_vm2211

a_vm2212

a_vm2213

a_vm2214

a_vm2215

a_vm2216

a_vm2217

a_vm2218

a_vm2219

a_vm2220

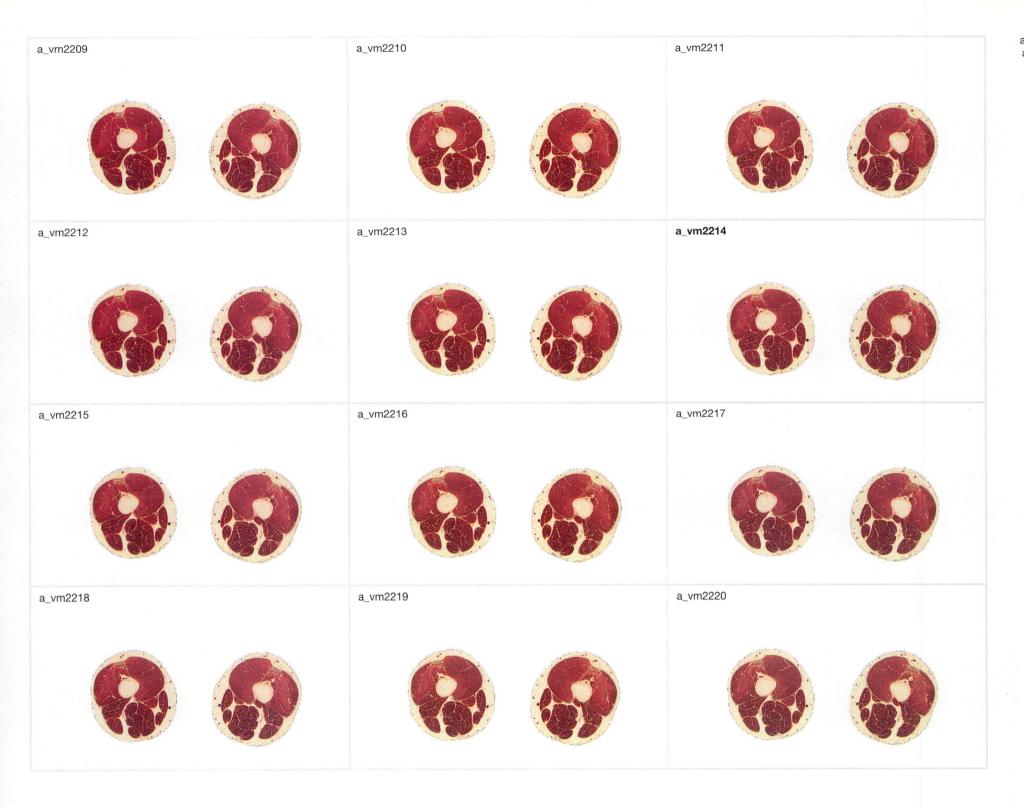

anterior

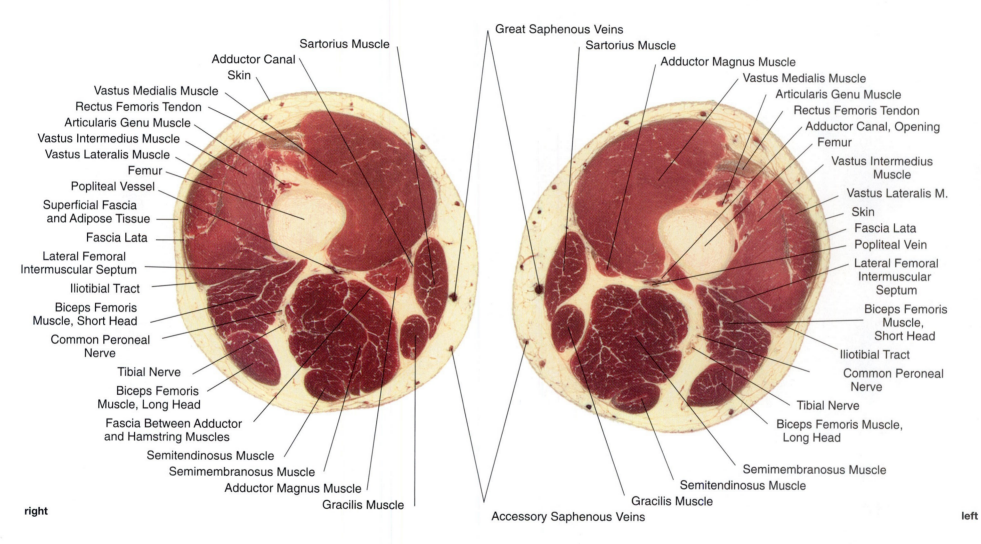

Great Saphenous Veins

Sartorius Muscle

Adductor Magnus Muscle

Vastus Medialis Muscle

Articularis Genu Muscle

Rectus Femoris Tendon

Adductor Canal, Opening

Femur

Vastus Intermedius
Muscle

Vastus Lateralis M.

Skin

Fascia Lata

Popliteal Vein

Lateral Femoral
Intermuscular
Septum

Biceps Femoris
Muscle,
Short Head

Iliotibial Tract

Common Peroneal
Nerve

Tibial Nerve

Biceps Femoris Muscle,
Long Head

Semimembranosus Muscle

Semitendinosus Muscle

Gracilis Muscle

Accessory Saphenous Veins

Sartorius Muscle

Adductor Canal

Skin

Vastus Medialis Muscle

Rectus Femoris Tendon

Articularis Genu Muscle

Vastus Intermedius Muscle

Vastus Lateralis Muscle

Femur

Popliteal Vessel

Superficial Fascia
and Adipose Tissue

Fascia Lata

Lateral Femoral
Intermuscular Septum

Iliotibial Tract

Biceps Femoris
Muscle, Short Head

Common Peroneal
Nerve

Tibial Nerve

Biceps Femoris
Muscle, Long Head

Fascia Between Adductor
and Hamstring Muscles

Semitendinosus Muscle

Semimembranosus Muscle

Adductor Magnus Muscle

Gracilis Muscle

right

left

posterior

204

a_vm2221

a_vm2222

a_vm2223

a_vm2224

a_vm2225

a_vm2226

a_vm2227

a_vm2228

a_vm2229

a_vm2230

a_vm2231

a_vm2232

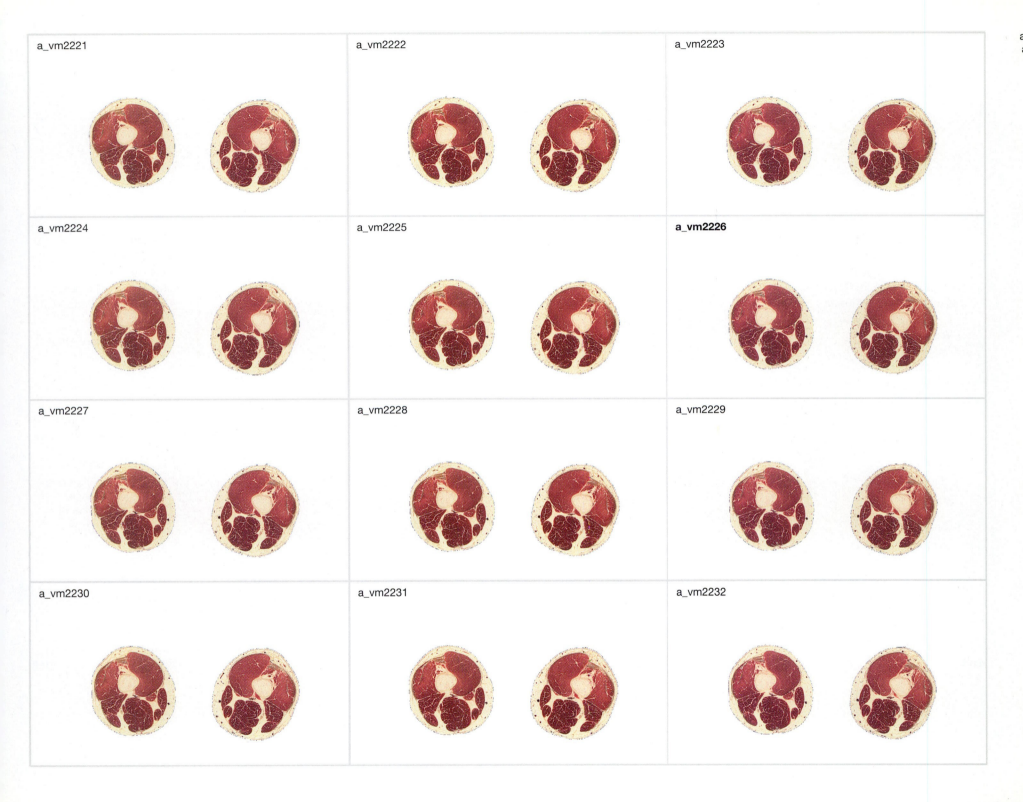

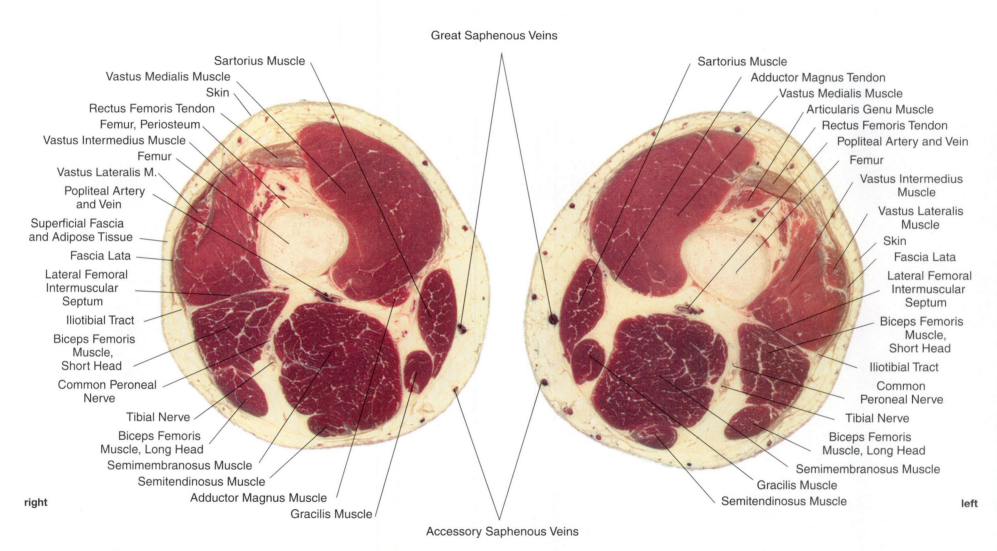

anterior

Great Saphenous Veins

Sartorius Muscle

Sartorius Muscle

Vastus Medialis Muscle

Adductor Magnus Tendon

Skin

Vastus Medialis Muscle

Rectus Femoris Tendon

Articularis Genu Muscle

Femur, Periosteum

Rectus Femoris Tendon

Vastus Intermedius Muscle

Popliteal Artery and Vein

Femur

Femur

Vastus Lateralis M.

Vastus Intermedius
Muscle

Popliteal Artery
and Vein

Vastus Lateralis
Muscle

Superficial Fascia
and Adipose Tissue

Skin

Fascia Lata

Fascia Lata

Lateral Femoral
Intermuscular
Septum

Lateral Femoral
Intermuscular
Septum

Iliotibial Tract

Biceps Femoris
Muscle,
Short Head

Biceps Femoris
Muscle,
Short Head

Common Peroneal
Nerve

Iliotibial Tract

Tibial Nerve

Common
Peroneal Nerve

Biceps Femoris
Muscle, Long Head

Tibial Nerve

Semimembranosus Muscle

Biceps Femoris
Muscle, Long Head

Semitendinosus Muscle

Semimembranosus Muscle

Adductor Magnus Muscle

Gracilis Muscle

Gracilis Muscle

Semitendinosus Muscle

Accessory Saphenous Veins

right

left

posterior

a_vm2233

a_vm2234

a_vm2235

a_vm2236

a_vm2237

a_vm2238

a_vm2239

a_vm2240

a_vm2241

a_vm2242

a_vm2243

a_vm2244

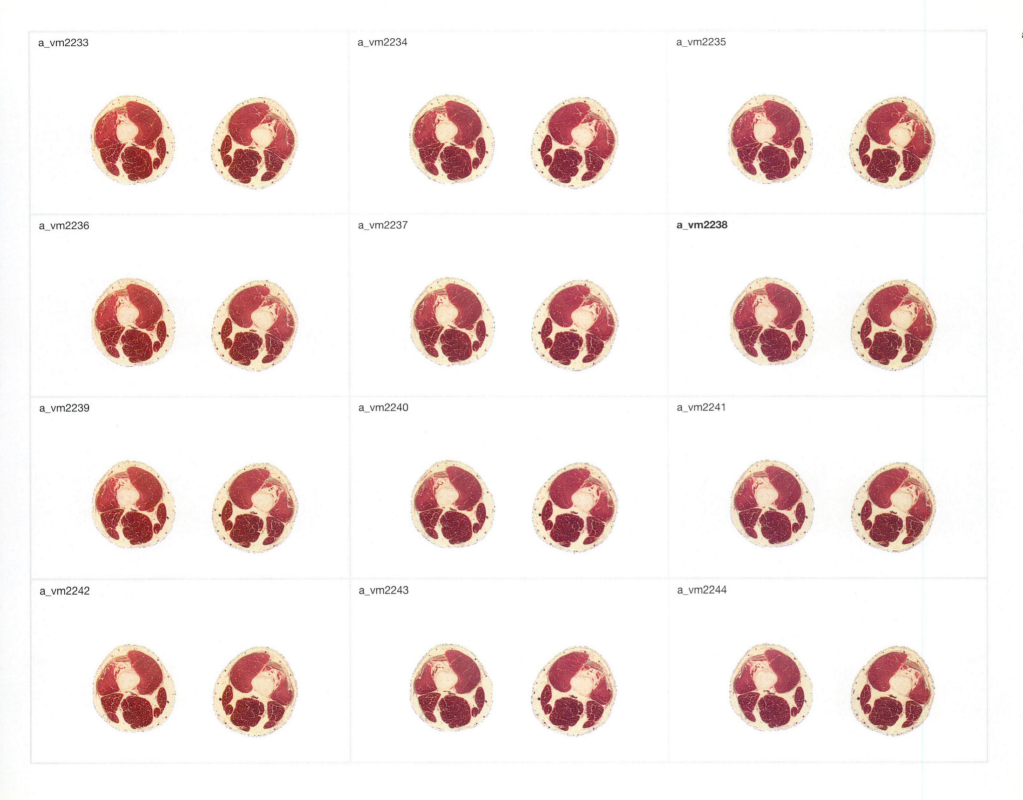

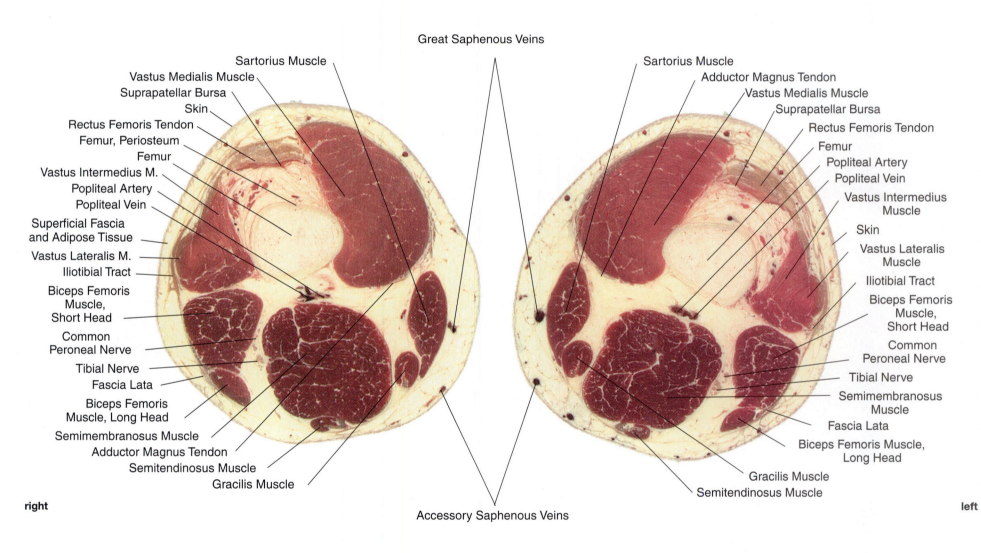

anterior

Great Saphenous Veins

Sartorius Muscle

Vastus Medialis Muscle

Suprapatellar Bursa

Skin

Rectus Femoris Tendon

Femur, Periosteum

Femur

Vastus Intermedius M.

Popliteal Artery

Popliteal Vein

Superficial Fascia
and Adipose Tissue

Vastus Lateralis M.

Iliotibial Tract

Biceps Femoris
Muscle,
Short Head

Common
Peroneal Nerve

Tibial Nerve

Fascia Lata

Biceps Femoris
Muscle, Long Head

Semimembranosus Muscle

Adductor Magnus Tendon

Semitendinosus Muscle

Gracilis Muscle

Sartorius Muscle

Adductor Magnus Tendon

Vastus Medialis Muscle

Suprapatellar Bursa

Rectus Femoris Tendon

Femur

Popliteal Artery

Popliteal Vein

Vastus Intermedius
Muscle

Skin

Vastus Lateralis
Muscle

Iliotibial Tract

Biceps Femoris
Muscle,
Short Head

Common
Peroneal Nerve

Tibial Nerve

Semimembranosus
Muscle

Fascia Lata

Biceps Femoris Muscle,
Long Head

Gracilis Muscle

Semitendinosus Muscle

Accessory Saphenous Veins

right

left

posterior

208

a_vm2245

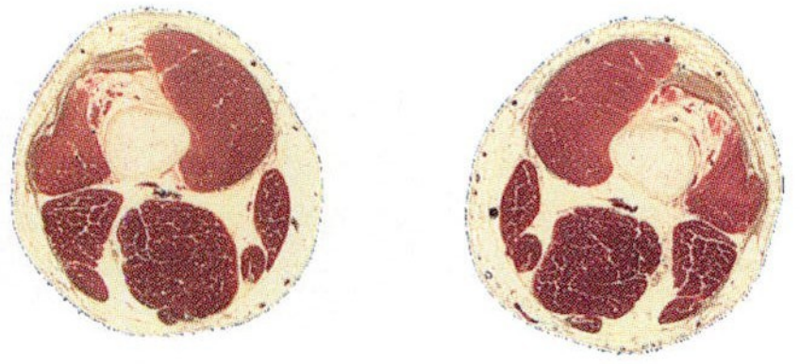

a_vm2246

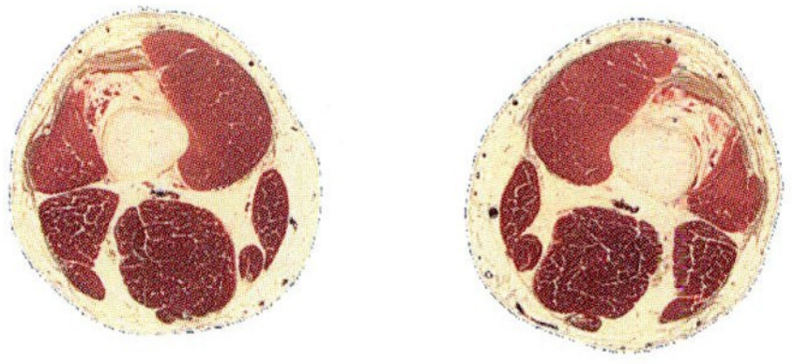

a_vm2247

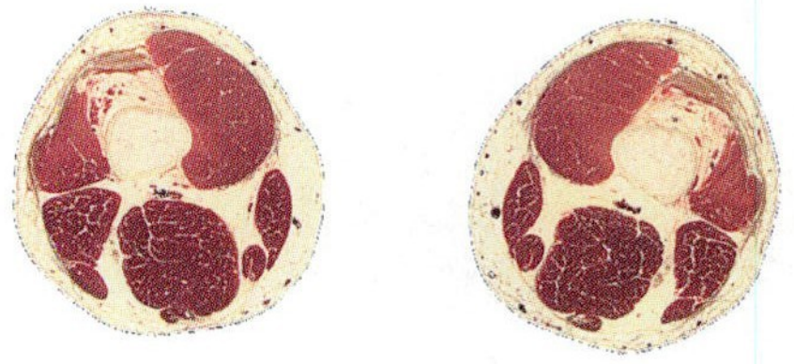

a_vm2248

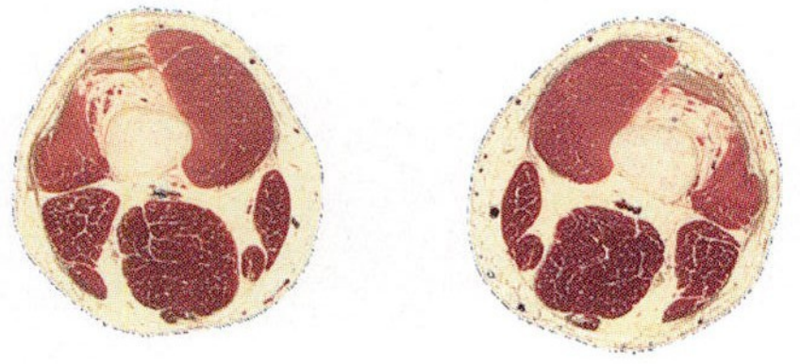

a_vm2249

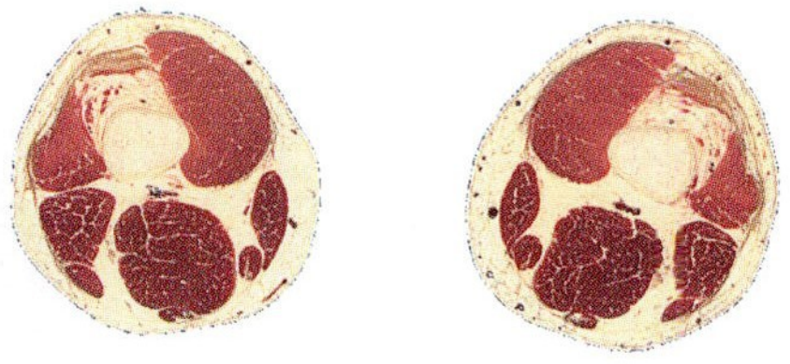

a_vm2250

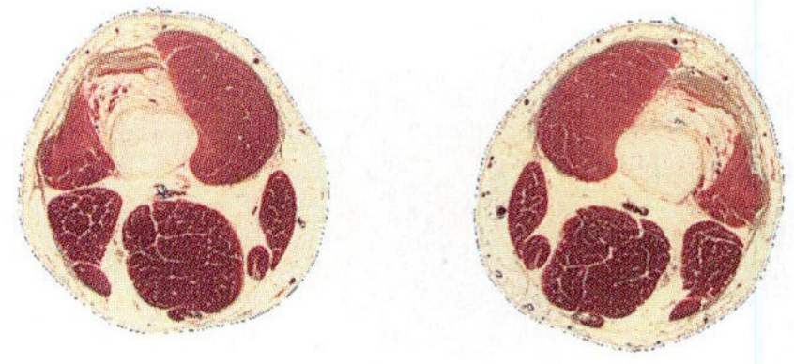

a_vm2251

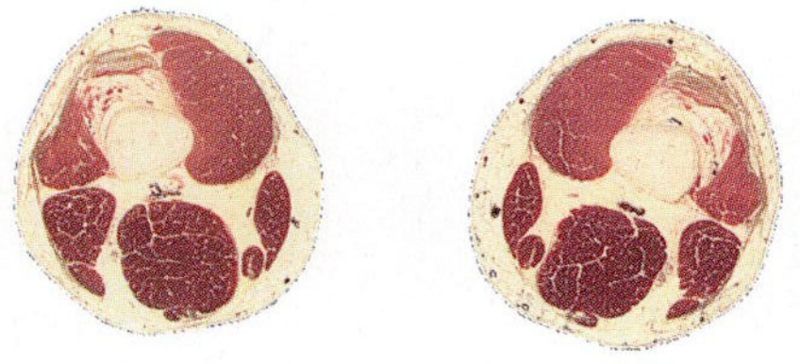

a_vm2252

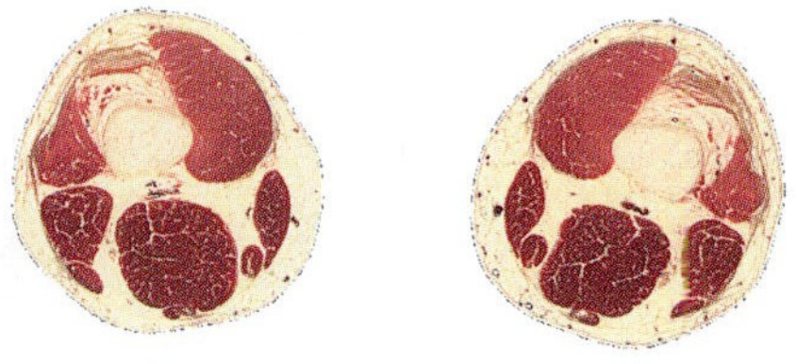

a_vm2253

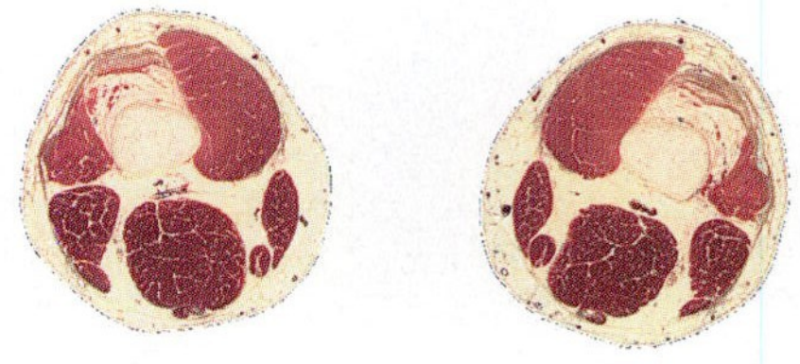

a_vm2254

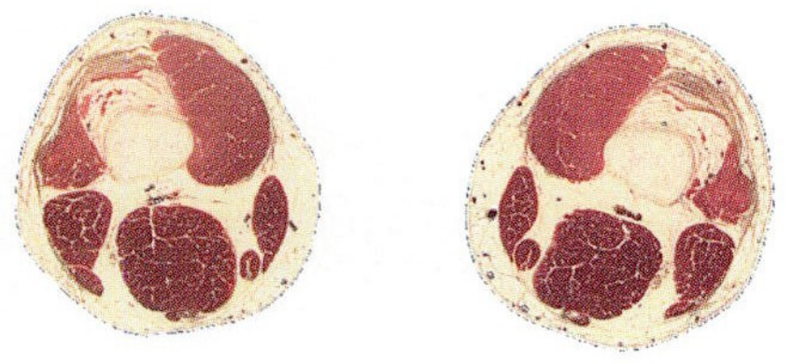

a_vm2255

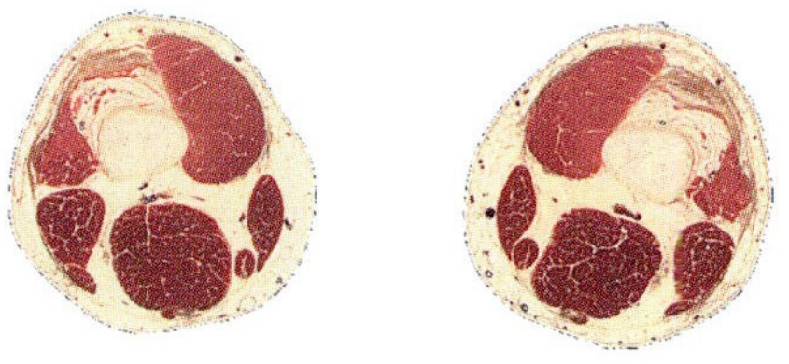

a_vm2256

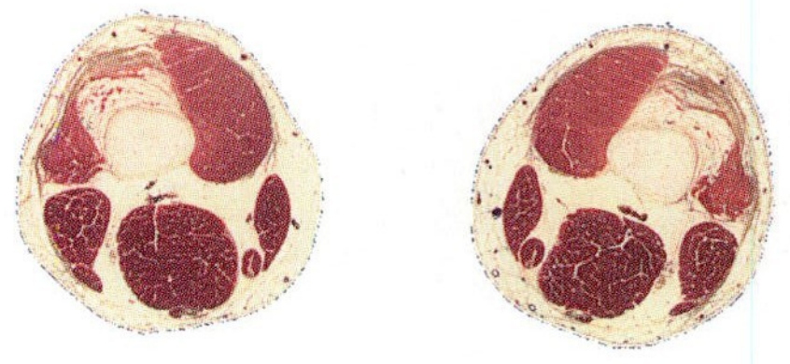

anterior

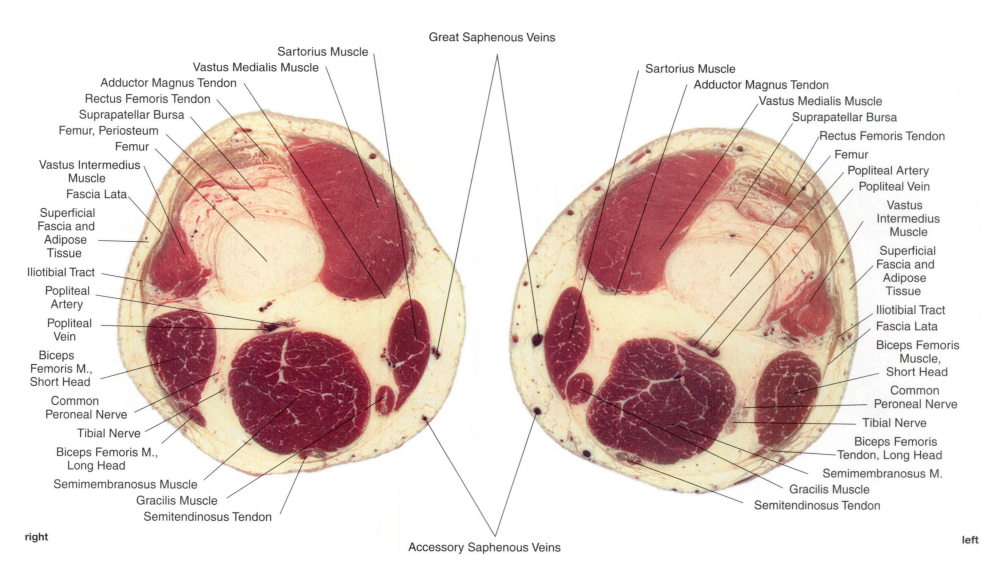

Great Saphenous Veins

Sartorius Muscle

Vastus Medialis Muscle

Sartorius Muscle

Adductor Magnus Tendon

Adductor Magnus Tendon

Rectus Femoris Tendon

Vastus Medialis Muscle

Suprapatellar Bursa

Suprapatellar Bursa

Femur, Periosteum

Rectus Femoris Tendon

Femur

Femur

Vastus Intermedius
Muscle

Popliteal Artery

Popliteal Vein

Fascia Lata

Vastus
Intermedius
Muscle

Superficial
Fascia and
Adipose
Tissue

Superficial
Fascia and
Adipose
Tissue

Iliotibial Tract

Iliotibial Tract

Popliteal
Artery

Fascia Lata

Popliteal
Vein

Biceps Femoris
Muscle,
Short Head

Biceps
Femoris M.,
Short Head

Common
Peroneal Nerve

Common
Peroneal Nerve

Tibial Nerve

Tibial Nerve

Biceps Femoris M.,
Long Head

Biceps Femoris
Tendon, Long Head

Semimembranosus Muscle

Semimembranosus M.

Gracilis Muscle

Gracilis Muscle

Semitendinosus Tendon

Semitendinosus Tendon

right

left

Accessory Saphenous Veins

posterior

210

a_vm2257

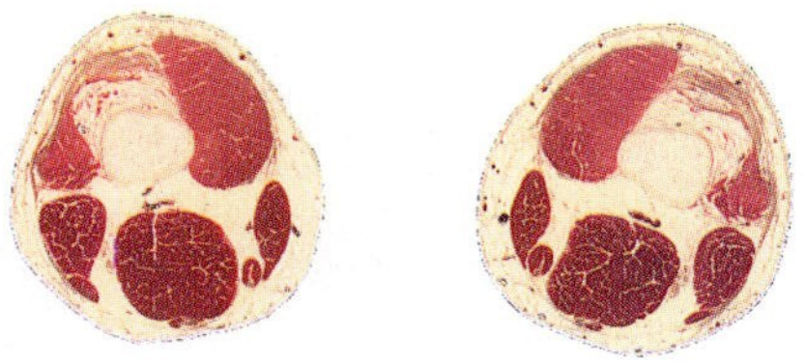

a_vm2258

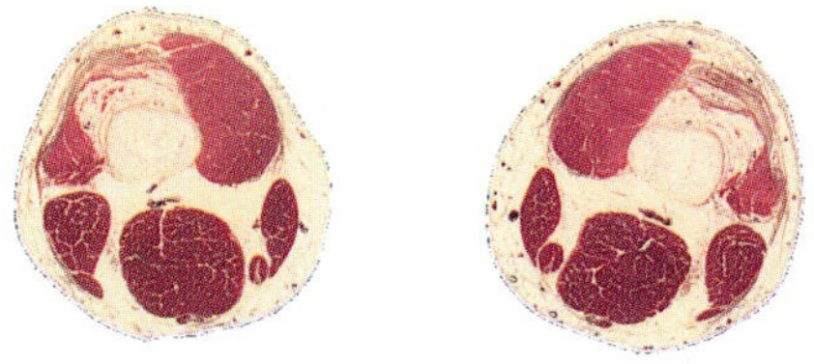

a_vm2259

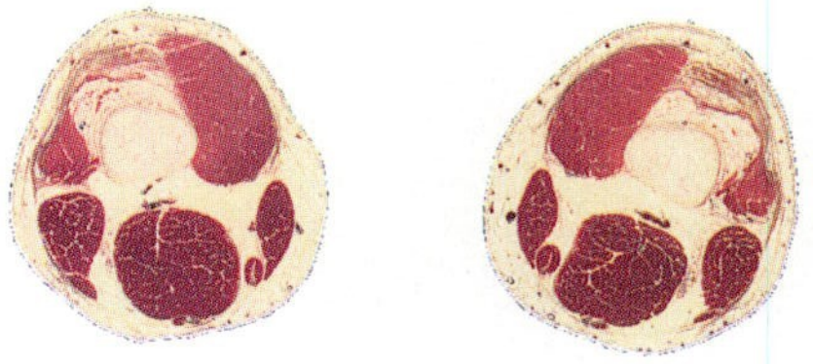

a_vm2260

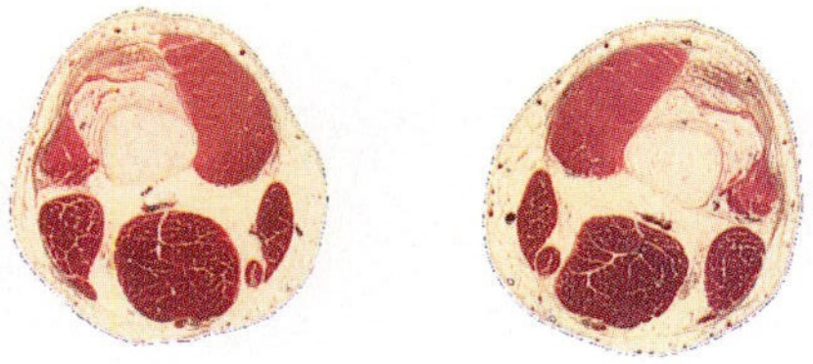

a_vm2261

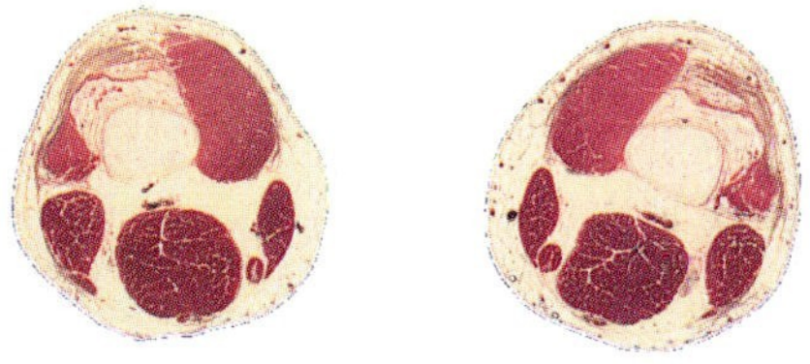

a_vm2262

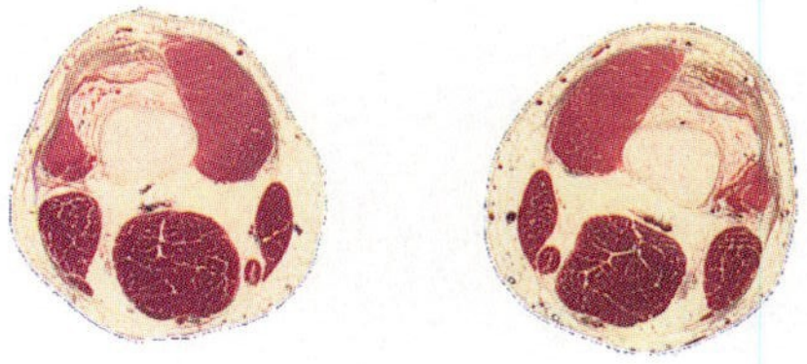

a_vm2263

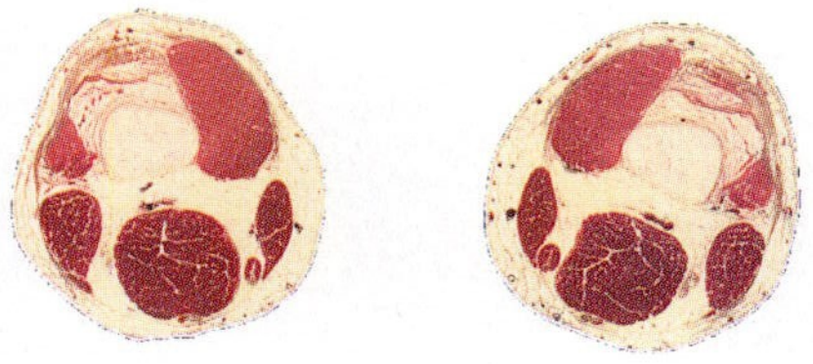

a_vm2264

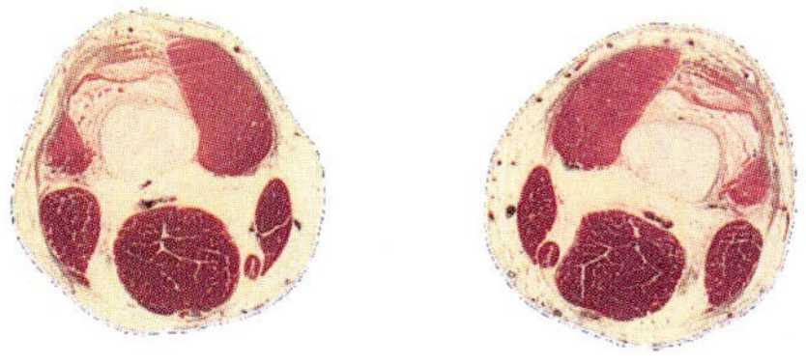

a_vm2265

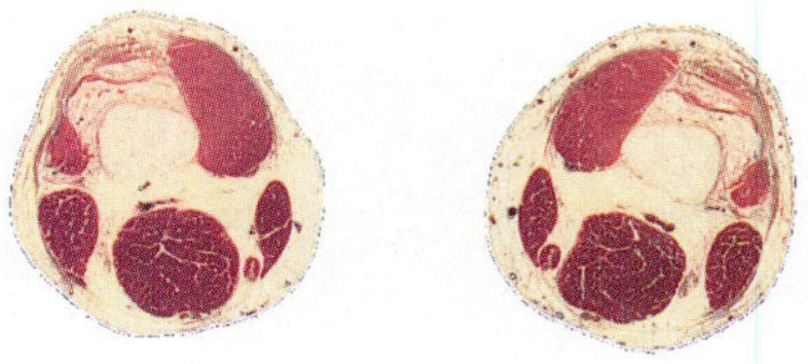

a_vm2266

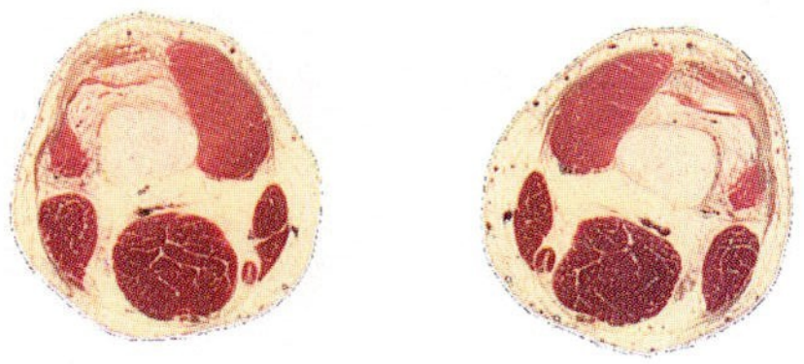

a_vm2267

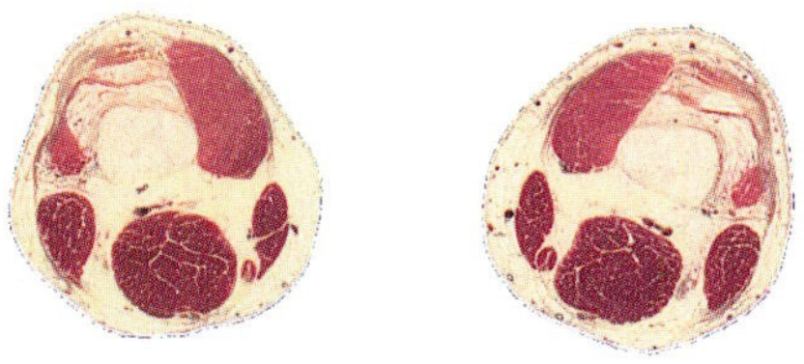

a_vm2268

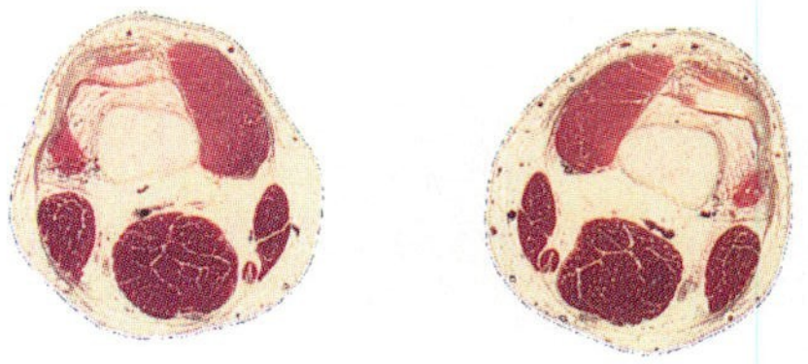

anterior

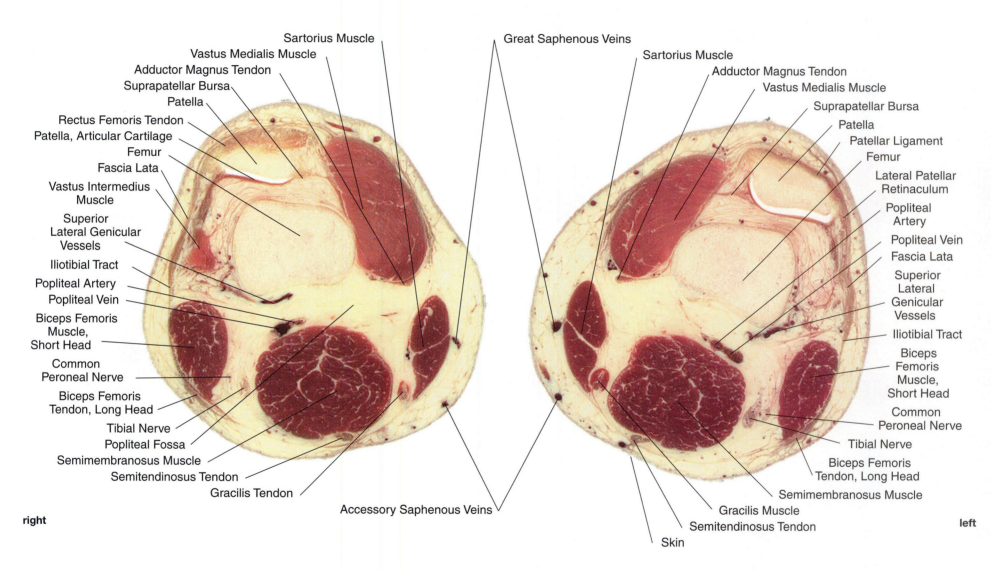

Sartorius Muscle

Vastus Medialis Muscle

Adductor Magnus Tendon

Suprapatellar Bursa

Patella

Rectus Femoris Tendon

Patella, Articular Cartilage

Femur

Fascia Lata

Vastus Intermedius Muscle

Superior Lateral Genicular Vessels

Iliotibial Tract

Popliteal Artery

Popliteal Vein

Biceps Femoris Muscle, Short Head

Common Peroneal Nerve

Biceps Femoris Tendon, Long Head

Tibial Nerve

Popliteal Fossa

Semimembranosus Muscle

Semitendinosus Tendon

Gracilis Tendon

Great Saphenous Veins

Sartorius Muscle

Adductor Magnus Tendon

Vastus Medialis Muscle

Suprapatellar Bursa

Patella

Patellar Ligament

Femur

Lateral Patellar Retinaculum

Popliteal Artery

Popliteal Vein

Fascia Lata

Superior Lateral Genicular Vessels

Iliotibial Tract

Biceps Femoris Muscle, Short Head

Common Peroneal Nerve

Tibial Nerve

Biceps Femoris Tendon, Long Head

Semimembranosus Muscle

Accessory Saphenous Veins

Gracilis Muscle

Semitendinosus Tendon

Skin

right

left

posterior

212

a_vm2269

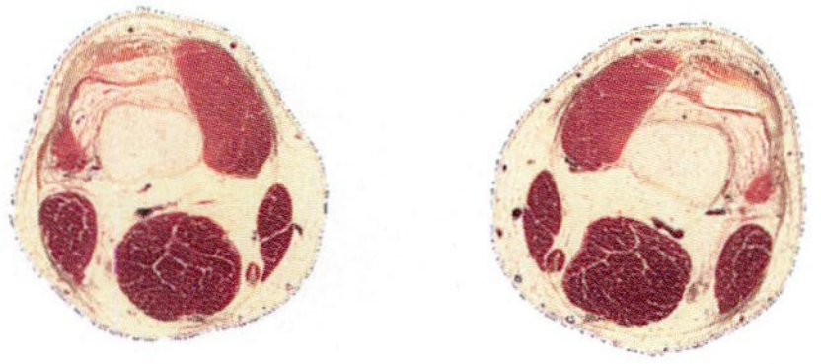

a_vm2270

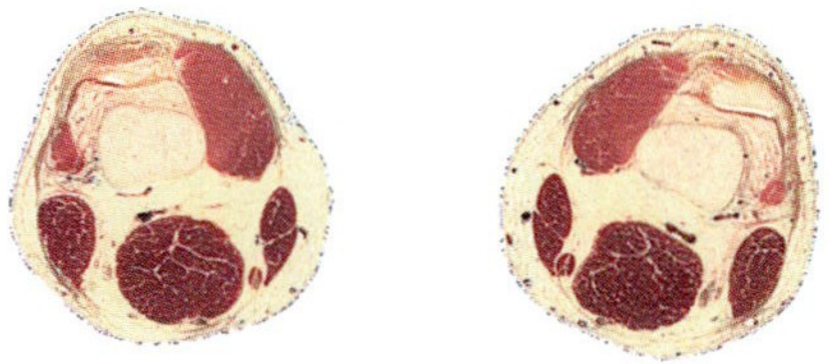

a_vm2271

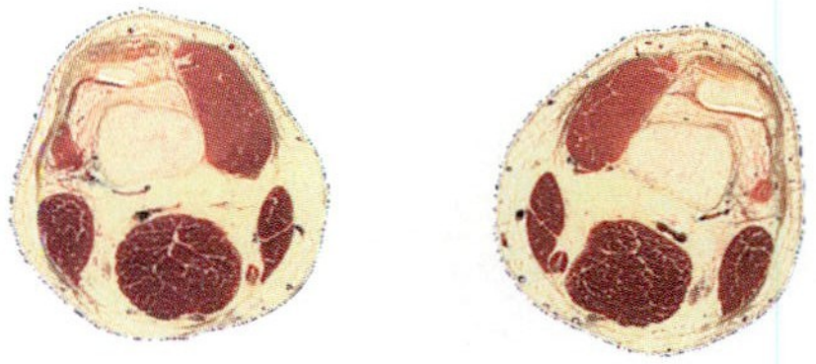

a_vm2272

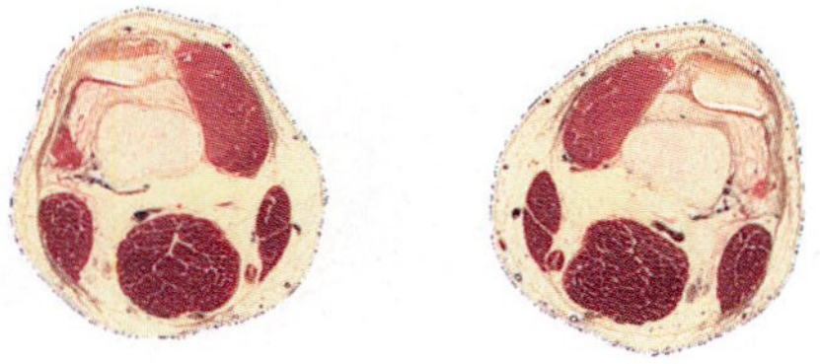

a_vm2273

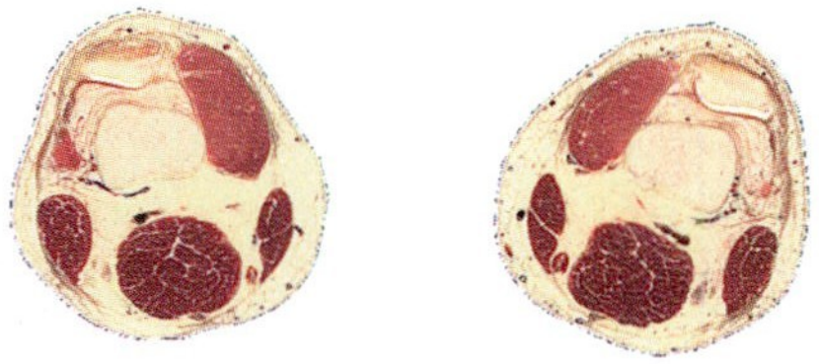

a_vm2274

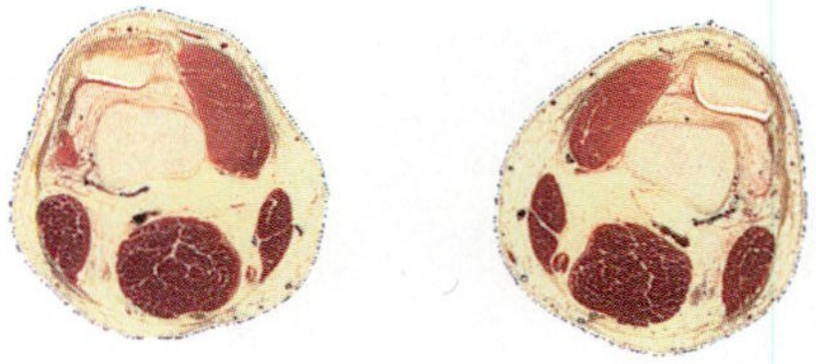

a_vm2275

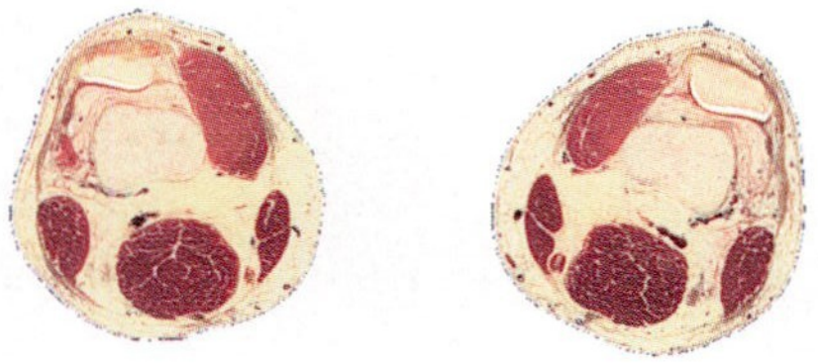

a_vm2276

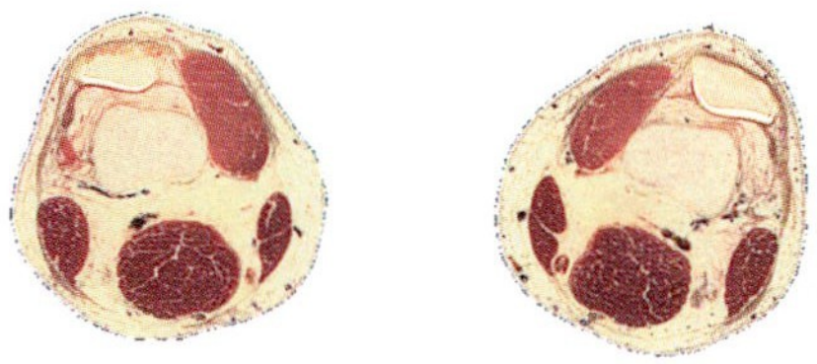

a_vm2277

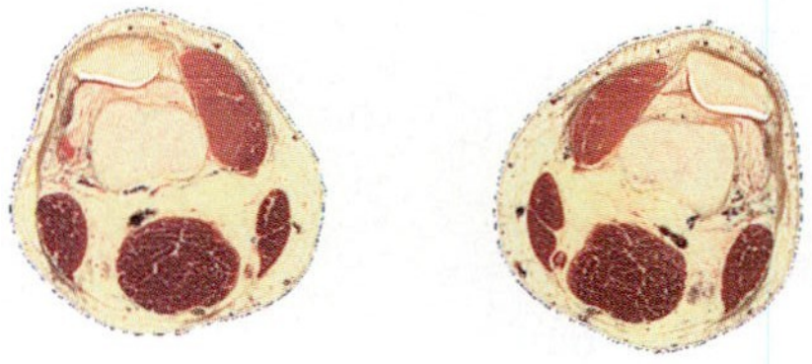

a_vm2278

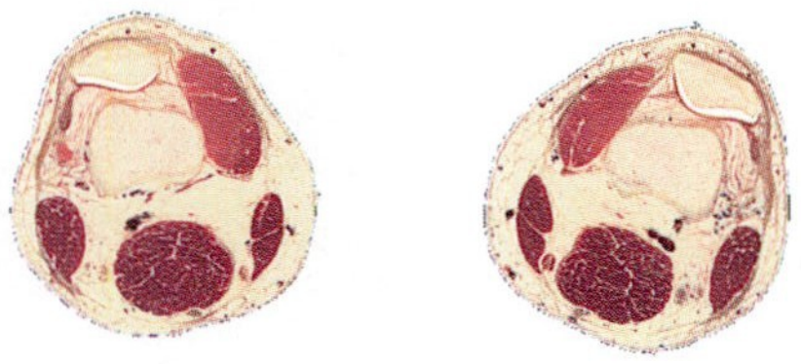

a_vm2279

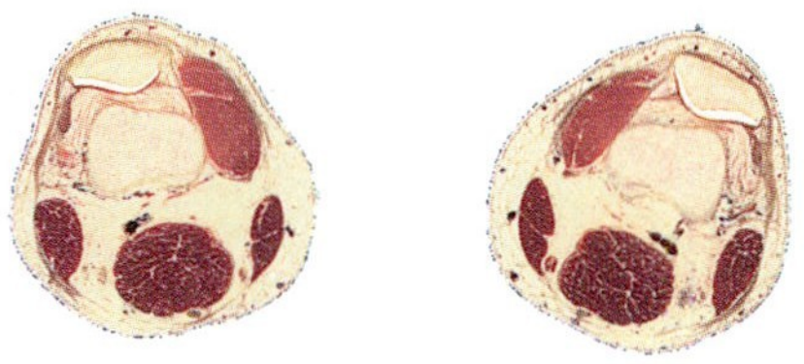

a_vm2280

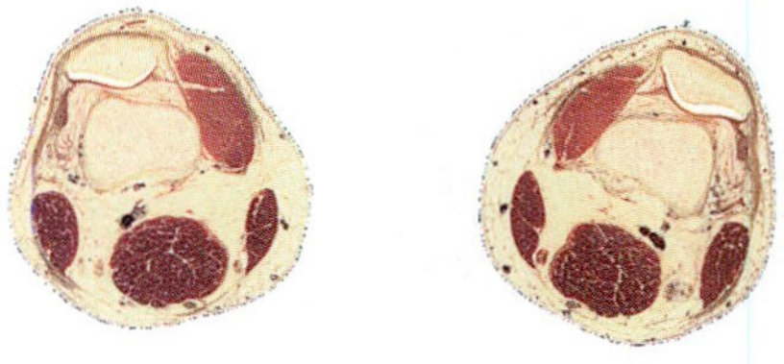

anterior

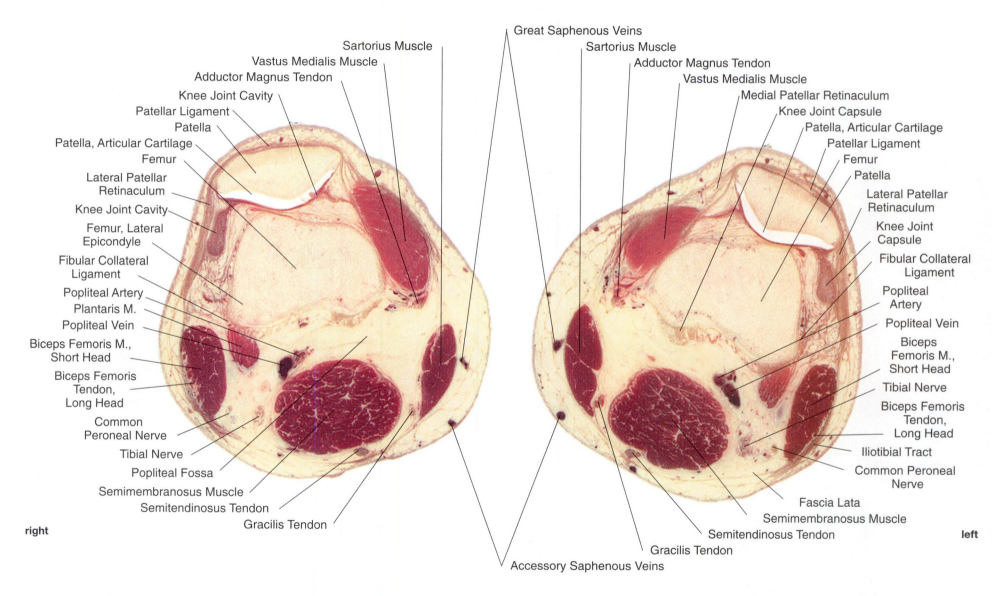

Great Saphenous Veins

Sartorius Muscle

Sartorius Muscle

Vastus Medialis Muscle

Adductor Magnus Tendon

Adductor Magnus Tendon

Vastus Medialis Muscle

Knee Joint Cavity

Medial Patellar Retinaculum

Patellar Ligament

Knee Joint Capsule

Patella

Patella, Articular Cartilage

Patella, Articular Cartilage

Patellar Ligament

Femur

Femur

Lateral Patellar
Retinaculum

Patella

Knee Joint Cavity

Lateral Patellar
Retinaculum

Femur, Lateral
Epicondyle

Knee Joint
Capsule

Fibular Collateral
Ligament

Fibular Collateral
Ligament

Popliteal Artery

Popliteal
Artery

Plantaris M.

Popliteal Vein

Popliteal Vein

Biceps Femoris M.,
Short Head

Biceps
Femoris
M.,
Short Head

Biceps Femoris
Tendon,
Long Head

Tibial Nerve

Biceps Femoris
Tendon,
Long Head

Common
Peroneal Nerve

Iliotibial Tract

Tibial Nerve

Common Peroneal
Nerve

Popliteal Fossa

Semimembranosus Muscle

Fascia Lata

Semitendinosus Tendon

Semimembranosus Muscle

Gracilis Tendon

Semitendinosus Tendon

Gracilis Tendon

Accessory Saphenous Veins

right

left

posterior

a_vm2281

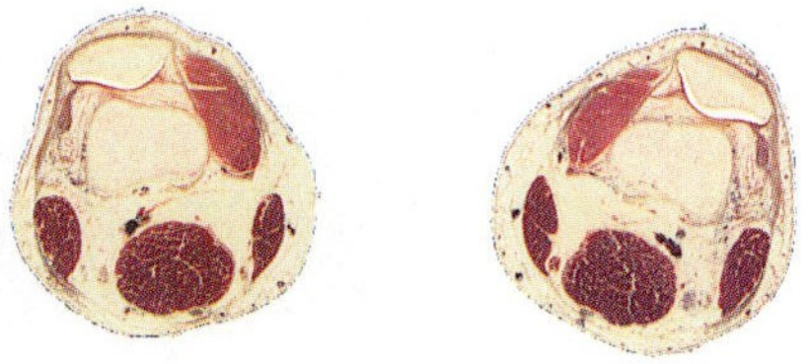

a_vm2282

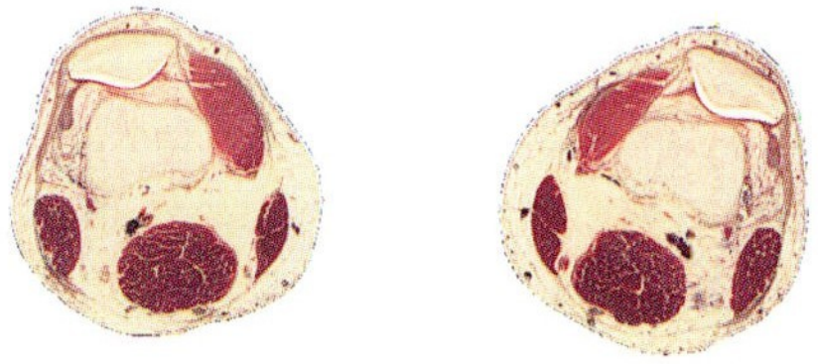

a_vm2283

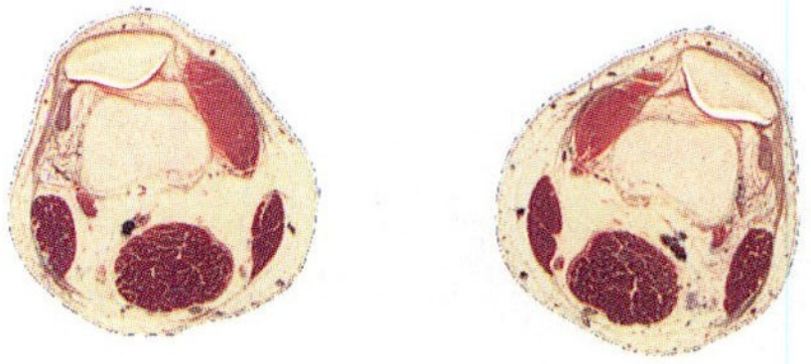

a_vm2281-
a_vm2292

a_vm2284

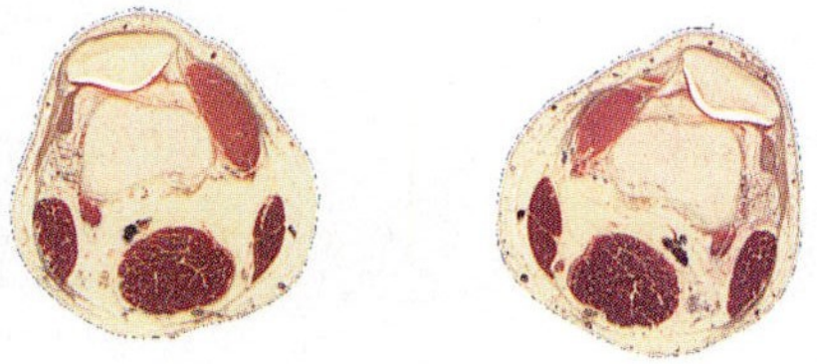

a_vm2285

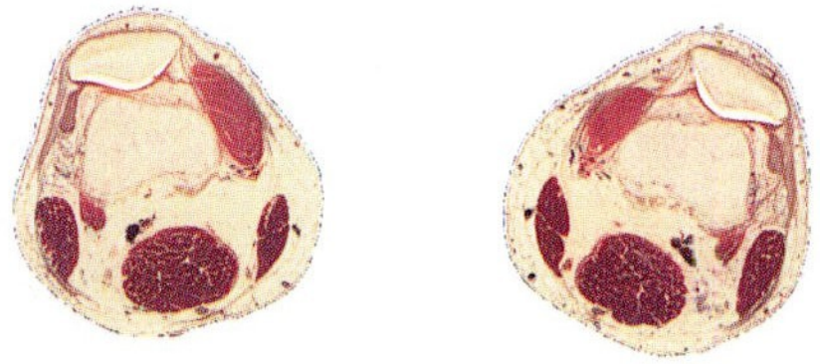

a_vm2286

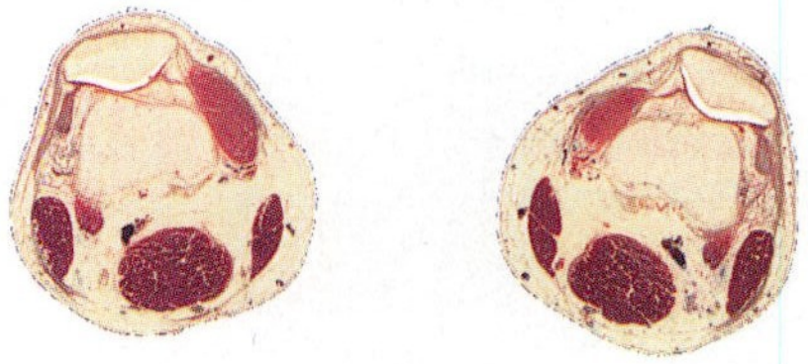

a_vm2287

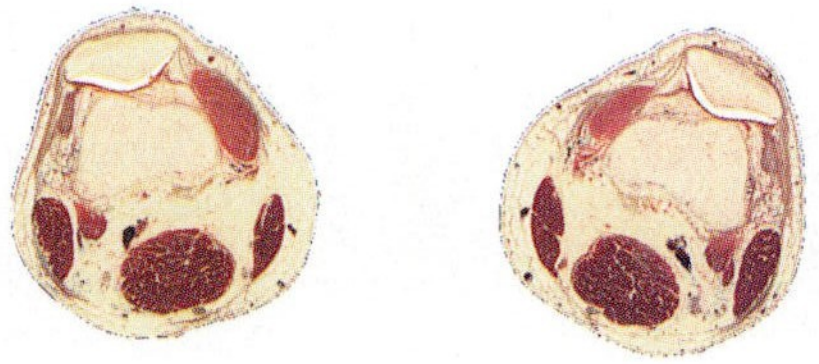

a_vm2288

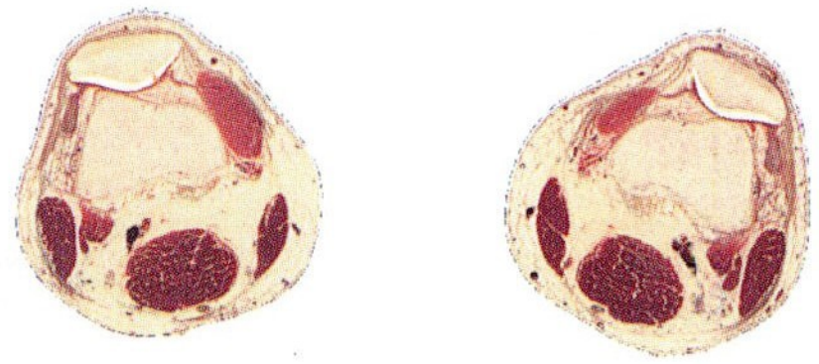

a_vm2289

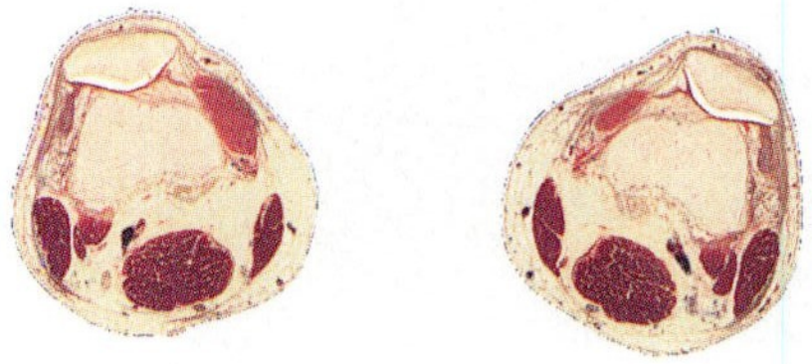

a_vm2290

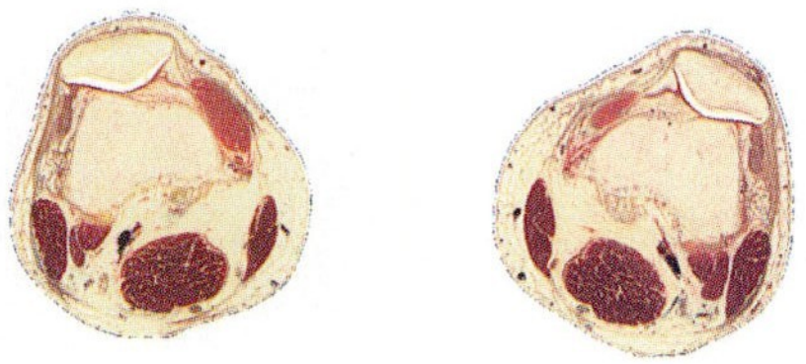

a_vm2291

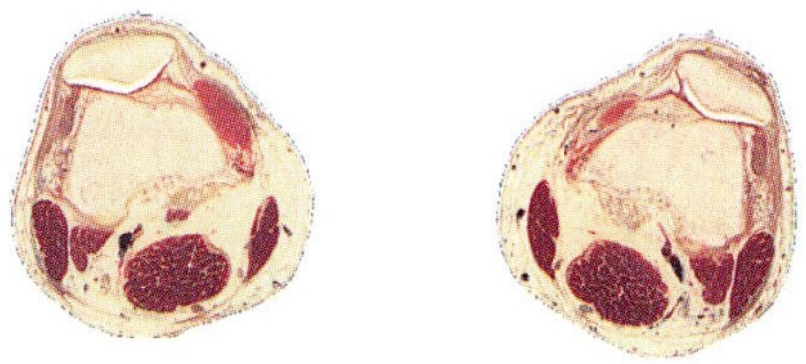

a_vm2292

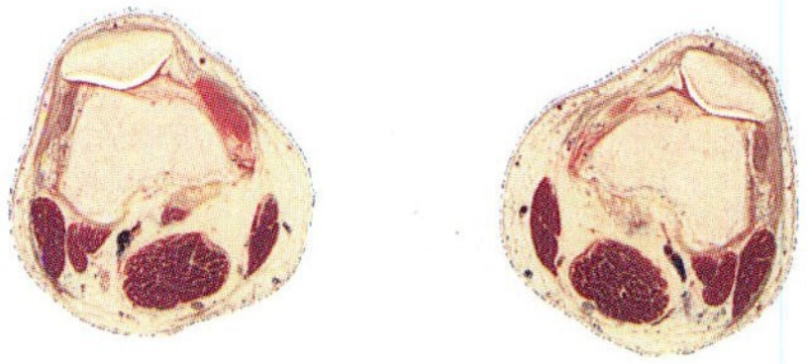

anterior

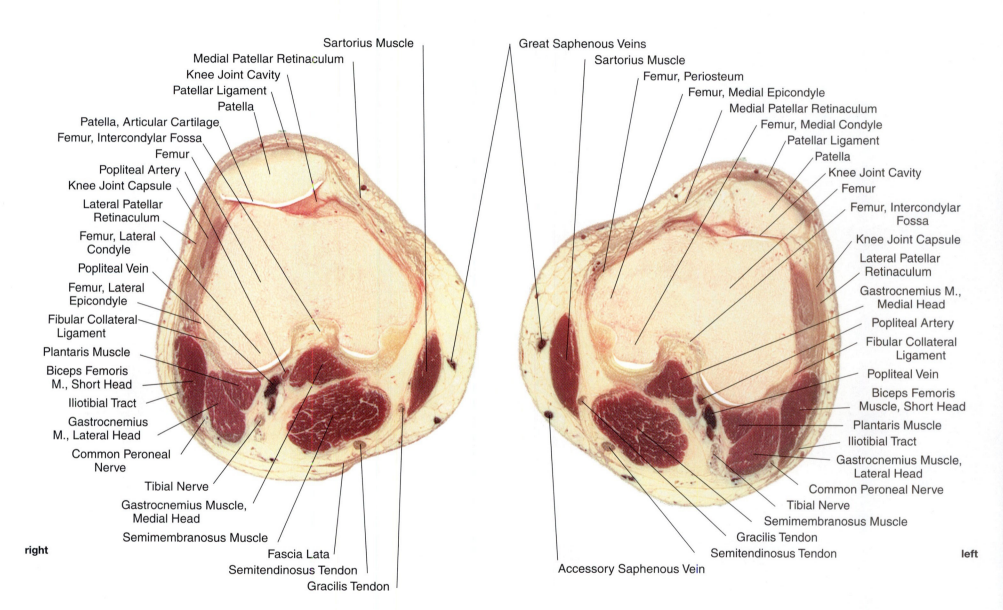

Sartorius Muscle

Medial Patellar Retinaculum
Knee Joint Cavity
Patellar Ligament
Patella

Patella, Articular Cartilage
Femur, Intercondylar Fossa
Femur
Popliteal Artery
Knee Joint Capsule
Lateral Patellar
Retinaculum
Femur, Lateral
Condyle
Popliteal Vein
Femur, Lateral
Epicondyle
Fibular Collateral
Ligament
Plantaris Muscle
Biceps Femoris
M., Short Head
Iliotibial Tract
Gastrocnemius
M., Lateral Head
Common Peroneal
Nerve
Tibial Nerve
Gastrocnemius Muscle,
Medial Head
Semimembranosus Muscle
Fascia Lata
Semitendinosus Tendon
Gracilis Tendon

Great Saphenous Veins
Sartorius Muscle
Femur, Periosteum
Femur, Medial Epicondyle
Medial Patellar Retinaculum
Femur, Medial Condyle
Patellar Ligament
Patella
Knee Joint Cavity
Femur
Femur, Intercondylar
Fossa
Knee Joint Capsule
Lateral Patellar
Retinaculum
Gastrocnemius M.,
Medial Head
Popliteal Artery
Fibular Collateral
Ligament
Popliteal Vein
Biceps Femoris
Muscle, Short Head
Plantaris Muscle
Iliotibial Tract
Gastrocnemius Muscle,
Lateral Head
Common Peroneal Nerve
Tibial Nerve
Semimembranosus Muscle
Gracilis Tendon
Semitendinosus Tendon

Accessory Saphenous Vein

right

left

posterior

216

a_vm2293

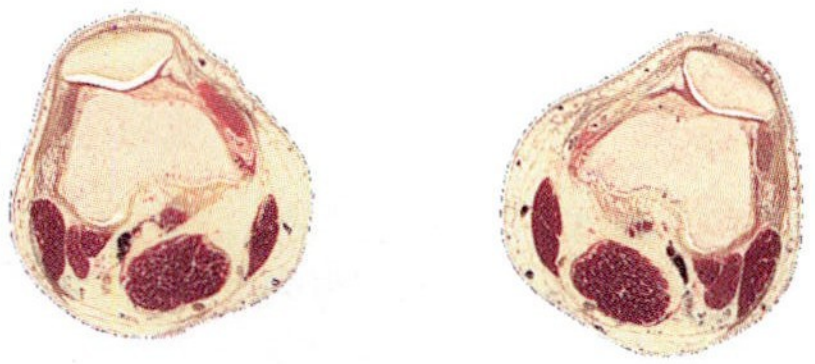

a_vm2294

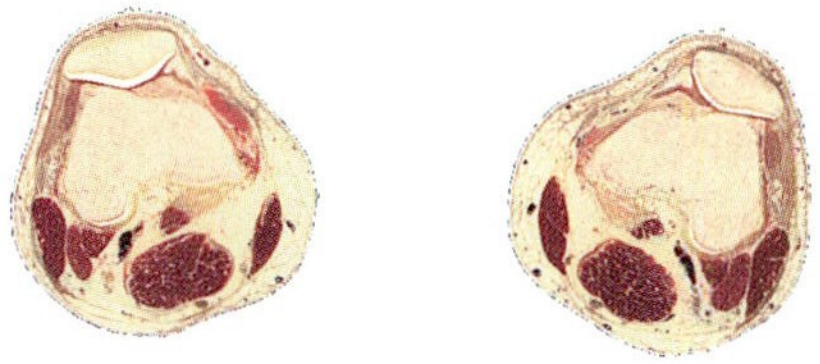

a_vm2295

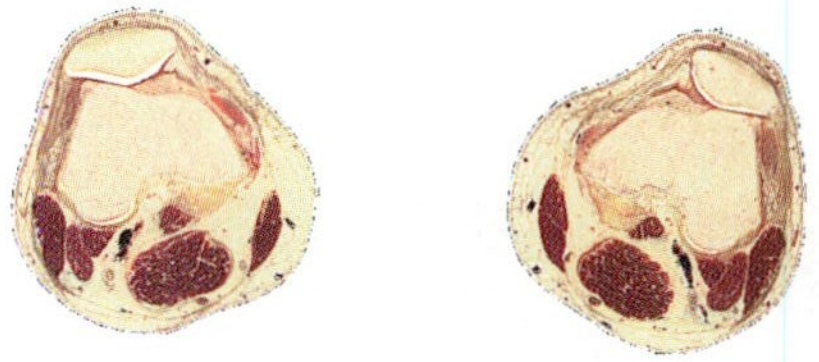

a_vm2296

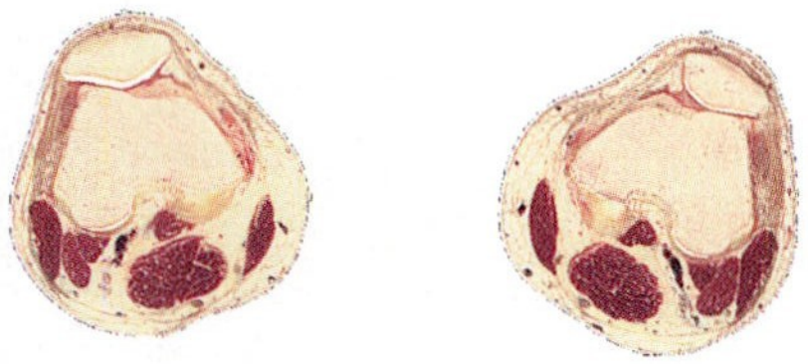

a_vm2297

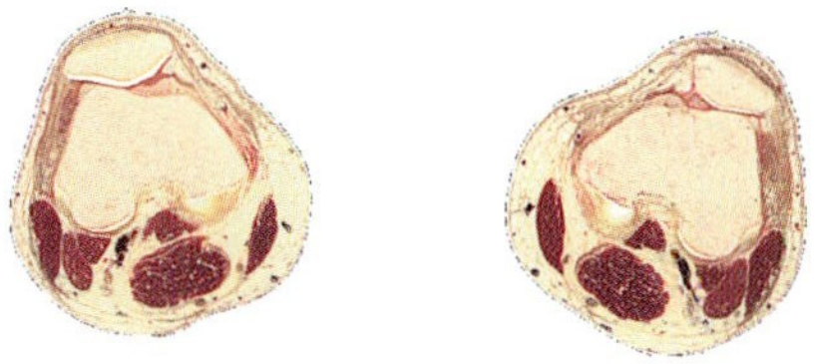

a_vm2298

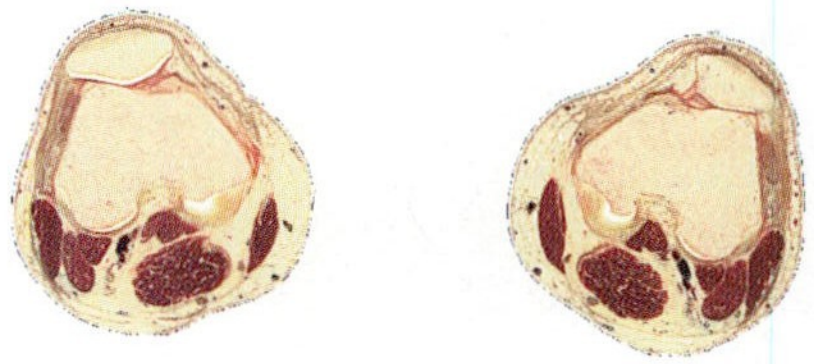

a_vm2299

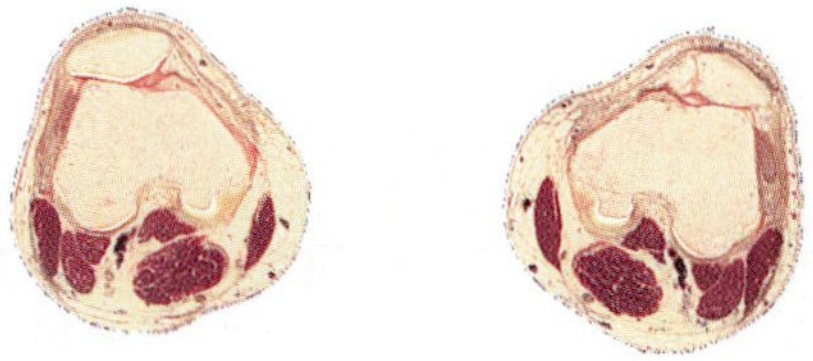

a_vm2300

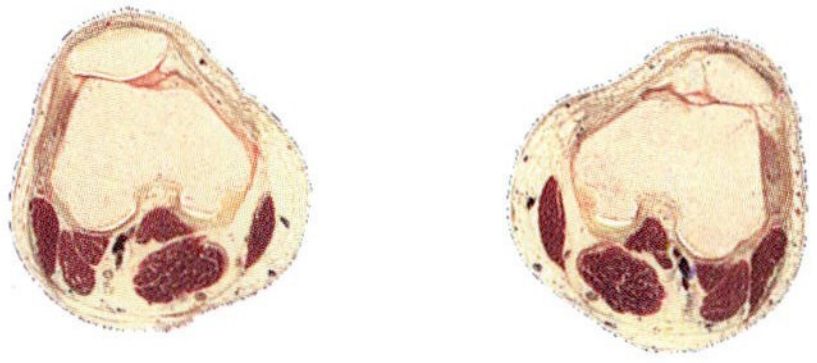

a_vm2301

a_vm2302

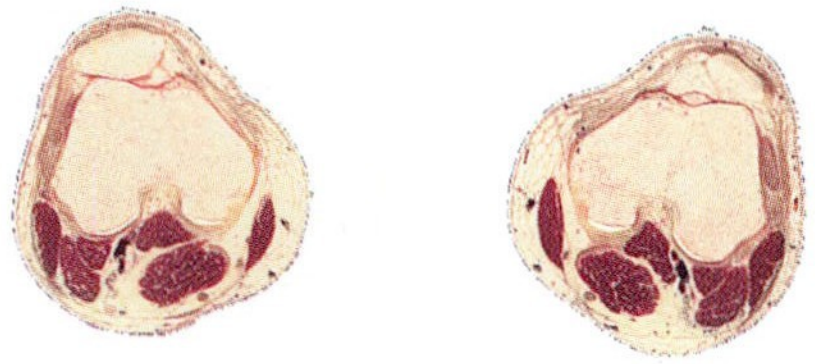

a_vm2303

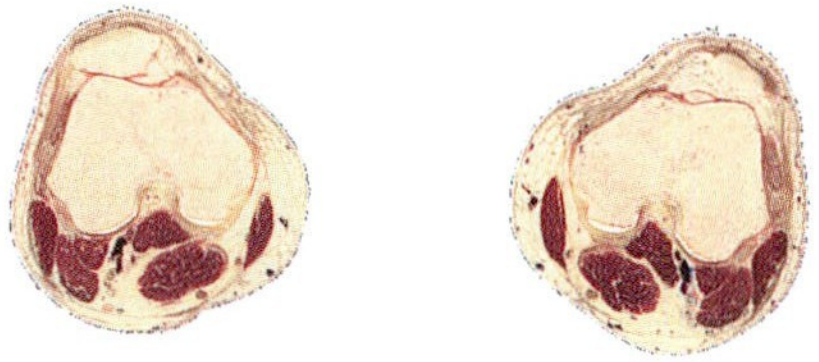

a_vm2304

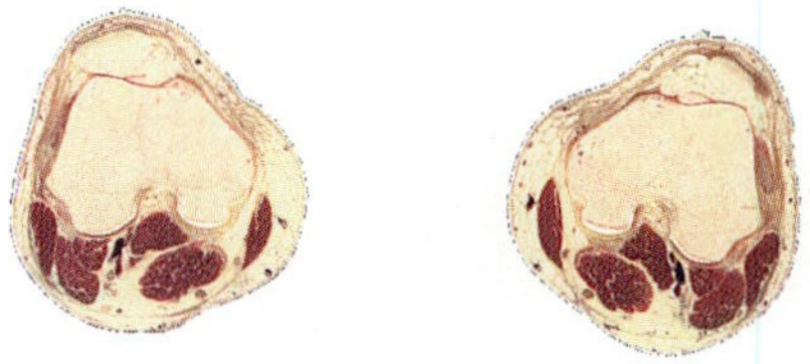

anterior

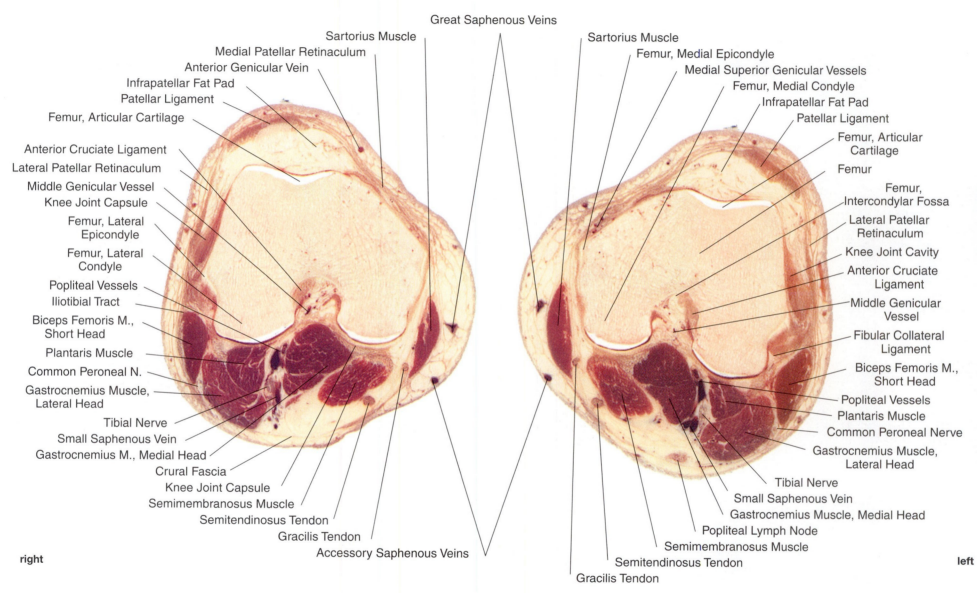

Great Saphenous Veins

Sartorius Muscle

Sartorius Muscle

Medial Patellar Retinaculum

Femur, Medial Epicondyle

Anterior Genicular Vein

Medial Superior Genicular Vessels

Infrapatellar Fat Pad

Femur, Medial Condyle

Patellar Ligament

Infrapatellar Fat Pad

Femur, Articular Cartilage

Patellar Ligament

Femur, Articular
Cartilage

Anterior Cruciate Ligament

Lateral Patellar Retinaculum

Femur

Middle Genicular Vessel

Femur,
Intercondylar Fossa

Knee Joint Capsule

Lateral Patellar
Retinaculum

Femur, Lateral
Epicondyle

Knee Joint Cavity

Femur, Lateral
Condyle

Anterior Cruciate
Ligament

Popliteal Vessels

Middle Genicular
Vessel

Iliotibial Tract

Biceps Femoris M.,
Short Head

Fibular Collateral
Ligament

Plantaris Muscle

Biceps Femoris M.,
Short Head

Common Peroneal N.

Popliteal Vessels

Gastrocnemius Muscle,
Lateral Head

Plantaris Muscle

Tibial Nerve

Common Peroneal Nerve

Small Saphenous Vein

Gastrocnemius Muscle,
Lateral Head

Gastrocnemius M., Medial Head

Crural Fascia

Tibial Nerve

Knee Joint Capsule

Small Saphenous Vein

Semimembranosus Muscle

Gastrocnemius Muscle, Medial Head

Semitendinosus Tendon

Popliteal Lymph Node

Gracilis Tendon

Semimembranosus Muscle

Accessory Saphenous Veins

Semitendinosus Tendon

Gracilis Tendon

right

left

posterior

a_vm2305

a_vm2306

a_vm2307

a_vm2308

a_vm2309

a_vm2310

a_vm2311

a_vm2312

a_vm2313

a_vm2314

a_vm2315

a_vm2316

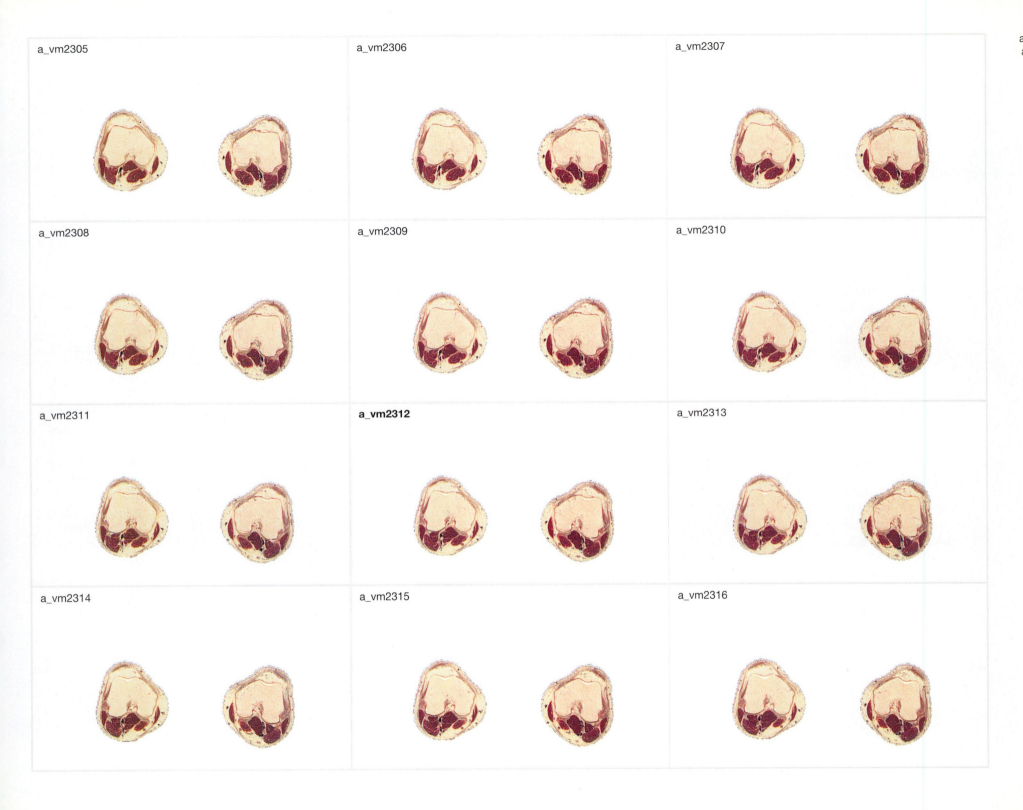

anterior

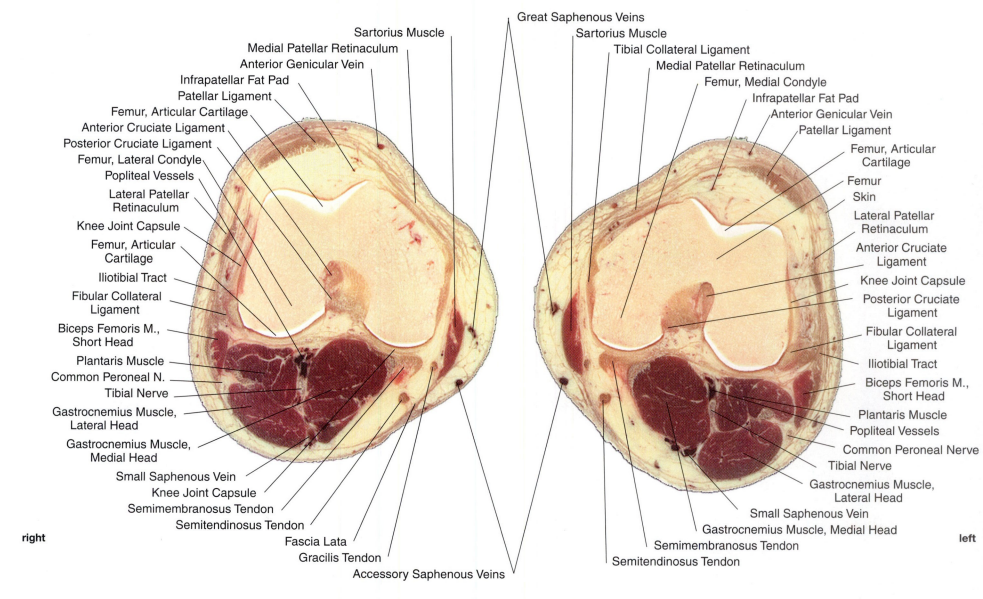

Great Saphenous Veins
Sartorius Muscle
Tibial Collateral Ligament
Medial Patellar Retinaculum
Femur, Medial Condyle
Infrapatellar Fat Pad
Anterior Genicular Vein
Patellar Ligament
Femur, Articular Cartilage
Femur
Skin
Lateral Patellar Retinaculum
Anterior Cruciate Ligament
Knee Joint Capsule
Posterior Cruciate Ligament
Fibular Collateral Ligament
Iliotibial Tract
Biceps Femoris M., Short Head
Plantaris Muscle
Popliteal Vessels
Common Peroneal Nerve
Tibial Nerve
Gastrocnemius Muscle, Lateral Head
Small Saphenous Vein
Gastrocnemius Muscle, Medial Head
Semimembranosus Tendon
Semitendinosus Tendon

Sartorius Muscle
Medial Patellar Retinaculum
Anterior Genicular Vein
Infrapatellar Fat Pad
Patellar Ligament
Femur, Articular Cartilage
Anterior Cruciate Ligament
Posterior Cruciate Ligament
Femur, Lateral Condyle
Popliteal Vessels
Lateral Patellar Retinaculum
Knee Joint Capsule
Femur, Articular Cartilage
Iliotibial Tract
Fibular Collateral Ligament
Biceps Femoris M., Short Head
Plantaris Muscle
Common Peroneal N.
Tibial Nerve
Gastrocnemius Muscle, Lateral Head
Gastrocnemius Muscle, Medial Head
Small Saphenous Vein
Knee Joint Capsule
Semimembranosus Tendon
Semitendinosus Tendon
Fascia Lata
Gracilis Tendon
Accessory Saphenous Veins

right

left

posterior

a_vm2317

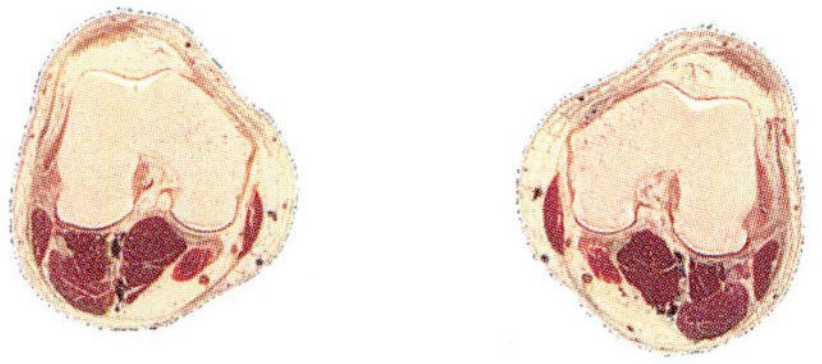

a_vm2318

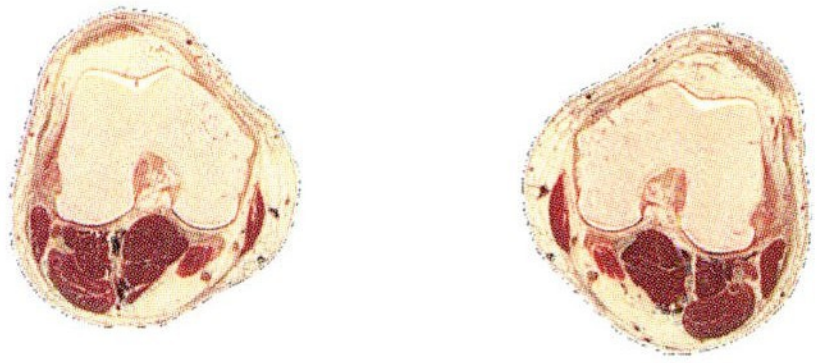

a_vm2319

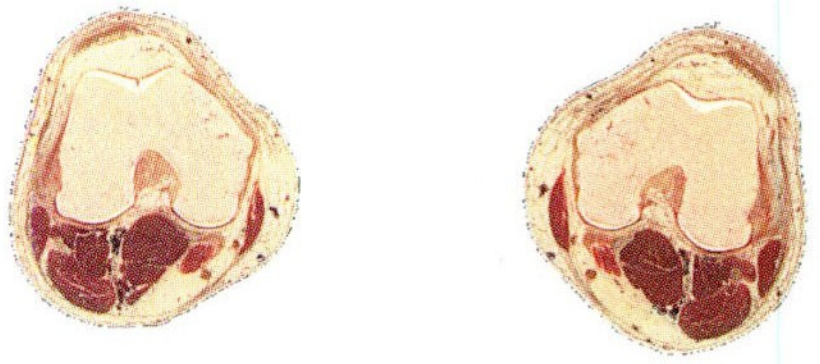

a_vm2320

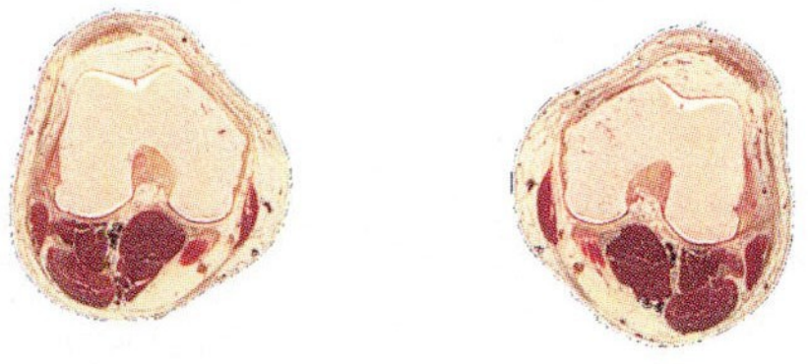

a_vm2321

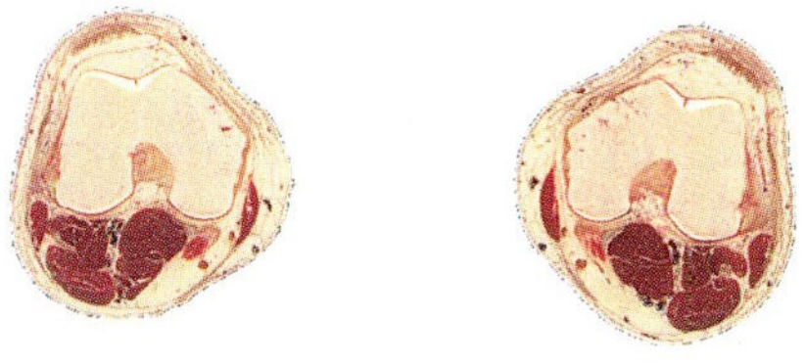

a_vm2322

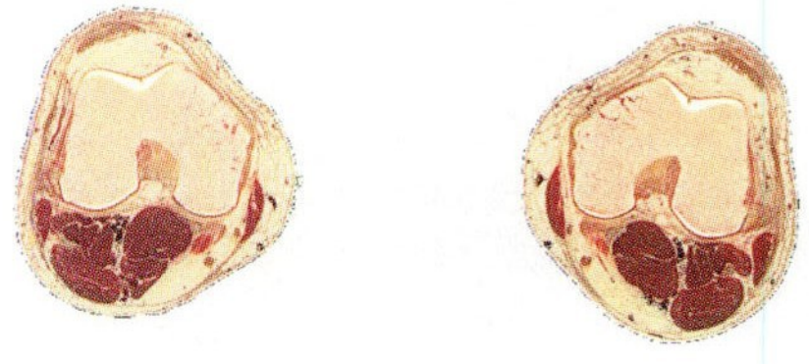

a_vm2323

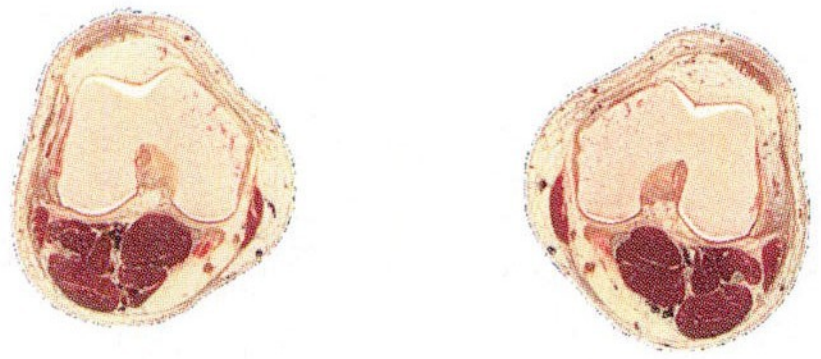

a_vm2324

a_vm2325

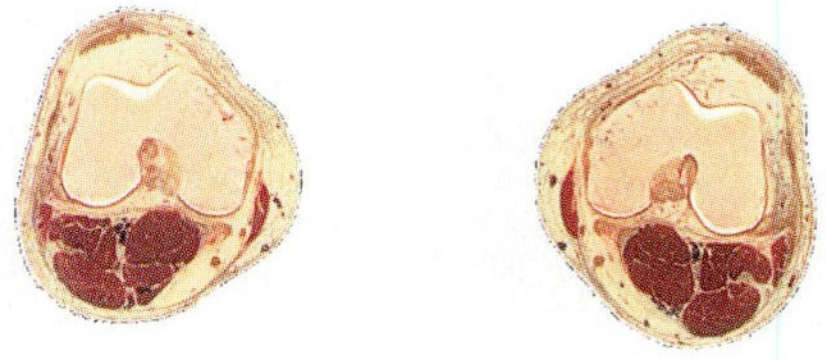

a_vm2326

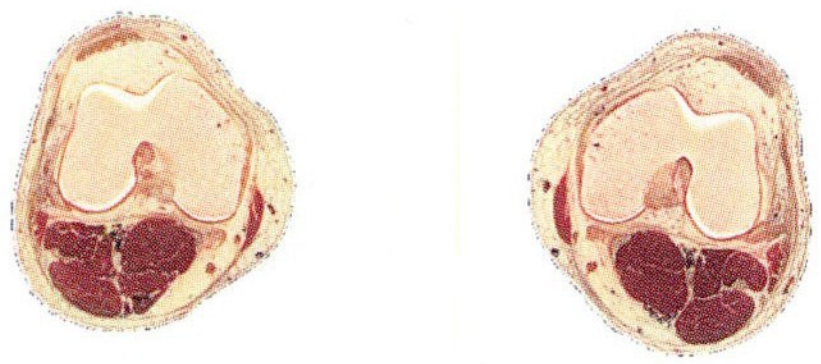

a_vm2327

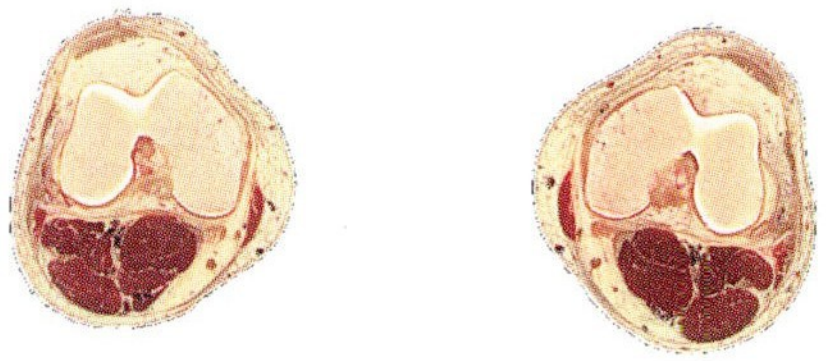

a_vm2328

Transverse
a_vm2330

anterior

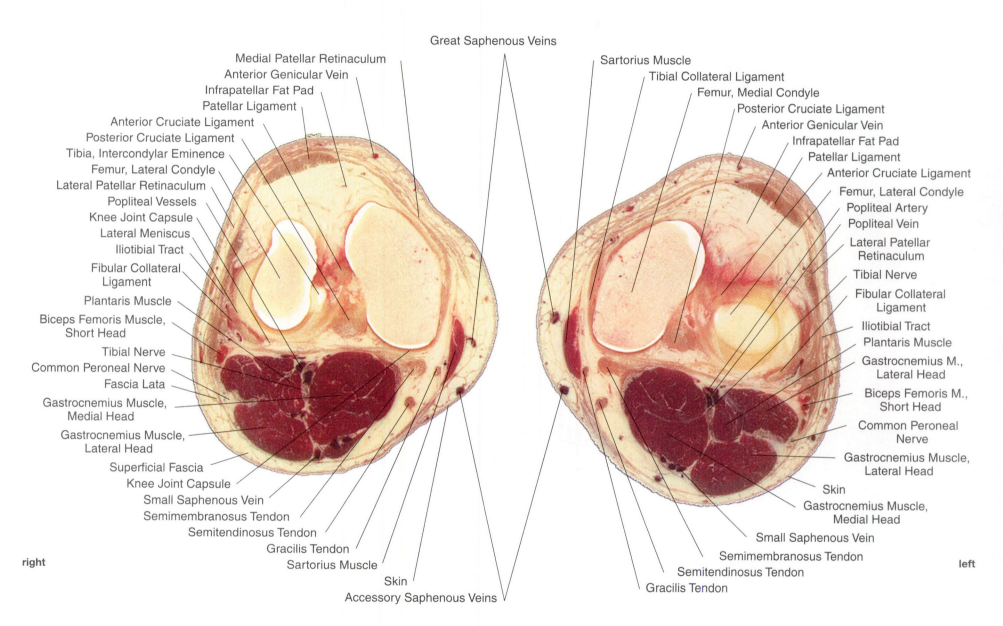

Great Saphenous Veins

Medial Patellar Retinaculum
Anterior Genicular Vein
Infrapatellar Fat Pad
Patellar Ligament
Anterior Cruciate Ligament
Posterior Cruciate Ligament
Tibia, Intercondylar Eminence
Femur, Lateral Condyle
Lateral Patellar Retinaculum
Popliteal Vessels
Knee Joint Capsule
Lateral Meniscus
Iliotibial Tract
Fibular Collateral Ligament
Plantaris Muscle
Biceps Femoris Muscle, Short Head
Tibial Nerve
Common Peroneal Nerve
Fascia Lata
Gastrocnemius Muscle, Medial Head
Gastrocnemius Muscle, Lateral Head
Superficial Fascia
Knee Joint Capsule
Small Saphenous Vein
Semimembranosus Tendon
Semitendinosus Tendon
Gracilis Tendon
Sartorius Muscle
Skin
Accessory Saphenous Veins

Sartorius Muscle
Tibial Collateral Ligament
Femur, Medial Condyle
Posterior Cruciate Ligament
Anterior Genicular Vein
Infrapatellar Fat Pad
Patellar Ligament
Anterior Cruciate Ligament
Femur, Lateral Condyle
Popliteal Artery
Popliteal Vein
Lateral Patellar Retinaculum
Tibial Nerve
Fibular Collateral Ligament
Iliotibial Tract
Plantaris Muscle
Gastrocnemius M., Lateral Head
Biceps Femoris M., Short Head
Common Peroneal Nerve
Gastrocnemius Muscle, Lateral Head
Skin
Gastrocnemius Muscle, Medial Head
Small Saphenous Vein
Semimembranosus Tendon
Semitendinosus Tendon
Gracilis Tendon

right

left

posterior

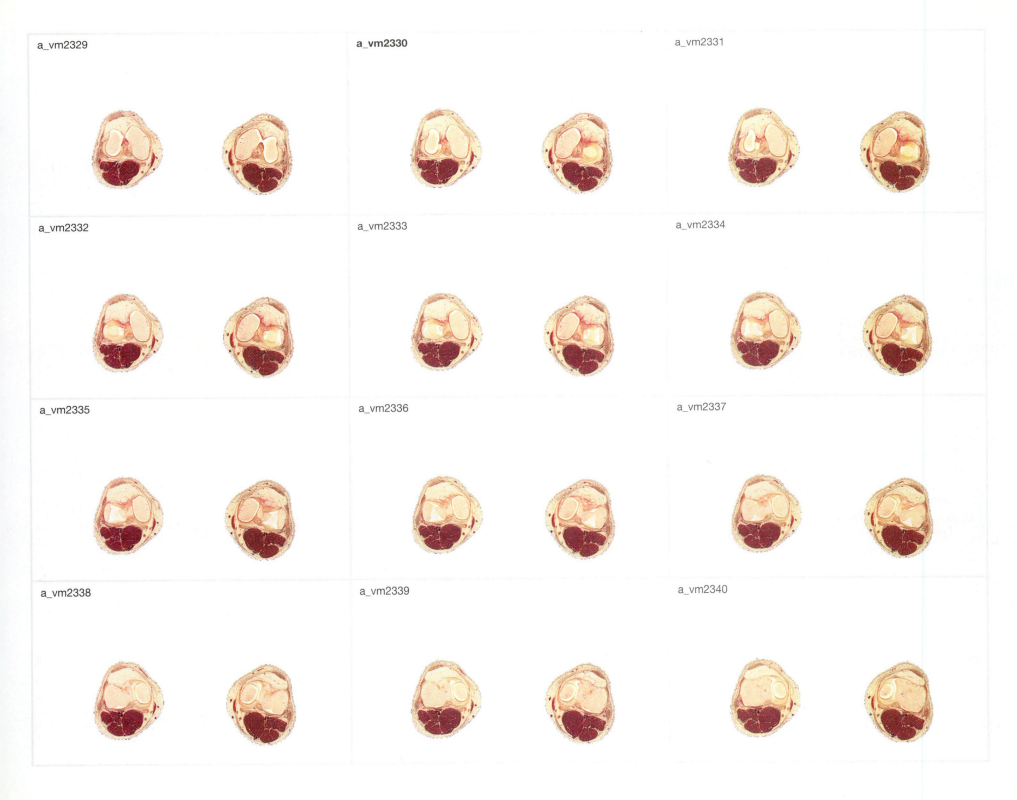

a_vm2329

a_vm2330

a_vm2331

a_vm2332

a_vm2333

a_vm2334

a_vm2335

a_vm2336

a_vm2337

a_vm2338

a_vm2339

a_vm2340

anterior

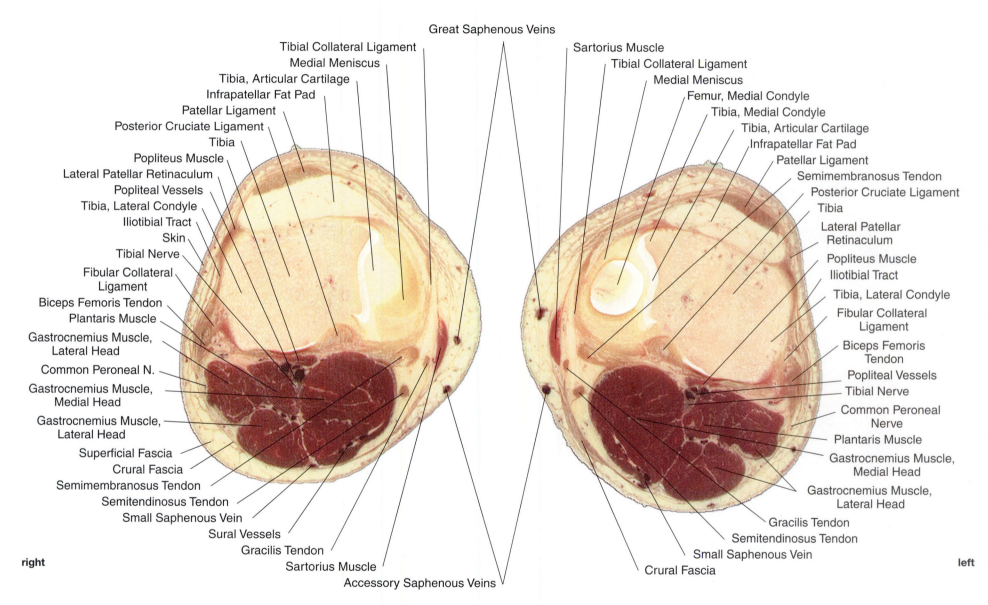

Great Saphenous Veins

Tibial Collateral Ligament
Medial Meniscus
Tibia, Articular Cartilage
Infrapatellar Fat Pad
Patellar Ligament
Posterior Cruciate Ligament
Tibia
Popliteus Muscle
Lateral Patellar Retinaculum
Popliteal Vessels
Tibia, Lateral Condyle
Iliotibial Tract
Skin
Tibial Nerve
Fibular Collateral Ligament
Biceps Femoris Tendon
Plantaris Muscle
Gastrocnemius Muscle, Lateral Head
Common Peroneal N.
Gastrocnemius Muscle, Medial Head
Gastrocnemius Muscle, Lateral Head
Superficial Fascia
Crural Fascia
Semimembranosus Tendon
Semitendinosus Tendon
Small Saphenous Vein
Sural Vessels
Gracilis Tendon
Sartorius Muscle
Accessory Saphenous Veins

Sartorius Muscle
Tibial Collateral Ligament
Medial Meniscus
Femur, Medial Condyle
Tibia, Medial Condyle
Tibia, Articular Cartilage
Infrapatellar Fat Pad
Patellar Ligament
Semimembranosus Tendon
Posterior Cruciate Ligament
Tibia
Lateral Patellar Retinaculum
Popliteus Muscle
Iliotibial Tract
Tibia, Lateral Condyle
Fibular Collateral Ligament
Biceps Femoris Tendon
Popliteal Vessels
Tibial Nerve
Common Peroneal Nerve
Plantaris Muscle
Gastrocnemius Muscle, Medial Head
Gastrocnemius Muscle, Lateral Head
Gracilis Tendon
Semitendinosus Tendon
Small Saphenous Vein
Crural Fascia

right

left

posterior

a_vm2341

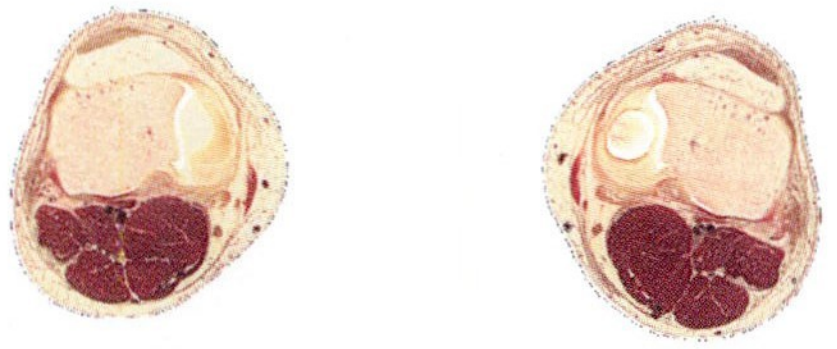

a_vm2342

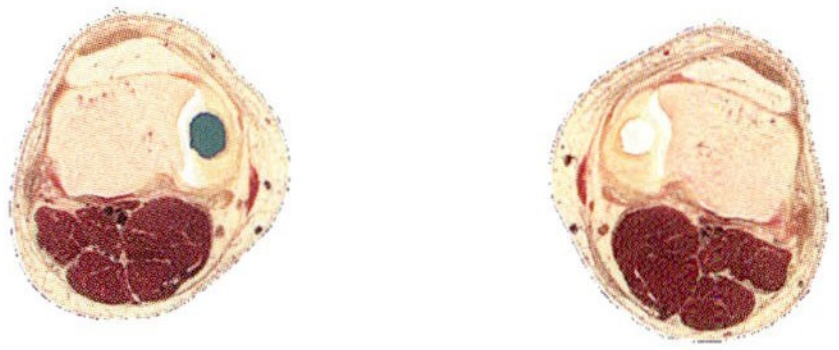

a_vm2343

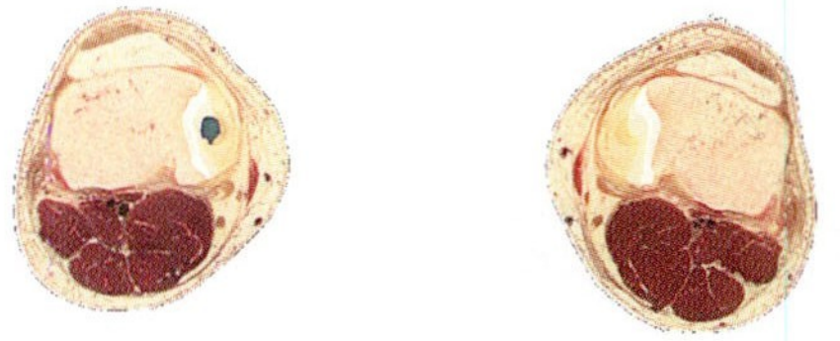

a_vm2344

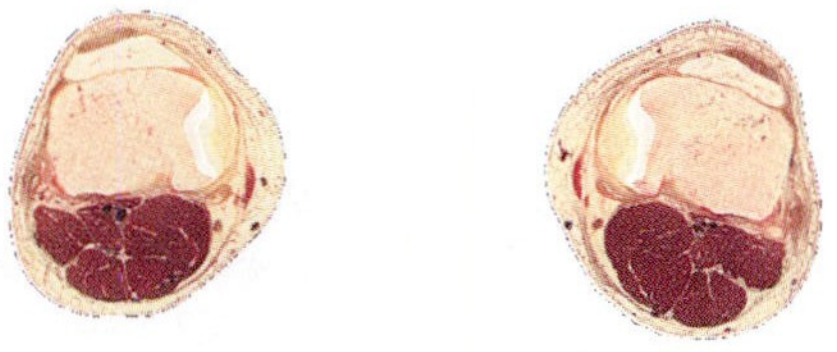

a_vm2345

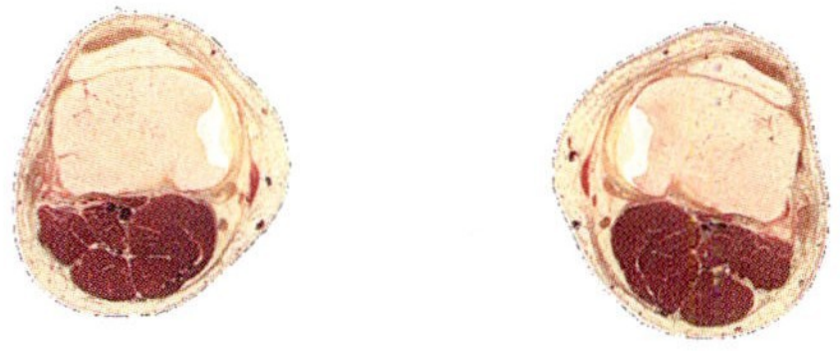

a_vm2346

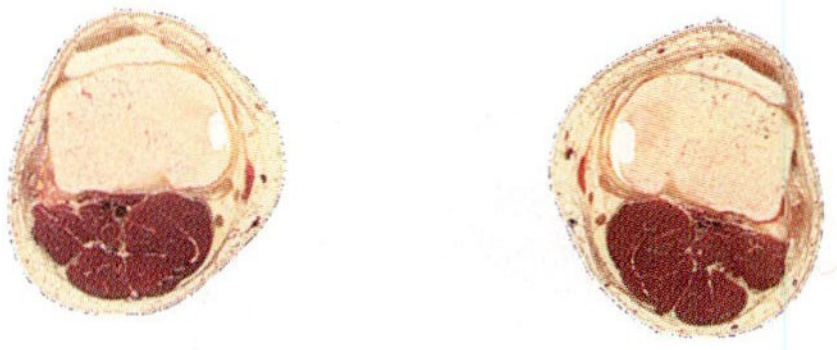

a_vm2347

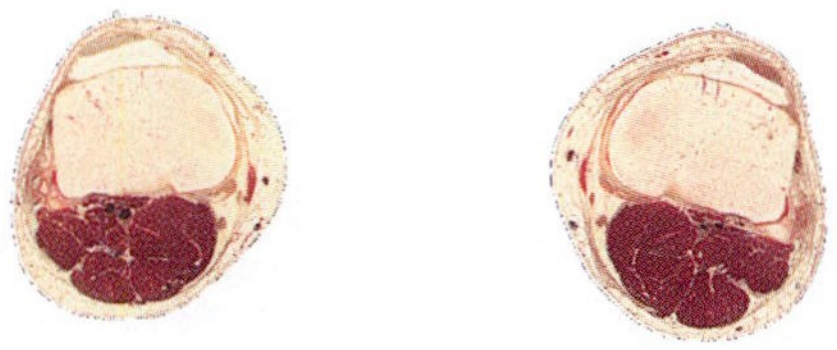

a_vm2348

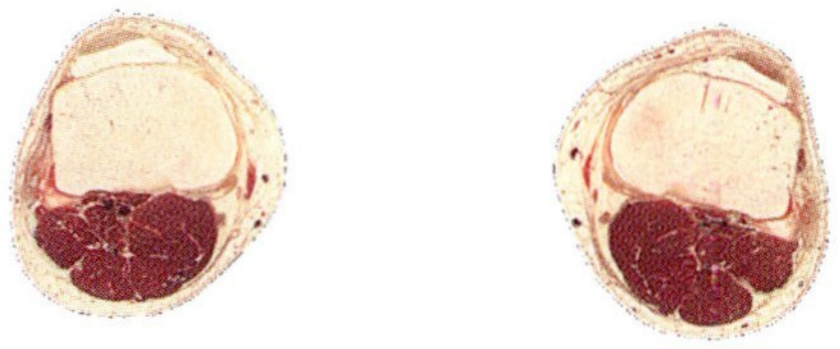

a_vm2349

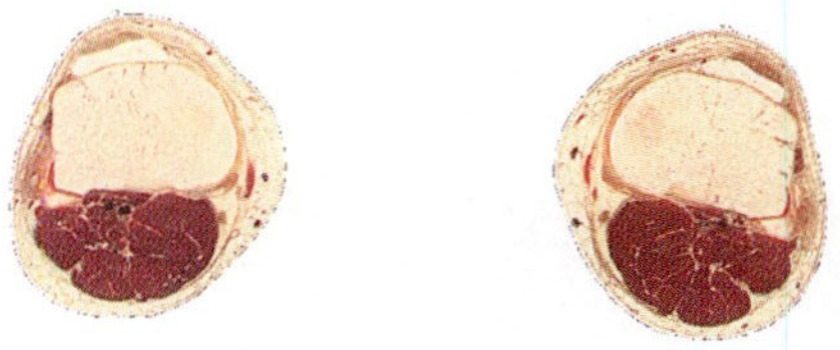

a_vm2350

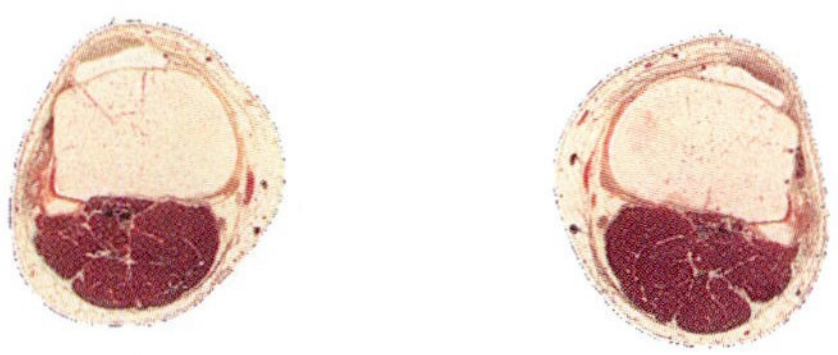

a_vm2351

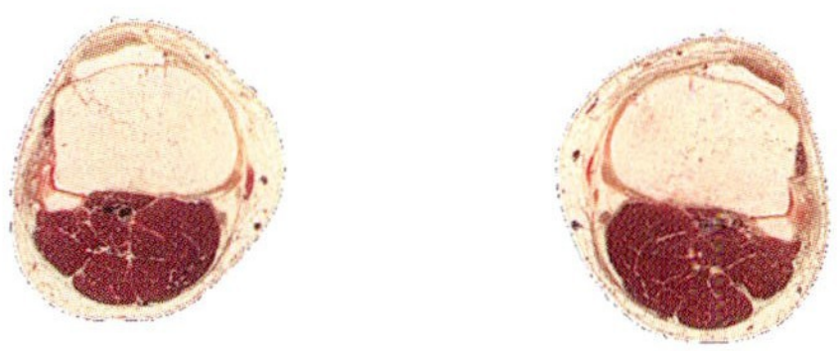

a_vm2352

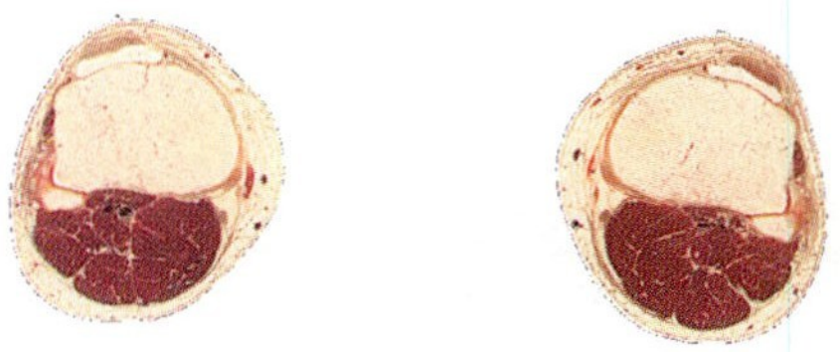

anterior

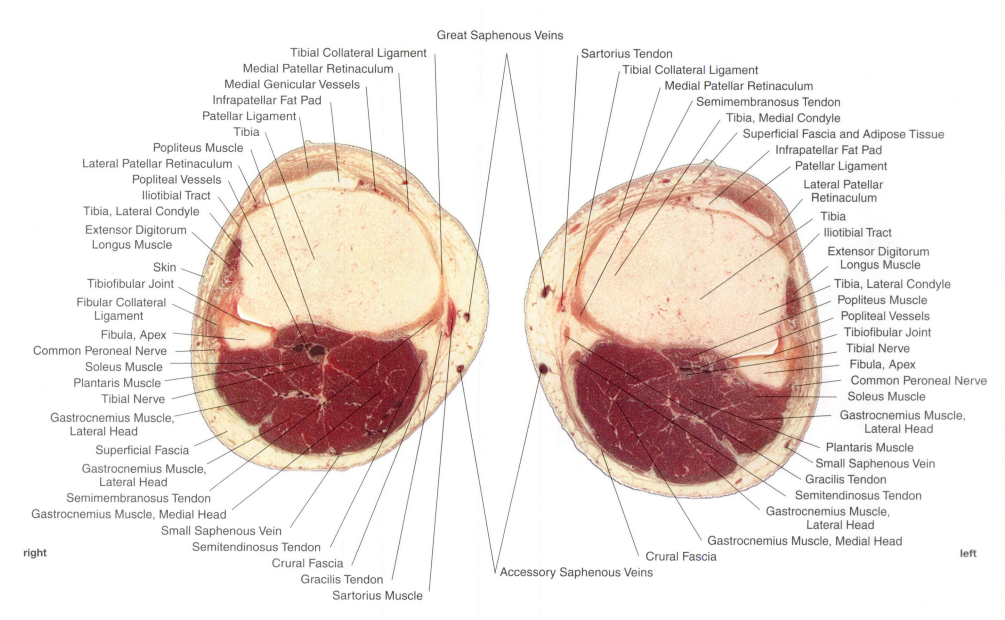

Great Saphenous Veins

Tibial Collateral Ligament
Medial Patellar Retinaculum
Medial Genicular Vessels
Infrapatellar Fat Pad
Patellar Ligament
Tibia
Popliteus Muscle
Lateral Patellar Retinaculum
Popliteal Vessels
Iliotibial Tract
Tibia, Lateral Condyle
Extensor Digitorum
Longus Muscle
Skin
Tibiofibular Joint
Fibular Collateral
Ligament
Fibula, Apex
Common Peroneal Nerve
Soleus Muscle
Plantaris Muscle
Tibial Nerve
Gastrocnemius Muscle,
Lateral Head
Superficial Fascia
Gastrocnemius Muscle,
Lateral Head
Semimembranosus Tendon
Gastrocnemius Muscle, Medial Head
Small Saphenous Vein
Semitendinosus Tendon
Crural Fascia
Gracilis Tendon
Sartorius Muscle

Sartorius Tendon
Tibial Collateral Ligament
Medial Patellar Retinaculum
Semimembranosus Tendon
Tibia, Medial Condyle
Superficial Fascia and Adipose Tissue
Infrapatellar Fat Pad
Patellar Ligament
Lateral Patellar
Retinaculum
Tibia
Iliotibial Tract
Extensor Digitorum
Longus Muscle
Tibia, Lateral Condyle
Popliteus Muscle
Popliteal Vessels
Tibiofibular Joint
Tibial Nerve
Fibula, Apex
Common Peroneal Nerve
Soleus Muscle
Gastrocnemius Muscle,
Lateral Head
Plantaris Muscle
Small Saphenous Vein
Gracilis Tendon
Semitendinosus Tendon
Gastrocnemius Muscle,
Lateral Head
Gastrocnemius Muscle, Medial Head
Crural Fascia
Accessory Saphenous Veins

right

left

posterior

a_vm2354

a_vm2355

a_vm2356

a_vm2357

a_vm2358

a_vm2359

a_vm2360

a_vm2361

a_vm2362

a_vm2363

a_vm2364

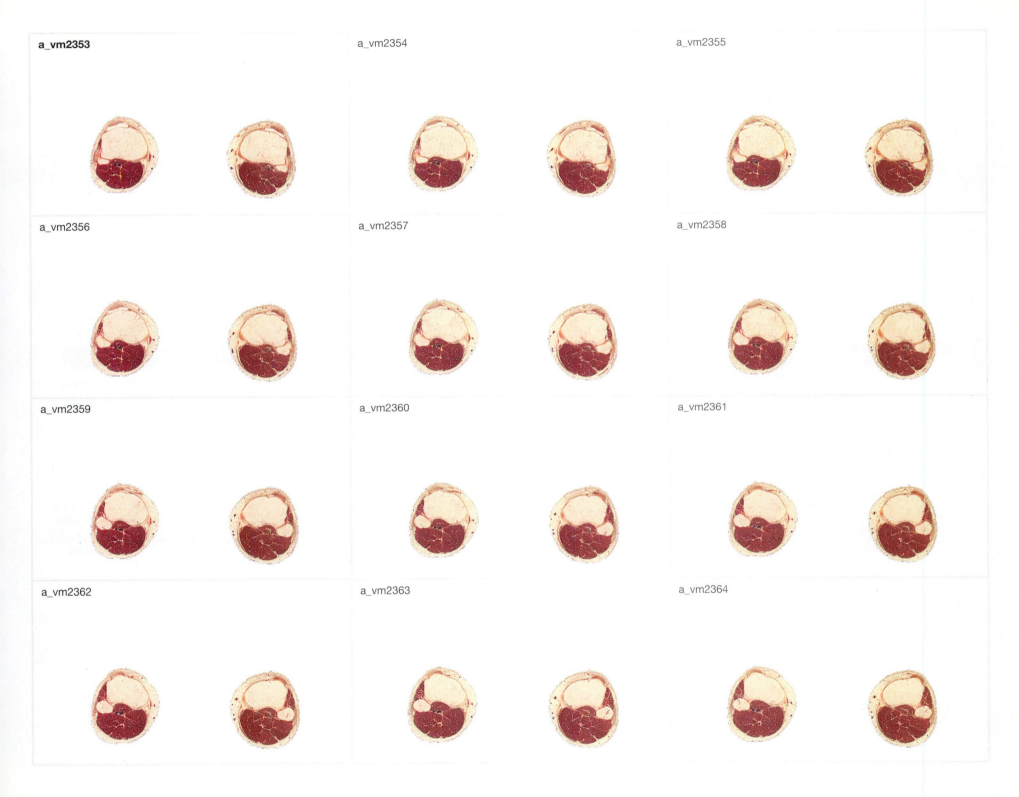

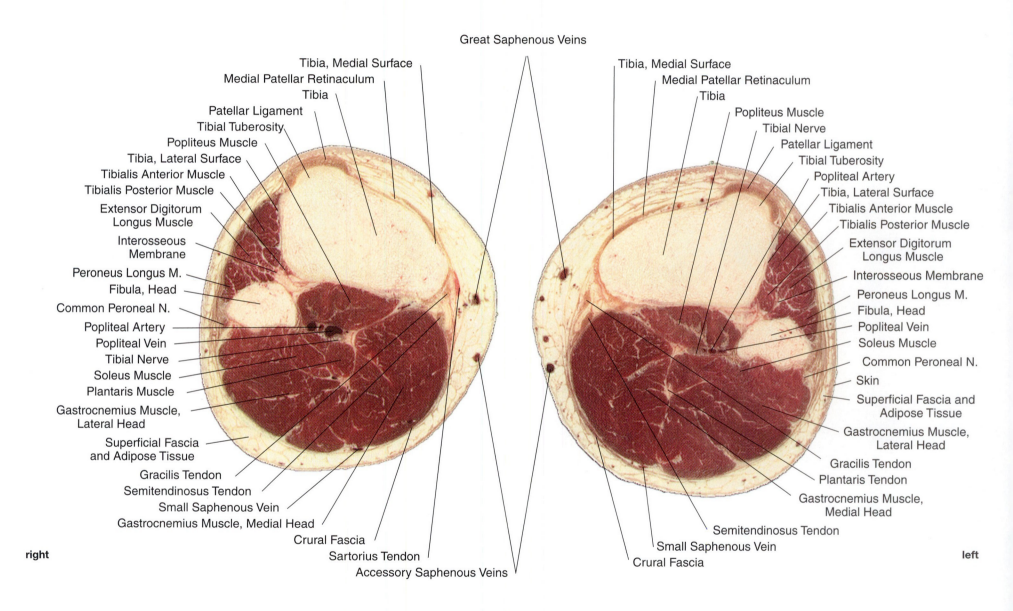

anterior

Great Saphenous Veins

Tibia, Medial Surface
Medial Patellar Retinaculum
Tibia
Patellar Ligament
Tibial Tuberosity
Popliteus Muscle
Tibia, Lateral Surface
Tibialis Anterior Muscle
Tibialis Posterior Muscle
Extensor Digitorum
Longus Muscle
Interosseous
Membrane
Peroneus Longus M.
Fibula, Head
Common Peroneal N.
Popliteal Artery
Popliteal Vein
Tibial Nerve
Soleus Muscle
Plantaris Muscle
Gastrocnemius Muscle,
Lateral Head
Superficial Fascia
and Adipose Tissue
Gracilis Tendon
Semitendinosus Tendon
Small Saphenous Vein
Gastrocnemius Muscle, Medial Head
Crural Fascia
Sartorius Tendon
Accessory Saphenous Veins

Tibia, Medial Surface
Medial Patellar Retinaculum
Tibia
Popliteus Muscle
Tibial Nerve
Patellar Ligament
Tibial Tuberosity
Popliteal Artery
Tibia, Lateral Surface
Tibialis Anterior Muscle
Tibialis Posterior Muscle
Extensor Digitorum
Longus Muscle
Interosseous Membrane
Peroneus Longus M.
Fibula, Head
Popliteal Vein
Soleus Muscle
Common Peroneal N.
Skin
Superficial Fascia and
Adipose Tissue
Gastrocnemius Muscle,
Lateral Head
Gracilis Tendon
Plantaris Tendon
Gastrocnemius Muscle,
Medial Head
Semitendinosus Tendon
Small Saphenous Vein
Crural Fascia

right

left

posterior

228

a_vm2365

a_vm2366

a_vm2367

a_vm2368

a_vm2369

a_vm2370

a_vm2371

a_vm2372

a_vm2373

a_vm2374

a_vm2375

a_vm2376

anterior

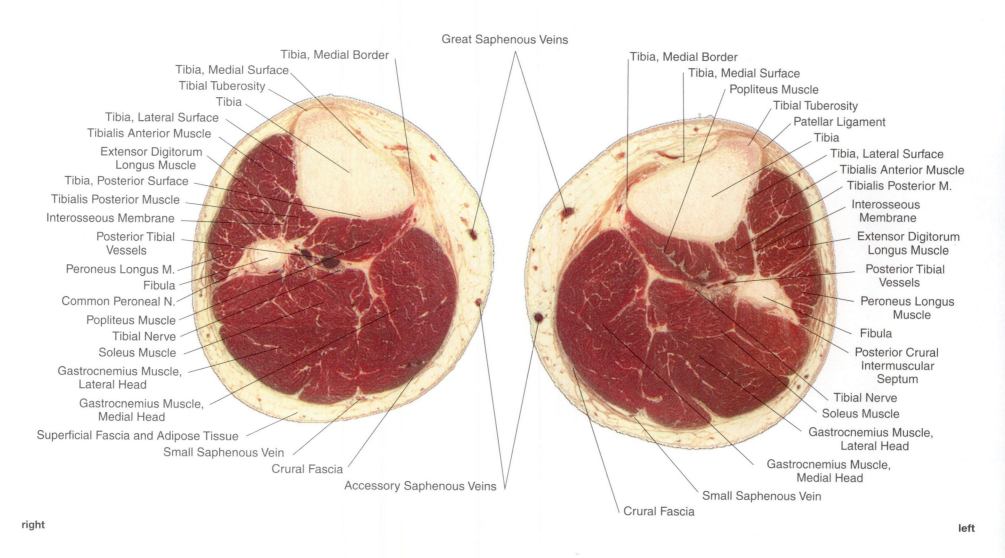

Great Saphenous Veins

Tibia, Medial Border

Tibia, Medial Surface

Tibial Tuberosity

Tibia

Tibia, Lateral Surface

Tibialis Anterior Muscle

Extensor Digitorum
Longus Muscle

Tibia, Posterior Surface

Tibialis Posterior Muscle

Interosseous Membrane

Posterior Tibial
Vessels

Peroneus Longus M.

Fibula

Common Peroneal N.

Popliteus Muscle

Tibial Nerve

Soleus Muscle

Gastrocnemius Muscle,
Lateral Head

Gastrocnemius Muscle,
Medial Head

Superficial Fascia and Adipose Tissue

Small Saphenous Vein

Crural Fascia

Accessory Saphenous Veins

Tibia, Medial Border

Tibia, Medial Surface

Popliteus Muscle

Tibial Tuberosity

Patellar Ligament

Tibia

Tibia, Lateral Surface

Tibialis Anterior Muscle

Tibialis Posterior M.

Interosseous
Membrane

Extensor Digitorum
Longus Muscle

Posterior Tibial
Vessels

Peroneus Longus
Muscle

Fibula

Posterior Crural
Intermuscular
Septum

Tibial Nerve

Soleus Muscle

Gastrocnemius Muscle,
Lateral Head

Gastrocnemius Muscle,
Medial Head

Small Saphenous Vein

Crural Fascia

right

left

posterior

a_vm2377

a_vm2378

a_vm2379

a_vm2380

a_vm2381

a_vm2382

a_vm2383

a_vm2384

a_vm2385

a_vm2386

a_vm2387

a_vm2388

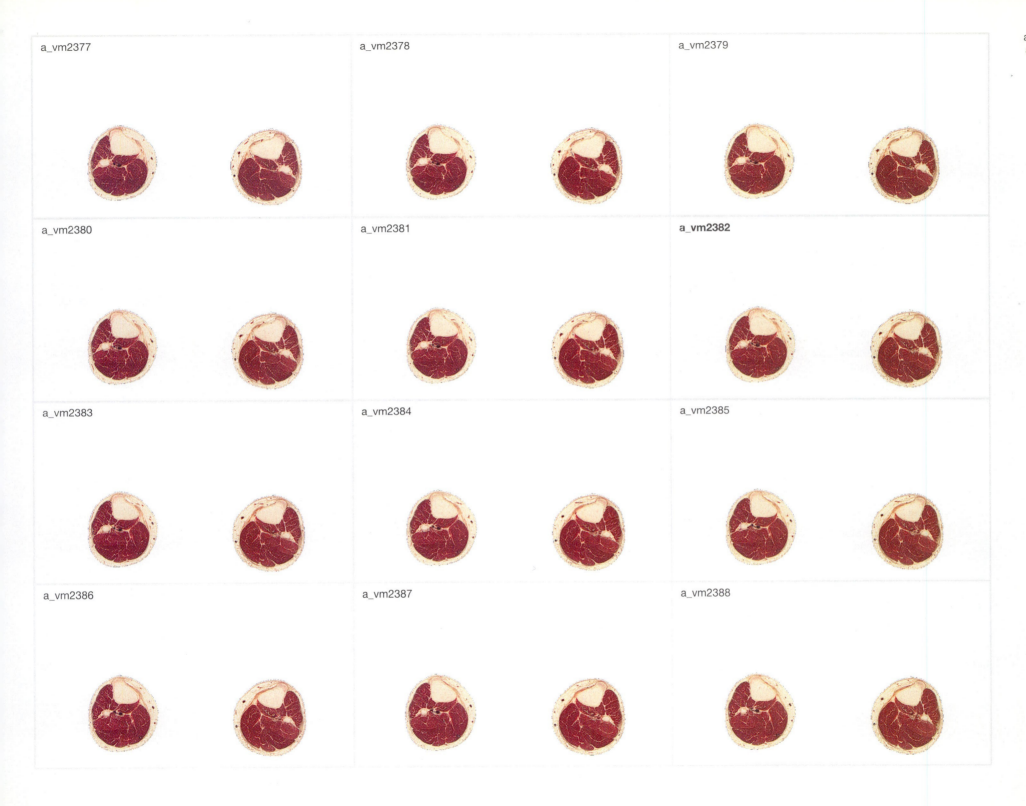

anterior

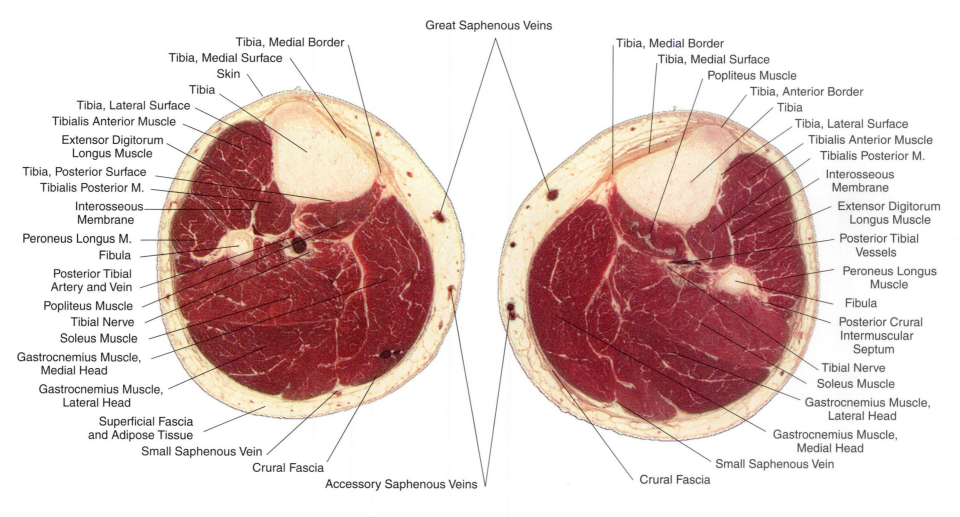

Great Saphenous Veins

Tibia, Medial Border

Tibia, Medial Surface

Skin

Tibia

Tibia, Lateral Surface

Tibialis Anterior Muscle

Extensor Digitorum
Longus Muscle

Tibia, Posterior Surface

Tibialis Posterior M.

Interosseous
Membrane

Peroneus Longus M.

Fibula

Posterior Tibial
Artery and Vein

Popliteus Muscle

Tibial Nerve

Soleus Muscle

Gastrocnemius Muscle,
Medial Head

Gastrocnemius Muscle,
Lateral Head

Superficial Fascia
and Adipose Tissue

Small Saphenous Vein

Crural Fascia

Accessory Saphenous Veins

Tibia, Medial Border

Tibia, Medial Surface

Popliteus Muscle

Tibia, Anterior Border

Tibia

Tibia, Lateral Surface

Tibialis Anterior Muscle

Tibialis Posterior M.

Interosseous
Membrane

Extensor Digitorum
Longus Muscle

Posterior Tibial
Vessels

Peroneus Longus
Muscle

Fibula

Posterior Crural
Intermuscular
Septum

Tibial Nerve

Soleus Muscle

Gastrocnemius Muscle,
Lateral Head

Gastrocnemius Muscle,
Medial Head

Small Saphenous Vein

Crural Fascia

right

left

posterior

a_vm2389

a_vm2390

a_vm2391

a_vm2392

a_vm2393

a_vm2394

a_vm2395

a_vm2396

a_vm2397

a_vm2398

a_vm2399

a_vm2400

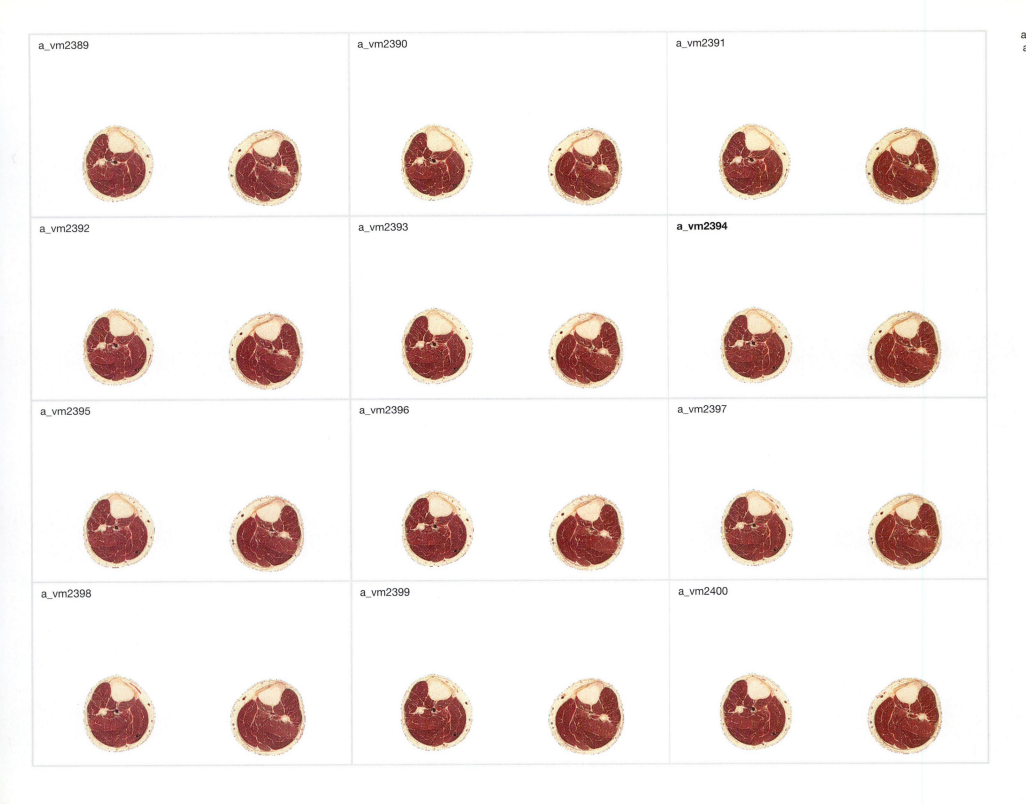

anterior

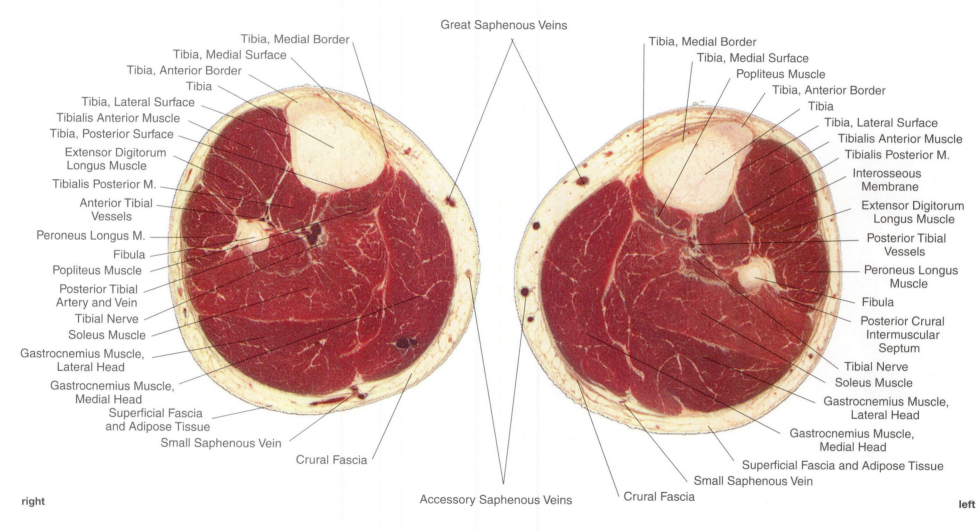

Great Saphenous Veins

Tibia, Medial Border

Tibia, Medial Surface

Tibia, Anterior Border

Tibia

Tibia, Lateral Surface

Tibialis Anterior Muscle

Tibia, Posterior Surface

Extensor Digitorum
Longus Muscle

Tibialis Posterior M.

Anterior Tibial
Vessels

Peroneus Longus M.

Fibula

Popliteus Muscle

Posterior Tibial
Artery and Vein

Tibial Nerve

Soleus Muscle

Gastrocnemius Muscle,
Lateral Head

Gastrocnemius Muscle,
Medial Head

Superficial Fascia
and Adipose Tissue

Small Saphenous Vein

Crural Fascia

Tibia, Medial Border

Tibia, Medial Surface

Popliteus Muscle

Tibia, Anterior Border

Tibia

Tibia, Lateral Surface

Tibialis Anterior Muscle

Tibialis Posterior M.

Interosseous
Membrane

Extensor Digitorum
Longus Muscle

Posterior Tibial
Vessels

Peroneus Longus
Muscle

Fibula

Posterior Crural
Intermuscular
Septum

Tibial Nerve

Soleus Muscle

Gastrocnemius Muscle,
Lateral Head

Gastrocnemius Muscle,
Medial Head

Superficial Fascia and Adipose Tissue

Small Saphenous Vein

Crural Fascia

Accessory Saphenous Veins

right

left

posterior

a_vm2401

a_vm2402

a_vm2403

a_vm2404

a_vm2405

a_vm2406

a_vm2407

a_vm2408

a_vm2409

a_vm2410

a_vm2411

a_vm2412

anterior

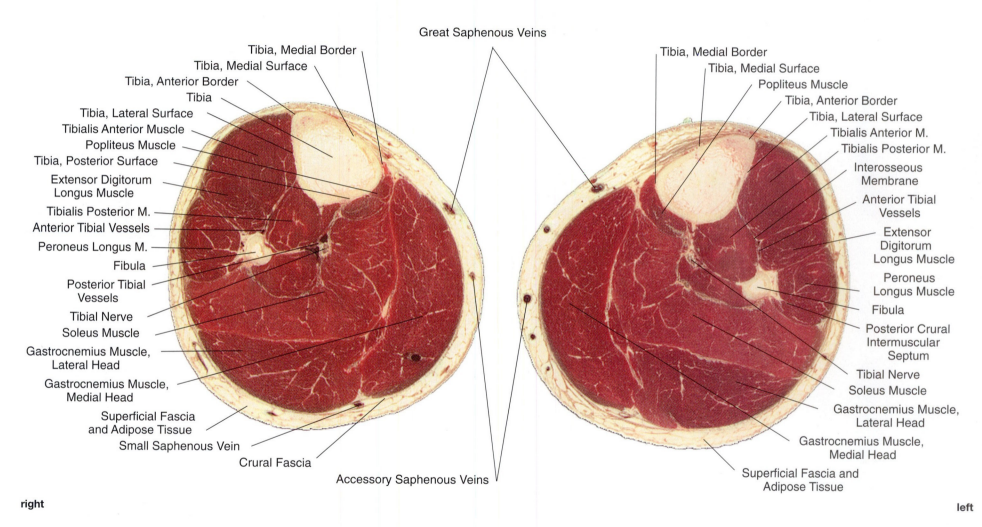

Great Saphenous Veins

Tibia, Medial Border

Tibia, Medial Surface

Tibia, Anterior Border

Tibia

Tibia, Lateral Surface

Tibialis Anterior Muscle

Popliteus Muscle

Tibia, Posterior Surface

Extensor Digitorum
Longus Muscle

Tibialis Posterior M.

Anterior Tibial Vessels

Peroneus Longus M.

Fibula

Posterior Tibial
Vessels

Tibial Nerve

Soleus Muscle

Gastrocnemius Muscle,
Lateral Head

Gastrocnemius Muscle,
Medial Head

Superficial Fascia
and Adipose Tissue

Small Saphenous Vein

Crural Fascia

Accessory Saphenous Veins

Tibia, Medial Border

Tibia, Medial Surface

Popliteus Muscle

Tibia, Anterior Border

Tibia, Lateral Surface

Tibialis Anterior M.

Tibialis Posterior M.

Interosseous
Membrane

Anterior Tibial
Vessels

Extensor
Digitorum
Longus Muscle

Peroneus
Longus Muscle

Fibula

Posterior Crural
Intermuscular
Septum

Tibial Nerve

Soleus Muscle

Gastrocnemius Muscle,
Lateral Head

Gastrocnemius Muscle,
Medial Head

Superficial Fascia and
Adipose Tissue

right

left

posterior

236

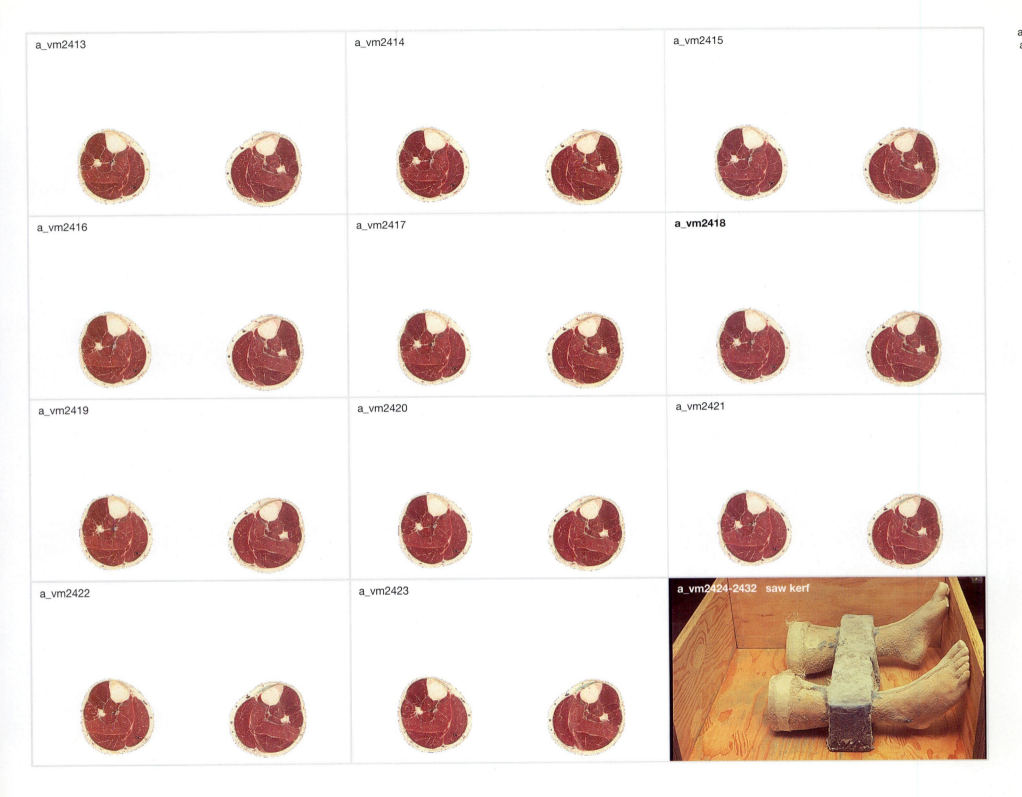

a_vm2413

a_vm2414

a_vm2415

a_vm2416

a_vm2417

a_vm2418

a_vm2419

a_vm2420

a_vm2421

a_vm2422

a_vm2423

a_vm2424-2432 saw kerf

anterior

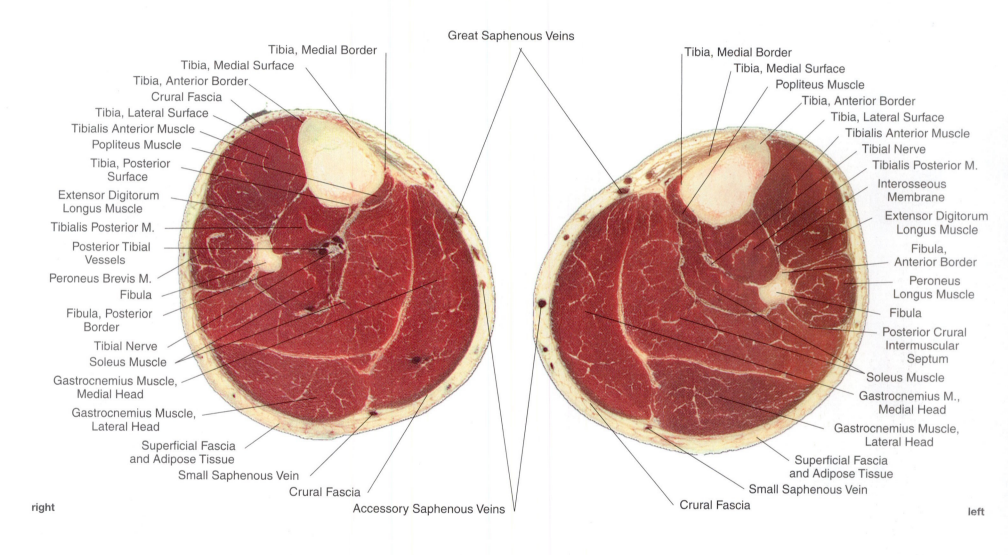

Great Saphenous Veins

Tibia, Medial Border

Tibia, Medial Surface

Tibia, Anterior Border

Crural Fascia

Tibia, Lateral Surface

Tibialis Anterior Muscle

Popliteus Muscle

Tibia, Posterior Surface

Extensor Digitorum Longus Muscle

Tibialis Posterior M.

Posterior Tibial Vessels

Peroneus Brevis M.

Fibula

Fibula, Posterior Border

Tibial Nerve

Soleus Muscle

Gastrocnemius Muscle, Medial Head

Gastrocnemius Muscle, Lateral Head

Superficial Fascia and Adipose Tissue

Small Saphenous Vein

Crural Fascia

Accessory Saphenous Veins

Tibia, Medial Border

Tibia, Medial Surface

Popliteus Muscle

Tibia, Anterior Border

Tibia, Lateral Surface

Tibialis Anterior Muscle

Tibial Nerve

Tibialis Posterior M.

Interosseous Membrane

Extensor Digitorum Longus Muscle

Fibula, Anterior Border

Peroneus Longus Muscle

Fibula

Posterior Crural Intermuscular Septum

Soleus Muscle

Gastrocnemius M., Medial Head

Gastrocnemius Muscle, Lateral Head

Superficial Fascia and Adipose Tissue

Small Saphenous Vein

Crural Fascia

right

left

posterior

238

a_vm2433

a_vm2434

a_vm2435

a_vm2436

a_vm2437

a_vm2438

a_vm2439

a_vm2440

a_vm2441

a_vm2442

a_vm2443

a_vm2444

anterior

Great Saphenous Veins

Tibia, Medial Border

Tibia, Medial Surface

Tibia, Anterior Border

Tibia

Tibialis Anterior Muscle

Popliteus Muscle

Tibia, Posterior Surface

Extensor Digitorum
Longus Muscle

Tibialis Posterior M.

Posterior Tibial
Vessels

Peroneus Brevis M.

Peroneus Longus M.

Fibula

Fibula, Posterior
Border

Tibial Nerve

Soleus Muscle

Gastrocnemius M.,
Medial Head

Gastrocnemius Muscle,
Lateral Head

Plantaris Tendon

Small Saphenous Vein

Crural Fascia

Accessory Saphenous Veins

Tibia, Medial Border

Popliteus Muscle

Tibia, Medial Surface

Tibia, Anterior Border

Tibia, Lateral Surface

Tibialis Anterior Muscle

Interosseous
Membrane

Tibialis Posterior
Muscle

Extensor
Digitorum
Longus Muscle

Fibula, Anterior
Border

Peroneus
Longus Muscle

Peroneus
Brevis Muscle

Fibula

Tibial Nerve

Soleus Muscle

Gastrocnemius M.,
Medial Head

Gastrocnemius Muscle,
Lateral Head

Plantaris Tendon

Small Saphenous Vein

Crural Fascia

right

left

posterior

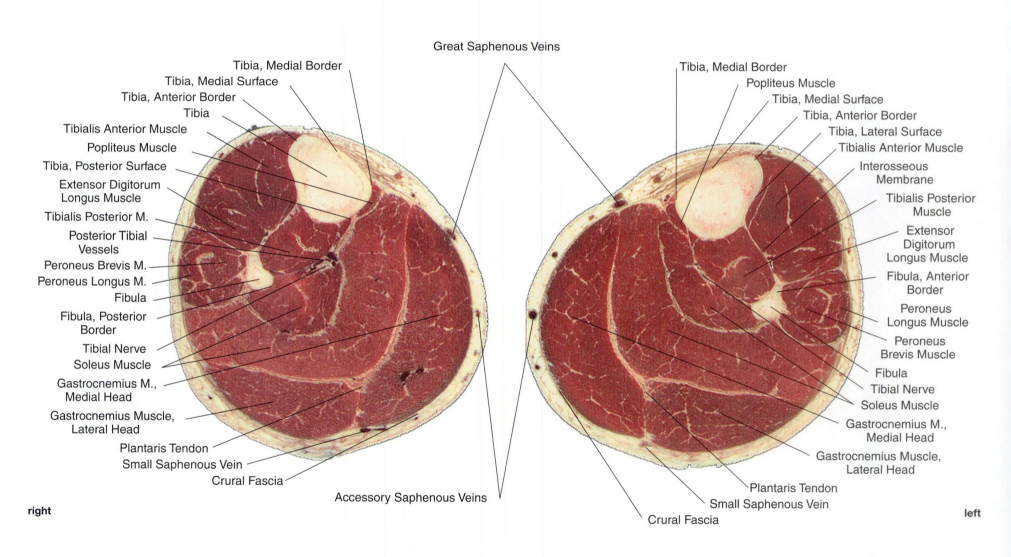

240

a_vm2445

a_vm2446

a_vm2447

a_vm2448

a_vm2449

a_vm2450

a_vm2451

a_vm2452

a_vm2453

a_vm2454

a_vm2455

a_vm2456

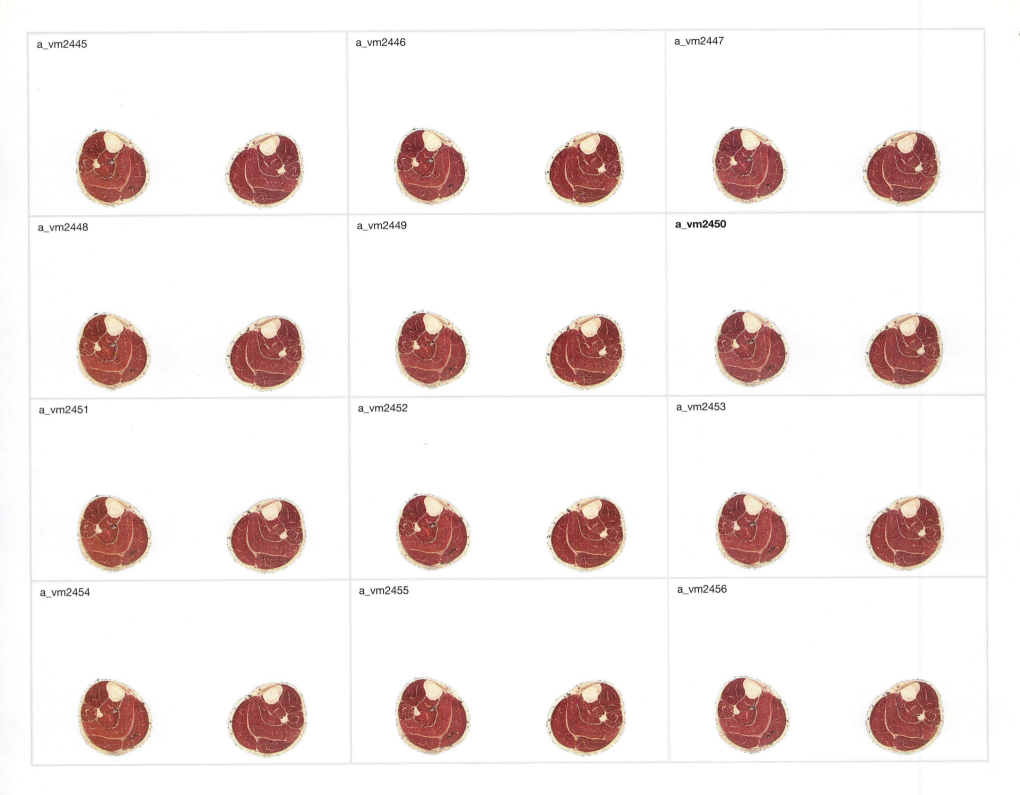

anterior

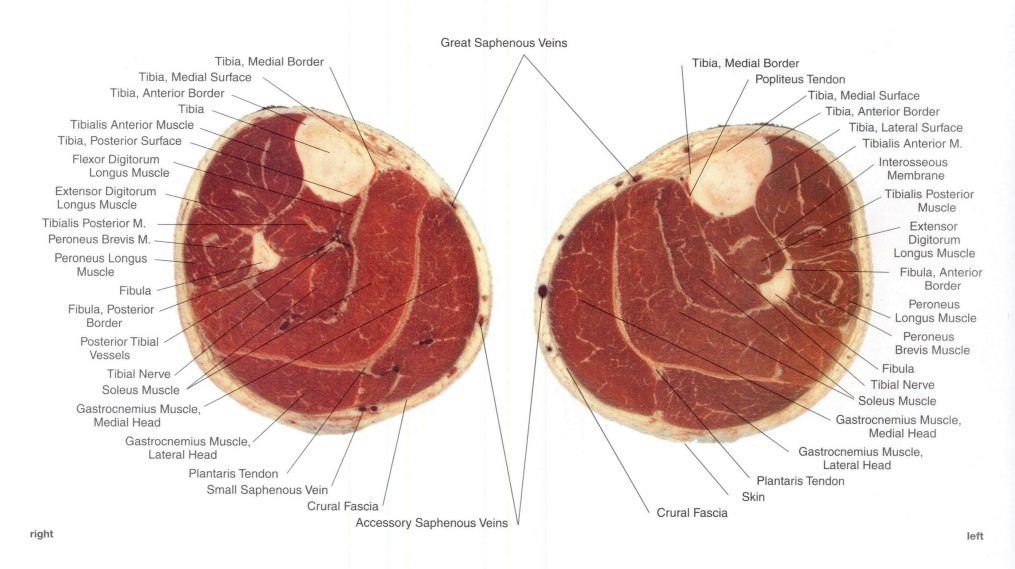

Great Saphenous Veins

Tibia, Medial Border

Tibia, Medial Surface

Tibia, Anterior Border

Tibia

Tibialis Anterior Muscle

Tibia, Posterior Surface

Flexor Digitorum
Longus Muscle

Extensor Digitorum
Longus Muscle

Tibialis Posterior M.

Peroneus Brevis M.

Peroneus Longus
Muscle

Fibula

Fibula, Posterior
Border

Posterior Tibial
Vessels

Tibial Nerve

Soleus Muscle

Gastrocnemius Muscle,
Medial Head

Gastrocnemius Muscle,
Lateral Head

Plantaris Tendon

Small Saphenous Vein

Crural Fascia

Accessory Saphenous Veins

Tibia, Medial Border

Popliteus Tendon

Tibia, Medial Surface

Tibia, Anterior Border

Tibia, Lateral Surface

Tibialis Anterior M.

Interosseous
Membrane

Tibialis Posterior
Muscle

Extensor
Digitorum
Longus Muscle

Fibula, Anterior
Border

Peroneus
Longus Muscle

Peroneus
Brevis Muscle

Fibula

Tibial Nerve

Soleus Muscle

Gastrocnemius Muscle,
Medial Head

Gastrocnemius Muscle,
Lateral Head

Plantaris Tendon

Skin

Crural Fascia

right

left

posterior

242

a_vm2457

a_vm2458

a_vm2459

a_vm2460

a_vm2461

a_vm2462

a_vm2463

a_vm2464

a_vm2465

a_vm2466

a_vm2467

a_vm2468

anterior

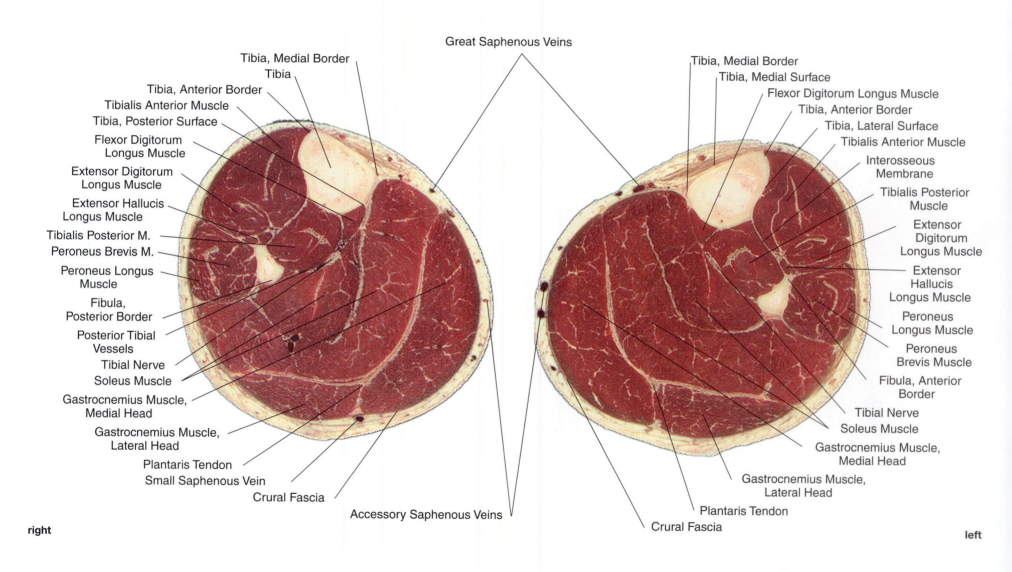

Great Saphenous Veins

Tibia, Medial Border
Tibia

Tibia, Anterior Border
Tibialis Anterior Muscle
Tibia, Posterior Surface
Flexor Digitorum
Longus Muscle

Extensor Digitorum
Longus Muscle
Extensor Hallucis
Longus Muscle

Tibialis Posterior M.
Peroneus Brevis M.

Peroneus Longus
Muscle

Fibula,
Posterior Border

Posterior Tibial
Vessels

Tibial Nerve

Soleus Muscle

Gastrocnemius Muscle,
Medial Head

Gastrocnemius Muscle,
Lateral Head

Plantaris Tendon
Small Saphenous Vein
Crural Fascia

Tibia, Medial Border
Tibia, Medial Surface
Flexor Digitorum Longus Muscle
Tibia, Anterior Border
Tibia, Lateral Surface
Tibialis Anterior Muscle

Interosseous
Membrane

Tibialis Posterior
Muscle

Extensor
Digitorum
Longus Muscle

Extensor
Hallucis
Longus Muscle

Peroneus
Longus Muscle

Peroneus
Brevis Muscle

Fibula, Anterior
Border

Tibial Nerve
Soleus Muscle

Gastrocnemius Muscle,
Medial Head

Gastrocnemius Muscle,
Lateral Head

Plantaris Tendon

Crural Fascia

Accessory Saphenous Veins

right

left

posterior

a_vm2469

a_vm2470

a_vm2471

a_vm2472

a_vm2473

a_vm2474

a_vm2475

a_vm2476

a_vm2477

a_vm2478

a_vm2479

a_vm2480

anterior

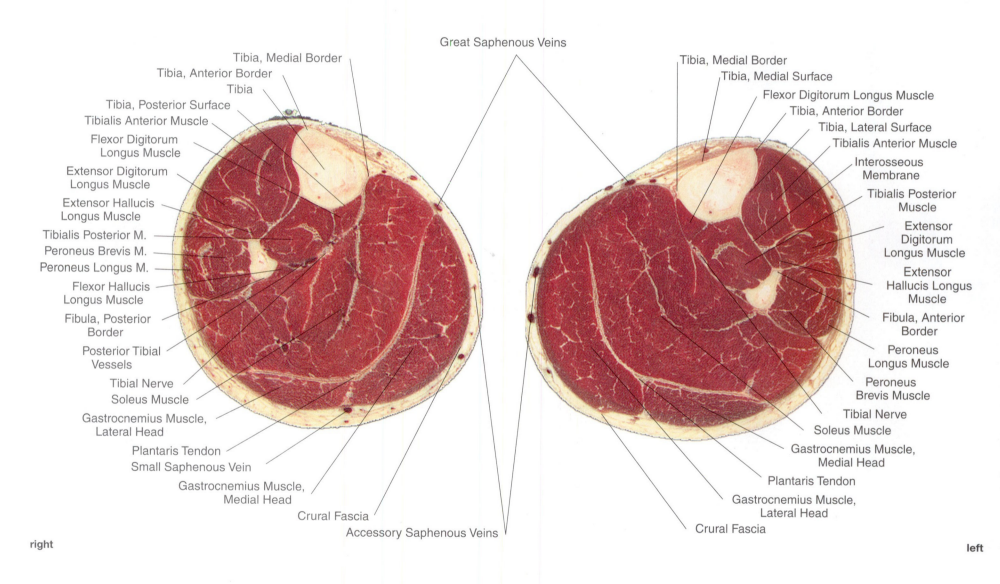

Great Saphenous Veins

Tibia, Medial Border
Tibia, Anterior Border
Tibia
Tibia, Posterior Surface
Tibialis Anterior Muscle
Flexor Digitorum
Longus Muscle
Extensor Digitorum
Longus Muscle
Extensor Hallucis
Longus Muscle
Tibialis Posterior M.
Peroneus Brevis M.
Peroneus Longus M.
Flexor Hallucis
Longus Muscle
Fibula, Posterior
Border
Posterior Tibial
Vessels
Tibial Nerve
Soleus Muscle
Gastrocnemius Muscle,
Lateral Head
Plantaris Tendon
Small Saphenous Vein
Gastrocnemius Muscle,
Medial Head
Crural Fascia
Accessory Saphenous Veins

Tibia, Medial Border
Tibia, Medial Surface
Flexor Digitorum Longus Muscle
Tibia, Anterior Border
Tibia, Lateral Surface
Tibialis Anterior Muscle
Interosseous
Membrane
Tibialis Posterior
Muscle
Extensor
Digitorum
Longus Muscle
Extensor
Hallucis Longus
Muscle
Fibula, Anterior
Border
Peroneus
Longus Muscle
Peroneus
Brevis Muscle
Tibial Nerve
Soleus Muscle
Gastrocnemius Muscle,
Medial Head
Plantaris Tendon
Gastrocnemius Muscle,
Lateral Head
Crural Fascia

right

left

posterior

a_vm2481

a_vm2482

a_vm2483

a_vm2484

a_vm2485

a_vm2486

a_vm2487

a_vm2488

a_vm2489

a_vm2490

a_vm2491

a_vm2492

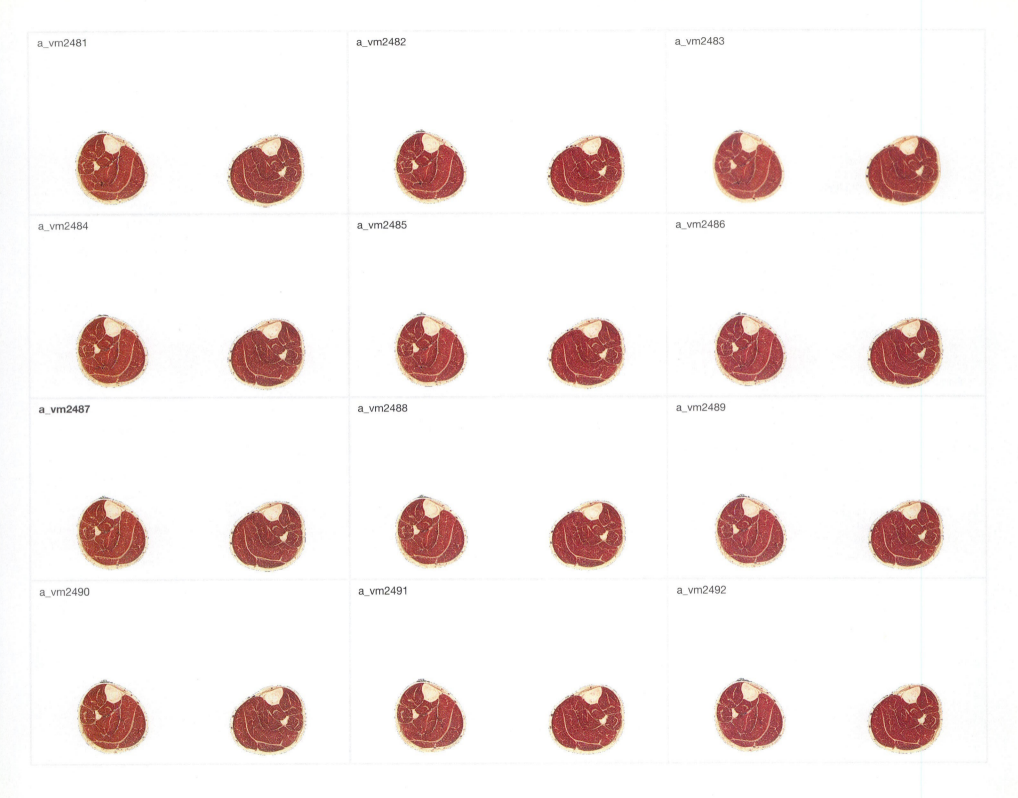

anterior

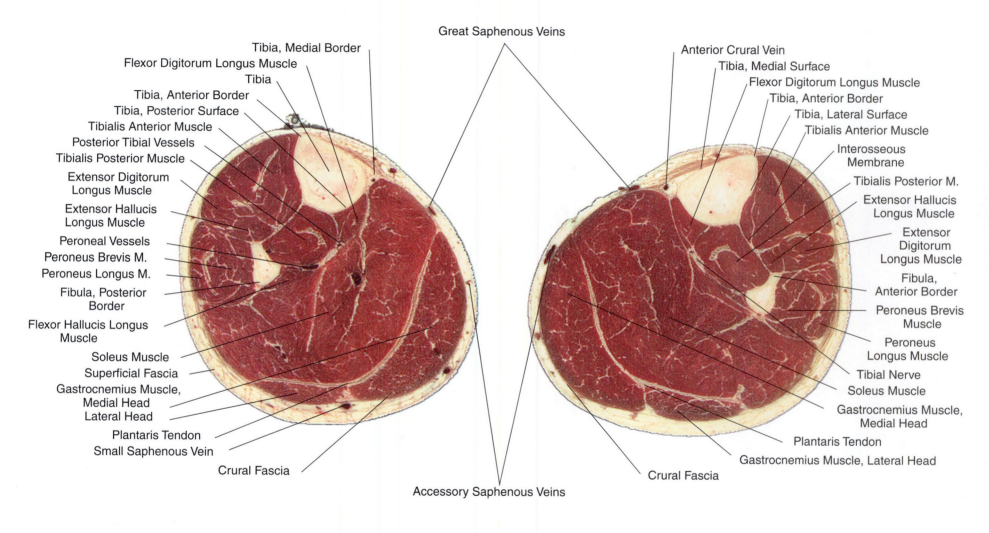

Great Saphenous Veins

Tibia, Medial Border

Flexor Digitorum Longus Muscle

Tibia

Tibia, Anterior Border

Tibia, Posterior Surface

Tibialis Anterior Muscle

Posterior Tibial Vessels

Tibialis Posterior Muscle

Extensor Digitorum
Longus Muscle

Extensor Hallucis
Longus Muscle

Peroneal Vessels

Peroneus Brevis M.

Peroneus Longus M.

Fibula, Posterior
Border

Flexor Hallucis Longus
Muscle

Soleus Muscle

Superficial Fascia

Gastrocnemius Muscle,
Medial Head
Lateral Head

Plantaris Tendon

Small Saphenous Vein

Crural Fascia

Anterior Crural Vein

Tibia, Medial Surface

Flexor Digitorum Longus Muscle

Tibia, Anterior Border

Tibia, Lateral Surface

Tibialis Anterior Muscle

Interosseous
Membrane

Tibialis Posterior M.

Extensor Hallucis
Longus Muscle

Extensor
Digitorum
Longus Muscle

Fibula,
Anterior Border

Peroneus Brevis
Muscle

Peroneus
Longus Muscle

Tibial Nerve

Soleus Muscle

Gastrocnemius Muscle,
Medial Head

Plantaris Tendon

Gastrocnemius Muscle, Lateral Head

Crural Fascia

Accessory Saphenous Veins

right

left

posterior

248

a_vm2493

a_vm2494

a_vm2495

a_vm2496

a_vm2497

a_vm2498

a_vm2499

a_vm2500

a_vm2501

a_vm2502

a_vm2503

a_vm2504

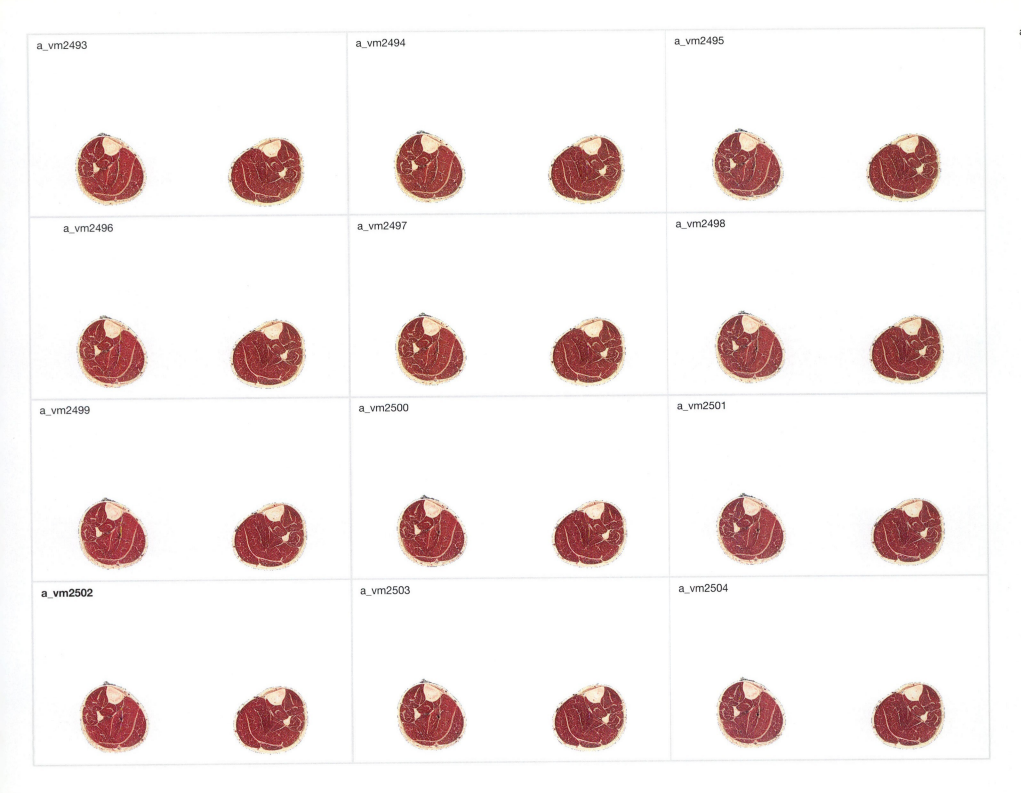

anterior

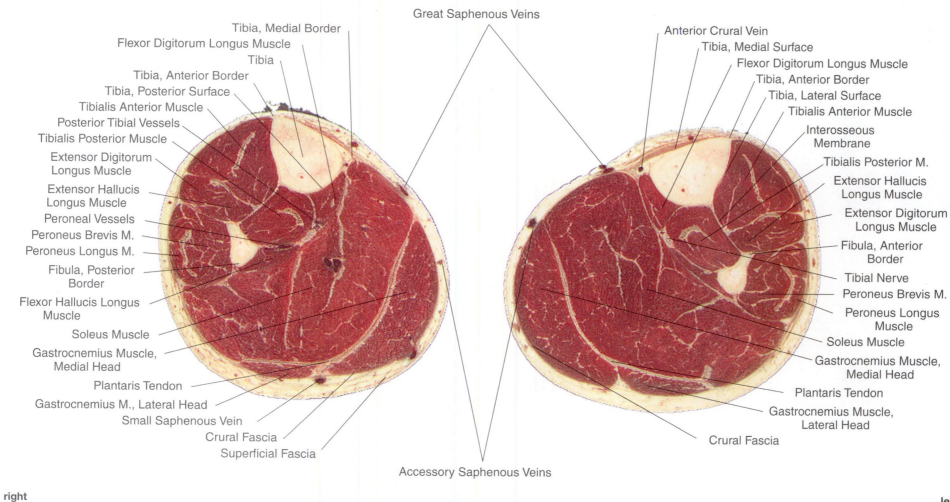

Great Saphenous Veins

Tibia, Medial Border

Flexor Digitorum Longus Muscle

Tibia

Tibia, Anterior Border

Tibia, Posterior Surface

Tibialis Anterior Muscle

Posterior Tibial Vessels

Tibialis Posterior Muscle

Extensor Digitorum
Longus Muscle

Extensor Hallucis
Longus Muscle

Peroneal Vessels

Peroneus Brevis M.

Peroneus Longus M.

Fibula, Posterior
Border

Flexor Hallucis Longus
Muscle

Soleus Muscle

Gastrocnemius Muscle,
Medial Head

Plantaris Tendon

Gastrocnemius M., Lateral Head

Small Saphenous Vein

Crural Fascia

Superficial Fascia

Anterior Crural Vein

Tibia, Medial Surface

Flexor Digitorum Longus Muscle

Tibia, Anterior Border

Tibia, Lateral Surface

Tibialis Anterior Muscle

Interosseous
Membrane

Tibialis Posterior M.

Extensor Hallucis
Longus Muscle

Extensor Digitorum
Longus Muscle

Fibula, Anterior
Border

Tibial Nerve

Peroneus Brevis M.

Peroneus Longus
Muscle

Soleus Muscle

Gastrocnemius Muscle,
Medial Head

Plantaris Tendon

Gastrocnemius Muscle,
Lateral Head

Crural Fascia

Accessory Saphenous Veins

right

left

posterior

a_vm2505

a_vm2506

a_vm2507

a_vm2508

a_vm2509

a_vm2510

a_vm2511

a_vm2512

a_vm2513

a_vm2514

a_vm2515

a_vm2516

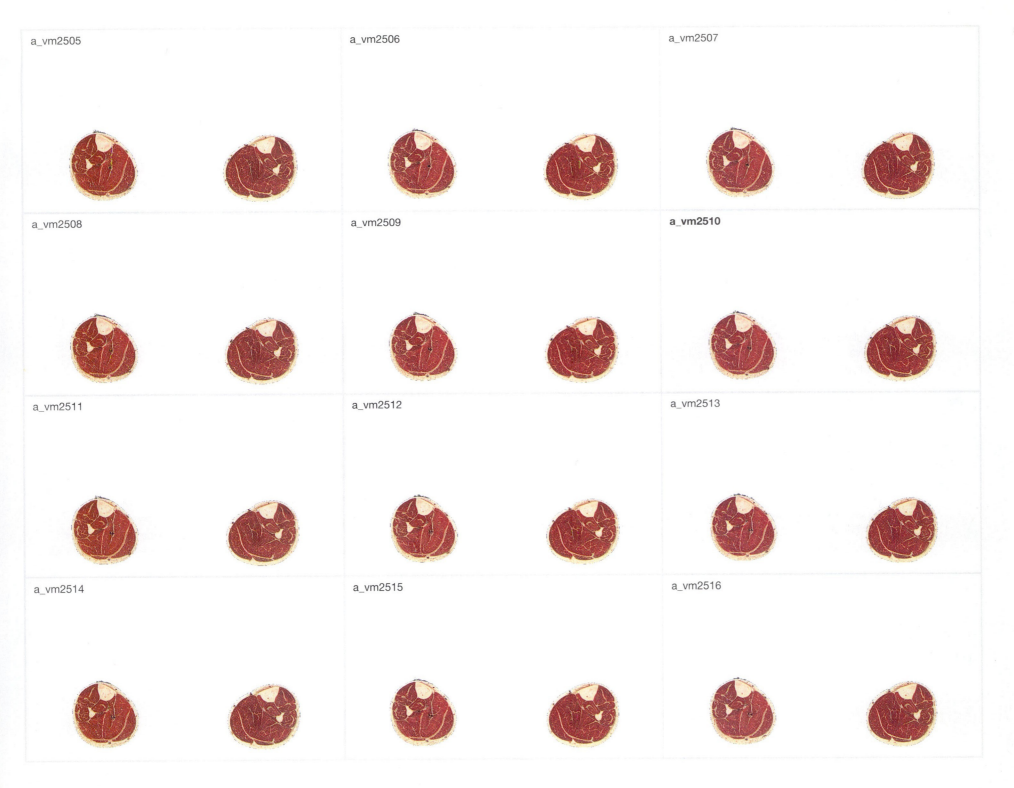

anterior

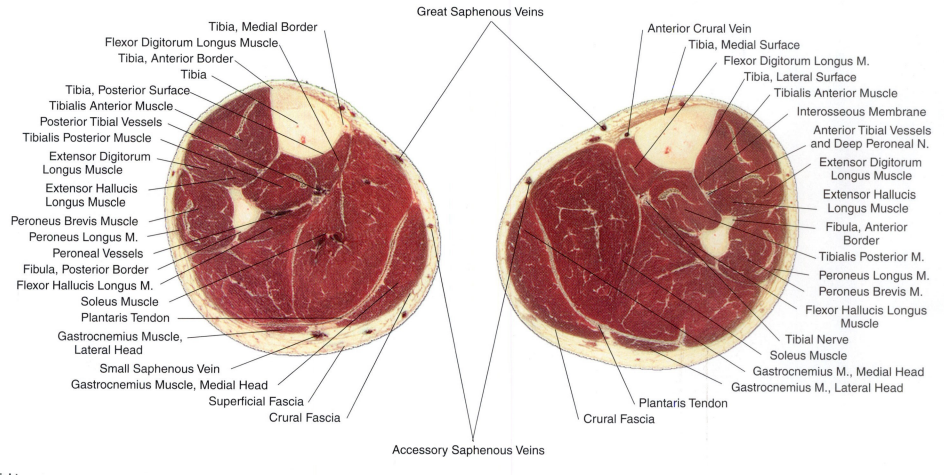

Great Saphenous Veins

Tibia, Medial Border

Flexor Digitorum Longus Muscle

Tibia, Anterior Border

Tibia

Tibia, Posterior Surface

Tibialis Anterior Muscle

Posterior Tibial Vessels

Tibialis Posterior Muscle

Extensor Digitorum
Longus Muscle

Extensor Hallucis
Longus Muscle

Peroneus Brevis Muscle

Peroneus Longus M.

Peroneal Vessels

Fibula, Posterior Border

Flexor Hallucis Longus M.

Soleus Muscle

Plantaris Tendon

Gastrocnemius Muscle,
Lateral Head

Small Saphenous Vein

Gastrocnemius Muscle, Medial Head

Superficial Fascia

Crural Fascia

Anterior Crural Vein

Tibia, Medial Surface

Flexor Digitorum Longus M.

Tibia, Lateral Surface

Tibialis Anterior Muscle

Interosseous Membrane

Anterior Tibial Vessels
and Deep Peroneal N.

Extensor Digitorum
Longus Muscle

Extensor Hallucis
Longus Muscle

Fibula, Anterior
Border

Tibialis Posterior M.

Peroneus Longus M.

Peroneus Brevis M.

Flexor Hallucis Longus
Muscle

Tibial Nerve

Soleus Muscle

Gastrocnemius M., Medial Head

Gastrocnemius M., Lateral Head

Plantaris Tendon

Crural Fascia

Accessory Saphenous Veins

right

left

posterior

252

a_vm2517

a_vm2518

a_vm2519

a_vm2520

a_vm2521

a_vm2522

a_vm2523

a_vm2524

a_vm2525

a_vm2526

a_vm2527

a_vm2528

anterior

Great Saphenous Veins

Tibia, Medial Border

Tibia, Anterior Border

Flexor Digitorum Longus Muscle

Tibia, Posterior Surface

Tibialis Anterior Muscle

Posterior Tibial Vessels

Tibialis Posterior M.

Extensor Digitorum
Longus Muscle

Extensor Hallucis
Longus Muscle

Peroneus Brevis
Muscle

Peroneus Longus
Tendon

Peroneal Vessels

Fibula, Post. Border

Flexor Hallucis
Longus Muscle

Muscular Vessel

Soleus Muscle

Calcaneal Tendon

Small Saphenous Vein

Superficial Fascia

Gastrocnemius Muscle, Medial Head

Crural Fascia

Anterior Crural Vein

Tibia, Medial Surface

Flexor Digitorum Longus Muscle

Tibia, Anterior Border

Tibia, Lateral Surface

Tibialis Anterior Muscle

Interosseous
Membrane

Anterior Tibial
Vessels and Deep
Peroneal Nerve

Extensor
Digitorum
Longus Muscle

Extensor Hallucis
Longus Muscle

Fibula, Ant. Border

Tibialis Posterior M.

Peroneus Longus M.

Peroneus Brevis M.

Tibial Nerve

Flexor Hallucis Longus M.

Soleus Muscle

Gastrocnemius M., Medial Head

Gastrocnemius Muscle, Lateral Head

Plantaris Tendon

Crural Fascia

Accessory Saphenous Veins

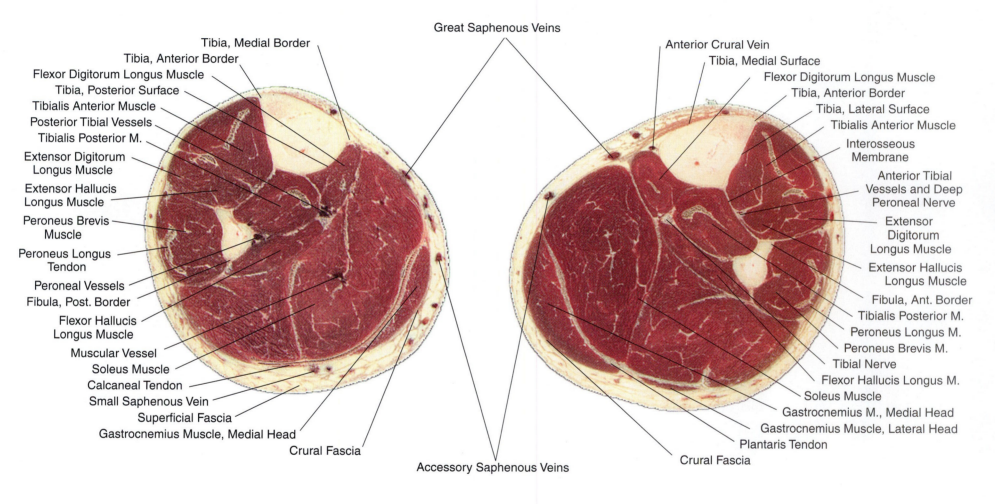

right

left

posterior

a_vm2529

a_vm2530

a_vm2531

a_vm2532

a_vm2533

a_vm2534

a_vm2535

a_vm2536

a_vm2537

a_vm2538

a_vm2539

a_vm2540

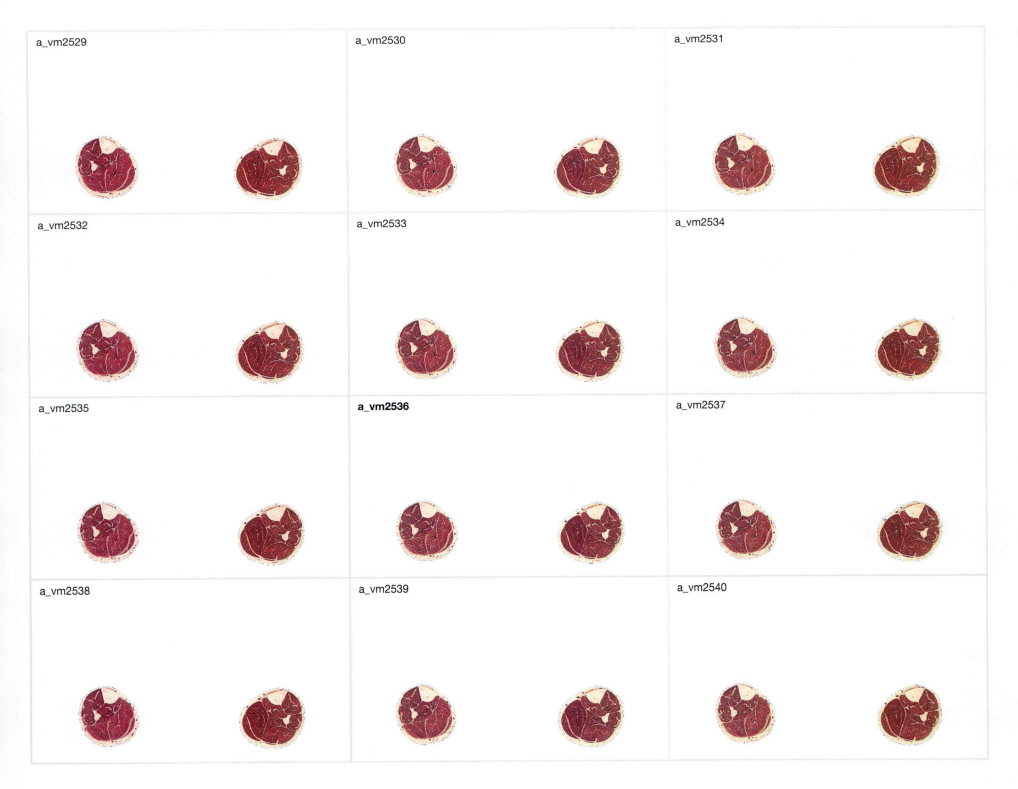

anterior

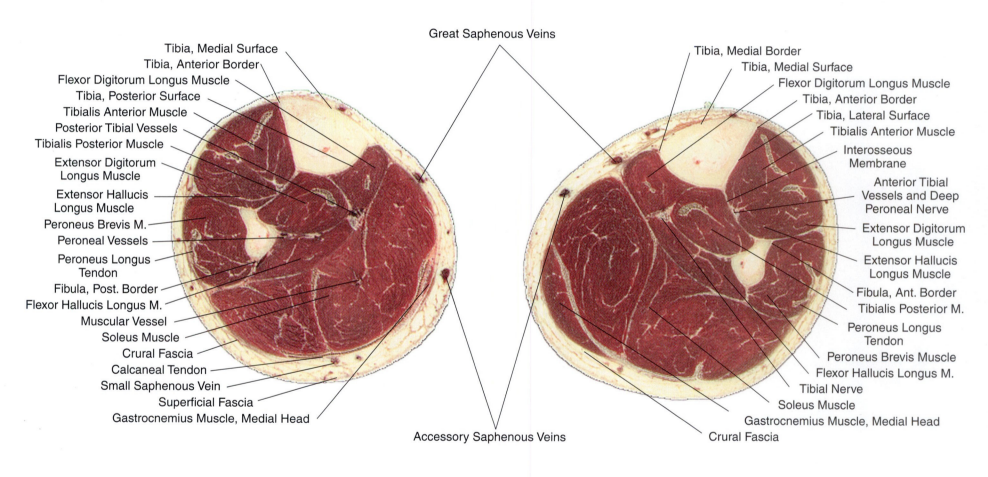

Great Saphenous Veins

Tibia, Medial Surface

Tibia, Anterior Border

Flexor Digitorum Longus Muscle

Tibia, Posterior Surface

Tibialis Anterior Muscle

Posterior Tibial Vessels

Tibialis Posterior Muscle

Extensor Digitorum Longus Muscle

Extensor Hallucis Longus Muscle

Peroneus Brevis M.

Peroneal Vessels

Peroneus Longus Tendon

Fibula, Post. Border

Flexor Hallucis Longus M.

Muscular Vessel

Soleus Muscle

Crural Fascia

Calcaneal Tendon

Small Saphenous Vein

Superficial Fascia

Gastrocnemius Muscle, Medial Head

Tibia, Medial Border

Tibia, Medial Surface

Flexor Digitorum Longus Muscle

Tibia, Anterior Border

Tibia, Lateral Surface

Tibialis Anterior Muscle

Interosseous Membrane

Anterior Tibial Vessels and Deep Peroneal Nerve

Extensor Digitorum Longus Muscle

Extensor Hallucis Longus Muscle

Fibula, Ant. Border

Tibialis Posterior M.

Peroneus Longus Tendon

Peroneus Brevis Muscle

Flexor Hallucis Longus M.

Tibial Nerve

Soleus Muscle

Gastrocnemius Muscle, Medial Head

Crural Fascia

Accessory Saphenous Veins

right

left

posterior

a_vm2541

a_vm2542

a_vm2543

a_vm2544

a_vm2545

a_vm2546

a_vm2547

a_vm2548

a_vm2549

a_vm2550

a_vm2551

a_vm2552

anterior

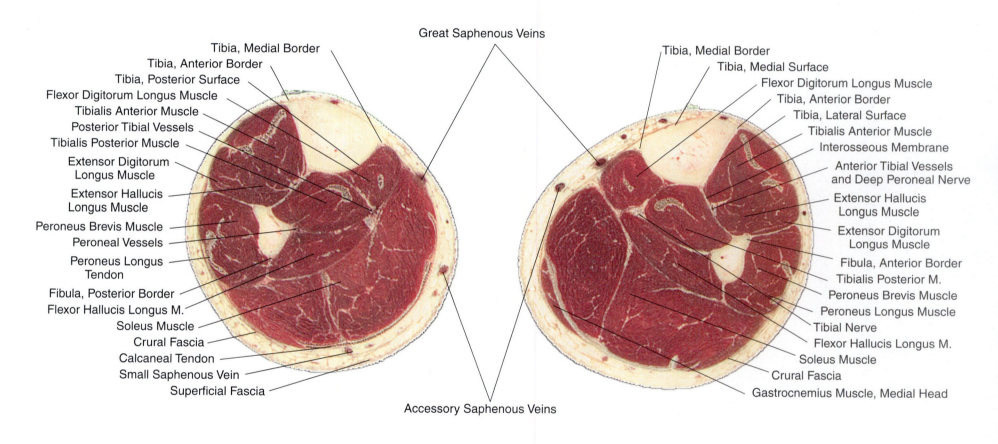

Great Saphenous Veins

Tibia, Medial Border

Tibia, Anterior Border

Tibia, Posterior Surface

Flexor Digitorum Longus Muscle

Tibialis Anterior Muscle

Posterior Tibial Vessels

Tibialis Posterior Muscle

Extensor Digitorum
Longus Muscle

Extensor Hallucis
Longus Muscle

Peroneus Brevis Muscle

Peroneal Vessels

Peroneus Longus
Tendon

Fibula, Posterior Border

Flexor Hallucis Longus M.

Soleus Muscle

Crural Fascia

Calcaneal Tendon

Small Saphenous Vein

Superficial Fascia

Tibia, Medial Border

Tibia, Medial Surface

Flexor Digitorum Longus Muscle

Tibia, Anterior Border

Tibia, Lateral Surface

Tibialis Anterior Muscle

Interosseous Membrane

Anterior Tibial Vessels
and Deep Peroneal Nerve

Extensor Hallucis
Longus Muscle

Extensor Digitorum
Longus Muscle

Fibula, Anterior Border

Tibialis Posterior M.

Peroneus Brevis Muscle

Peroneus Longus Muscle

Tibial Nerve

Flexor Hallucis Longus M.

Soleus Muscle

Crural Fascia

Gastrocnemius Muscle, Medial Head

Accessory Saphenous Veins

right

left

posterior

258

a_vm2553

a_vm2554

a_vm2555

a_vm2556

a_vm2557

a_vm2558

a_vm2559

a_vm2560

a_vm2561

a_vm2562

a_vm2563

a_vm2564

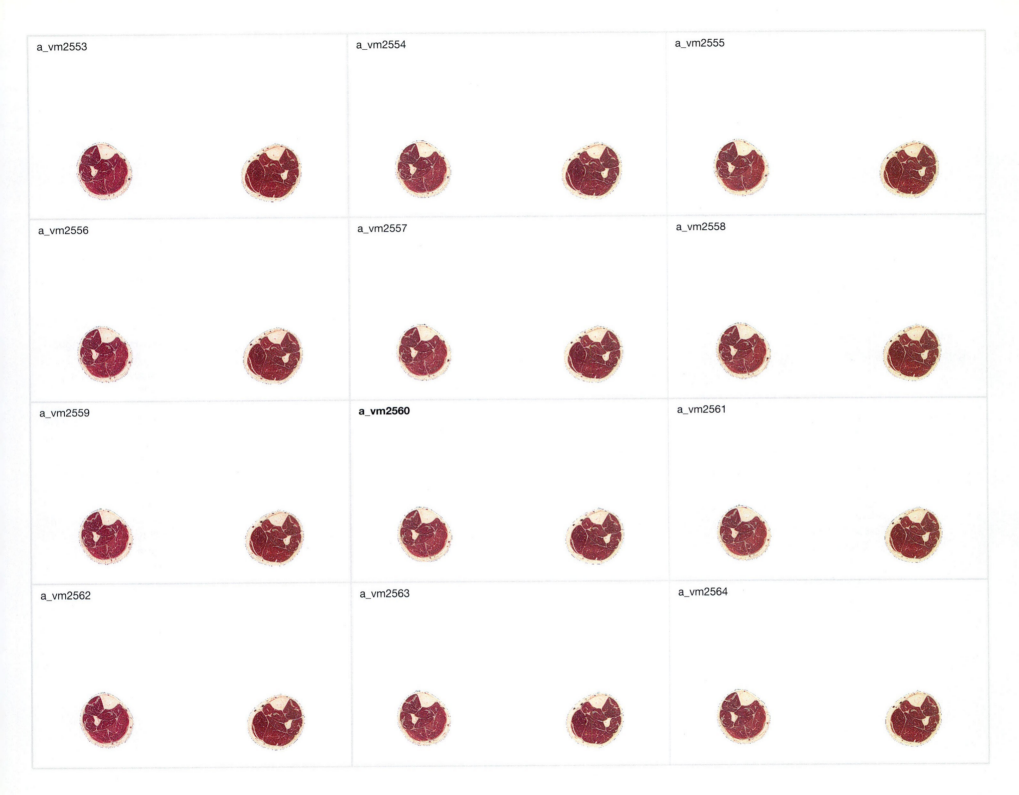

anterior

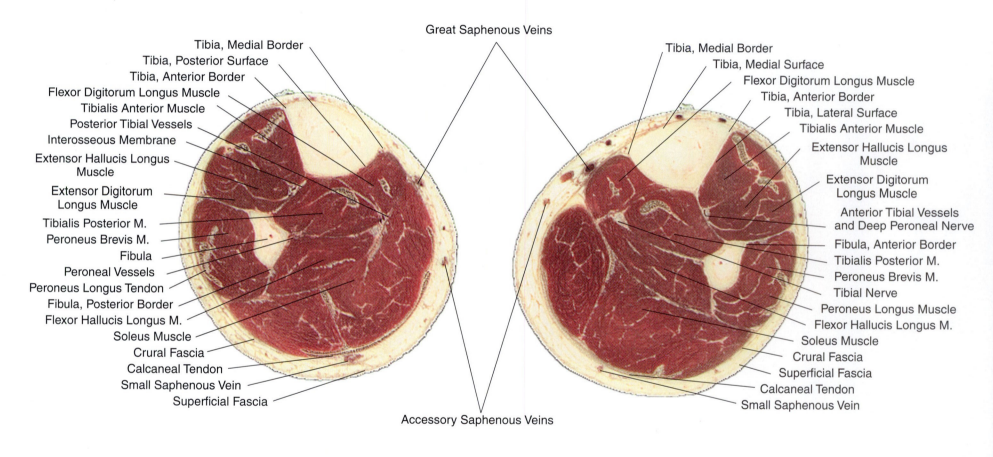

Great Saphenous Veins

Tibia, Medial Border
Tibia, Posterior Surface
Tibia, Anterior Border
Flexor Digitorum Longus Muscle
Tibialis Anterior Muscle
Posterior Tibial Vessels
Interosseous Membrane
Extensor Hallucis Longus Muscle
Extensor Digitorum Longus Muscle
Tibialis Posterior M.
Peroneus Brevis M.
Fibula
Peroneal Vessels
Peroneus Longus Tendon
Fibula, Posterior Border
Flexor Hallucis Longus M.
Soleus Muscle
Crural Fascia
Calcaneal Tendon
Small Saphenous Vein
Superficial Fascia

Tibia, Medial Border
Tibia, Medial Surface
Flexor Digitorum Longus Muscle
Tibia, Anterior Border
Tibia, Lateral Surface
Tibialis Anterior Muscle
Extensor Hallucis Longus Muscle
Extensor Digitorum Longus Muscle
Anterior Tibial Vessels and Deep Peroneal Nerve
Fibula, Anterior Border
Tibialis Posterior M.
Peroneus Brevis M.
Tibial Nerve
Peroneus Longus Muscle
Flexor Hallucis Longus M.
Soleus Muscle
Crural Fascia
Superficial Fascia
Calcaneal Tendon
Small Saphenous Vein

Accessory Saphenous Veins

right

left

posterior

a_vm2565

a_vm2566

a_vm2567

a_vm2568

a_vm2569

a_vm2570

a_vm2571

a_vm2572

a_vm2573

a_vm2574

a_vm2575

a_vm2576

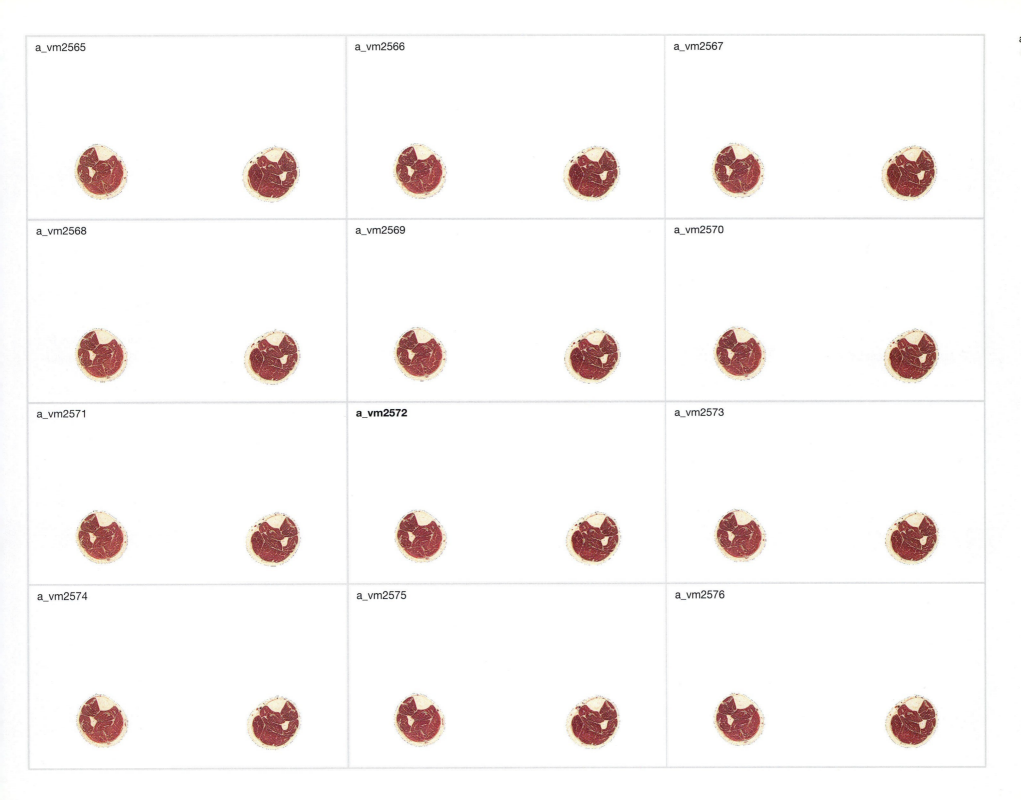

anterior

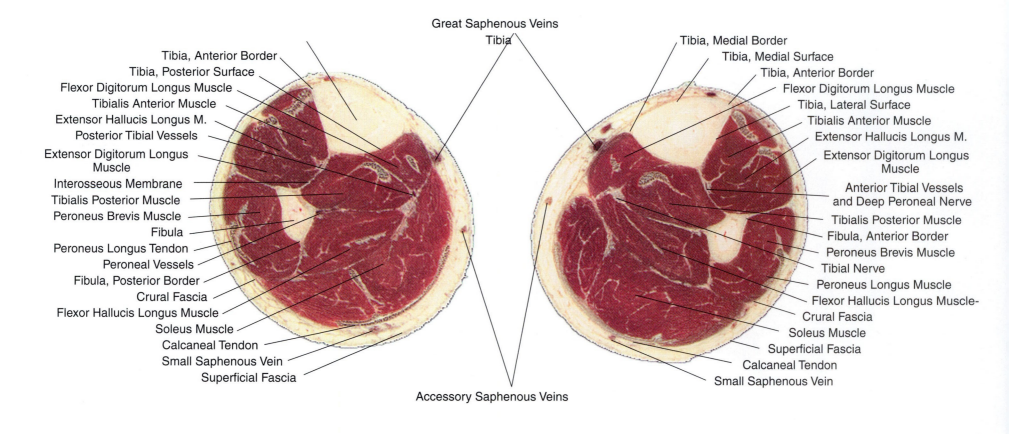

Great Saphenous Veins
Tibia

Tibia, Anterior Border
Tibia, Posterior Surface
Flexor Digitorum Longus Muscle
Tibialis Anterior Muscle
Extensor Hallucis Longus M.
Posterior Tibial Vessels
Extensor Digitorum Longus Muscle
Interosseous Membrane
Tibialis Posterior Muscle
Peroneus Brevis Muscle
Fibula
Peroneus Longus Tendon
Peroneal Vessels
Fibula, Posterior Border
Crural Fascia
Flexor Hallucis Longus Muscle
Soleus Muscle
Calcaneal Tendon
Small Saphenous Vein
Superficial Fascia

Tibia, Medial Border
Tibia, Medial Surface
Tibia, Anterior Border
Flexor Digitorum Longus Muscle
Tibia, Lateral Surface
Tibialis Anterior Muscle
Extensor Hallucis Longus M.
Extensor Digitorum Longus Muscle
Anterior Tibial Vessels and Deep Peroneal Nerve
Tibialis Posterior Muscle
Fibula, Anterior Border
Peroneus Brevis Muscle
Tibial Nerve
Peroneus Longus Muscle
Flexor Hallucis Longus Muscle
Crural Fascia
Soleus Muscle
Superficial Fascia
Calcaneal Tendon
Small Saphenous Vein

Accessory Saphenous Veins

right

left

posterior

a_vm2577

a_vm2578

a_vm2579

a_vm2580

a_vm2581

a_vm2582

a_vm2583

a_vm2584

a_vm2585

a_vm2586

a_vm2587

a_vm2588

anterior

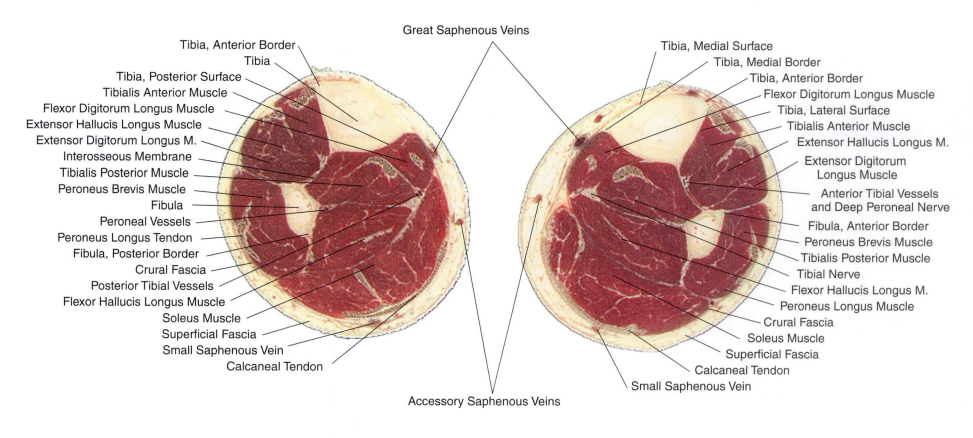

Great Saphenous Veins

Tibia, Anterior Border
Tibia
Tibia, Posterior Surface
Tibialis Anterior Muscle
Flexor Digitorum Longus Muscle
Extensor Hallucis Longus Muscle
Extensor Digitorum Longus M.
Interosseous Membrane
Tibialis Posterior Muscle
Peroneus Brevis Muscle
Fibula
Peroneal Vessels
Peroneus Longus Tendon
Fibula, Posterior Border
Crural Fascia
Posterior Tibial Vessels
Flexor Hallucis Longus Muscle
Soleus Muscle
Superficial Fascia
Small Saphenous Vein
Calcaneal Tendon

Tibia, Medial Surface
Tibia, Medial Border
Tibia, Anterior Border
Flexor Digitorum Longus Muscle
Tibia, Lateral Surface
Tibialis Anterior Muscle
Extensor Hallucis Longus M.
Extensor Digitorum
Longus Muscle
Anterior Tibial Vessels
and Deep Peroneal Nerve
Fibula, Anterior Border
Peroneus Brevis Muscle
Tibialis Posterior Muscle
Tibial Nerve
Flexor Hallucis Longus M.
Peroneus Longus Muscle
Crural Fascia
Soleus Muscle
Superficial Fascia
Calcaneal Tendon
Small Saphenous Vein

Accessory Saphenous Veins

right

left

posterior

a_vm2589

a_vm2590

a_vm2591

a_vm2592

a_vm2593

a_vm2594

a_vm2595

a_vm2596

a_vm2597

a_vm2598

a_vm2599

a_vm2600

anterior

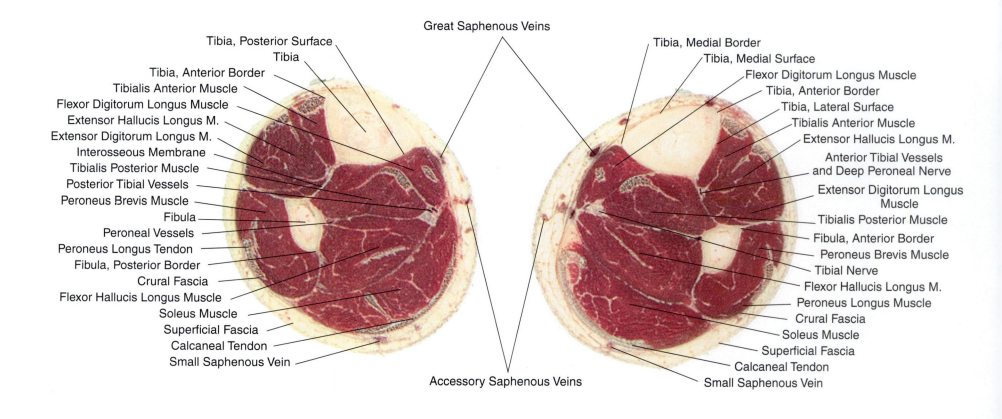

Great Saphenous Veins

Tibia, Posterior Surface
Tibia

Tibia, Anterior Border
Tibialis Anterior Muscle
Flexor Digitorum Longus Muscle
Extensor Hallucis Longus M.
Extensor Digitorum Longus M.
Interosseous Membrane
Tibialis Posterior Muscle
Posterior Tibial Vessels
Peroneus Brevis Muscle
Fibula
Peroneal Vessels
Peroneus Longus Tendon
Fibula, Posterior Border
Crural Fascia
Flexor Hallucis Longus Muscle
Soleus Muscle
Superficial Fascia
Calcaneal Tendon
Small Saphenous Vein

Tibia, Medial Border
Tibia, Medial Surface
Flexor Digitorum Longus Muscle
Tibia, Anterior Border
Tibia, Lateral Surface
Tibialis Anterior Muscle
Extensor Hallucis Longus M.
Anterior Tibial Vessels
and Deep Peroneal Nerve
Extensor Digitorum Longus
Muscle
Tibialis Posterior Muscle
Fibula, Anterior Border
Peroneus Brevis Muscle
Tibial Nerve
Flexor Hallucis Longus M.
Peroneus Longus Muscle
Crural Fascia
Soleus Muscle
Superficial Fascia
Calcaneal Tendon
Small Saphenous Vein

Accessory Saphenous Veins

right

left

posterior

266

a_vm2601

a_vm2602

a_vm2603

a_vm2604

a_vm2605

a_vm2606

a_vm2607

a_vm2608

a_vm2609

a_vm2610

a_vm2611

a_vm2612

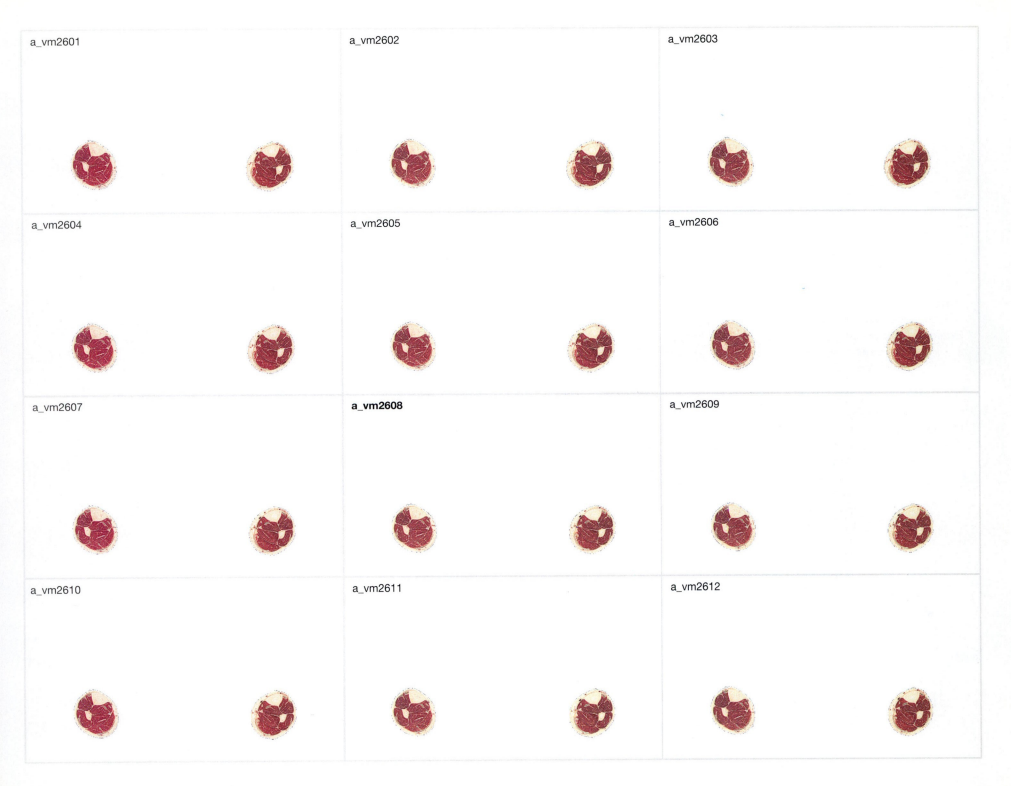

anterior

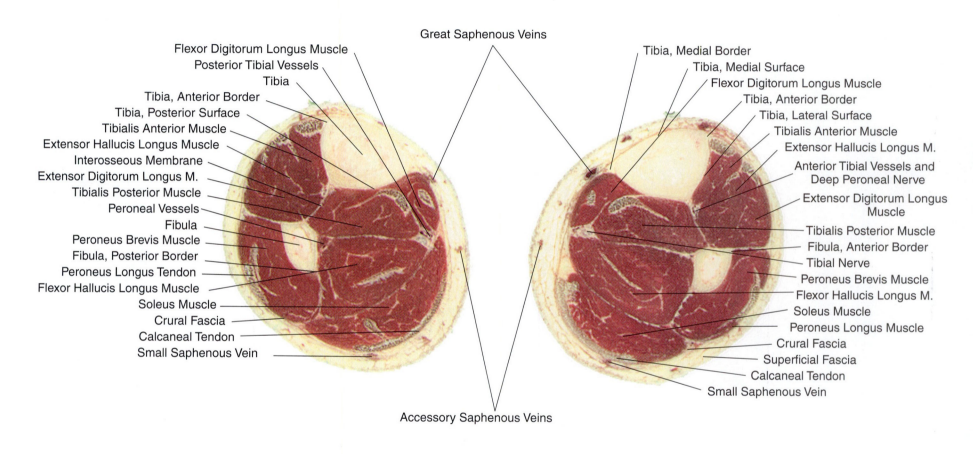

Great Saphenous Veins

Flexor Digitorum Longus Muscle
Posterior Tibial Vessels
Tibia
Tibia, Anterior Border
Tibia, Posterior Surface
Tibialis Anterior Muscle
Extensor Hallucis Longus Muscle
Interosseous Membrane
Extensor Digitorum Longus M.
Tibialis Posterior Muscle
Peroneal Vessels
Fibula
Peroneus Brevis Muscle
Fibula, Posterior Border
Peroneus Longus Tendon
Flexor Hallucis Longus Muscle
Soleus Muscle
Crural Fascia
Calcaneal Tendon
Small Saphenous Vein

Tibia, Medial Border
Tibia, Medial Surface
Flexor Digitorum Longus Muscle
Tibia, Anterior Border
Tibia, Lateral Surface
Tibialis Anterior Muscle
Extensor Hallucis Longus M.
Anterior Tibial Vessels and
Deep Peroneal Nerve
Extensor Digitorum Longus
Muscle
Tibialis Posterior Muscle
Fibula, Anterior Border
Tibial Nerve
Peroneus Brevis Muscle
Flexor Hallucis Longus M.
Soleus Muscle
Peroneus Longus Muscle
Crural Fascia
Superficial Fascia
Calcaneal Tendon
Small Saphenous Vein

Accessory Saphenous Veins

right

left

posterior

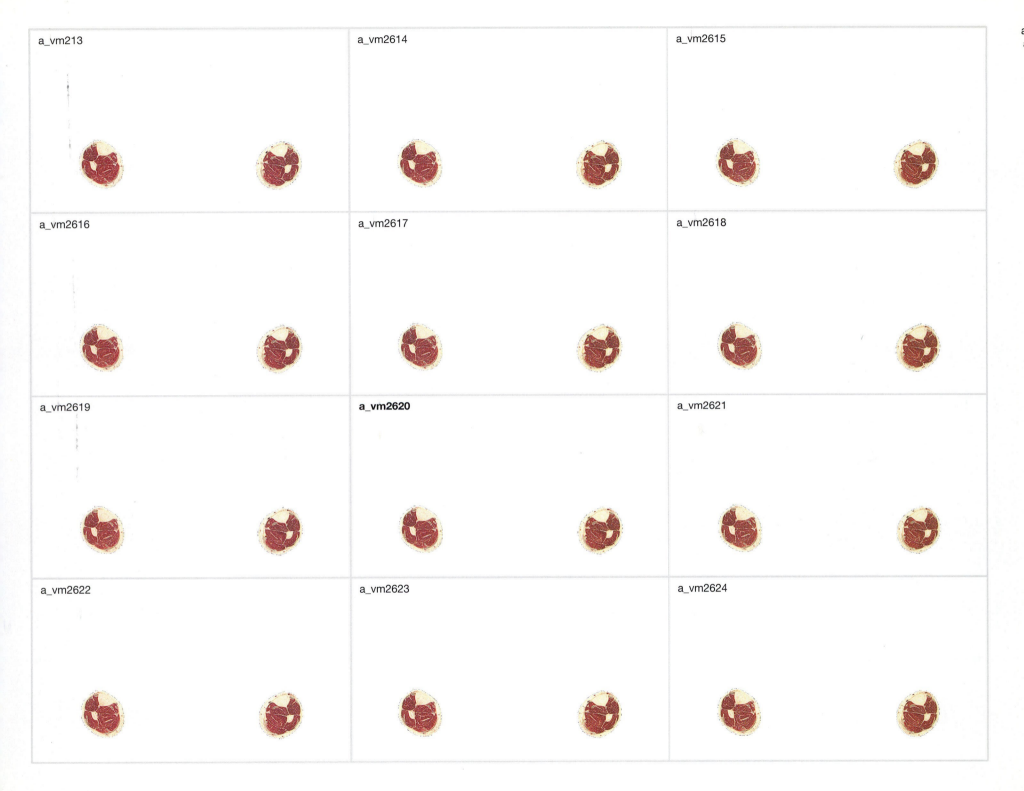

a_vm213

a_vm2614

a_vm2615

a_vm2616

a_vm2617

a_vm2618

a_vm2619

a_vm2620

a_vm2621

a_vm2622

a_vm2623

a_vm2624

anterior

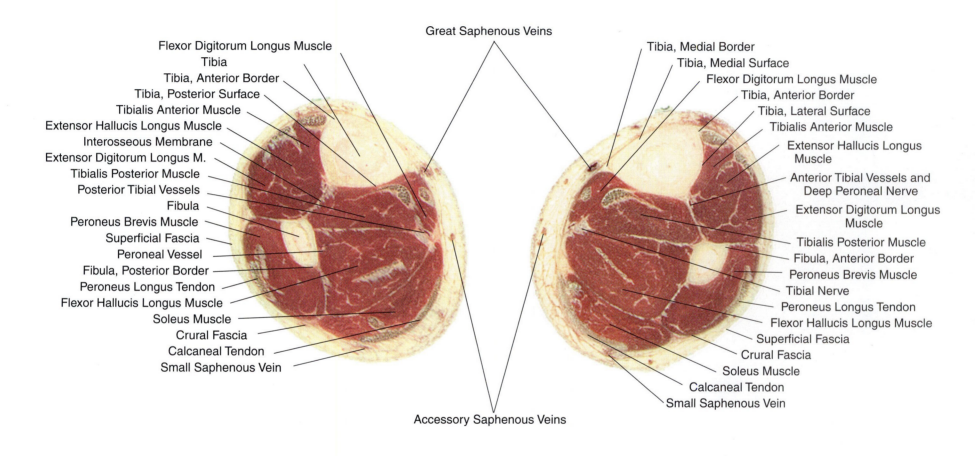

Great Saphenous Veins

Flexor Digitorum Longus Muscle
Tibia
Tibia, Anterior Border
Tibia, Posterior Surface
Tibialis Anterior Muscle
Extensor Hallucis Longus Muscle
Interosseous Membrane
Extensor Digitorum Longus M.
Tibialis Posterior Muscle
Posterior Tibial Vessels
Fibula
Peroneus Brevis Muscle
Superficial Fascia
Peroneal Vessel
Fibula, Posterior Border
Peroneus Longus Tendon
Flexor Hallucis Longus Muscle
Soleus Muscle
Crural Fascia
Calcaneal Tendon
Small Saphenous Vein

Tibia, Medial Border
Tibia, Medial Surface
Flexor Digitorum Longus Muscle
Tibia, Anterior Border
Tibia, Lateral Surface
Tibialis Anterior Muscle
Extensor Hallucis Longus Muscle
Anterior Tibial Vessels and Deep Peroneal Nerve
Extensor Digitorum Longus Muscle
Tibialis Posterior Muscle
Fibula, Anterior Border
Peroneus Brevis Muscle
Tibial Nerve
Peroneus Longus Tendon
Flexor Hallucis Longus Muscle
Superficial Fascia
Crural Fascia
Soleus Muscle
Calcaneal Tendon
Small Saphenous Vein

Accessory Saphenous Veins

right

left

posterior

270

a_vm2625

a_vm2626

a_vm2627

a_vm2628

a_vm2629

a_vm2630

a_vm2631

a_vm2632

a_vm2633

a_vm2634

a_vm2635

a_vm2636

anterior

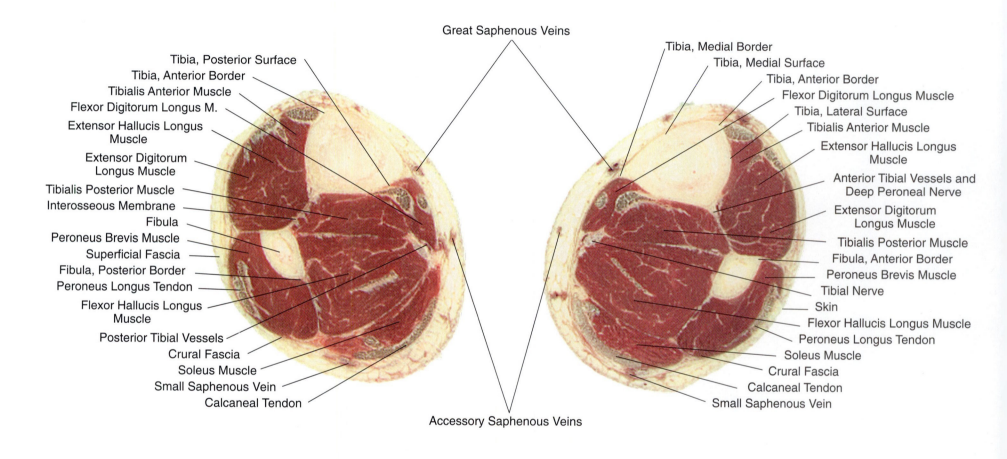

Great Saphenous Veins

Tibia, Posterior Surface
Tibia, Anterior Border
Tibialis Anterior Muscle
Flexor Digitorum Longus M.
Extensor Hallucis Longus Muscle
Extensor Digitorum Longus Muscle
Tibialis Posterior Muscle
Interosseous Membrane
Fibula
Peroneus Brevis Muscle
Superficial Fascia
Fibula, Posterior Border
Peroneus Longus Tendon
Flexor Hallucis Longus Muscle
Posterior Tibial Vessels
Crural Fascia
Soleus Muscle
Small Saphenous Vein
Calcaneal Tendon

Tibia, Medial Border
Tibia, Medial Surface
Tibia, Anterior Border
Flexor Digitorum Longus Muscle
Tibia, Lateral Surface
Tibialis Anterior Muscle
Extensor Hallucis Longus Muscle
Anterior Tibial Vessels and Deep Peroneal Nerve
Extensor Digitorum Longus Muscle
Tibialis Posterior Muscle
Fibula, Anterior Border
Peroneus Brevis Muscle
Tibial Nerve
Skin
Flexor Hallucis Longus Muscle
Peroneus Longus Tendon
Soleus Muscle
Crural Fascia
Calcaneal Tendon
Small Saphenous Vein

Accessory Saphenous Veins

right

left

posterior

a_vm2637

a_vm2638

a_vm2639

a_vm2640

a_vm2641

a_vm2642

a_vm2643

a_vm2644

a_vm2645

a_vm2646

a_vm2647

a_vm2648

anterior

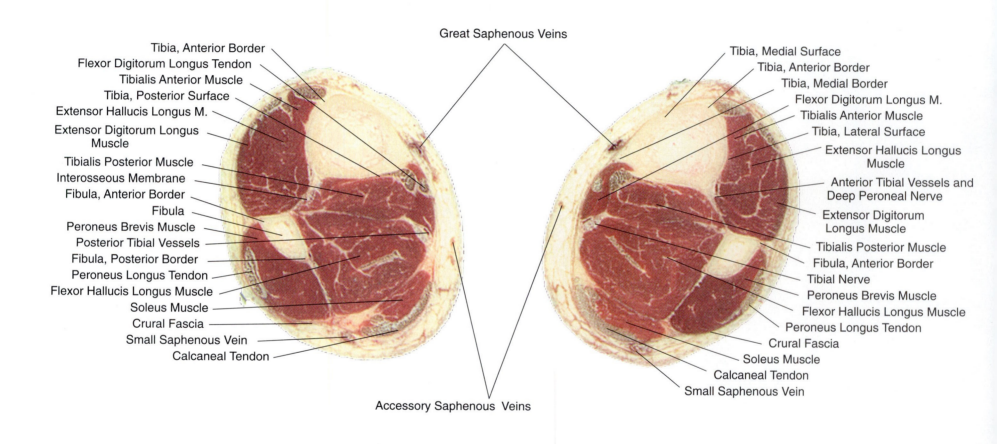

Great Saphenous Veins

Tibia, Anterior Border
Flexor Digitorum Longus Tendon
Tibialis Anterior Muscle
Tibia, Posterior Surface
Extensor Hallucis Longus M.
Extensor Digitorum Longus Muscle
Tibialis Posterior Muscle
Interosseous Membrane
Fibula, Anterior Border
Fibula
Peroneus Brevis Muscle
Posterior Tibial Vessels
Fibula, Posterior Border
Peroneus Longus Tendon
Flexor Hallucis Longus Muscle
Soleus Muscle
Crural Fascia
Small Saphenous Vein
Calcaneal Tendon

Tibia, Medial Surface
Tibia, Anterior Border
Tibia, Medial Border
Flexor Digitorum Longus M.
Tibialis Anterior Muscle
Tibia, Lateral Surface
Extensor Hallucis Longus Muscle
Anterior Tibial Vessels and Deep Peroneal Nerve
Extensor Digitorum Longus Muscle
Tibialis Posterior Muscle
Fibula, Anterior Border
Tibial Nerve
Peroneus Brevis Muscle
Flexor Hallucis Longus Muscle
Peroneus Longus Tendon
Crural Fascia
Soleus Muscle
Calcaneal Tendon
Small Saphenous Vein

Accessory Saphenous Veins

right

left

posterior

a_vm2649

a_vm2650

a_vm2651

a_vm2652

a_vm2653

a_vm2654

a_vm2655

a_vm2656

a_vm2657

a_vm2658

a_vm2659

a_vm2660

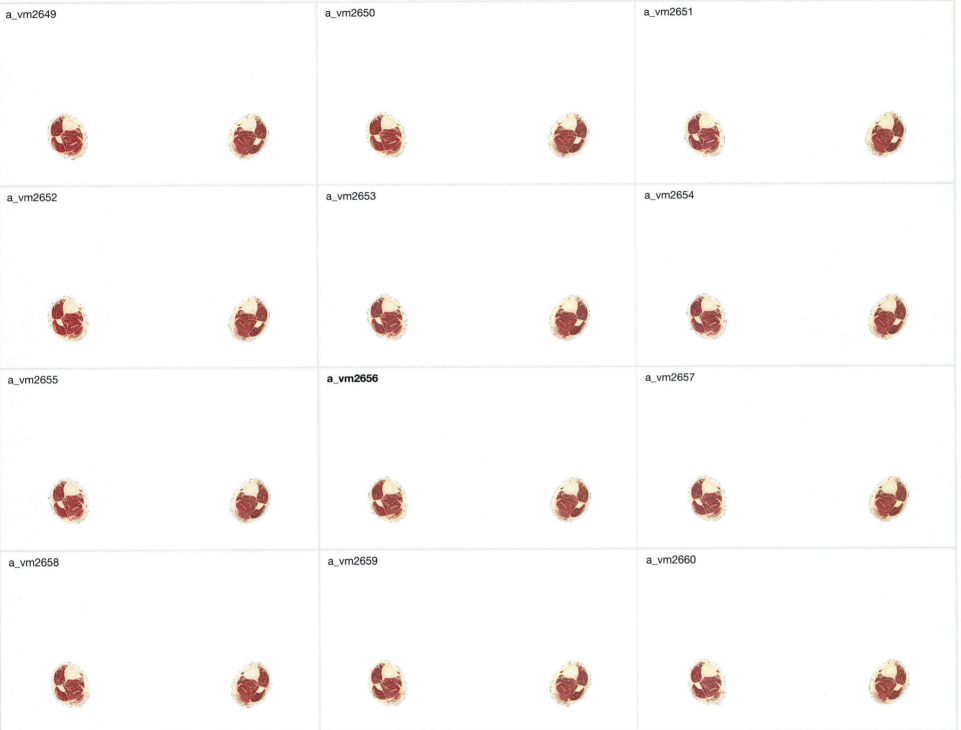

anterior

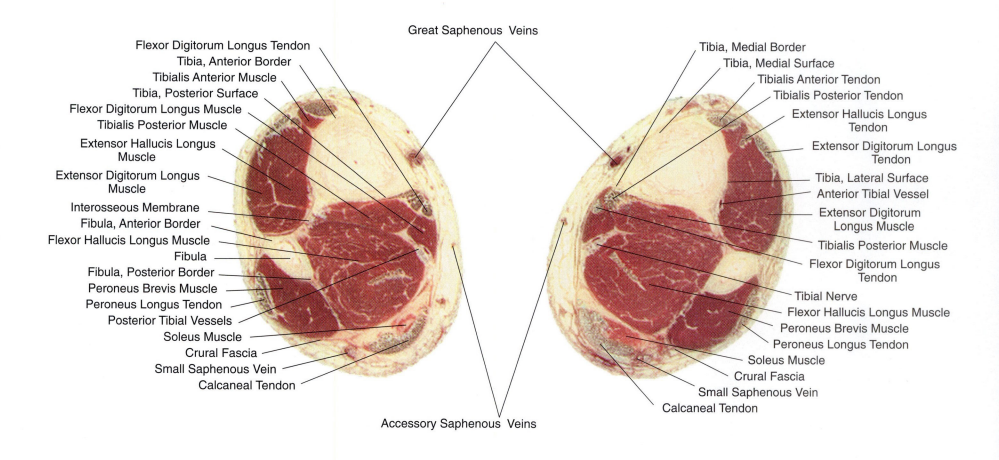

Great Saphenous Veins

Flexor Digitorum Longus Tendon
Tibia, Anterior Border
Tibialis Anterior Muscle
Tibia, Posterior Surface
Flexor Digitorum Longus Muscle
Tibialis Posterior Muscle
Extensor Hallucis Longus Muscle
Extensor Digitorum Longus Muscle
Interosseous Membrane
Fibula, Anterior Border
Flexor Hallucis Longus Muscle
Fibula
Fibula, Posterior Border
Peroneus Brevis Muscle
Peroneus Longus Tendon
Posterior Tibial Vessels
Soleus Muscle
Crural Fascia
Small Saphenous Vein
Calcaneal Tendon

Tibia, Medial Border
Tibia, Medial Surface
Tibialis Anterior Tendon
Tibialis Posterior Tendon
Extensor Hallucis Longus Tendon
Extensor Digitorum Longus Tendon
Tibia, Lateral Surface
Anterior Tibial Vessel
Extensor Digitorum Longus Muscle
Tibialis Posterior Muscle
Flexor Digitorum Longus Tendon
Tibial Nerve
Flexor Hallucis Longus Muscle
Peroneus Brevis Muscle
Peroneus Longus Tendon
Soleus Muscle
Crural Fascia
Small Saphenous Vein
Calcaneal Tendon

Accessory Saphenous Veins

right

left

posterior

a_vm2661

a_vm2662

a_vm2663

a_vm2664

a_vm2665

a_vm2666

a_vm2667

a_vm2668

a_vm2669

a_vm2670

a_vm2671

a_vm2572

anterior

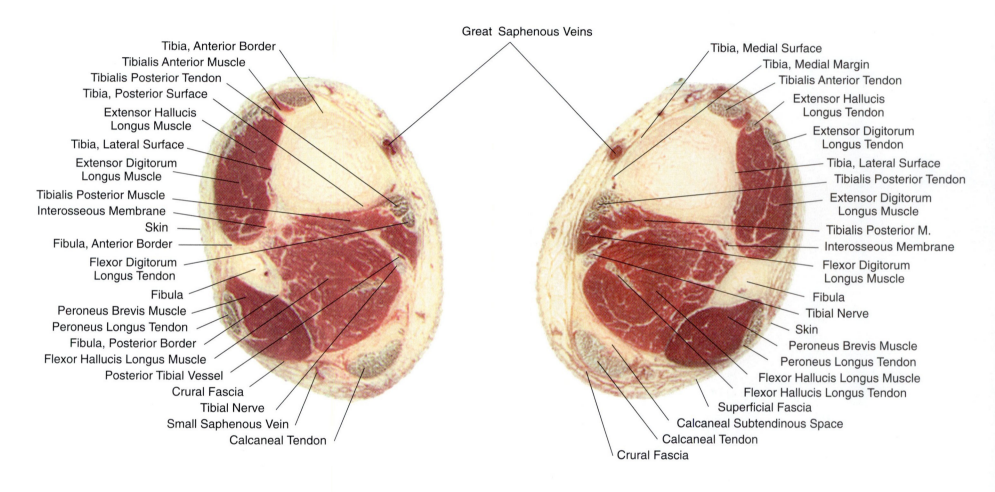

Great Saphenous Veins

Tibia, Anterior Border
Tibialis Anterior Muscle
Tibialis Posterior Tendon
Tibia, Posterior Surface
Extensor Hallucis
Longus Muscle
Tibia, Lateral Surface
Extensor Digitorum
Longus Muscle
Tibialis Posterior Muscle
Interosseous Membrane
Skin
Fibula, Anterior Border
Flexor Digitorum
Longus Tendon
Fibula
Peroneus Brevis Muscle
Peroneus Longus Tendon
Fibula, Posterior Border
Flexor Hallucis Longus Muscle
Posterior Tibial Vessel
Crural Fascia
Tibial Nerve
Small Saphenous Vein
Calcaneal Tendon

Tibia, Medial Surface
Tibia, Medial Margin
Tibialis Anterior Tendon
Extensor Hallucis
Longus Tendon
Extensor Digitorum
Longus Tendon
Tibia, Lateral Surface
Tibialis Posterior Tendon
Extensor Digitorum
Longus Muscle
Tibialis Posterior M.
Interosseous Membrane
Flexor Digitorum
Longus Muscle
Fibula
Tibial Nerve
Skin
Peroneus Brevis Muscle
Peroneus Longus Tendon
Flexor Hallucis Longus Muscle
Flexor Hallucis Longus Tendon
Superficial Fascia
Calcaneal Subtendinous Space
Calcaneal Tendon
Crural Fascia

right

left

posterior

278

a_vm2673

a_vm2674

a_vm2675

a_vm2676

a_vm2677

a_vm2678

a_vm2679

a_vm2680

a_vm2681

a_vm2682

a_vm2683

a_vm2684

anterior

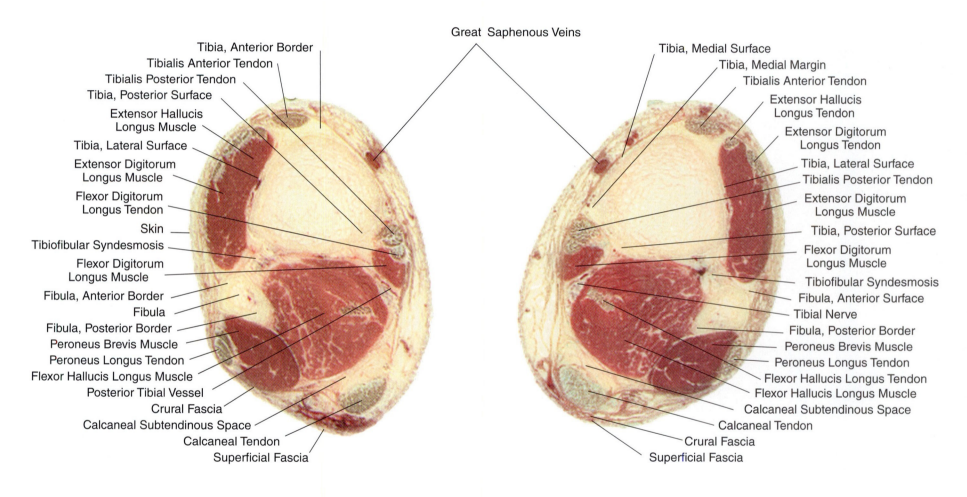

Great Saphenous Veins

Tibia, Anterior Border

Tibialis Anterior Tendon

Tibialis Posterior Tendon

Tibia, Posterior Surface

Extensor Hallucis
Longus Muscle

Tibia, Lateral Surface

Extensor Digitorum
Longus Muscle

Flexor Digitorum
Longus Tendon

Skin

Tibiofibular Syndesmosis

Flexor Digitorum
Longus Muscle

Fibula, Anterior Border

Fibula

Fibula, Posterior Border

Peroneus Brevis Muscle

Peroneus Longus Tendon

Flexor Hallucis Longus Muscle

Posterior Tibial Vessel

Crural Fascia

Calcaneal Subtendinous Space

Calcaneal Tendon

Superficial Fascia

Tibia, Medial Surface

Tibia, Medial Margin

Tibialis Anterior Tendon

Extensor Hallucis
Longus Tendon

Extensor Digitorum
Longus Tendon

Tibia, Lateral Surface

Tibialis Posterior Tendon

Extensor Digitorum
Longus Muscle

Tibia, Posterior Surface

Flexor Digitorum
Longus Muscle

Tibiofibular Syndesmosis

Fibula, Anterior Surface

Tibial Nerve

Fibula, Posterior Border

Peroneus Brevis Muscle

Peroneus Longus Tendon

Flexor Hallucis Longus Tendon

Flexor Hallucis Longus Muscle

Calcaneal Subtendinous Space

Calcaneal Tendon

Crural Fascia

Superficial Fascia

right

left

posterior

a_vm2685

a_vm2686

a_vm2687

a_vm2688

a_vm2689

a_vm2690

a_vm2691

a_vm2692

a_vm2693

a_vm2694

a_vm2695

a_vm2696

anterior

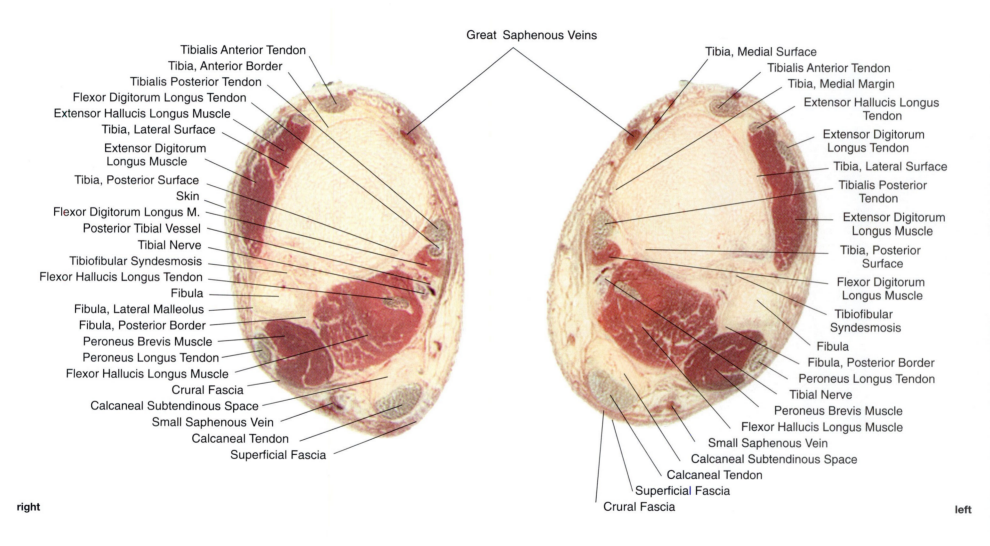

Great Saphenous Veins

Tibialis Anterior Tendon
Tibia, Anterior Border
Tibialis Posterior Tendon
Flexor Digitorum Longus Tendon
Extensor Hallucis Longus Muscle
Tibia, Lateral Surface
Extensor Digitorum Longus Muscle
Tibia, Posterior Surface
Skin
Flexor Digitorum Longus M.
Posterior Tibial Vessel
Tibial Nerve
Tibiofibular Syndesmosis
Flexor Hallucis Longus Tendon
Fibula
Fibula, Lateral Malleolus
Fibula, Posterior Border
Peroneus Brevis Muscle
Peroneus Longus Tendon
Flexor Hallucis Longus Muscle
Crural Fascia
Calcaneal Subtendinous Space
Small Saphenous Vein
Calcaneal Tendon
Superficial Fascia

Tibia, Medial Surface
Tibialis Anterior Tendon
Tibia, Medial Margin
Extensor Hallucis Longus Tendon
Extensor Digitorum Longus Tendon
Tibia, Lateral Surface
Tibialis Posterior Tendon
Extensor Digitorum Longus Muscle
Tibia, Posterior Surface
Flexor Digitorum Longus Muscle
Tibiofibular Syndesmosis
Fibula
Fibula, Posterior Border
Peroneus Longus Tendon
Tibial Nerve
Peroneus Brevis Muscle
Flexor Hallucis Longus Muscle
Small Saphenous Vein
Calcaneal Subtendinous Space
Calcaneal Tendon
Superficial Fascia
Crural Fascia

right

left

posterior

a_vm2697

a_vm2698

a_vm2699

a_vm2700

a_vm2701

a_vm2702

a_vm2703

a_vm2704

a_vm2705

a_vm2706

a_vm2707

a_vm2708

anterior

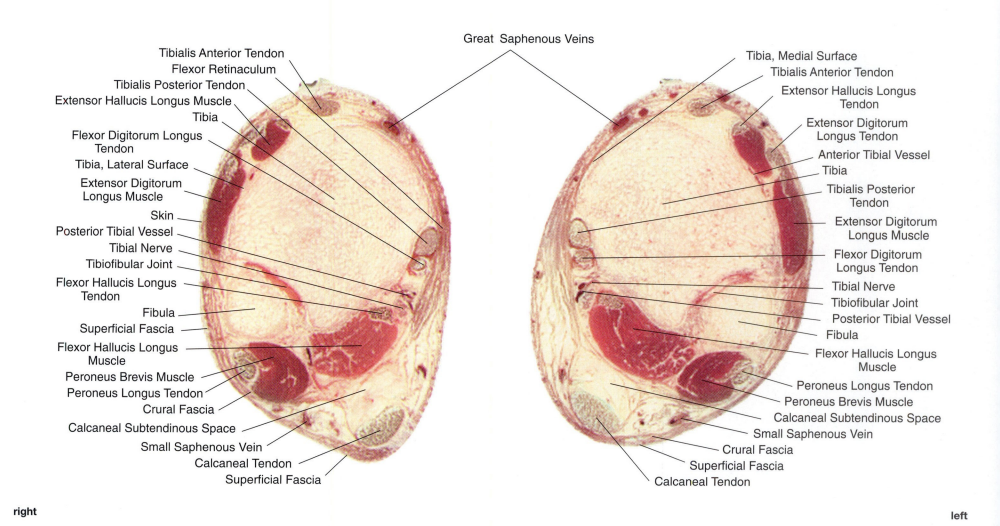

Great Saphenous Veins

Tibialis Anterior Tendon
Flexor Retinaculum
Tibialis Posterior Tendon
Extensor Hallucis Longus Muscle
Tibia
Flexor Digitorum Longus Tendon
Tibia, Lateral Surface
Extensor Digitorum Longus Muscle
Skin
Posterior Tibial Vessel
Tibial Nerve
Tibiofibular Joint
Flexor Hallucis Longus Tendon
Fibula
Superficial Fascia
Flexor Hallucis Longus Muscle
Peroneus Brevis Muscle
Peroneus Longus Tendon
Crural Fascia
Calcaneal Subtendinous Space
Small Saphenous Vein
Calcaneal Tendon
Superficial Fascia

Tibia, Medial Surface
Tibialis Anterior Tendon
Extensor Hallucis Longus Tendon
Extensor Digitorum Longus Tendon
Anterior Tibial Vessel
Tibia
Tibialis Posterior Tendon
Extensor Digitorum Longus Muscle
Flexor Digitorum Longus Tendon
Tibial Nerve
Tibiofibular Joint
Posterior Tibial Vessel
Fibula
Flexor Hallucis Longus Muscle
Peroneus Longus Tendon
Peroneus Brevis Muscle
Calcaneal Subtendinous Space
Small Saphenous Vein
Crural Fascia
Superficial Fascia
Calcaneal Tendon

right

left

posterior

284

a_vm2709

a_vm2710

a_vm2711

a_vm2712

a_vm2713

a_vm2714

a_vm2715

a_vm2716

a_vm2717

a_vm2718

a_vm2719

a_vm2720

anterior

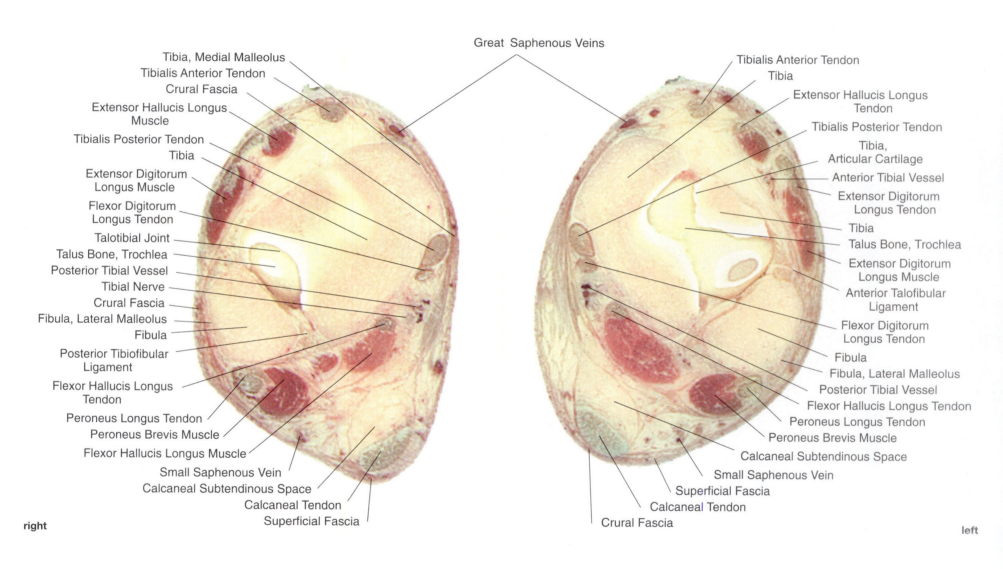

Great Saphenous Veins

Tibia, Medial Malleolus
Tibialis Anterior Tendon
Crural Fascia
Extensor Hallucis Longus
Muscle
Tibialis Posterior Tendon
Tibia
Extensor Digitorum
Longus Muscle
Flexor Digitorum
Longus Tendon
Talotibial Joint
Talus Bone, Trochlea
Posterior Tibial Vessel
Tibial Nerve
Crural Fascia
Fibula, Lateral Malleolus
Fibula
Posterior Tibiofibular
Ligament
Flexor Hallucis Longus
Tendon
Peroneus Longus Tendon
Peroneus Brevis Muscle
Flexor Hallucis Longus Muscle
Small Saphenous Vein
Calcaneal Subtendinous Space
Calcaneal Tendon
Superficial Fascia

Tibialis Anterior Tendon
Tibia
Extensor Hallucis Longus
Tendon
Tibialis Posterior Tendon
Tibia,
Articular Cartilage
Anterior Tibial Vessel
Extensor Digitorum
Longus Tendon
Tibia
Talus Bone, Trochlea
Extensor Digitorum
Longus Muscle
Anterior Talofibular
Ligament
Flexor Digitorum
Longus Tendon
Fibula
Fibula, Lateral Malleolus
Posterior Tibial Vessel
Flexor Hallucis Longus Tendon
Peroneus Longus Tendon
Peroneus Brevis Muscle
Calcaneal Subtendinous Space
Small Saphenous Vein
Superficial Fascia
Calcaneal Tendon
Crural Fascia

right

left

posterior

a_vm2721

a_vm2722

a_vm2723

a_vm2724

a_vm2725

a_vm2726

a_vm2727

a_vm2728

a_vm2729

a_vm2730

a_vm2731

a_vm2732

anterior

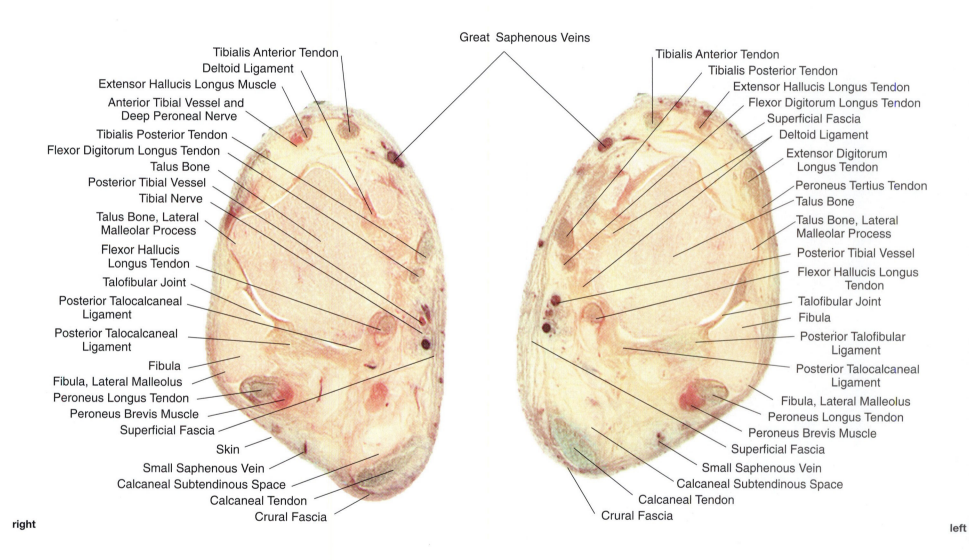

Great Saphenous Veins

Tibialis Anterior Tendon
Deltoid Ligament
Extensor Hallucis Longus Muscle
Anterior Tibial Vessel and
Deep Peroneal Nerve
Tibialis Posterior Tendon
Flexor Digitorum Longus Tendon
Talus Bone
Posterior Tibial Vessel
Tibial Nerve
Talus Bone, Lateral
Malleolar Process
Flexor Hallucis
Longus Tendon
Talofibular Joint
Posterior Talocalcaneal
Ligament
Posterior Talocalcaneal
Ligament
Fibula
Fibula, Lateral Malleolus
Peroneus Longus Tendon
Peroneus Brevis Muscle
Superficial Fascia
Skin
Small Saphenous Vein
Calcaneal Subtendinous Space
Calcaneal Tendon
Crural Fascia

Tibialis Anterior Tendon
Tibialis Posterior Tendon
Extensor Hallucis Longus Tendon
Flexor Digitorum Longus Tendon
Superficial Fascia
Deltoid Ligament
Extensor Digitorum
Longus Tendon
Peroneus Tertius Tendon
Talus Bone
Talus Bone, Lateral
Malleolar Process
Posterior Tibial Vessel
Flexor Hallucis Longus
Tendon
Talofibular Joint
Fibula
Posterior Talofibular
Ligament
Posterior Talocalcaneal
Ligament
Fibula, Lateral Malleolus
Peroneus Longus Tendon
Peroneus Brevis Muscle
Superficial Fascia
Small Saphenous Vein
Calcaneal Subtendinous Space
Calcaneal Tendon
Crural Fascia

right

left

posterior

a_vm2733

a_vm2734

a_vm2735

a_vm2736

a_vm2737

a_vm2738

a_vm2739

a_vm2740

a_vm2741

a_vm2742

a_vm2743

a_vm2744

anterior

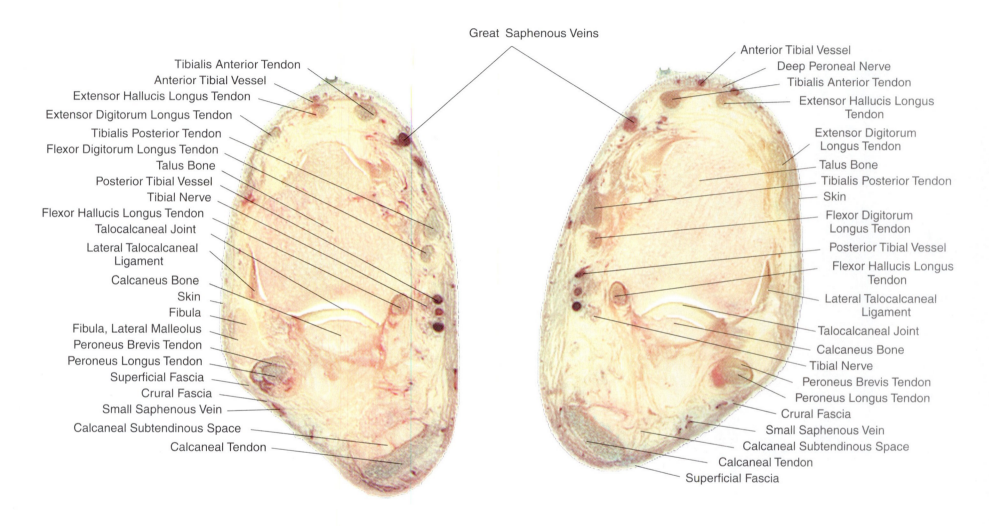

Great Saphenous Veins

Tibialis Anterior Tendon
Anterior Tibial Vessel
Extensor Hallucis Longus Tendon
Extensor Digitorum Longus Tendon
Tibialis Posterior Tendon
Flexor Digitorum Longus Tendon
Talus Bone
Posterior Tibial Vessel
Tibial Nerve
Flexor Hallucis Longus Tendon
Talocalcaneal Joint
Lateral Talocalcaneal Ligament
Calcaneus Bone
Skin
Fibula
Fibula, Lateral Malleolus
Peroneus Brevis Tendon
Peroneus Longus Tendon
Superficial Fascia
Crural Fascia
Small Saphenous Vein
Calcaneal Subtendinous Space
Calcaneal Tendon

Anterior Tibial Vessel
Deep Peroneal Nerve
Tibialis Anterior Tendon
Extensor Hallucis Longus Tendon
Extensor Digitorum Longus Tendon
Talus Bone
Tibialis Posterior Tendon
Skin
Flexor Digitorum Longus Tendon
Posterior Tibial Vessel
Flexor Hallucis Longus Tendon
Lateral Talocalcaneal Ligament
Talocalcaneal Joint
Calcaneus Bone
Tibial Nerve
Peroneus Brevis Tendon
Peroneus Longus Tendon
Crural Fascia
Small Saphenous Vein
Calcaneal Subtendinous Space
Calcaneal Tendon
Superficial Fascia

right

left

posterior

a_vm2748

a_vm2749

a_vm2750

a_vm2751

a_vm2752

a_vm2753

a_vm2754

a_vm2755

a_vm2756

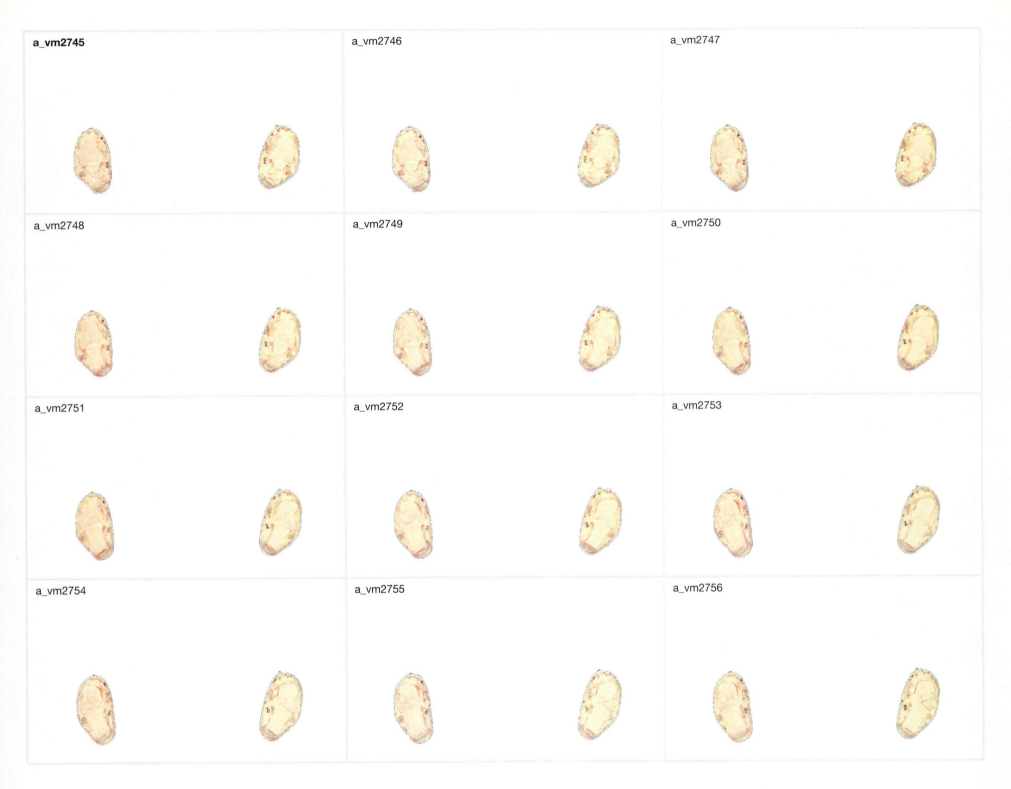

anterior

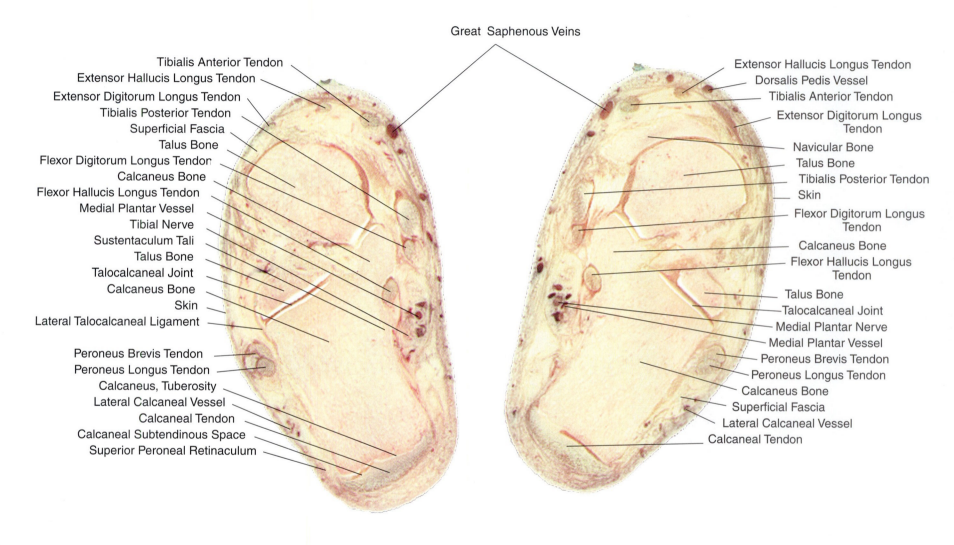

Great Saphenous Veins

Tibialis Anterior Tendon
Extensor Hallucis Longus Tendon
Extensor Digitorum Longus Tendon
Tibialis Posterior Tendon
Superficial Fascia
Talus Bone
Flexor Digitorum Longus Tendon
Calcaneus Bone
Flexor Hallucis Longus Tendon
Medial Plantar Vessel
Tibial Nerve
Sustentaculum Tali
Talus Bone
Talocalcaneal Joint
Calcaneus Bone
Skin
Lateral Talocalcaneal Ligament

Peroneus Brevis Tendon
Peroneus Longus Tendon
Calcaneus, Tuberosity
Lateral Calcaneal Vessel
Calcaneal Tendon
Calcaneal Subtendinous Space
Superior Peroneal Retinaculum

Extensor Hallucis Longus Tendon
Dorsalis Pedis Vessel
Tibialis Anterior Tendon
Extensor Digitorum Longus Tendon
Navicular Bone
Talus Bone
Tibialis Posterior Tendon
Skin
Flexor Digitorum Longus Tendon
Calcaneus Bone
Flexor Hallucis Longus Tendon
Talus Bone
Talocalcaneal Joint
Medial Plantar Nerve
Medial Plantar Vessel
Peroneus Brevis Tendon
Peroneus Longus Tendon
Calcaneus Bone
Superficial Fascia
Lateral Calcaneal Vessel
Calcaneal Tendon

right

left

posterior

292

a_vm2757

a_vm2758

a_vm2759

a_vm2760

a_vm2761

a_vm2762

a_vm2763

a_vm2764

a_vm2765

a_vm2766

a_vm2767

a_vm2768

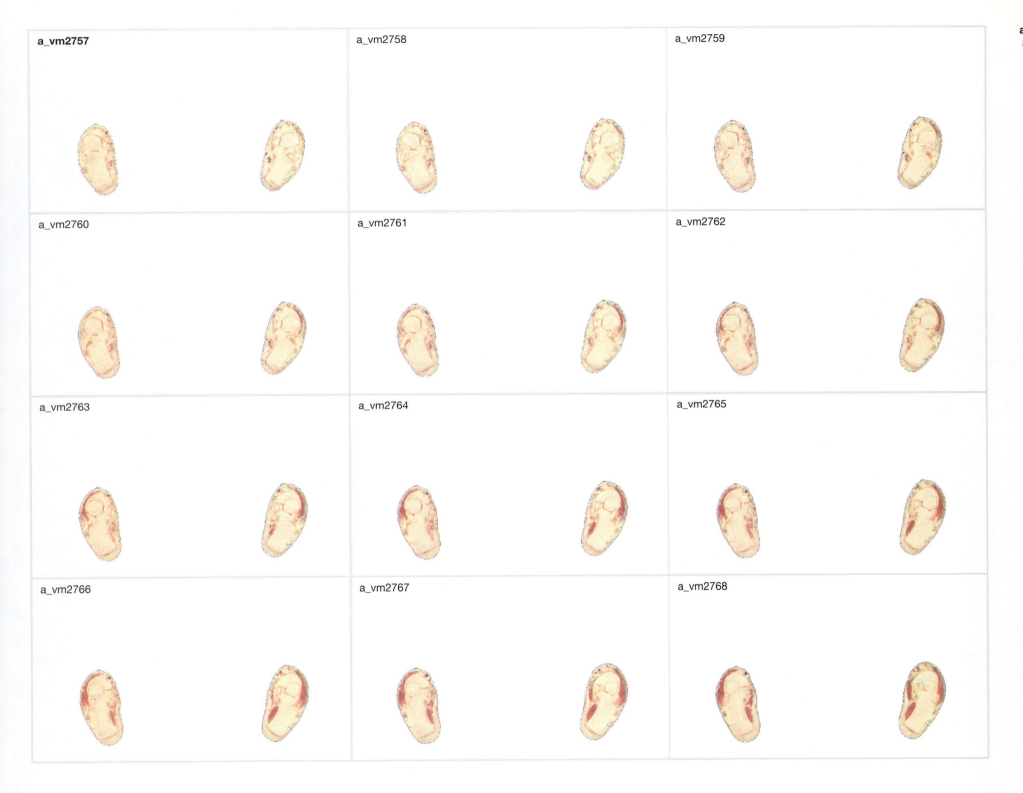

anterior

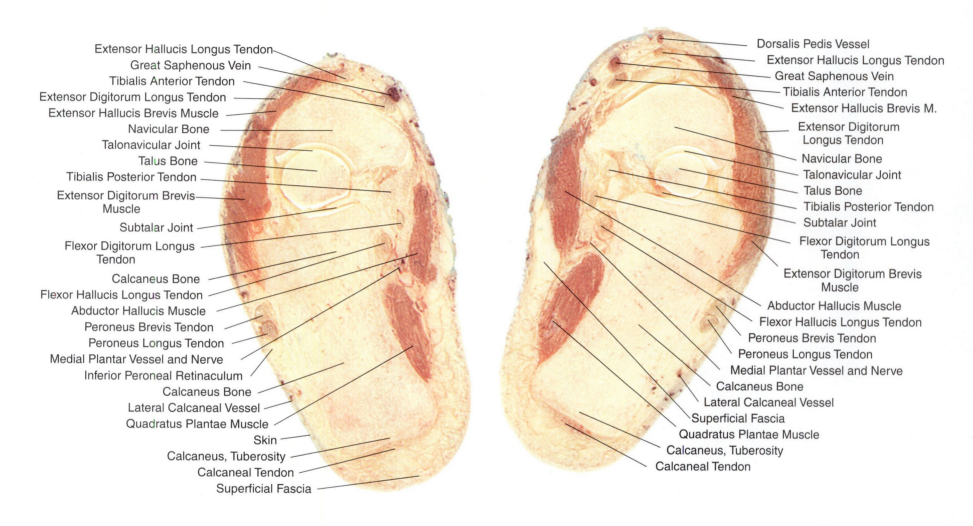

Extensor Hallucis Longus Tendon

Great Saphenous Vein

Tibialis Anterior Tendon

Extensor Digitorum Longus Tendon

Extensor Hallucis Brevis Muscle

Navicular Bone

Talonavicular Joint

Talus Bone

Tibialis Posterior Tendon

Extensor Digitorum Brevis Muscle

Subtalar Joint

Flexor Digitorum Longus Tendon

Calcaneus Bone

Flexor Hallucis Longus Tendon

Abductor Hallucis Muscle

Peroneus Brevis Tendon

Peroneus Longus Tendon

Medial Plantar Vessel and Nerve

Inferior Peroneal Retinaculum

Calcaneus Bone

Lateral Calcaneal Vessel

Quadratus Plantae Muscle

Skin

Calcaneus, Tuberosity

Calcaneal Tendon

Superficial Fascia

Dorsalis Pedis Vessel

Extensor Hallucis Longus Tendon

Great Saphenous Vein

Tibialis Anterior Tendon

Extensor Hallucis Brevis M.

Extensor Digitorum Longus Tendon

Navicular Bone

Talonavicular Joint

Talus Bone

Tibialis Posterior Tendon

Subtalar Joint

Flexor Digitorum Longus Tendon

Extensor Digitorum Brevis Muscle

Abductor Hallucis Muscle

Flexor Hallucis Longus Tendon

Peroneus Brevis Tendon

Peroneus Longus Tendon

Medial Plantar Vessel and Nerve

Calcaneus Bone

Lateral Calcaneal Vessel

Superficial Fascia

Quadratus Plantae Muscle

Calcaneus, Tuberosity

Calcaneal Tendon

right

left

posterior

a_vm2769

a_vm2770

a_vm2771

a_vm2772

a_vm2773

a_vm2774

a_vm2775

a_vm2776

a_vm2777

a_vm2778

a_vm2779

a_vm2780

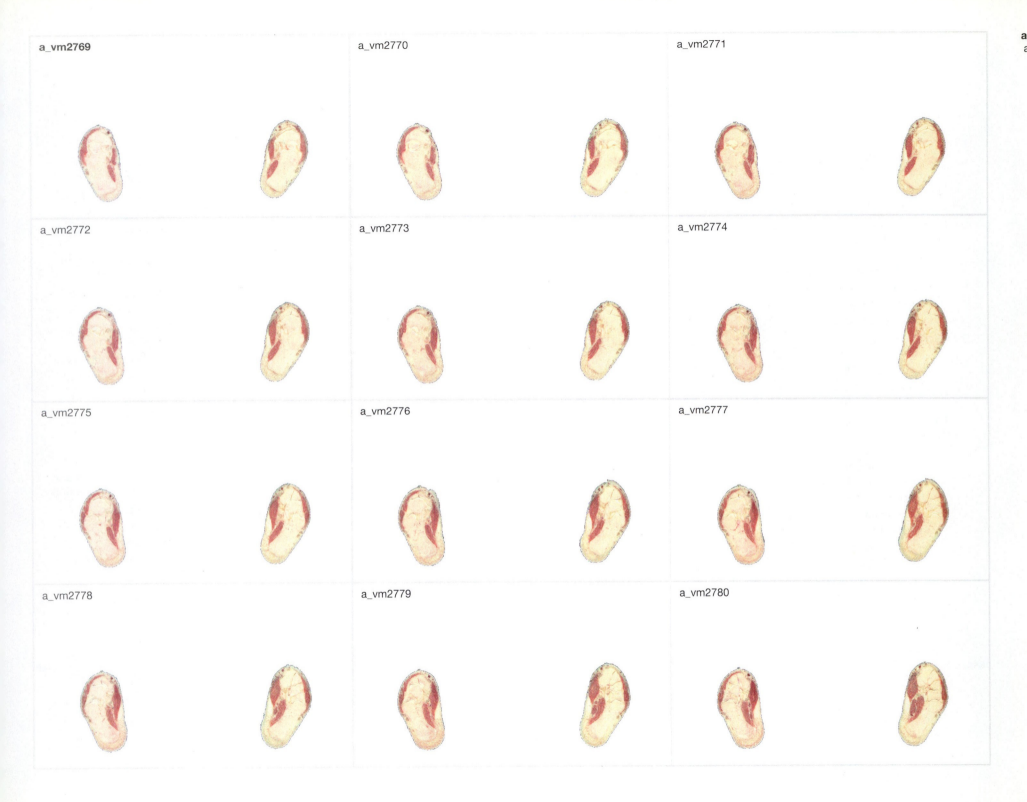

anterior

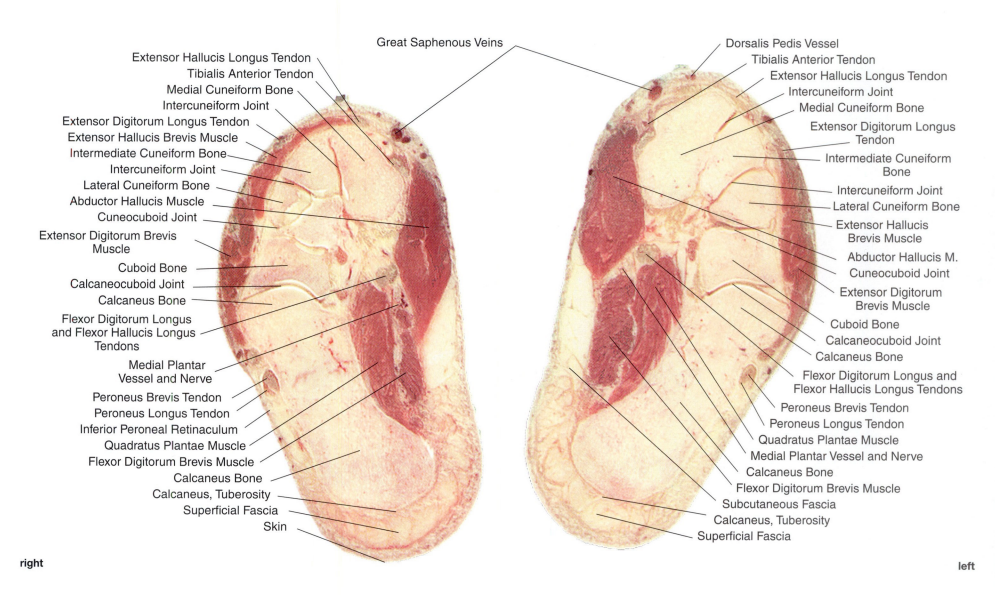

Great Saphenous Veins

Extensor Hallucis Longus Tendon
Tibialis Anterior Tendon
Medial Cuneiform Bone
Intercuneiform Joint
Extensor Digitorum Longus Tendon
Extensor Hallucis Brevis Muscle
Intermediate Cuneiform Bone
Intercuneiform Joint
Lateral Cuneiform Bone
Abductor Hallucis Muscle
Cuneocuboid Joint
Extensor Digitorum Brevis
Muscle
Cuboid Bone
Calcaneocuboid Joint
Calcaneus Bone
Flexor Digitorum Longus
and Flexor Hallucis Longus
Tendons
Medial Plantar
Vessel and Nerve
Peroneus Brevis Tendon
Peroneus Longus Tendon
Inferior Peroneal Retinaculum
Quadratus Plantae Muscle
Flexor Digitorum Brevis Muscle
Calcaneus Bone
Calcaneus, Tuberosity
Superficial Fascia
Skin

Dorsalis Pedis Vessel
Tibialis Anterior Tendon
Extensor Hallucis Longus Tendon
Intercuneiform Joint
Medial Cuneiform Bone
Extensor Digitorum Longus
Tendon
Intermediate Cuneiform
Bone
Intercuneiform Joint
Lateral Cuneiform Bone
Extensor Hallucis
Brevis Muscle
Abductor Hallucis M.
Cuneocuboid Joint
Extensor Digitorum
Brevis Muscle
Cuboid Bone
Calcaneocuboid Joint
Calcaneus Bone
Flexor Digitorum Longus and
Flexor Hallucis Longus Tendons
Peroneus Brevis Tendon
Peroneus Longus Tendon
Quadratus Plantae Muscle
Medial Plantar Vessel and Nerve
Calcaneus Bone
Flexor Digitorum Brevis Muscle
Subcutaneous Fascia
Calcaneus, Tuberosity
Superficial Fascia

right

left

posterior

a_vm2781

a_vm2782

a_vm2783

a_vm2784

a_vm2785

a_vm2786

a_vm2787

a_vm2788

a_vm2789

a_vm2790

a_vm2791

a_vm2792

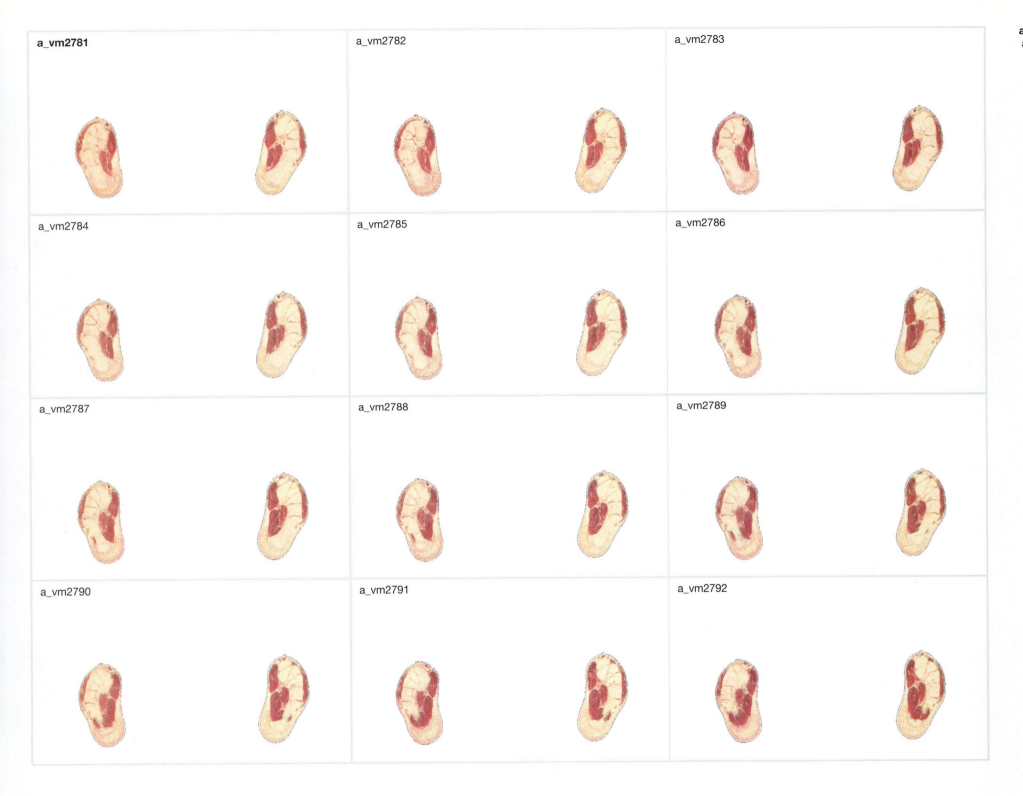

anterior

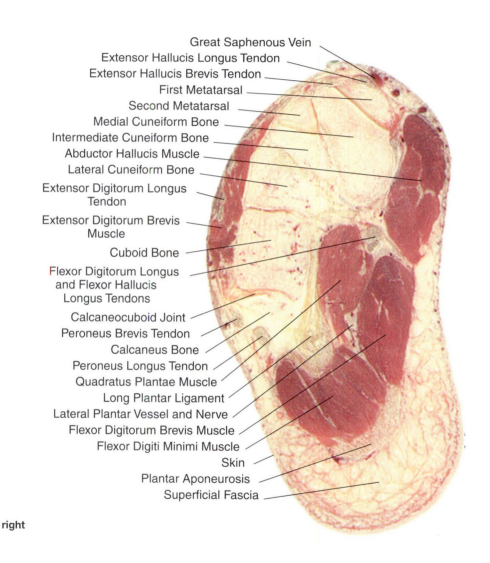

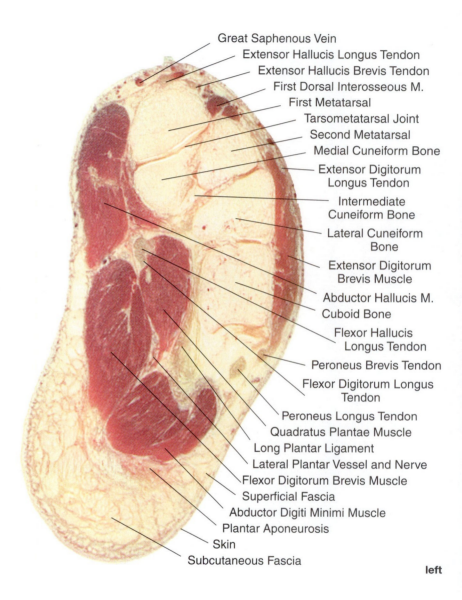

Great Saphenous Vein

Extensor Hallucis Longus Tendon

Extensor Hallucis Brevis Tendon

First Metatarsal

Second Metatarsal

Medial Cuneiform Bone

Intermediate Cuneiform Bone

Abductor Hallucis Muscle

Lateral Cuneiform Bone

Extensor Digitorum Longus
Tendon

Extensor Digitorum Brevis
Muscle

Cuboid Bone

Flexor Digitorum Longus
and Flexor Hallucis
Longus Tendons

Calcaneocuboid Joint

Peroneus Brevis Tendon

Calcaneus Bone

Peroneus Longus Tendon

Quadratus Plantae Muscle

Long Plantar Ligament

Lateral Plantar Vessel and Nerve

Flexor Digitorum Brevis Muscle

Flexor Digiti Minimi Muscle

Skin

Plantar Aponeurosis

Superficial Fascia

Great Saphenous Vein

Extensor Hallucis Longus Tendon

Extensor Hallucis Brevis Tendon

First Dorsal Interosseous M.

First Metatarsal

Tarsometatarsal Joint

Second Metatarsal

Medial Cuneiform Bone

Extensor Digitorum
Longus Tendon

Intermediate
Cuneiform Bone

Lateral Cuneiform
Bone

Extensor Digitorum
Brevis Muscle

Abductor Hallucis M.

Cuboid Bone

Flexor Hallucis
Longus Tendon

Peroneus Brevis Tendon

Flexor Digitorum Longus
Tendon

Peroneus Longus Tendon

Quadratus Plantae Muscle

Long Plantar Ligament

Lateral Plantar Vessel and Nerve

Flexor Digitorum Brevis Muscle

Superficial Fascia

Abductor Digiti Minimi Muscle

Plantar Aponeurosis

Skin

Subcutaneous Fascia

right

left

posterior

298

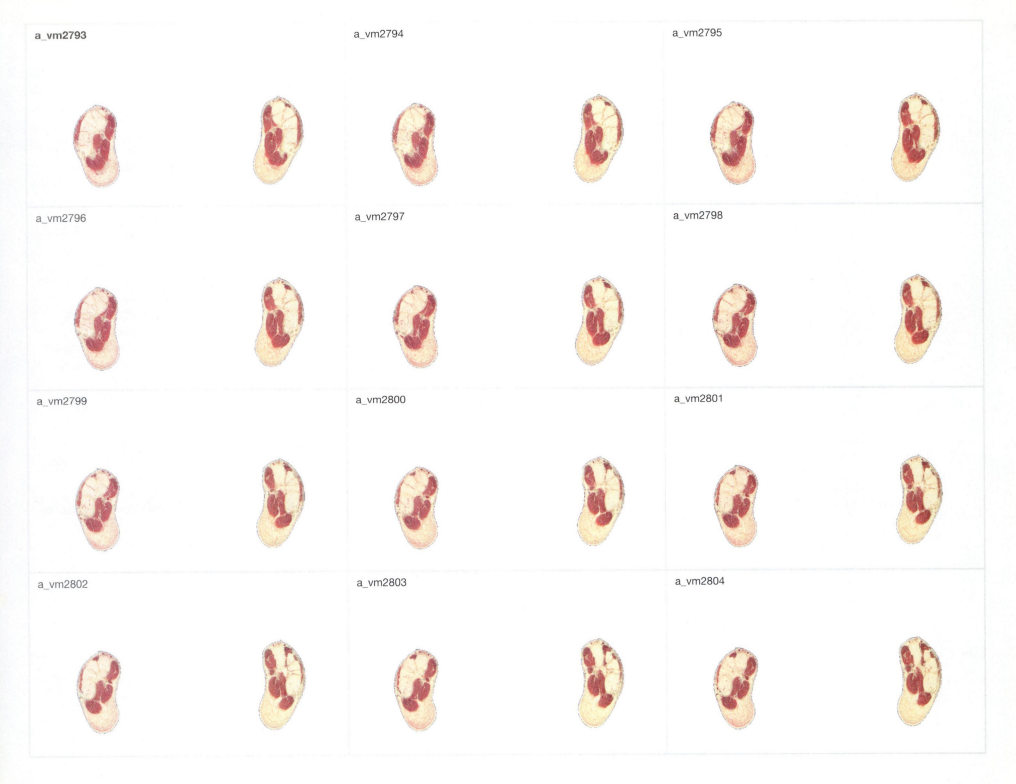

a_vm2793

a_vm2794

a_vm2795

a_vm2796

a_vm2797

a_vm2798

a_vm2799

a_vm2800

a_vm2801

a_vm2802

a_vm2803

a_vm2804

anterior

Great Saphenous Vein
Extensor Hallucis Longus Tendon
Extensor Hallucis Brevis Tendon
First Dorsal Interosseous Muscle
First Metatarsal
Second Metatarsal
Abductor Hallucis Muscle
Third Metatarsal
Extensor Digitorum Brevis M.
Long Plantar Ligament
Lateral Cuneiform Bone
Cuneocuboid Joint
Flexor Hallucis Brevis Muscle
Flexor Hallucis Longus
Tendon
Flexor Digitorum Longus
Tendon
Cuboid Bone
Quadratus Plantae Muscle
Skin
Peroneus Brevis Tendon
Peroneus Longus Tendon
Flexor Digitorum Brevis Muscle
Lateral Plantar Vessel and Nerve
Abductor Digiti Minimi Muscle
Flexor Digiti Minimi Brevis Muscle
Plantar Aponeurosis
Superficial Fascia

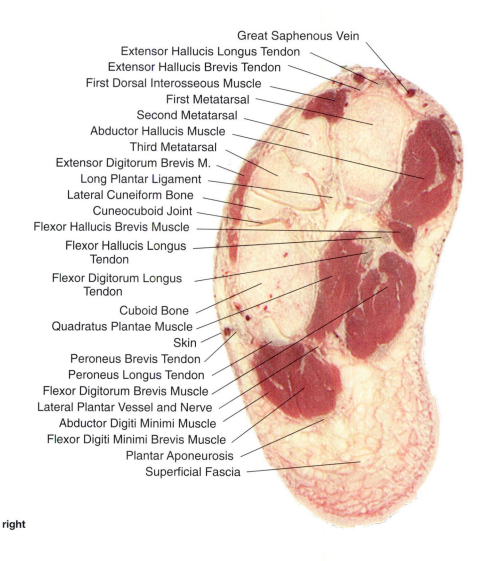

Great Saphenous Vein
Extensor Hallucis Longus Tendon
Extensor Hallucis Brevis Tendon
First Dorsal Interosseous Muscle
First Metatarsal
Abductor Hallucis Muscle
Second Metatarsal
Flexor Hallucis Brevis Muscle
Adductor Hallucis Muscle,
Transverse Head
Extensor Digitorum Brevis M.
Third Metatarsal
Long Plantar Ligament
Fourth Metatarsal
Tarsometatarsal Ligament
Flexor Hallucis Longus
Tendon
Flexor Digitorum Longus
Tendon
Cuboid Bone
Quadratus Plantae Muscle
Flexor Digitorum Brevis Muscle
Peroneus Brevis Tendon
Peroneus Longus Tendon
Lateral Plantar Vessel and Nerve
Skin
Abductor Digiti Minimi Muscle
Flexor Digiti Minimi Brevis Muscle
Plantar Aponeurosis
Superficial Fascia

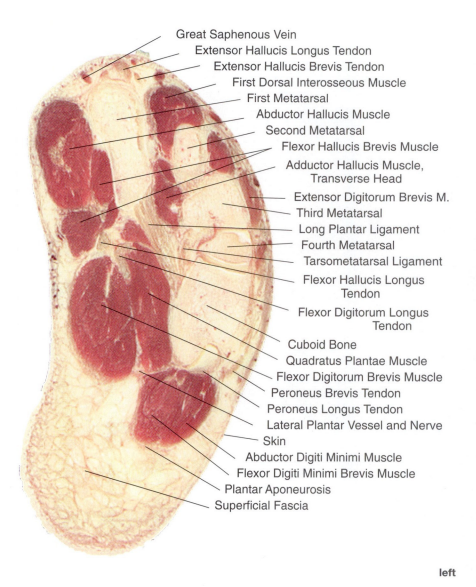

right

left

posterior

300

a_vm2805

a_vm2806

a_vm2807

a_vm2808

a_vm2809

a_vm2810

a_vm2811

a_vm2812

a_vm2813

a_vm2814

a_vm2815

a_vm2816

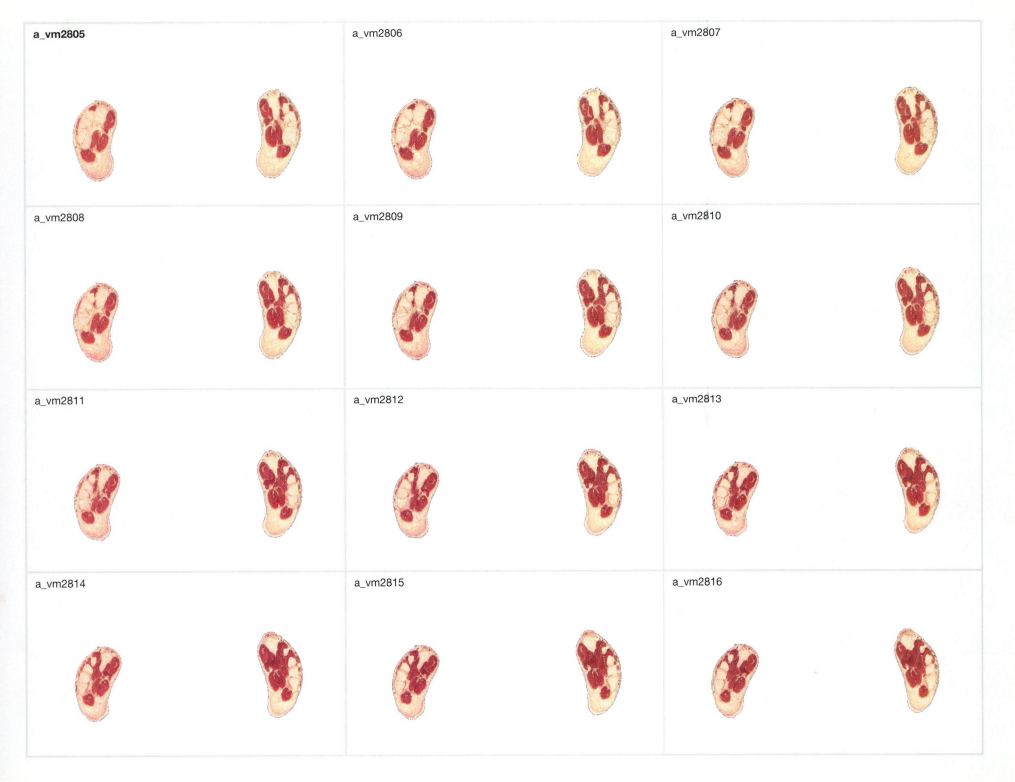

anterior

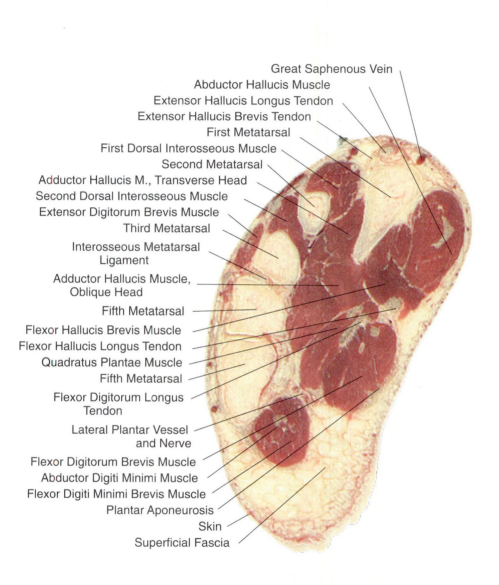

Great Saphenous Vein
Abductor Hallucis Muscle
Extensor Hallucis Longus Tendon
Extensor Hallucis Brevis Tendon
First Metatarsal
First Dorsal Interosseous Muscle
Second Metatarsal
Adductor Hallucis M., Transverse Head
Second Dorsal Interosseous Muscle
Extensor Digitorum Brevis Muscle
Third Metatarsal
Interosseous Metatarsal Ligament
Adductor Hallucis Muscle, Oblique Head
Fifth Metatarsal
Flexor Hallucis Brevis Muscle
Flexor Hallucis Longus Tendon
Quadratus Plantae Muscle
Fifth Metatarsal
Flexor Digitorum Longus Tendon
Lateral Plantar Vessel and Nerve
Flexor Digitorum Brevis Muscle
Abductor Digiti Minimi Muscle
Flexor Digiti Minimi Brevis Muscle
Plantar Aponeurosis
Skin
Superficial Fascia

Extensor Hallucis Longus Tendon
Extensor Hallucis Brevis Tendon
Abductor Hallucis Muscle
First Metatarsal
Great Saphenous Vein
Flexor Hallucis Brevis Muscle
First Dorsal Interosseous Muscle
Extensor Digitorum Brevis Tendon
Extensor Digitorum Longus Tendon
Second Metatarsal
Second Dorsal Interosseous M.
Adductor Hallucis Muscle, Transverse Head
Extensor Digitorum Brevis Tendon
Third Metatarsal
Third Dorsal Interosseous M.
Extensor Digitorum Longus Tendon
First Plantar Interosseous M.
Fourth Metatarsal
Adductor Hallucis Muscle, Oblique Head
Interosseous Metatarsal Ligament
Fifth Metatarsal
Quadratus Plantae Muscle
Lateral Plantar Vessel and Nerve
Skin
Flexor Digitorum Longus Tendon
Abductor Digiti Minimi Muscle
Flexor Digitorum Brevis Muscle
Superficial Fascia
Flexor Hallucis Longus Tendon
Plantar Aponeurosis

right

left

posterior

302

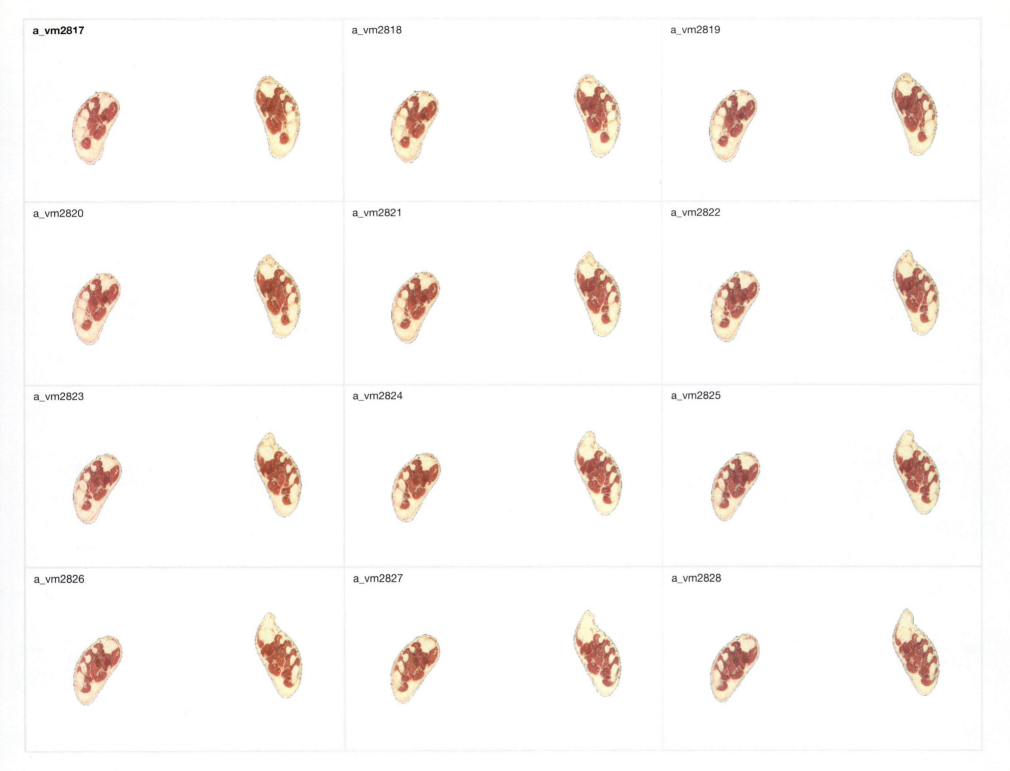

a_vm2817

a_vm2818

a_vm2819

a_vm2820

a_vm2821

a_vm2822

a_vm2823

a_vm2824

a_vm2825

a_vm2826

a_vm2827

a_vm2828

anterior

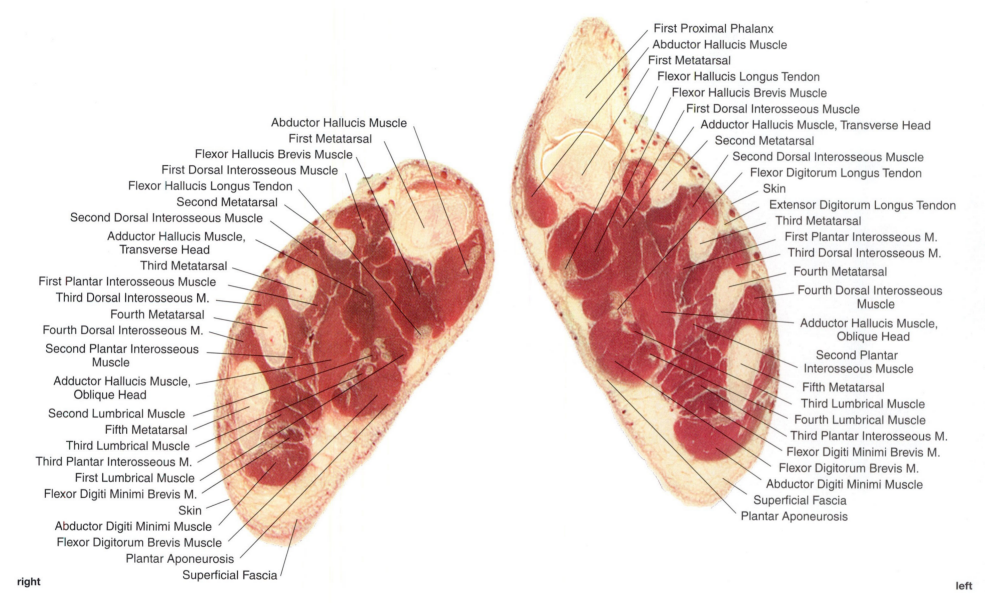

First Proximal Phalanx
Abductor Hallucis Muscle
First Metatarsal
Flexor Hallucis Longus Tendon
Flexor Hallucis Brevis Muscle
First Dorsal Interosseous Muscle
Adductor Hallucis Muscle, Transverse Head
Second Metatarsal
Second Dorsal Interosseous Muscle
Flexor Digitorum Longus Tendon
Skin
Extensor Digitorum Longus Tendon
Third Metatarsal
First Plantar Interosseous M.
Third Dorsal Interosseous M.
Fourth Metatarsal
Fourth Dorsal Interosseous Muscle
Adductor Hallucis Muscle, Oblique Head
Second Plantar Interosseous Muscle
Fifth Metatarsal
Third Lumbrical Muscle
Fourth Lumbrical Muscle
Third Plantar Interosseous M.
Flexor Digiti Minimi Brevis M.
Flexor Digitorum Brevis M.
Abductor Digiti Minimi Muscle
Superficial Fascia
Plantar Aponeurosis

Abductor Hallucis Muscle
First Metatarsal
Flexor Hallucis Brevis Muscle
First Dorsal Interosseous Muscle
Flexor Hallucis Longus Tendon
Second Metatarsal
Second Dorsal Interosseous Muscle
Adductor Hallucis Muscle, Transverse Head
Third Metatarsal
First Plantar Interosseous Muscle
Third Dorsal Interosseous M.
Fourth Metatarsal
Fourth Dorsal Interosseous M.
Second Plantar Interosseous Muscle
Adductor Hallucis Muscle, Oblique Head
Second Lumbrical Muscle
Fifth Metatarsal
Third Lumbrical Muscle
Third Plantar Interosseous M.
First Lumbrical Muscle
Flexor Digiti Minimi Brevis M.
Skin
Abductor Digiti Minimi Muscle
Flexor Digitorum Brevis Muscle
Plantar Aponeurosis
Superficial Fascia

right

left

posterior

a_vm2829

a_vm2830

a_vm2831

a_vm2832

a_vm2833

a_vm2834

a_vm2835

a_vm2836

a_vm2837

a_vm2838

a_vm2839

a_vm2840

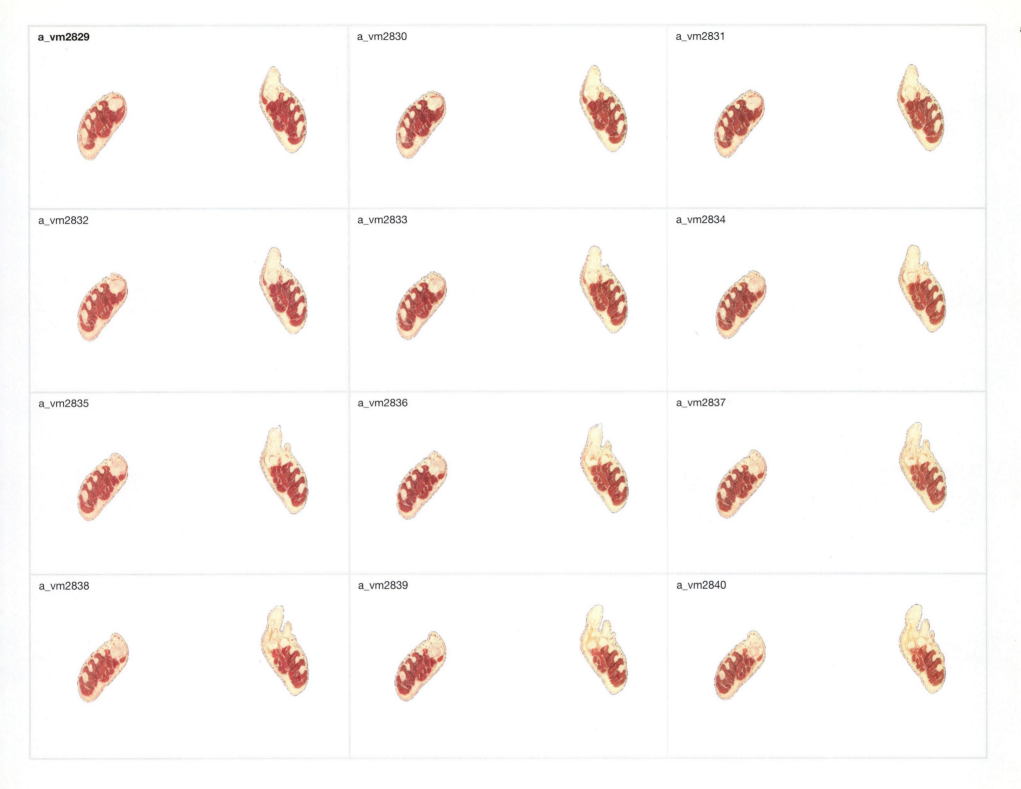

anterior

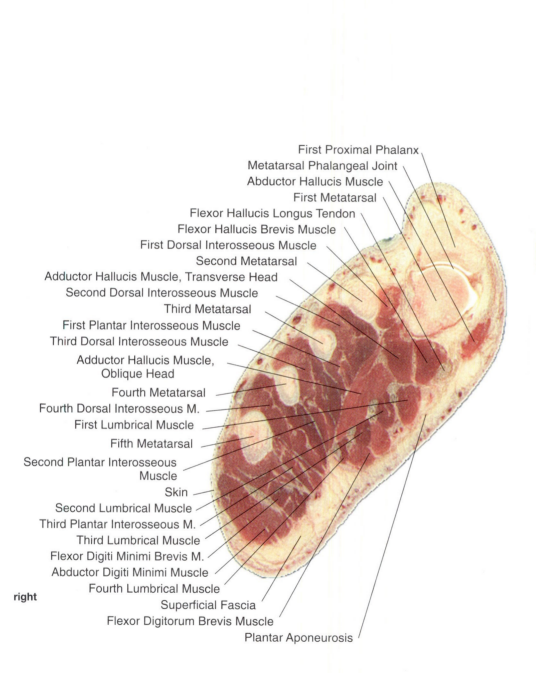

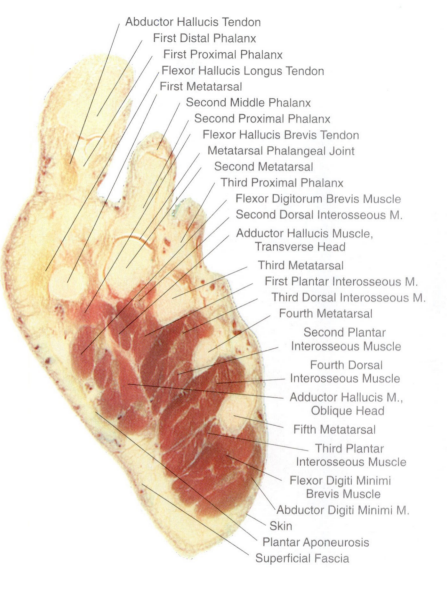

Abductor Hallucis Tendon

First Distal Phalanx

First Proximal Phalanx

Flexor Hallucis Longus Tendon

First Metatarsal

Second Middle Phalanx

Second Proximal Phalanx

Flexor Hallucis Brevis Tendon

Metatarsal Phalangeal Joint

Second Metatarsal

Third Proximal Phalanx

Flexor Digitorum Brevis Muscle

Second Dorsal Interosseous M.

Adductor Hallucis Muscle,
Transverse Head

Third Metatarsal

First Plantar Interosseous M.

Third Dorsal Interosseous M.

Fourth Metatarsal

Second Plantar
Interosseous Muscle

Fourth Dorsal
Interosseous Muscle

Adductor Hallucis M.,
Oblique Head

Fifth Metatarsal

Third Plantar
Interosseous Muscle

Flexor Digiti Minimi
Brevis Muscle

Abductor Digiti Minimi M.

Skin

Plantar Aponeurosis

Superficial Fascia

First Proximal Phalanx

Metatarsal Phalangeal Joint

Abductor Hallucis Muscle

First Metatarsal

Flexor Hallucis Longus Tendon

Flexor Hallucis Brevis Muscle

First Dorsal Interosseous Muscle

Second Metatarsal

Adductor Hallucis Muscle, Transverse Head

Second Dorsal Interosseous Muscle

Third Metatarsal

First Plantar Interosseous Muscle

Third Dorsal Interosseous Muscle

Adductor Hallucis Muscle,
Oblique Head

Fourth Metatarsal

Fourth Dorsal Interosseous M.

First Lumbrical Muscle

Fifth Metatarsal

Second Plantar Interosseous
Muscle

Skin

Second Lumbrical Muscle

Third Plantar Interosseous M.

Third Lumbrical Muscle

Flexor Digiti Minimi Brevis M.

Abductor Digiti Minimi Muscle

Fourth Lumbrical Muscle

Superficial Fascia

Flexor Digitorum Brevis Muscle

Plantar Aponeurosis

right

left

posterior

a_vm2841

a_vm2842

a_vm2843

a_vm2844

a_vm2845

a_vm2846

a_vm2847

a_vm2848

a_vm2849

a_vm2850

a_vm2851

a_vm2852

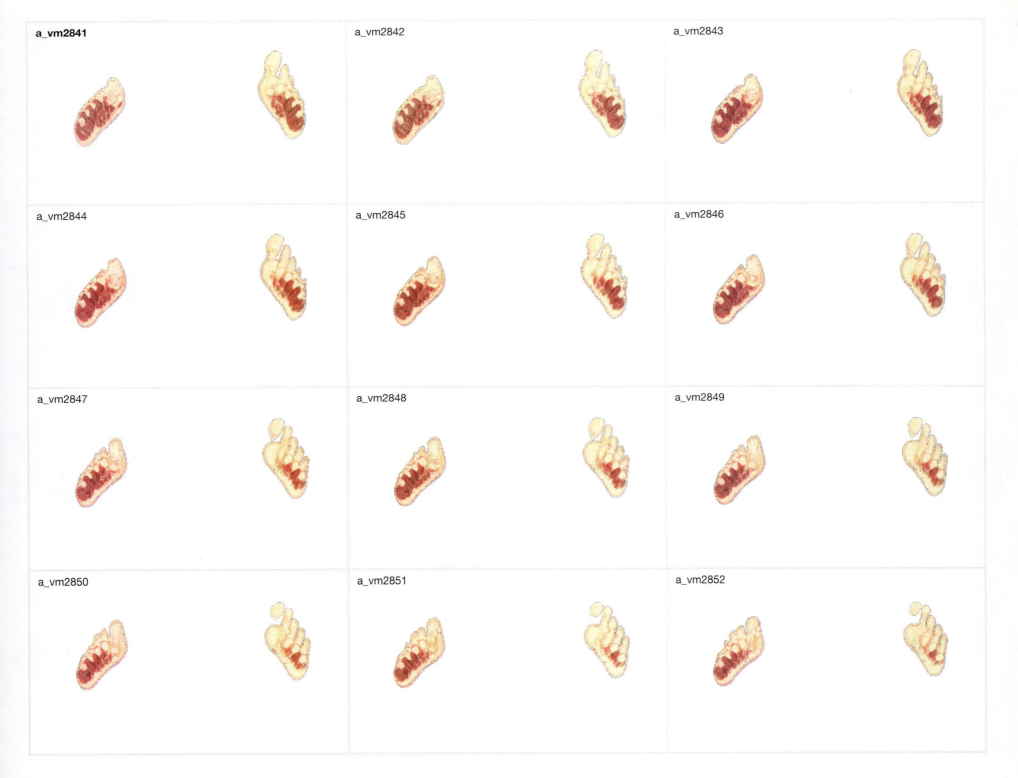

anterior

First Toe

Second Distal Phalanx

First Distal Phalanx

First Proximal Phalanx

First Lumbrical Muscle

Second Proximal Phalanx

Second Metatarsal

Second Lumbrical Muscle

Third Proximal Phalanx

Third Metatarsal

First Plantar Interosseous M.

Fourth Proximal Phalanx

Third Lumbrical Muscle

Second Plantar
Interosseous Muscle

Fourth Metatarsal

Fourth Lumbrical Muscle

Fourth Dorsal
Interosseous Muscle

Fifth Metatarsal

Third Plantar
Interosseous Muscle

Flexor Digiti Minimi
Brevis Muscle

Abductor Digiti
Minimi Muscle

Superficial Fascia

Skin

Third Distal Phalanx

Third Middle Phalanx

Third Proximal Phalanx

Third Metatarsal

Fourth Middle Phalanx

Third Dorsal Interosseous Tendon

Fourth Proximal Phalanx

Fourth Metatarsal

Fifth Middle Phalanx

Fifth Proximal Phalanx

Third Plantar Interosseous M.

Flexor Digiti Minimi Brevis M.

Fifth Metatarsal

Abductor Digiti Minimi Tendon

Superficial Fascia

Flexor Digitorum Longus and
Brevis Tendons

Skin

Plantar Aponeurosis

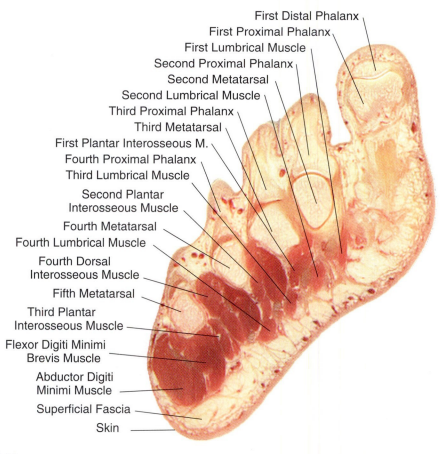

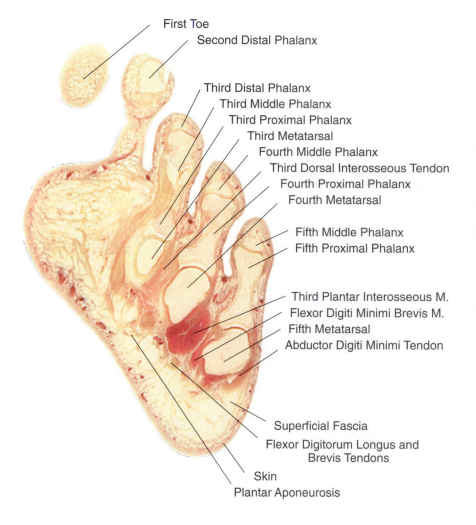

right

left

posterior

308

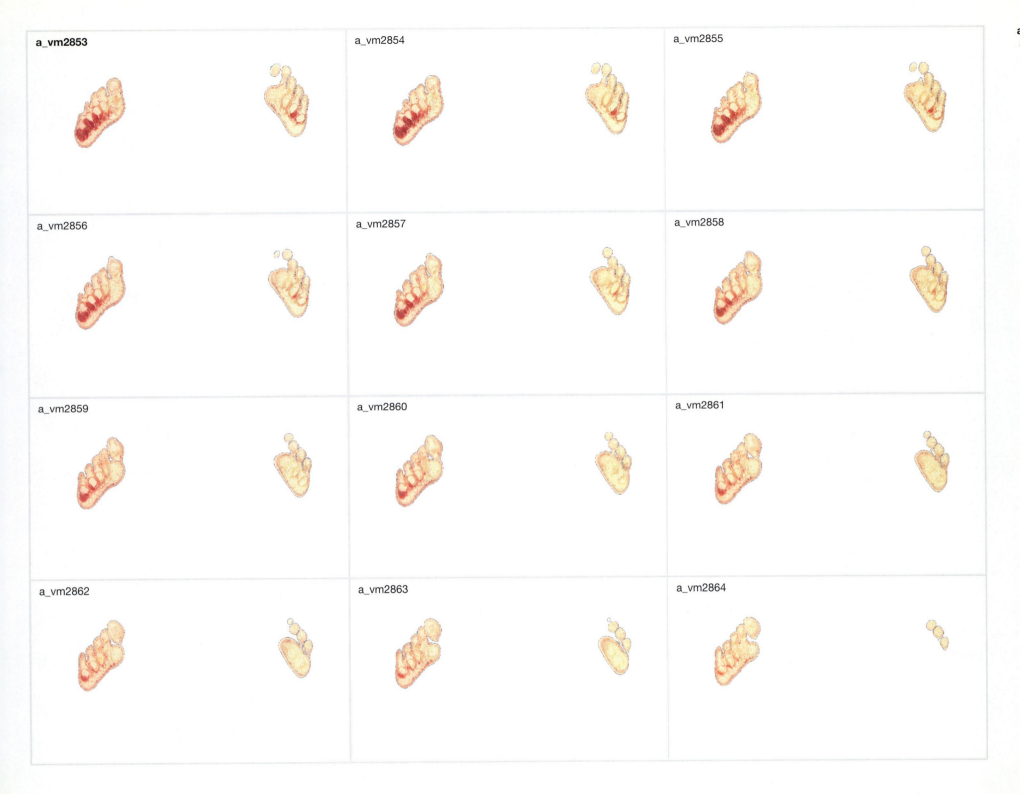

a_vm2853

a_vm2854

a_vm2855

a_vm2856

a_vm2857

a_vm2858

a_vm2859

a_vm2860

a_vm2861

a_vm2862

a_vm2863

a_vm2864

anterior

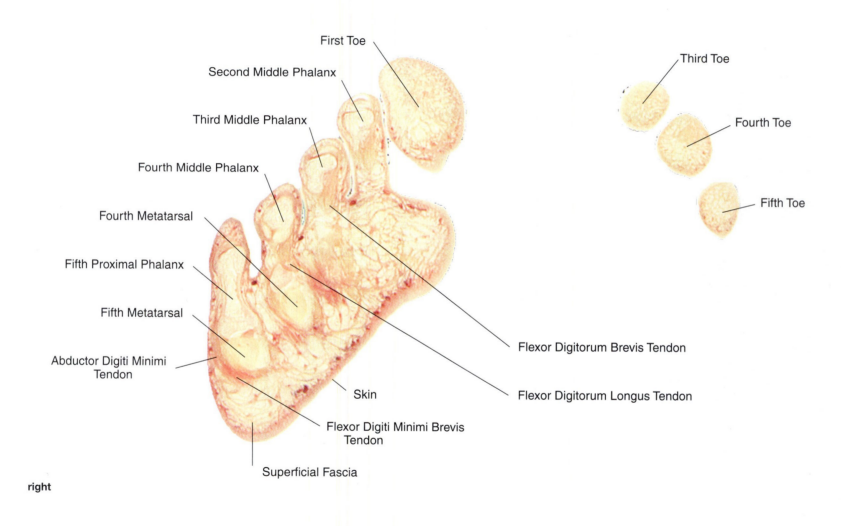

First Toe

Second Middle Phalanx

Third Middle Phalanx

Fourth Middle Phalanx

Fourth Metatarsal

Fifth Proximal Phalanx

Fifth Metatarsal

Abductor Digiti Minimi
Tendon

Third Toe

Fourth Toe

Fifth Toe

Flexor Digitorum Brevis Tendon

Flexor Digitorum Longus Tendon

Skin

Flexor Digiti Minimi Brevis
Tendon

Superficial Fascia

right

left

posterior

a_vm2865

a_vm2866

a_vm2867

a_vm2868

a_vm2869

a_vm2870

a_vm2871

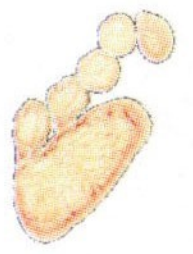

a_vm2872

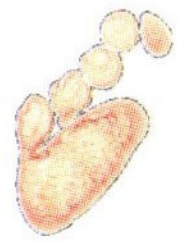

a_vm2873

a_vm2874

a_vm2875

a_vm2876

Transverse
a_vm2877-
a_vm2878

right

anterior

Fourth Toe

Fifth Toe

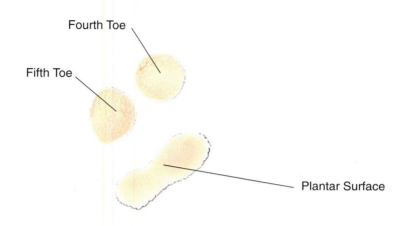

Plantar Surface

left

posterior

a_vm2877	a_vm2878

PART TWO

RECONSTRUCTED IMAGES
Coronal

The coronal images are of vertical planes through the long axis of the body and parallel to or through the coronal suture of the skull. They are presented in one-millimeter increments in this atlas starting at the front. These are standard radiological planes used in direct magnetic resonance and ultrasound imaging. Coronal clinical images are often reconstructed from transverse computed tomography and single photon emission computed tomography images. This plane, like the transverse, is particulary useful for comparison of bilateral anatomical structures.

These images are virtual slices—the Visible Human Male was never physically cut in coronal planes. Each coronal image was computer reconstructed from an edge view of all slices in the transverse image collection as seen from the anterior perspective and each is of a different depth at one-millimeter intervals through that collection.

The maximum resolution of images reconstructed in the coronal plane is determined by the one-millimeter thickness of the original transverse sections. Therefore, the level of detail in the coronal images (one-millimeter pixels) is only one third that seen in the transverse images of the previous part (0.32 millimeter pixels), in each dimension.

The labeling format for coronal images is the same as that used for the transverse images. In the same fashion as the transverse image display, the labeled slice is always repeated from the six coronal images on the same and facing page. The magnification remains the same, however, for all coronal images.

Nose, Tip

Rectus Abdominis Muscle

Superficial Fascia

Umbilicus

First Proximal Phalanx

First Metacarpal

First Metacarpal

Thenar Muscles

First Dorsal Interosseous Muscle

Second Metacarpal

Second Metacarpal

Third Metacarpal

Second Dorsal Interosseous Muscle

Palmar Interosseous Muscles

Third Metacarpal

Third Dorsal Interosseous Muscle

Dorsal Interosseous Muscles

Fourth Metacarpal

Thumb

Fourth Proximal Phalanx

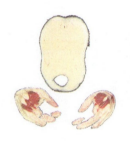

vmc3041

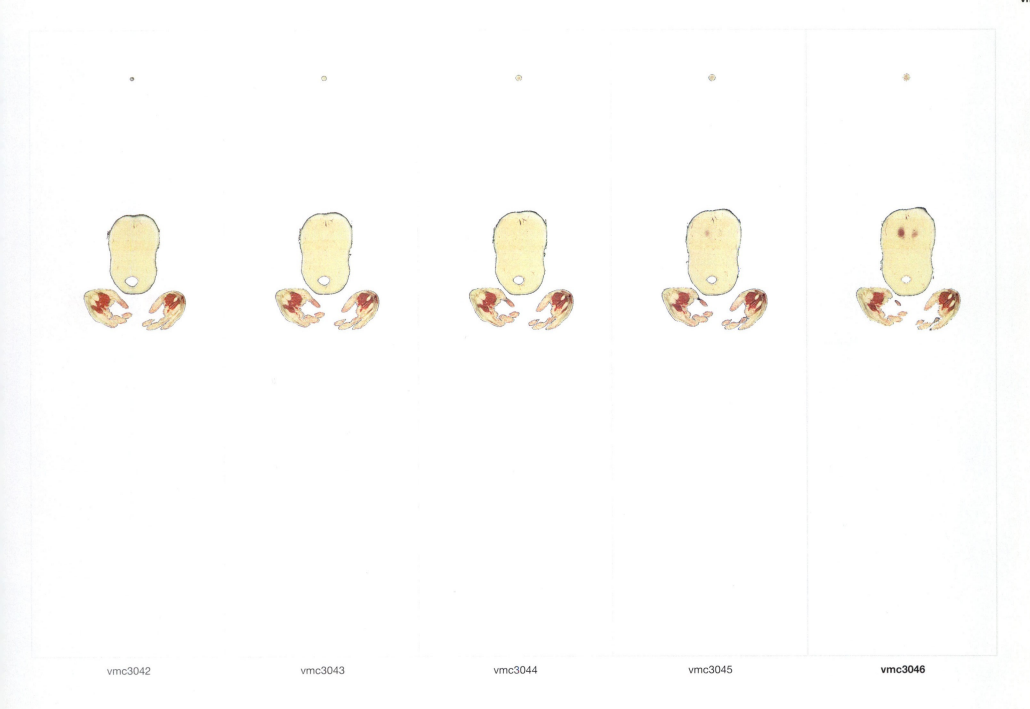

vmc3042   vmc3043   vmc3044   vmc3045   **vmc3046**

Nose ———— Naris

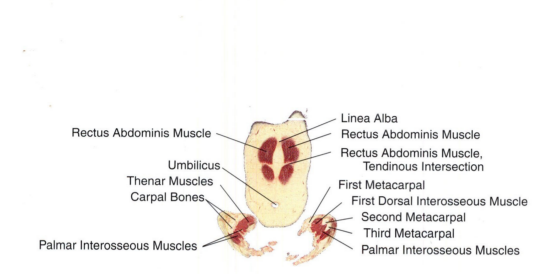

Rectus Abdominis Muscle

Linea Alba
Rectus Abdominis Muscle
Rectus Abdominis Muscle,
Tendinous Intersection

Umbilicus
Thenar Muscles
Carpal Bones

First Metacarpal
First Dorsal Interosseous Muscle
Second Metacarpal
Third Metacarpal
Palmar Interosseous Muscles

Palmar Interosseous Muscles

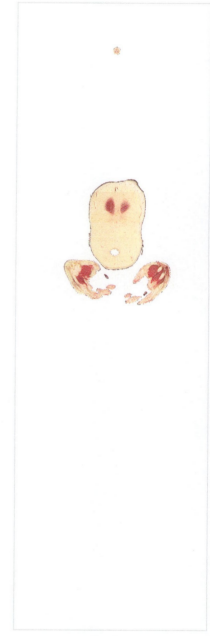

vmc3047

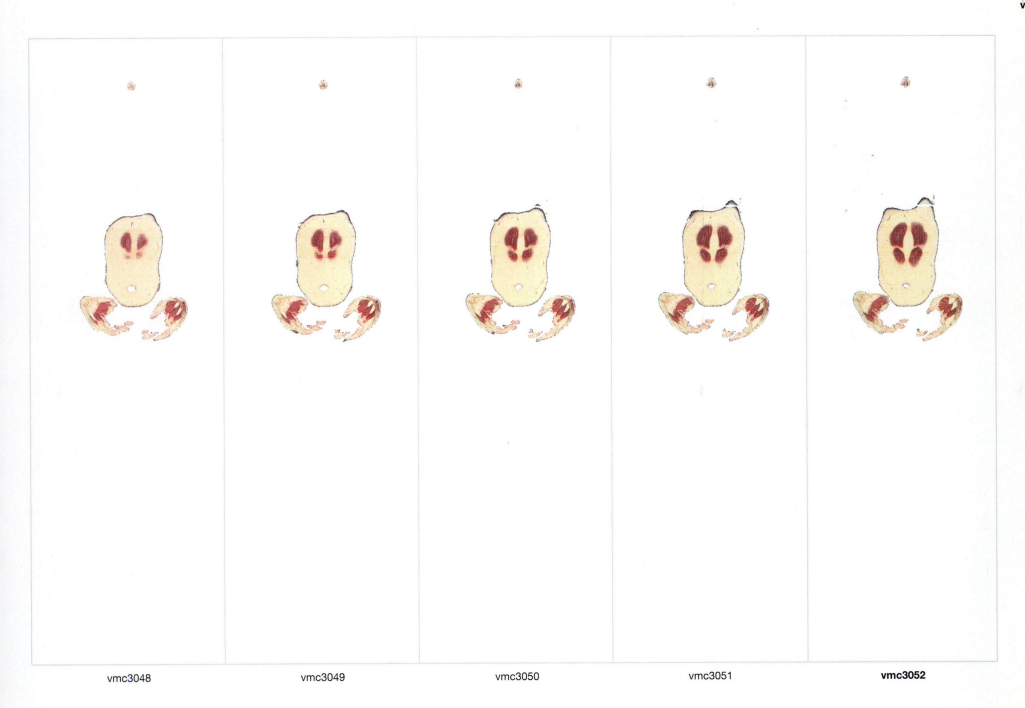

vmc3048   vmc3049   vmc3050   vmc3051   **vmc3052**

Nose — Nasal Septum
Upper Lip — Naris
Lower Lip

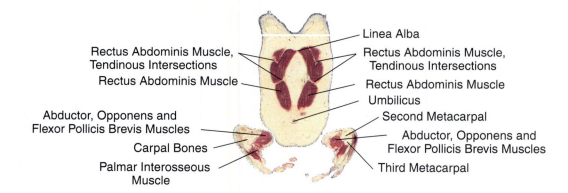

Linea Alba

Rectus Abdominis Muscle, — Rectus Abdominis Muscle,
Tendinous Intersections — Tendinous Intersections

Rectus Abdominis Muscle — Rectus Abdominis Muscle

Umbilicus

Abductor, Opponens and — Second Metacarpal
Flexor Pollicis Brevis Muscles

Carpal Bones — Abductor, Opponens and
Flexor Pollicis Brevis Muscles

Palmar Interosseous — Third Metacarpal
Muscle

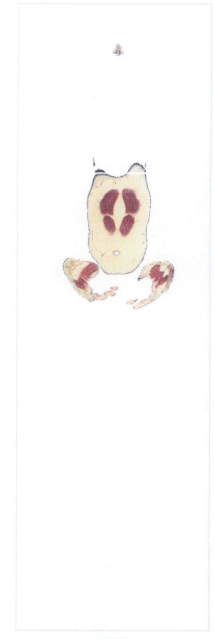

vmc3053

First Toe

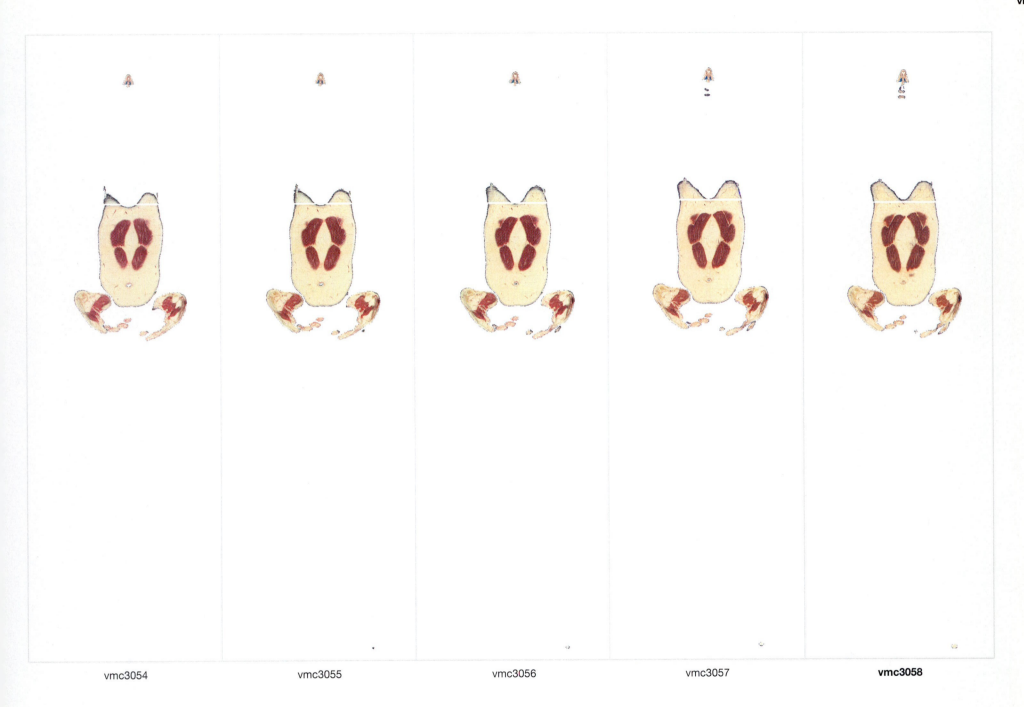

vmc3054   vmc3055   vmc3056   vmc3057   **vmc3058**

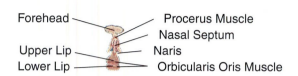

Forehead — Procerus Muscle
Nasal Septum
Upper Lip — Naris
Lower Lip — Orbicularis Oris Muscle

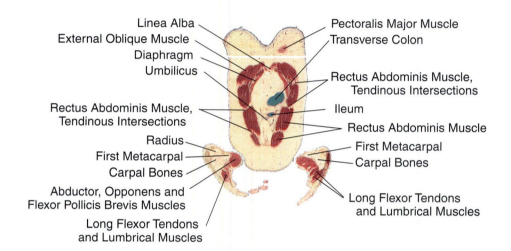

Linea Alba — Pectoralis Major Muscle
External Oblique Muscle — Transverse Colon
Diaphragm
Umbilicus — Rectus Abdominis Muscle,
Tendinous Intersections
Rectus Abdominis Muscle,
Tendinous Intersections — Ileum
Rectus Abdominis Muscle
Radius — First Metacarpal
First Metacarpal — Carpal Bones
Carpal Bones
Abductor, Opponens and — Long Flexor Tendons
Flexor Pollicis Brevis Muscles — and Lumbrical Muscles
Long Flexor Tendons
and Lumbrical Muscles

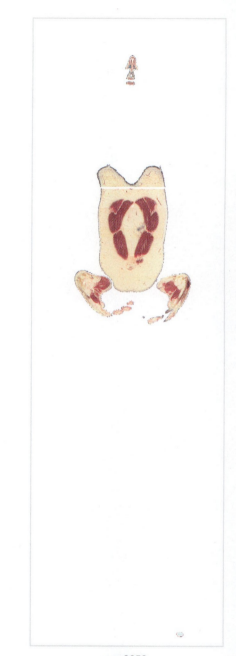

vmc3059

First Distal Phalanx

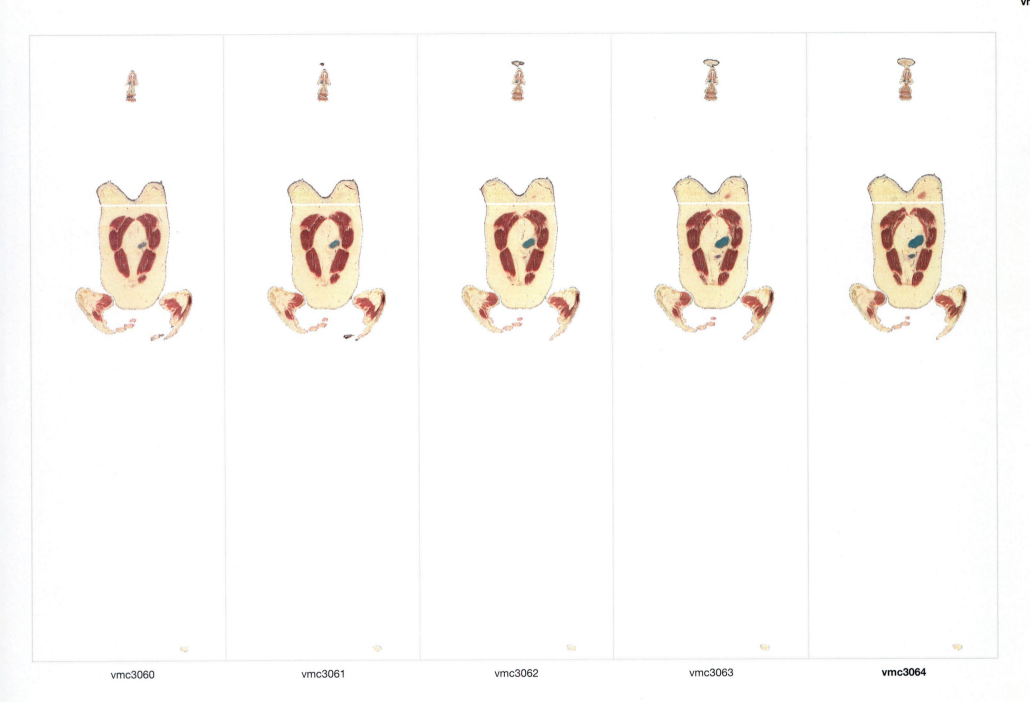

vmc3060 vmc3061 vmc3062 vmc3063 **vmc3064**

Frontal Bone —————— Procerus Muscle

Upper Lip ———— Nasal Septum
Chin ————— Lower Lip

Superficial Fascia —————— Pectoralis Major Muscle
Rib 5, Costal Cartilage

Xiphoid Process —————— External Oblique Muscle
Diaphragm
Diaphragm —————— Ileum
Transverse Colon —————
Rectus Abdominis Muscle —————

Ileum —————— Rectus Abdominis Muscle,
Mesentery —————— Tendinous Intersection
Radius —————— Thenar Muscles
Carpal Bones —————— Carpal Bones
Thenar Muscles —————— Palmar Interosseous and Lumbrical
Hypothenar Muscles —————— Muscles with Long Flexor Tendons
Fifth Metacarpal

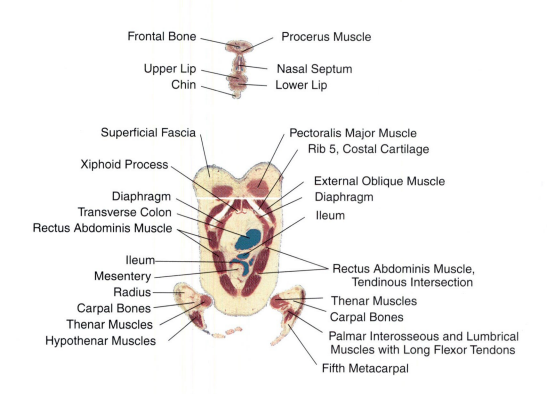

First Distal Phalanx
Second Toe

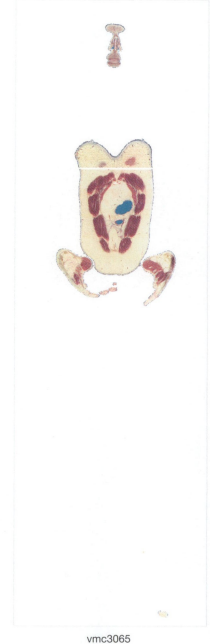

vmc3065

322

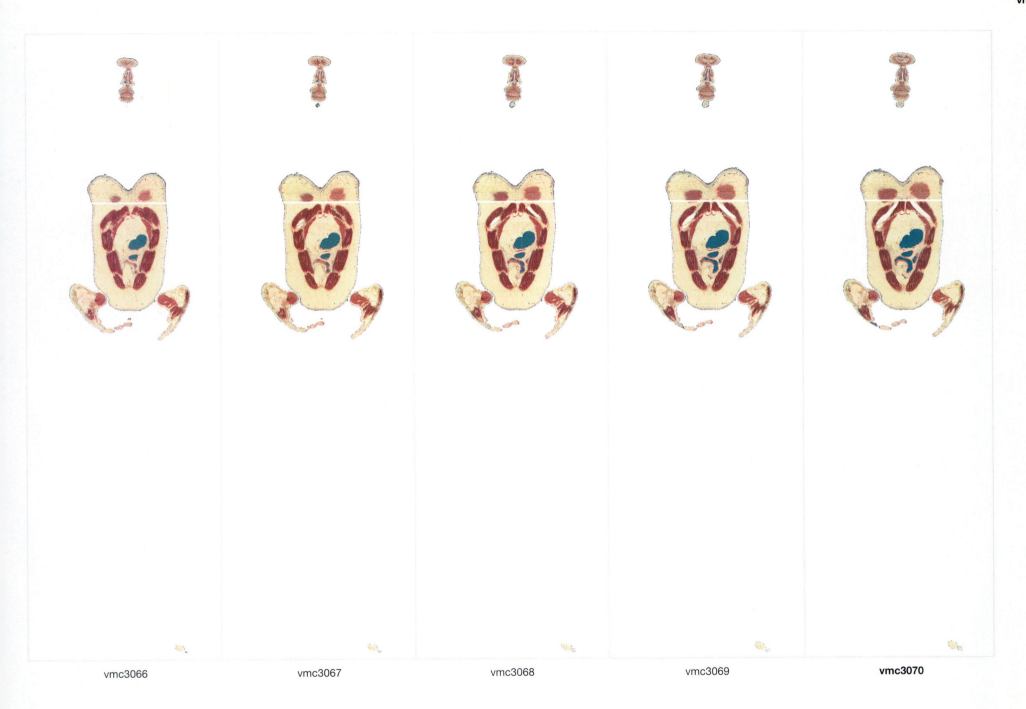

vmc3066 vmc3067 vmc3068 vmc3069 **vmc3070**

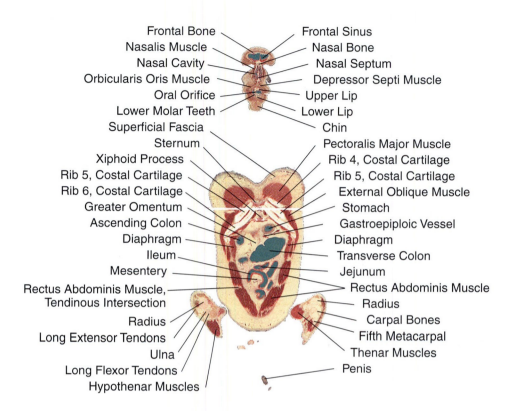

Frontal Bone — — Frontal Sinus
Nasalis Muscle — — Nasal Bone
Nasal Cavity — — Nasal Septum
Orbicularis Oris Muscle — — Depressor Septi Muscle
Oral Orifice — — Upper Lip
Lower Molar Teeth — — Lower Lip
Superficial Fascia — — Chin
Sternum — — Pectoralis Major Muscle
Xiphoid Process — — Rib 4, Costal Cartilage
Rib 5, Costal Cartilage — — Rib 5, Costal Cartilage
Rib 6, Costal Cartilage — — External Oblique Muscle
Greater Omentum — — Stomach
Ascending Colon — — Gastroepiploic Vessel
Diaphragm — — Diaphragm
Ileum — — Transverse Colon
Mesentery — — Jejunum
Rectus Abdominis Muscle, — — Rectus Abdominis Muscle
Tendinous Intersection — — Radius
Radius — — Carpal Bones
Long Extensor Tendons — — Fifth Metacarpal
Ulna — — Thenar Muscles
Long Flexor Tendons — — Penis
Hypothenar Muscles

First Toe

First Distal Phalanx
Second Distal Phalanx

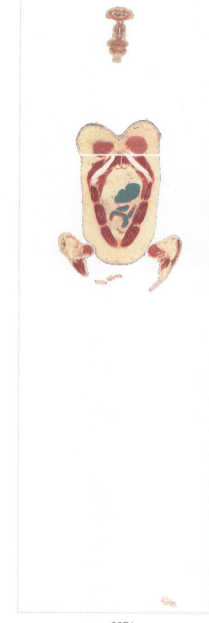

vmc3071

324

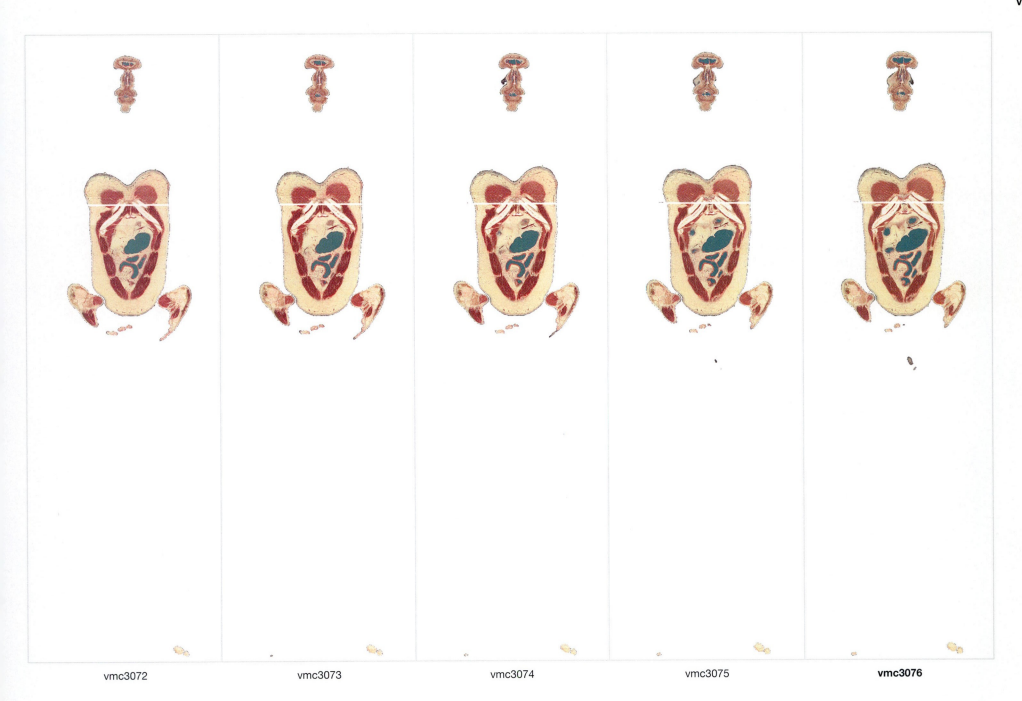

vmc3072   vmc3073   vmc3074   vmc3075   **vmc3076**

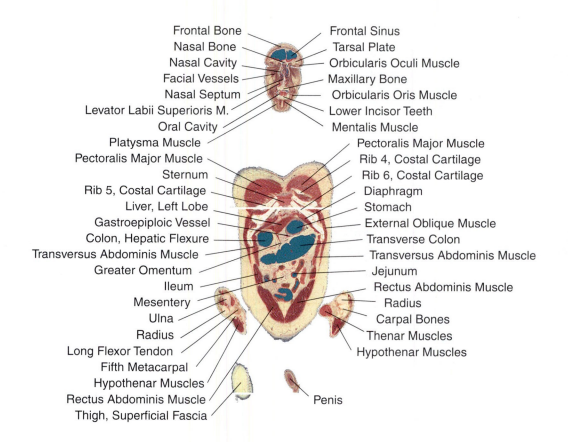

Frontal Bone — Frontal Sinus
Nasal Bone — Tarsal Plate
Nasal Cavity — Orbicularis Oculi Muscle
Facial Vessels — Maxillary Bone
Nasal Septum — Orbicularis Oris Muscle
Levator Labii Superioris M. — Lower Incisor Teeth
Oral Cavity — Mentalis Muscle
Platysma Muscle — Pectoralis Major Muscle
Pectoralis Major Muscle — Rib 4, Costal Cartilage
Sternum — Rib 6, Costal Cartilage
Rib 5, Costal Cartilage — Diaphragm
Liver, Left Lobe — Stomach
Gastroepiploic Vessel — External Oblique Muscle
Colon, Hepatic Flexure — Transverse Colon
Transversus Abdominis Muscle — Transversus Abdominis Muscle
Greater Omentum — Jejunum
Ileum — Rectus Abdominis Muscle
Mesentery — Radius
Ulna — Carpal Bones
Radius — Thenar Muscles
Long Flexor Tendon — Hypothenar Muscles
Fifth Metacarpal
Hypothenar Muscles
Rectus Abdominis Muscle
Thigh, Superficial Fascia

Penis

First Distal Phalanx
Second Distal Phalanx
First Distal Phalanx   Third Toe
Second Toe

vmc3077

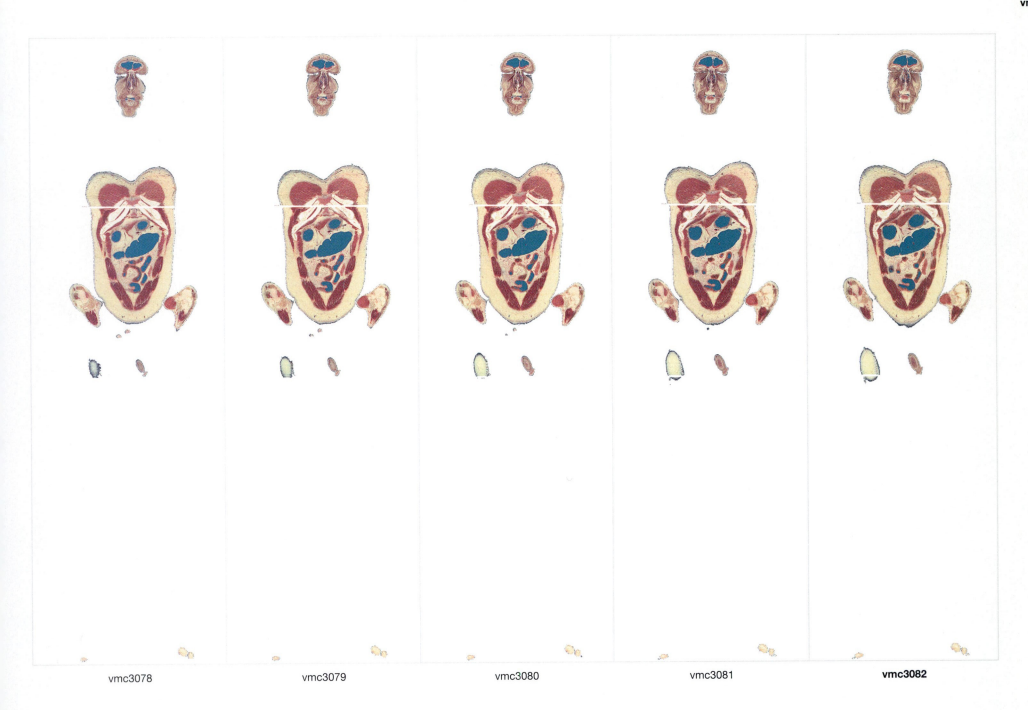

vmc3078   vmc3079   vmc3080   vmc3081   **vmc3082**

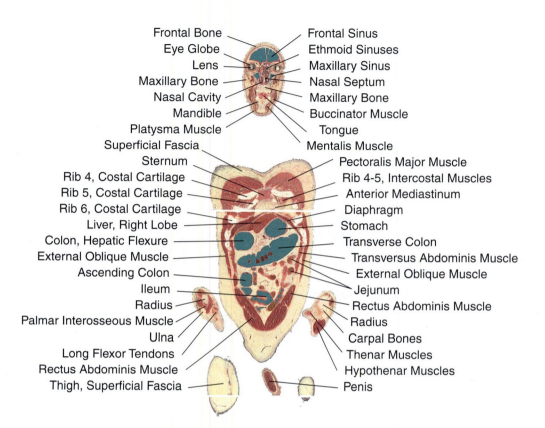

Frontal Bone — Frontal Sinus
Eye Globe — Ethmoid Sinuses
Lens — Maxillary Sinus
Maxillary Bone — Nasal Septum
Nasal Cavity — Maxillary Bone
Mandible — Buccinator Muscle
Platysma Muscle — Tongue
Superficial Fascia — Mentalis Muscle
Sternum — Pectoralis Major Muscle
Rib 4, Costal Cartilage — Rib 4-5, Intercostal Muscles
Rib 5, Costal Cartilage — Anterior Mediastinum
Rib 6, Costal Cartilage — Diaphragm
Liver, Right Lobe — Stomach
Colon, Hepatic Flexure — Transverse Colon
External Oblique Muscle — Transversus Abdominis Muscle
Ascending Colon — External Oblique Muscle
Ileum — Jejunum
Radius — Rectus Abdominis Muscle
Palmar Interosseous Muscle — Radius
Ulna — Carpal Bones
Long Flexor Tendons — Thenar Muscles
Rectus Abdominis Muscle — Hypothenar Muscles
Thigh, Superficial Fascia — Penis

First Distal Phalanx
First Distal Phalanx
Second Distal Phalanx
Second Distal Phalanx
Third Distal Phalanx

vmc3083

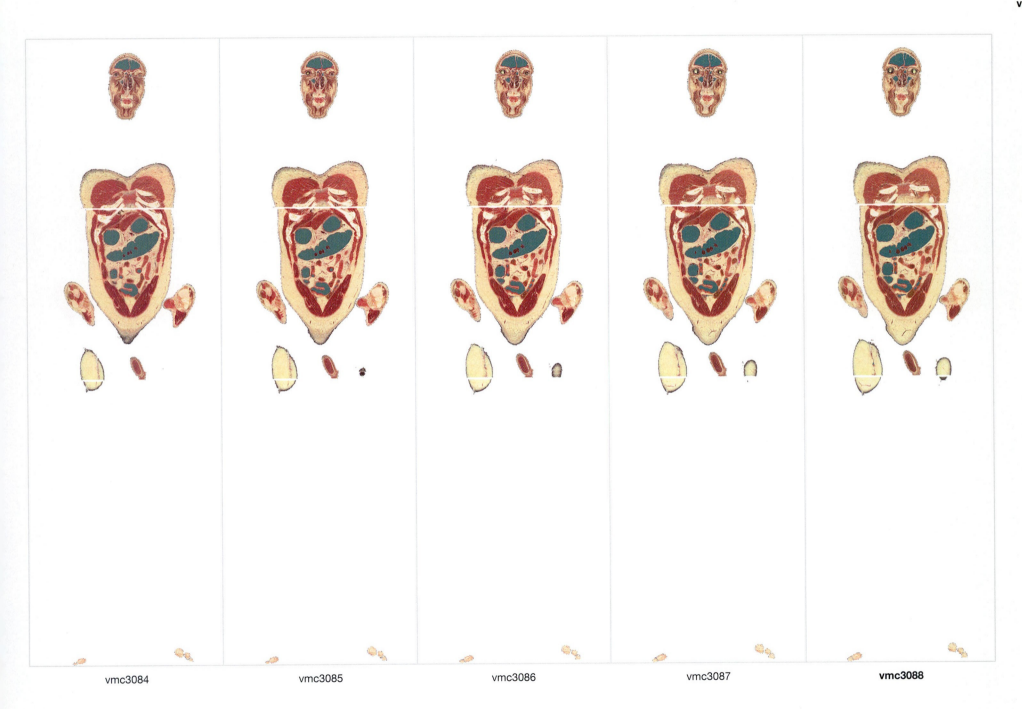

vmc3084　　　　　vmc3085　　　　　vmc3086　　　　　vmc3087　　　　　**vmc3088**

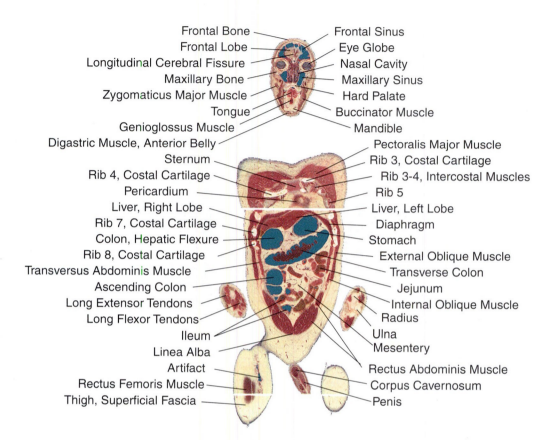

Frontal Bone — — Frontal Sinus
Frontal Lobe — — Eye Globe
Longitudinal Cerebral Fissure — — Nasal Cavity
Maxillary Bone — — Maxillary Sinus
Zygomaticus Major Muscle — — Hard Palate
Tongue — — Buccinator Muscle
Genioglossus Muscle — — Mandible
Digastric Muscle, Anterior Belly — — Pectoralis Major Muscle
Sternum — — Rib 3, Costal Cartilage
Rib 4, Costal Cartilage — — Rib 3-4, Intercostal Muscles
Pericardium — — Rib 5
Liver, Right Lobe — — Liver, Left Lobe
Rib 7, Costal Cartilage — — Diaphragm
Colon, Hepatic Flexure — — Stomach
Rib 8, Costal Cartilage — — External Oblique Muscle
Transversus Abdominis Muscle — — Transverse Colon
Ascending Colon — — Jejunum
Long Extensor Tendons — — Internal Oblique Muscle
Long Flexor Tendons — — Radius
Ileum — — Ulna
Linea Alba — — Mesentery
Artifact — — Rectus Abdominis Muscle
Rectus Femoris Muscle — — Corpus Cavernosum
Thigh, Superficial Fascia — — Penis

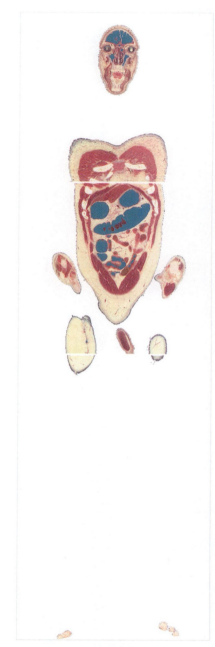

First Distal Phalanx
Second Distal Phalanx

First Distal Phalanx
Second Distal Phalanx
Third Distal Phalanx
Fourth Toe

vmc3089

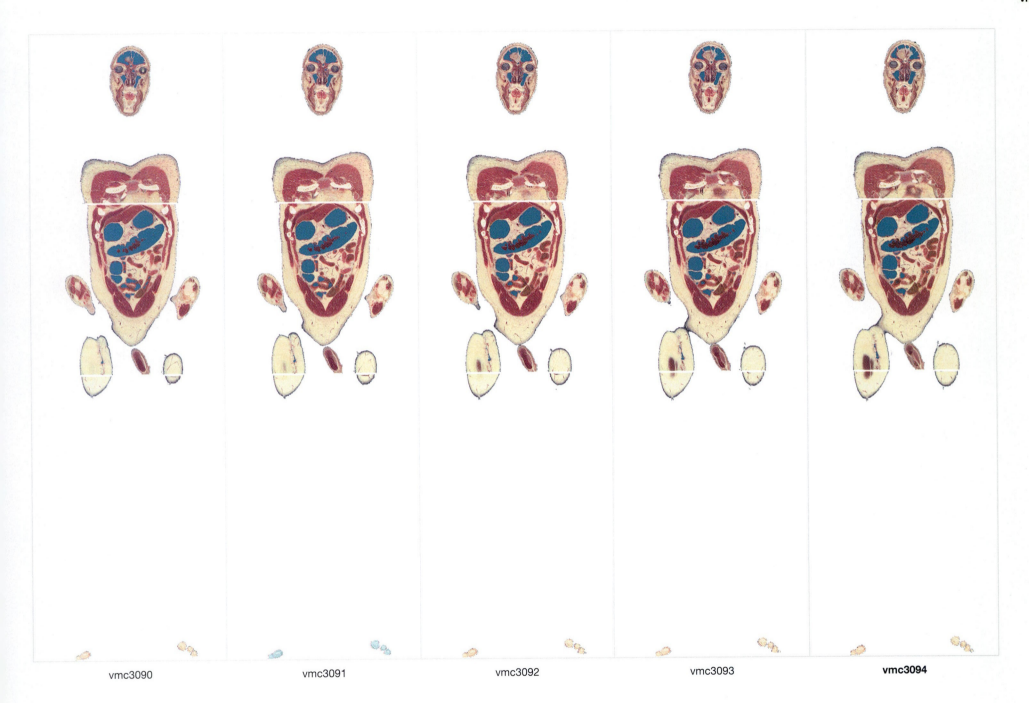

vmc3090   vmc3091   vmc3092   vmc3093   **vmc3094**

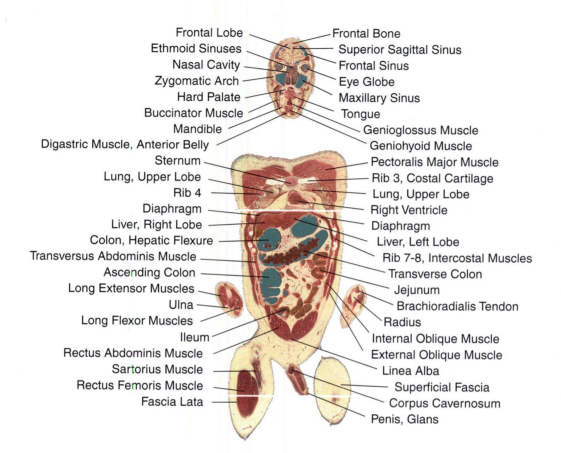

Frontal Lobe — Frontal Bone
Ethmoid Sinuses — Superior Sagittal Sinus
Nasal Cavity — Frontal Sinus
Zygomatic Arch — Eye Globe
Hard Palate — Maxillary Sinus
Buccinator Muscle — Tongue
Mandible — Genioglossus Muscle
Digastric Muscle, Anterior Belly — Geniohyoid Muscle
Sternum — Pectoralis Major Muscle
Lung, Upper Lobe — Rib 3, Costal Cartilage
Rib 4 — Lung, Upper Lobe
Diaphragm — Right Ventricle
Liver, Right Lobe — Diaphragm
Colon, Hepatic Flexure — Liver, Left Lobe
Transversus Abdominis Muscle — Rib 7-8, Intercostal Muscles
Ascending Colon — Transverse Colon
Long Extensor Muscles — Jejunum
Ulna — Brachioradialis Tendon
Long Flexor Muscles — Radius
Ileum — Internal Oblique Muscle
Rectus Abdominis Muscle — External Oblique Muscle
Sartorius Muscle — Linea Alba
Rectus Femoris Muscle — Superficial Fascia
Fascia Lata — Corpus Cavernosum
Penis, Glans

First Proximal Phalanx
First Proximal Phalanx — Second Middle Phalanx
Second Middle Phalanx — Third Middle Phalanx
Third Toe — Fourth Distal Phalanx

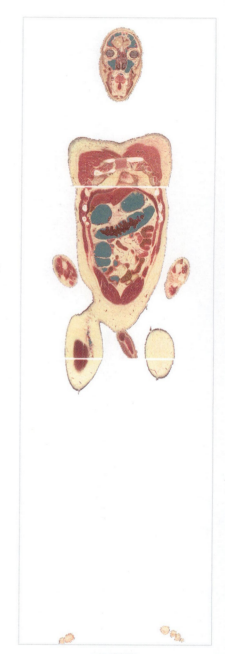

vmc3095

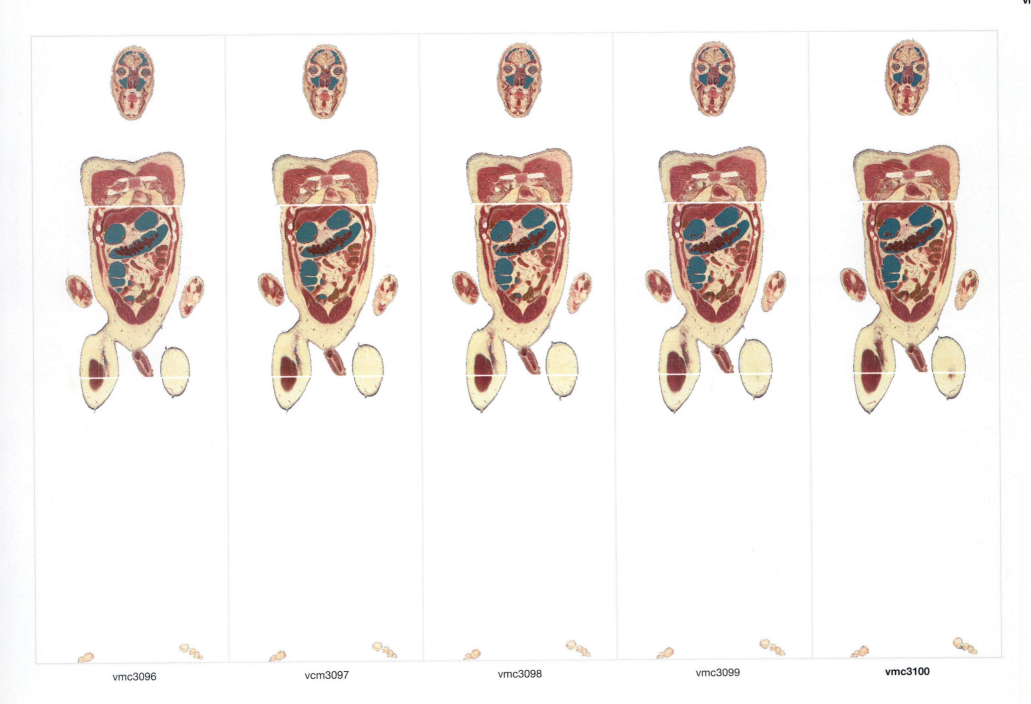

vmc3096   vcm3097   vmc3098   vmc3099   **vmc3100**

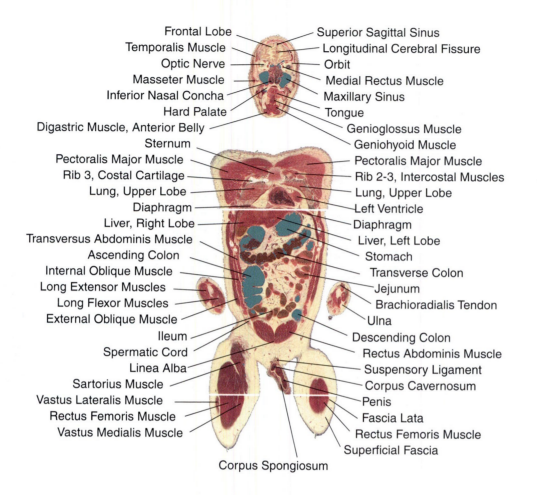

Frontal Lobe — Superior Sagittal Sinus
Temporalis Muscle — Longitudinal Cerebral Fissure
Optic Nerve — Orbit
Masseter Muscle — Medial Rectus Muscle
Inferior Nasal Concha — Maxillary Sinus
Hard Palate — Tongue
Digastric Muscle, Anterior Belly — Genioglossus Muscle
Sternum — Geniohyoid Muscle
Pectoralis Major Muscle — Pectoralis Major Muscle
Rib 3, Costal Cartilage — Rib 2-3, Intercostal Muscles
Lung, Upper Lobe — Lung, Upper Lobe
Diaphragm — Left Ventricle
Liver, Right Lobe — Diaphragm
Transversus Abdominis Muscle — Liver, Left Lobe
Ascending Colon — Stomach
Internal Oblique Muscle — Transverse Colon
Long Extensor Muscles — Jejunum
Long Flexor Muscles — Brachioradialis Tendon
External Oblique Muscle — Ulna
Ileum — Descending Colon
Spermatic Cord — Rectus Abdominis Muscle
Linea Alba — Suspensory Ligament
Sartorius Muscle — Corpus Cavernosum
Vastus Lateralis Muscle — Penis
Rectus Femoris Muscle — Fascia Lata
Vastus Medialis Muscle — Rectus Femoris Muscle
Superficial Fascia
Corpus Spongiosum

First Proximal Phalanx — First Proximal Phalanx
Second Middle Phalanx — Second Middle Phalanx
Third Distal Phalanx — Third Middle Phalanx
Fourth Distal Phalanx

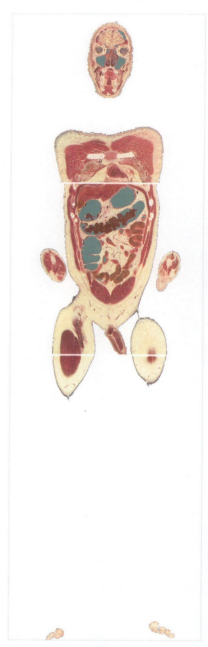

vmc3101

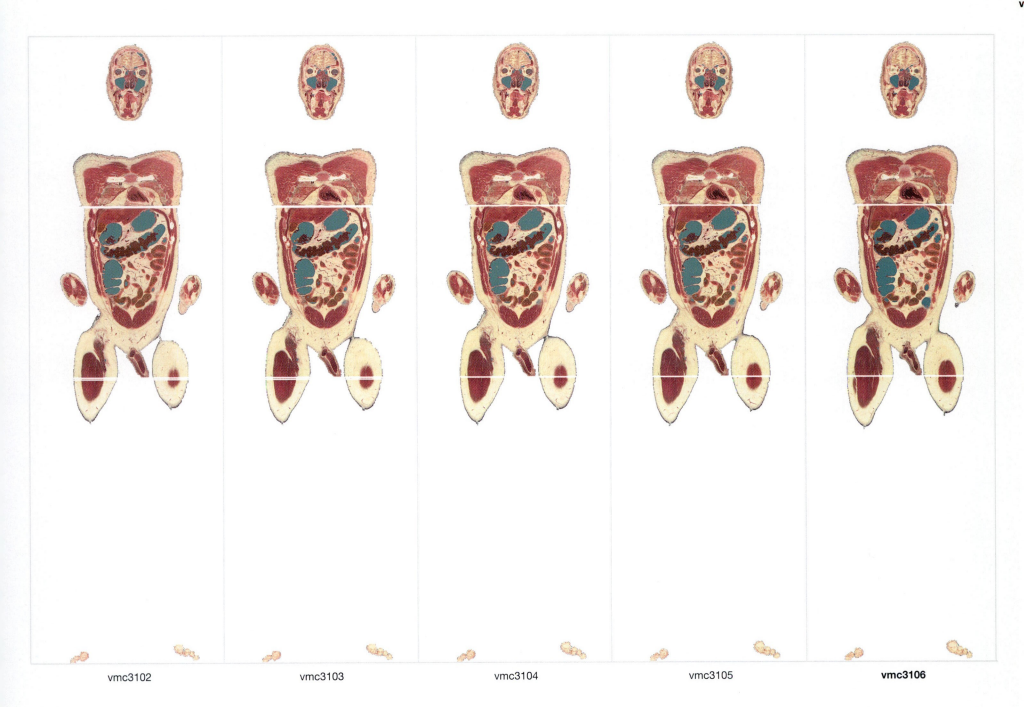

vmc3102   vmc3103   vmc3104   vmc3105   **vmc3106**

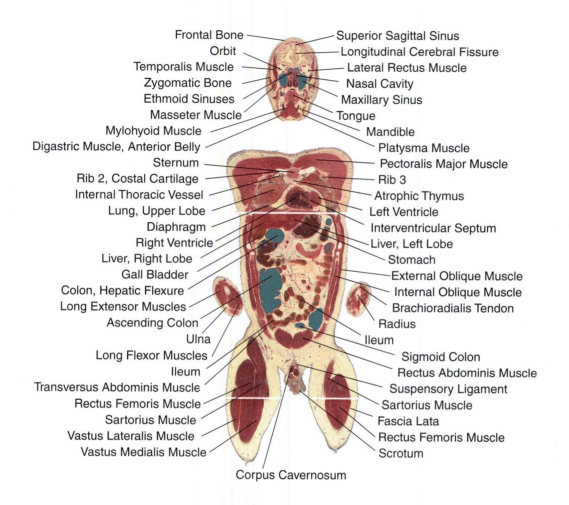

Frontal Bone — Superior Sagittal Sinus
Orbit — Longitudinal Cerebral Fissure
Temporalis Muscle — Lateral Rectus Muscle
Zygomatic Bone — Nasal Cavity
Ethmoid Sinuses — Maxillary Sinus
Masseter Muscle — Tongue
Mylohyoid Muscle — Mandible
Digastric Muscle, Anterior Belly — Platysma Muscle
Sternum — Pectoralis Major Muscle
Rib 2, Costal Cartilage — Rib 3
Internal Thoracic Vessel — Atrophic Thymus
Lung, Upper Lobe — Left Ventricle
Diaphragm — Interventricular Septum
Right Ventricle — Liver, Left Lobe
Liver, Right Lobe — Stomach
Gall Bladder — External Oblique Muscle
Colon, Hepatic Flexure — Internal Oblique Muscle
Long Extensor Muscles — Brachioradialis Tendon
Ascending Colon — Radius
Ulna — Ileum
Long Flexor Muscles — Sigmoid Colon
Ileum — Rectus Abdominis Muscle
Transversus Abdominis Muscle — Suspensory Ligament
Rectus Femoris Muscle — Sartorius Muscle
Sartorius Muscle — Fascia Lata
Vastus Lateralis Muscle — Rectus Femoris Muscle
Vastus Medialis Muscle — Scrotum
Corpus Cavernosum

First Proximal Phalanx — First Proximal Phalanx
Second Middle Phalanx — Second Proximal Phalanx
Third Middle Phalanx — Third Proximal Phalanx
Fourth Toe — Fourth Middle Phalanx
Fifth Toe

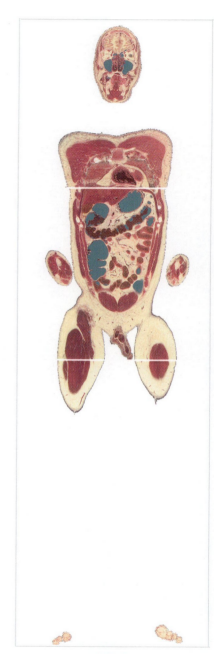

vmc3107

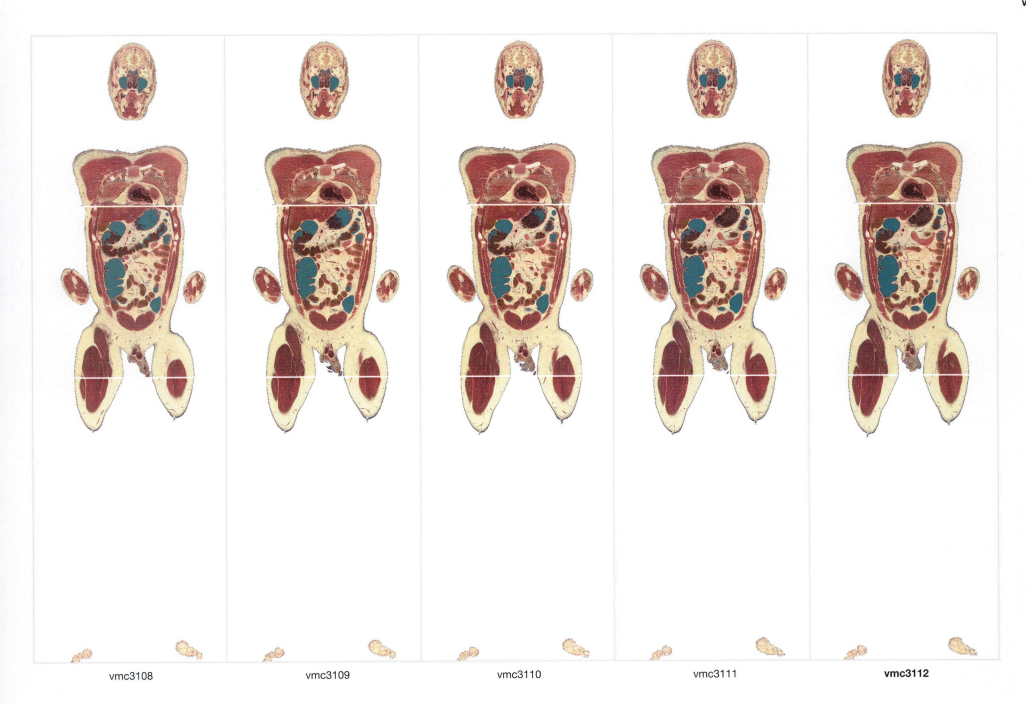

vmc3108   vmc3109   vmc3110   vmc3111   **vmc3112**

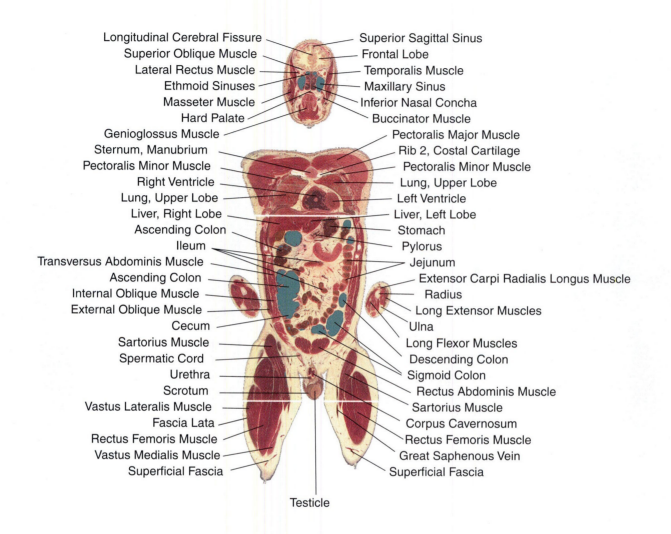

Longitudinal Cerebral Fissure — Superior Sagittal Sinus
Superior Oblique Muscle — Frontal Lobe
Lateral Rectus Muscle — Temporalis Muscle
Ethmoid Sinuses — Maxillary Sinus
Masseter Muscle — Inferior Nasal Concha
Hard Palate — Buccinator Muscle
Genioglossus Muscle — Pectoralis Major Muscle
Sternum, Manubrium — Rib 2, Costal Cartilage
Pectoralis Minor Muscle — Pectoralis Minor Muscle
Right Ventricle — Lung, Upper Lobe
Lung, Upper Lobe — Left Ventricle
Liver, Right Lobe — Liver, Left Lobe
Ascending Colon — Stomach
Ileum — Pylorus
Transversus Abdominis Muscle — Jejunum
Ascending Colon — Extensor Carpi Radialis Longus Muscle
Internal Oblique Muscle — Radius
External Oblique Muscle — Long Extensor Muscles
Cecum — Ulna
Sartorius Muscle — Long Flexor Muscles
Spermatic Cord — Descending Colon
Urethra — Sigmoid Colon
Scrotum — Rectus Abdominis Muscle
Vastus Lateralis Muscle — Sartorius Muscle
Fascia Lata — Corpus Cavernosum
Rectus Femoris Muscle — Rectus Femoris Muscle
Vastus Medialis Muscle — Great Saphenous Vein
Superficial Fascia — Superficial Fascia

Testicle

First Metatarsal

First Proximal Phalanx — Second Metatarsal
Second Proximal Phalanx — Third Proximal Phalanx
Third Middle Phalanx — Fourth Proximal Phalanx
Fourth Distal Phalanx — Fifth Middle Phalanx

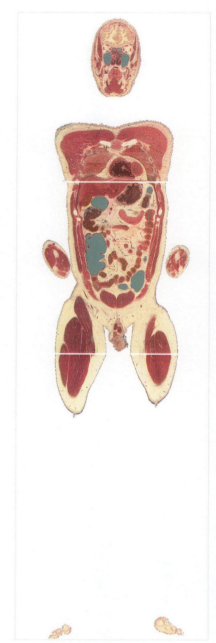

vmc3113

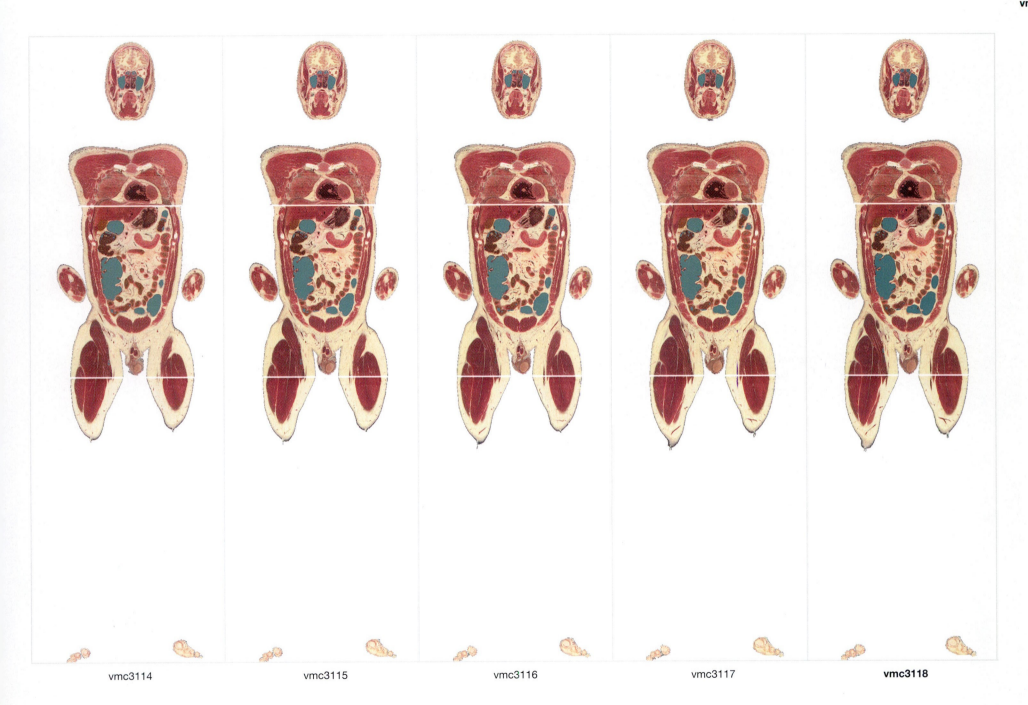

vmc3114   vmc3115   vmc3116   vmc3117   **vmc3118**

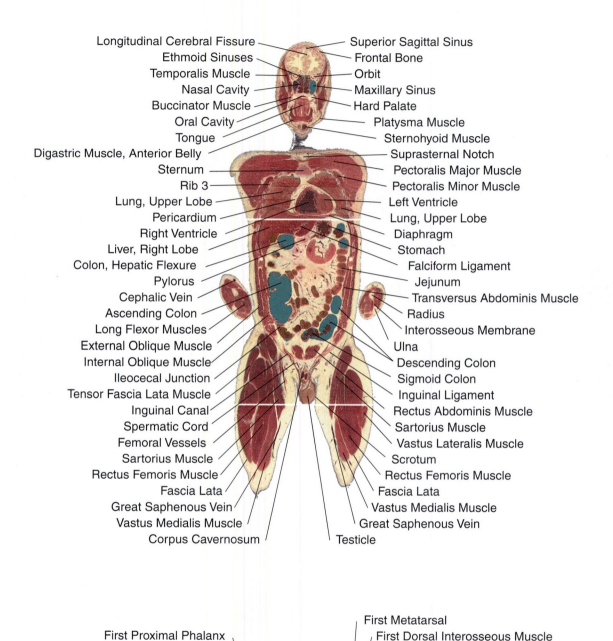

Longitudinal Cerebral Fissure — Superior Sagittal Sinus
Ethmoid Sinuses — Frontal Bone
Temporalis Muscle — Orbit
Nasal Cavity — Maxillary Sinus
Buccinator Muscle — Hard Palate
Oral Cavity — Platysma Muscle
Tongue — Sternohyoid Muscle
Digastric Muscle, Anterior Belly — Suprasternal Notch
Sternum — Pectoralis Major Muscle
Rib 3 — Pectoralis Minor Muscle
Lung, Upper Lobe — Left Ventricle
Pericardium — Lung, Upper Lobe
Right Ventricle — Diaphragm
Liver, Right Lobe — Stomach
Colon, Hepatic Flexure — Falciform Ligament
Pylorus — Jejunum
Cephalic Vein — Transversus Abdominis Muscle
Ascending Colon — Radius
Long Flexor Muscles — Interosseous Membrane
External Oblique Muscle — Ulna
Internal Oblique Muscle — Descending Colon
Ileocecal Junction — Sigmoid Colon
Tensor Fascia Lata Muscle — Inguinal Ligament
Inguinal Canal — Rectus Abdominis Muscle
Spermatic Cord — Sartorius Muscle
Femoral Vessels — Vastus Lateralis Muscle
Sartorius Muscle — Scrotum
Rectus Femoris Muscle — Rectus Femoris Muscle
Fascia Lata — Fascia Lata
Great Saphenous Vein — Vastus Medialis Muscle
Vastus Medialis Muscle — Great Saphenous Vein
Corpus Cavernosum — Testicle

First Proximal Phalanx — First Metatarsal
Second Proximal Phalanx — First Dorsal Interosseous Muscle
Third Middle Phalanx — Second Metatarsal
Fourth Middle Phalanx — Third Metatarsal
Fifth Distal Phalanx — Fourth Proximal Phalanx
— Fifth Middle Phalanx

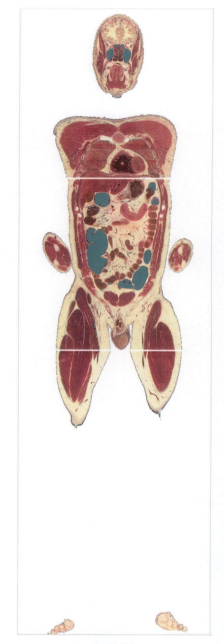

vmc3119

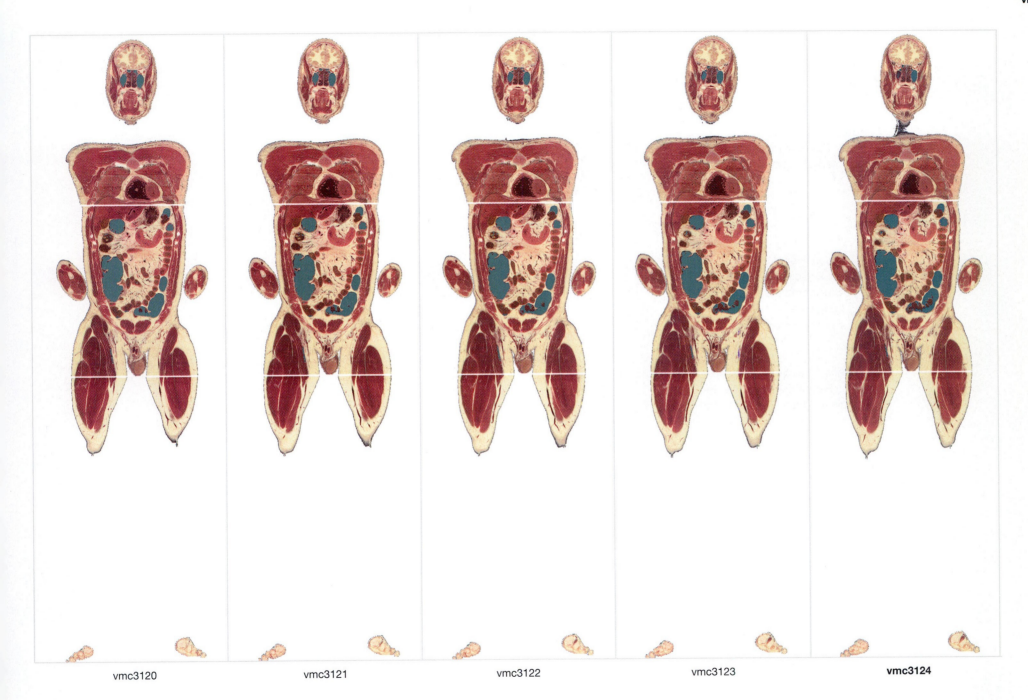

vmc3120   vmc3121   vmc3122   vmc3123   **vmc3124**

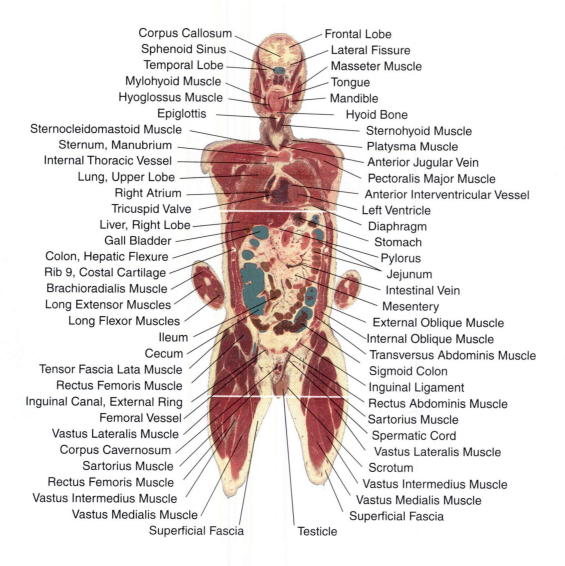

Corpus Callosum — Frontal Lobe
Sphenoid Sinus — Lateral Fissure
Temporal Lobe — Masseter Muscle
Mylohyoid Muscle — Tongue
Hyoglossus Muscle — Mandible
Epiglottis — Hyoid Bone
Sternocleidomastoid Muscle — Sternohyoid Muscle
Sternum, Manubrium — Platysma Muscle
Internal Thoracic Vessel — Anterior Jugular Vein
Lung, Upper Lobe — Pectoralis Major Muscle
Right Atrium — Anterior Interventricular Vessel
Tricuspid Valve — Left Ventricle
Liver, Right Lobe — Diaphragm
Gall Bladder — Stomach
Colon, Hepatic Flexure — Pylorus
Rib 9, Costal Cartilage — Jejunum
Brachioradialis Muscle — Intestinal Vein
Long Extensor Muscles — Mesentery
Long Flexor Muscles — External Oblique Muscle
Ileum — Internal Oblique Muscle
Cecum — Transversus Abdominis Muscle
Tensor Fascia Lata Muscle — Sigmoid Colon
Rectus Femoris Muscle — Inguinal Ligament
Inguinal Canal, External Ring — Rectus Abdominis Muscle
Femoral Vessel — Sartorius Muscle
Vastus Lateralis Muscle — Spermatic Cord
Corpus Cavernosum — Vastus Lateralis Muscle
Sartorius Muscle — Scrotum
Rectus Femoris Muscle — Vastus Intermedius Muscle
Vastus Intermedius Muscle — Vastus Medialis Muscle
Vastus Medialis Muscle — Superficial Fascia
Superficial Fascia — Testicle

First Metatarsal — First Metatarsal
Second Metatarsal — First Dorsal Interosseous Muscle
Third Proximal Phalanx — Third Metatarsal
Fourth Middle Phalanx — Fourth Metatarsal
Fifth Distal Phalanx — Fifth Proximal Phalanx

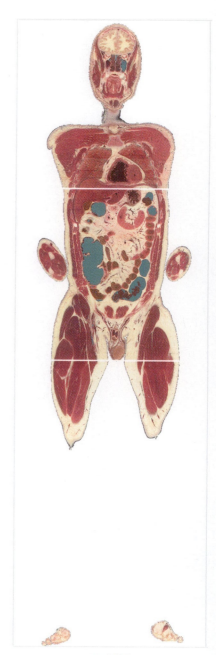

vmc3125

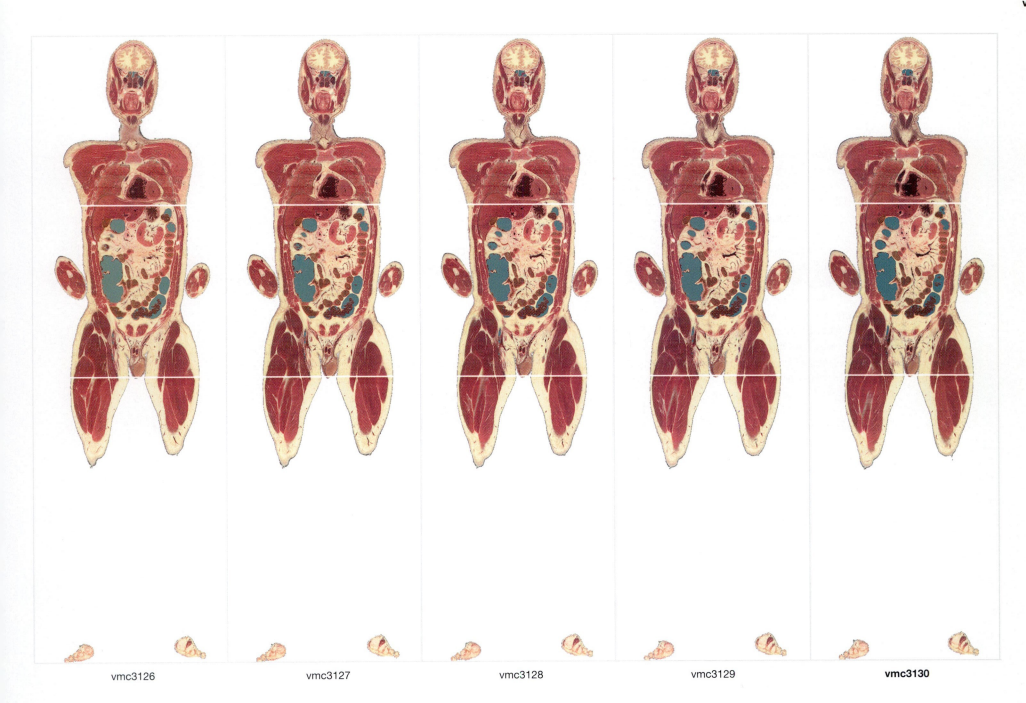

vmc3126   vmc3127   vmc3128   vmc3129   **vmc3130**

Corpus Callosum — Frontal Lobe

Temporalis Muscle — Lateral Ventricle

Zygomatic Arch — Sphenoid Sinus

Lateral Pterygoid Muscle — Nasopharynx

Medial Pterygoid Muscle — Masseter Muscle

Hyoid Bone — Hyoglossus Muscle

Thyrohyoid Muscle — Submandibular Gland

Sternocleidomastoid Muscle — Thyroid Cartilage

External Jugular Vein — Anterior Jugular Vein

Deltoid Muscle — Pectoralis Major Muscle

Rib 1, Costal Cartilage — Pectoralis Minor Muscle

Atrophic Thymus — Manubrium, Suprasternal Notch

Right Ventricle — Lung, Upper Lobe

Tricuspid Valve — Left Ventricle

Liver, Right Lobe — Liver, Left Lobe

Falciform Ligament — Stomach

Ascending Colon — Jejunum

Long Extensor Muscles — Radius

Long Flexor Muscles — Interosseous Membrane

Internal Oblique Muscle — Ulna

Transversus Abdominis Muscle — External Oblique Muscle

Ilium, Anterior Inferior Spine — Mesentery

Tensor Fascia Lata Muscle — Superior Mesenteric Vein

Iliacus Muscle — Sigmoid Colon

Rectus Femoris Muscle — Spermatic Cord and Inguinal Canal, Internal Ring

Vastus Lateralis Muscle — Femoral Vessels

Adductor Longus Muscle — Vastus Lateralis Muscle

Corpus Cavernosum — Inguinal Ligament

Corpus Spongiosum — Vastus Intermedius Muscle

Sartorius Muscle — Pectineus Muscle

Vastus Medialis Muscle — Sartorius Muscle

Patella — Vastus Medialis Muscle

Scrotum — Abductor Hallucis Muscle

Testicle — First Metatarsal

First Dorsal Interosseous Muscle

First Metatarsal — Second Metatarsal

Second Metatarsal — Second Dorsal Interosseous Muscle

Third Proximal Phalanx — Third Metatarsal

Fourth Proximal Phalanx — Fourth Metatarsal

Fifth Middle Phalanx — Fifth Proximal Phalanx

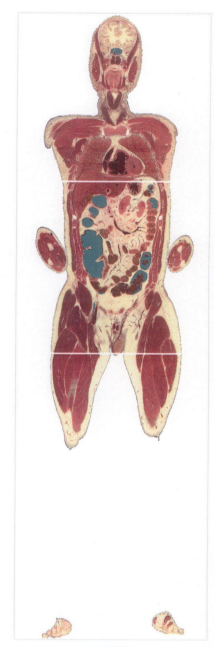

vmc3131

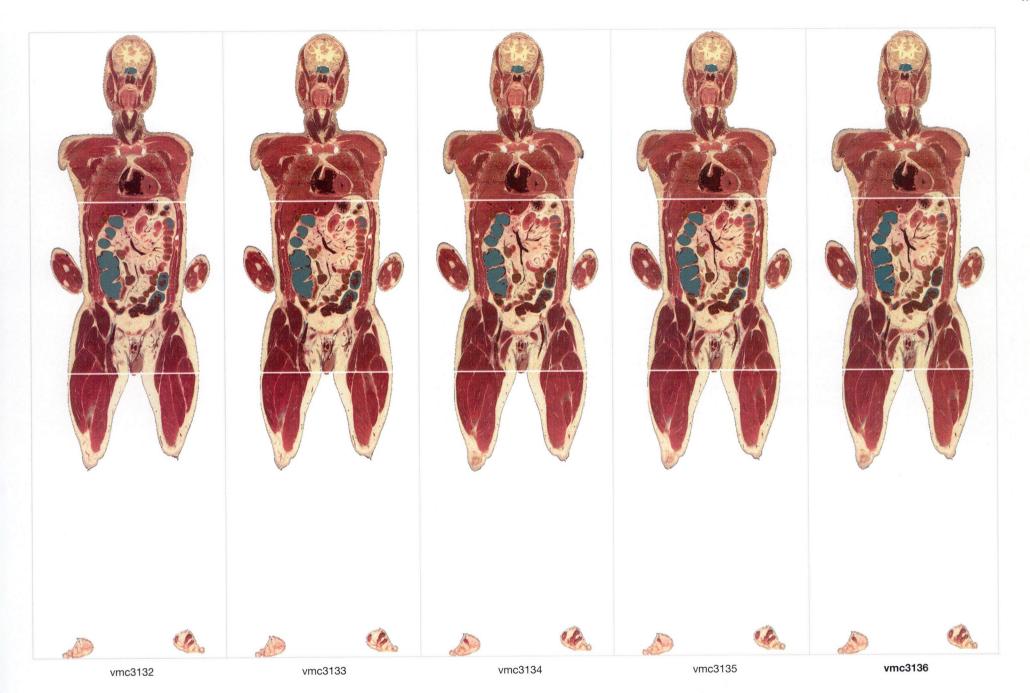

vmc3132   vmc3133   vmc3134   vmc3135   **vmc3136**

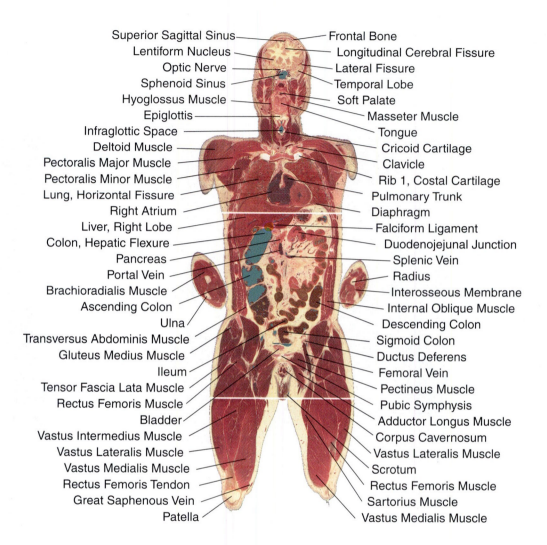

Superior Sagittal Sinus — — Frontal Bone
Lentiform Nucleus — — Longitudinal Cerebral Fissure
Optic Nerve — — Lateral Fissure
Sphenoid Sinus — — Temporal Lobe
Hyoglossus Muscle — — Soft Palate
Epiglottis — — Masseter Muscle
Infraglottic Space — — Tongue
Deltoid Muscle — — Cricoid Cartilage
Pectoralis Major Muscle — — Clavicle
Pectoralis Minor Muscle — — Rib 1, Costal Cartilage
Lung, Horizontal Fissure — — Pulmonary Trunk
Right Atrium — — Diaphragm
Liver, Right Lobe — — Falciform Ligament
Colon, Hepatic Flexure — — Duodenojejunal Junction
Pancreas — — Splenic Vein
Portal Vein — — Radius
Brachioradialis Muscle — — Interosseous Membrane
Ascending Colon — — Internal Oblique Muscle
Ulna — — Descending Colon
Transversus Abdominis Muscle — — Sigmoid Colon
Gluteus Medius Muscle — — Ductus Deferens
Ileum — — Femoral Vein
Tensor Fascia Lata Muscle — — Pectineus Muscle
Rectus Femoris Muscle — — Pubic Symphysis
Bladder — — Adductor Longus Muscle
Vastus Intermedius Muscle — — Corpus Cavernosum
Vastus Lateralis Muscle — — Vastus Lateralis Muscle
Vastus Medialis Muscle — — Scrotum
Rectus Femoris Tendon — — Rectus Femoris Muscle
Great Saphenous Vein — — Sartorius Muscle
Patella — — Vastus Medialis Muscle

First Metatarsal — Abductor Hallucis Muscle
First Dorsal Interosseous Muscle — First Metatarsal
Second Metatarsal — Second Metatarsal
Second Dorsal Interosseous Muscle — Third Metatarsal
Third Metatarsal — Flexor Digitorum Brevis Muscle
Fourth Metatarsal — Fourth Metatarsal
Fifth Proximal Phalanx — Fifth Proximal Phalanx

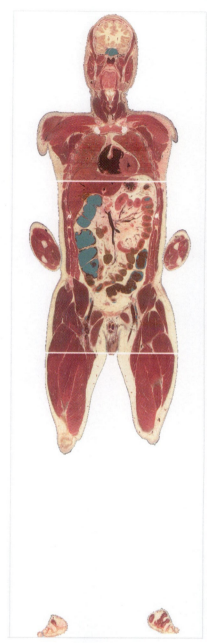

vmc3137

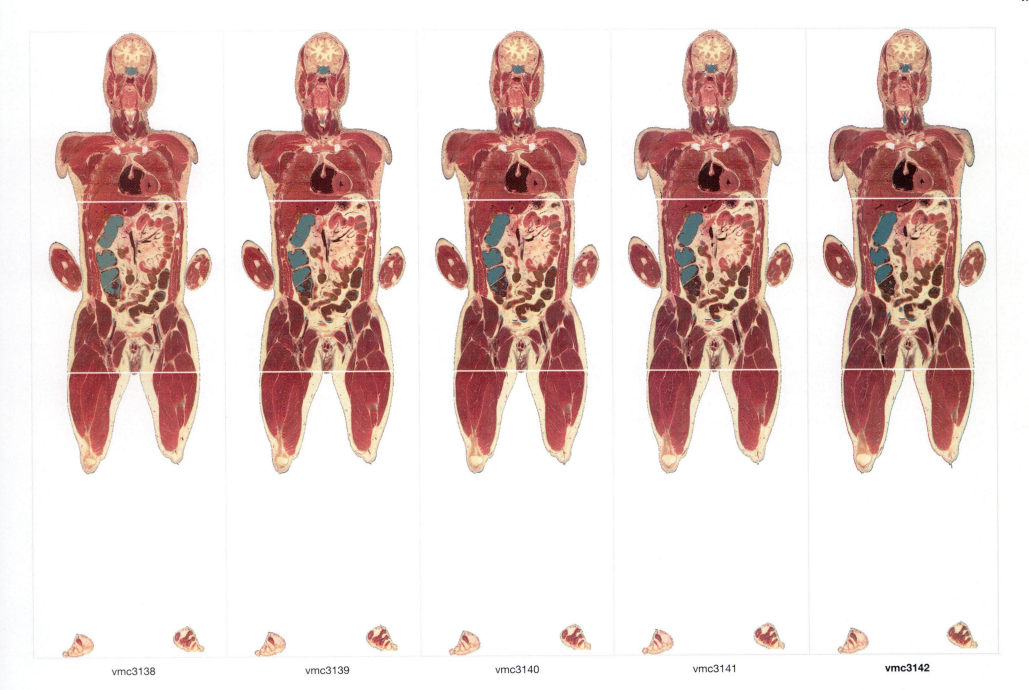

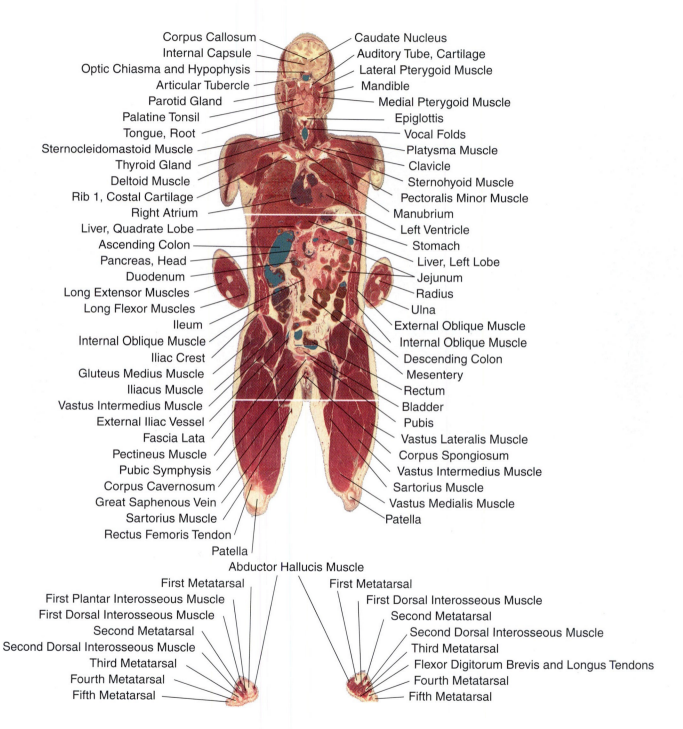

Corpus Callosum

Caudate Nucleus

Internal Capsule

Auditory Tube, Cartilage

Optic Chiasma and Hypophysis

Lateral Pterygoid Muscle

Articular Tubercle

Mandible

Parotid Gland

Medial Pterygoid Muscle

Palatine Tonsil

Epiglottis

Tongue, Root

Vocal Folds

Sternocleidomastoid Muscle

Platysma Muscle

Thyroid Gland

Clavicle

Deltoid Muscle

Sternohyoid Muscle

Rib 1, Costal Cartilage

Pectoralis Minor Muscle

Right Atrium

Manubrium

Liver, Quadrate Lobe

Left Ventricle

Ascending Colon

Stomach

Pancreas, Head

Liver, Left Lobe

Duodenum

Jejunum

Long Extensor Muscles

Radius

Long Flexor Muscles

Ulna

Ileum

External Oblique Muscle

Internal Oblique Muscle

Internal Oblique Muscle

Iliac Crest

Descending Colon

Gluteus Medius Muscle

Mesentery

Iliacus Muscle

Rectum

Vastus Intermedius Muscle

Bladder

External Iliac Vessel

Pubis

Fascia Lata

Vastus Lateralis Muscle

Pectineus Muscle

Corpus Spongiosum

Pubic Symphysis

Vastus Intermedius Muscle

Corpus Cavernosum

Sartorius Muscle

Great Saphenous Vein

Vastus Medialis Muscle

Sartorius Muscle

Patella

Rectus Femoris Tendon

Patella

Abductor Hallucis Muscle

First Metatarsal

First Metatarsal

First Plantar Interosseous Muscle

First Dorsal Interosseous Muscle

First Dorsal Interosseous Muscle

Second Metatarsal

Second Metatarsal

Second Dorsal Interosseous Muscle

Second Dorsal Interosseous Muscle

Third Metatarsal

Third Metatarsal

Flexor Digitorum Brevis and Longus Tendons

Fourth Metatarsal

Fourth Metatarsal

Fifth Metatarsal

Fifth Metatarsal

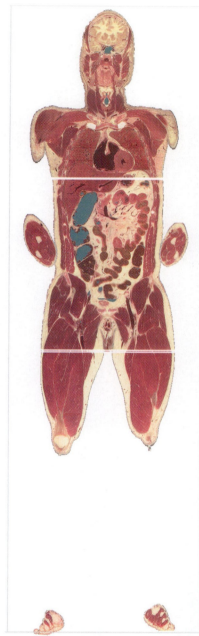

vmc3143

348

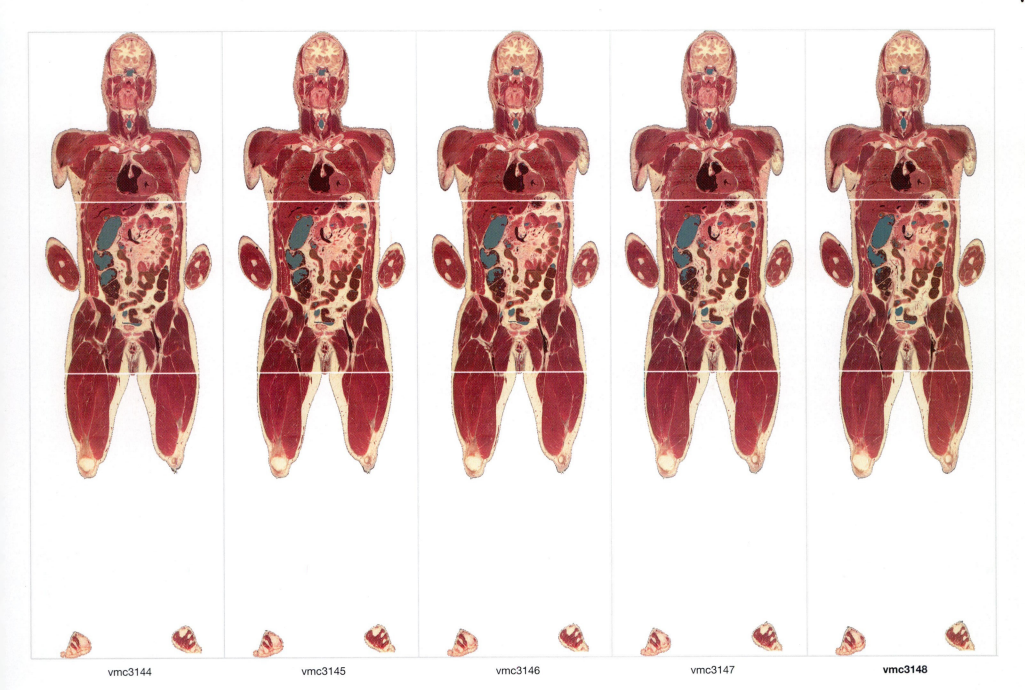

vmc3144   vmc3145   vmc3146   vmc3147   **vmc3148**

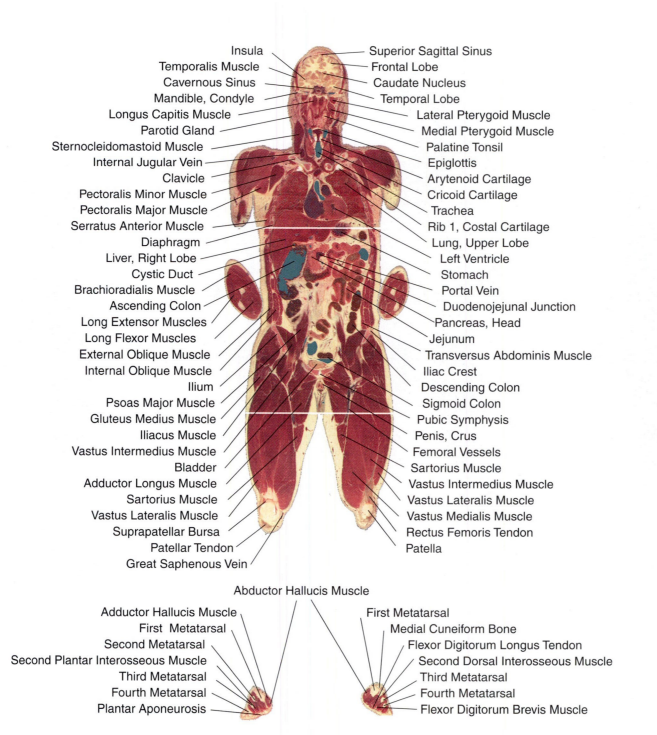

Insula
Temporalis Muscle
Cavernous Sinus
Mandible, Condyle
Longus Capitis Muscle
Parotid Gland
Sternocleidomastoid Muscle
Internal Jugular Vein
Clavicle
Pectoralis Minor Muscle
Pectoralis Major Muscle
Serratus Anterior Muscle
Diaphragm
Liver, Right Lobe
Cystic Duct
Brachioradialis Muscle
Ascending Colon
Long Extensor Muscles
Long Flexor Muscles
External Oblique Muscle
Internal Oblique Muscle
Ilium
Psoas Major Muscle
Gluteus Medius Muscle
Iliacus Muscle
Vastus Intermedius Muscle
Bladder
Adductor Longus Muscle
Sartorius Muscle
Vastus Lateralis Muscle
Suprapatellar Bursa
Patellar Tendon
Great Saphenous Vein

Superior Sagittal Sinus
Frontal Lobe
Caudate Nucleus
Temporal Lobe
Lateral Pterygoid Muscle
Medial Pterygoid Muscle
Palatine Tonsil
Epiglottis
Arytenoid Cartilage
Cricoid Cartilage
Trachea
Rib 1, Costal Cartilage
Lung, Upper Lobe
Left Ventricle
Stomach
Portal Vein
Duodenojejunal Junction
Pancreas, Head
Jejunum
Transversus Abdominis Muscle
Iliac Crest
Descending Colon
Sigmoid Colon
Pubic Symphysis
Penis, Crus
Femoral Vessels
Sartorius Muscle
Vastus Intermedius Muscle
Vastus Lateralis Muscle
Vastus Medialis Muscle
Rectus Femoris Tendon
Patella

Abductor Hallucis Muscle

Adductor Hallucis Muscle
First Metatarsal
Second Metatarsal
Second Plantar Interosseous Muscle
Third Metatarsal
Fourth Metatarsal
Plantar Aponeurosis

First Metatarsal
Medial Cuneiform Bone
Flexor Digitorum Longus Tendon
Second Dorsal Interosseous Muscle
Third Metatarsal
Fourth Metatarsal
Flexor Digitorum Brevis Muscle

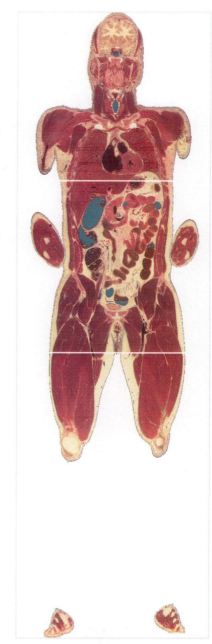

vmc3149

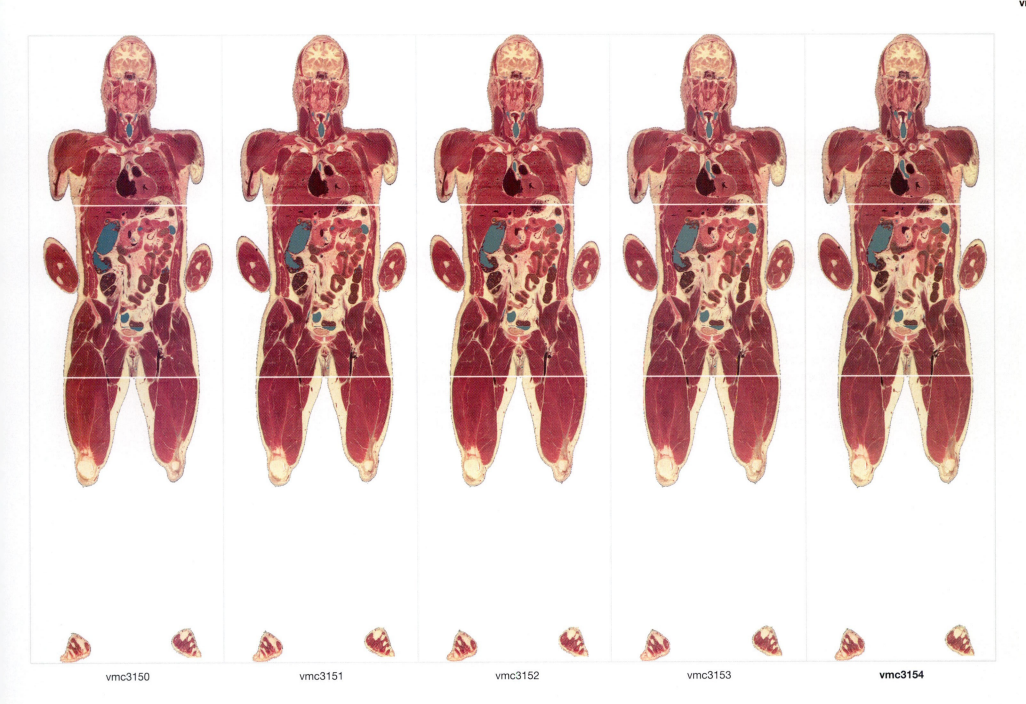

vmc3150   vmc3151   vmc3152   vmc3153   **vmc3154**

Longitudinal Cerebral Fissure — — Frontal Bone
Corpus Callosum — — Lateral Ventricle
Lentiform Nucleus — — Temporalis Muscle
Amygdala — — Pons
Tensor Veli Palatini Muscle — — Mandible, Condyle
Longus Capitis Muscle — — Retromandibular Vein
Digastric Muscle, Posterior Belly — — Pharyngeal Constrictor Muscle
Laryngopharynx — — Deltoid Muscle
Humerus, Head — — Cricoid Cartilage
Internal Jugular Vein — — Pectoralis Minor Muscle
Superior Vena Cava — — Cephalic Vein
Biceps Brachii Muscle — — Brachiocephalic Vein
Ascending Aorta — — Trachea
Right Atrium — — Pulmonary Trunk
Coronary Sinus — — Left Ventricle
Liver, Right Lobe — — Portal Vein
Colon, Hepatic Flexure — — Radius
Pancreas, Head — — Ulna
Long Flexor Muscles — — Jejunum
Ascending Colon — — External Oblique Muscle
Internal Oblique Muscle — — Sigmoid Colon
Psoas Major Muscle — — Inferior Mesenteric Vessel
Gluteus Medius Muscle — — Rectus Femoris Muscle
Iliacus Muscle — — Iliopsoas Muscle
Rectus Femoris Muscle — — Femoral Vessel
Pubis, Superior Ramus — — Pectineus Muscle
Obturator Externus Muscle — — Bladder
Adductor Longus Muscle — — Adductor Longus Muscle
Pubic Symphysis — — Penis, Crus
Vastus Intermedius Muscle — — Adductor Canal
Sartorius Muscle — — Articularis Genu Muscle
Suprapatellar Bursa — — Rectus Femoris Tendon
Great Saphenous Vein — — Patella
Patellar Tendon — — Vastus Medialis Muscle
Adductor Hallucis Muscle — — Abductor Hallucis Muscle
First Metatarsal — — First Metatarsal
Medial Cuneiform Bone — — Medial Cuneiform Bone
Second Metatarsal — — Second Metatarsal
Second Dorsal Interosseous Muscle — — Third Metatarsal
Third Dorsal Interosseous Muscle — — Flexor Digitorum Brevis and Longus Muscles
Plantar Aponeurosis — — Fourth Metatarsal
Fourth Dorsal Interosseous Muscle — — Fifth Metatarsal

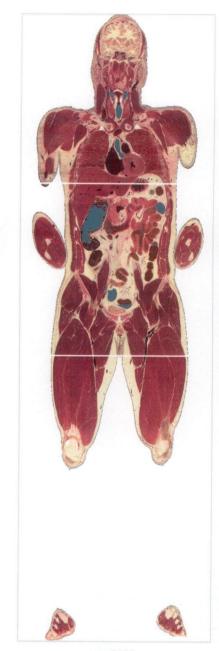

vmc3155

352

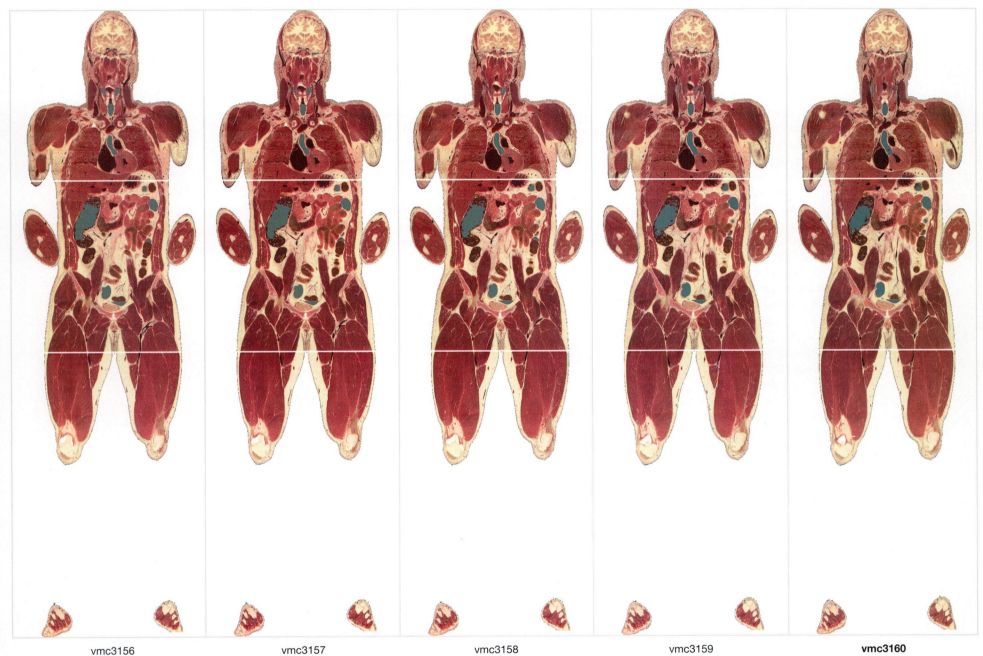

vmc3156   vmc3157   vmc3158   vmc3159   **vmc3160**

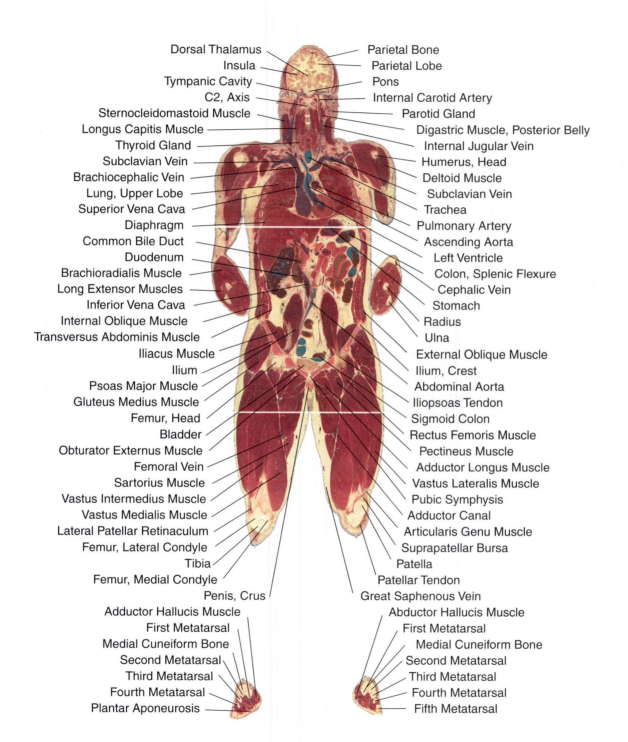

Dorsal Thalamus — Parietal Bone
Insula — Parietal Lobe
Tympanic Cavity — Pons
C2, Axis — Internal Carotid Artery
Sternocleidomastoid Muscle — Parotid Gland
Longus Capitis Muscle — Digastric Muscle, Posterior Belly
Thyroid Gland — Internal Jugular Vein
Subclavian Vein — Humerus, Head
Brachiocephalic Vein — Deltoid Muscle
Lung, Upper Lobe — Subclavian Vein
Superior Vena Cava — Trachea
Diaphragm — Pulmonary Artery
Common Bile Duct — Ascending Aorta
Duodenum — Left Ventricle
Brachioradialis Muscle — Colon, Splenic Flexure
Long Extensor Muscles — Cephalic Vein
Inferior Vena Cava — Stomach
Internal Oblique Muscle — Radius
Transversus Abdominis Muscle — Ulna
Iliacus Muscle — External Oblique Muscle
Ilium — Ilium, Crest
Psoas Major Muscle — Abdominal Aorta
Gluteus Medius Muscle — Iliopsoas Tendon
Femur, Head — Sigmoid Colon
Bladder — Rectus Femoris Muscle
Obturator Externus Muscle — Pectineus Muscle
Femoral Vein — Adductor Longus Muscle
Sartorius Muscle — Vastus Lateralis Muscle
Vastus Intermedius Muscle — Pubic Symphysis
Vastus Medialis Muscle — Adductor Canal
Lateral Patellar Retinaculum — Articularis Genu Muscle
Femur, Lateral Condyle — Suprapatellar Bursa
Tibia — Patella
Femur, Medial Condyle — Patellar Tendon
Penis, Crus — Great Saphenous Vein
Adductor Hallucis Muscle — Abductor Hallucis Muscle
First Metatarsal — First Metatarsal
Medial Cuneiform Bone — Medial Cuneiform Bone
Second Metatarsal — Second Metatarsal
Third Metatarsal — Third Metatarsal
Fourth Metatarsal — Fourth Metatarsal
Plantar Aponeurosis — Fifth Metatarsal

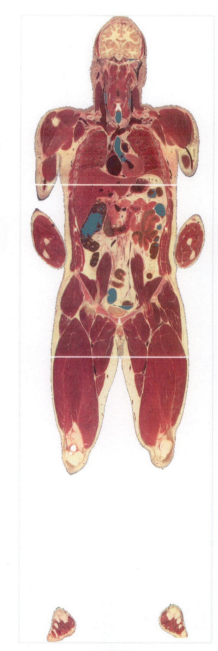

vmc3161

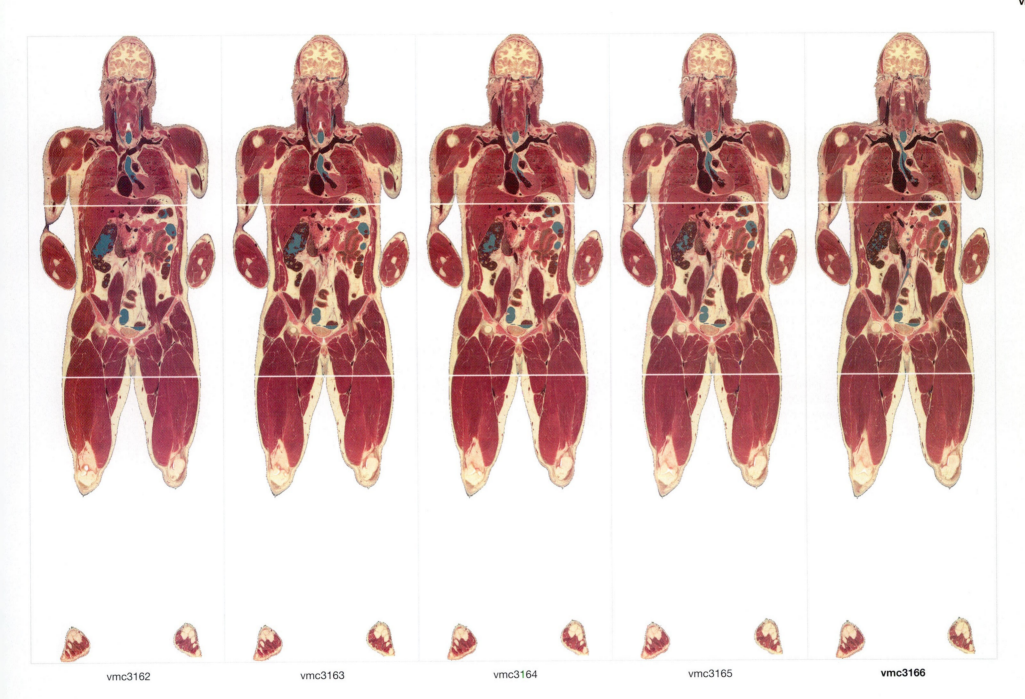

vmc3162   vmc3163   vmc3164   vmc3165   **vmc3166**

Corpus Callosum
Hippocampus
Temporal Lobe
Cerebral Peduncle
Internal Jugular Vein
Occipital Condyle
C1, Atlas
C2, Dens
External Jugular Vein
Brachial Plexus
Pectoralis Minor Muscle
Scalenus Anterior Muscle
Brachiocephalic Vein
Superior Vena Cava
Right Atrium
Liver, Right Lobe
Long Extensor Muscles
Colon, Hepatic Flexure
Long Flexor Muscles
Duodenum
Kidney
Inferior Vena Cava
Gluteus Medius Muscle
Gluteus Minimus Muscle
Iliofemoral Ligament
Iliopsoas Tendon
Obturator Internus Muscle
Bladder
Obturator Externus Muscle
Vastus Lateralis Muscle
Adductor Canal
Femur, Shaft
Tibial Tuberosity
Femur, Medial Condyle

Superior Sagittal Sinus
Longitudinal Cerebral Fissure
Temporalis Muscle
External Acoustic Meatus
Sternocleidomastoid Muscle
Longus Cervicis Muscle
C5, Body
Omohyoid Muscle, Inferior Belly
Humerus, Lesser Tubercle
Deltoid Muscle
Aorta, Arch
Pulmonary Artery
Left Auricle
Lung, Lower Lobe
Left Ventricle
Stomach
Jejunum
Ulna
Descending Colon
External Oblique Muscle
Internal Oblique Muscle
Psoas Major Muscle
Iliacus Muscle
Ilium
Femur, Head
Sigmoid Colon
Pubic Symphysis
Penis, Crus
Vastus Lateralis Muscle
Vastus Intermedius Muscle
Sartorius Muscle
Femur, Lateral Condyle
Patellar Ligament
Vastus Medialis Muscle
Abductor Hallucis Muscle

Great Saphenous Vein
First Metatarsal
Medial Cuneiform Bone
Intermediate Cuneiform Bone
Second Metatarsal
Adductor Hallucis Muscle
Plantar Aponeurosis

Medial Cuneiform Bone
Intermediate Cuneiform Bone
Lateral Cuneiform Bone
Third Metatarsal
Fourth Metatarsal
Fifth Metatarsal
Flexor Digitorum Brevis Muscle

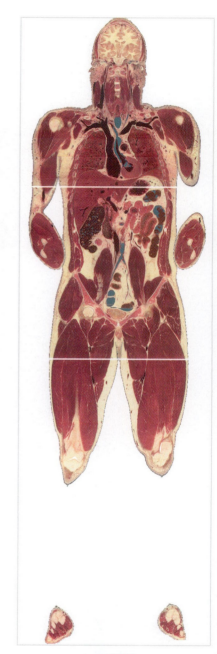

vmc3167

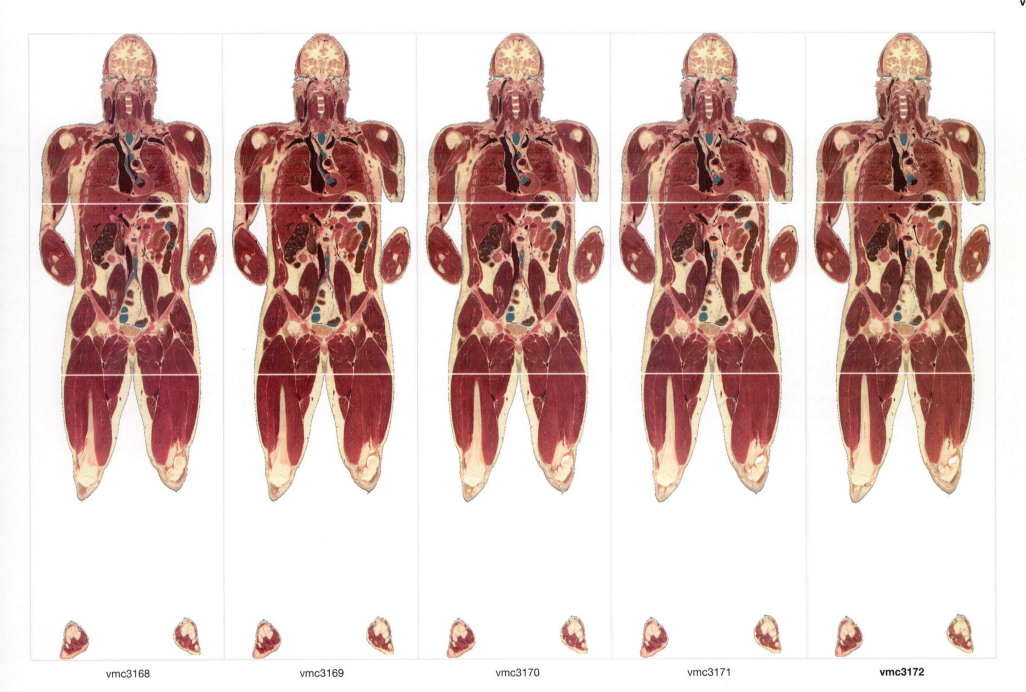

vmc3168   vmc3169   vmc3170   vmc3171   **vmc3172**

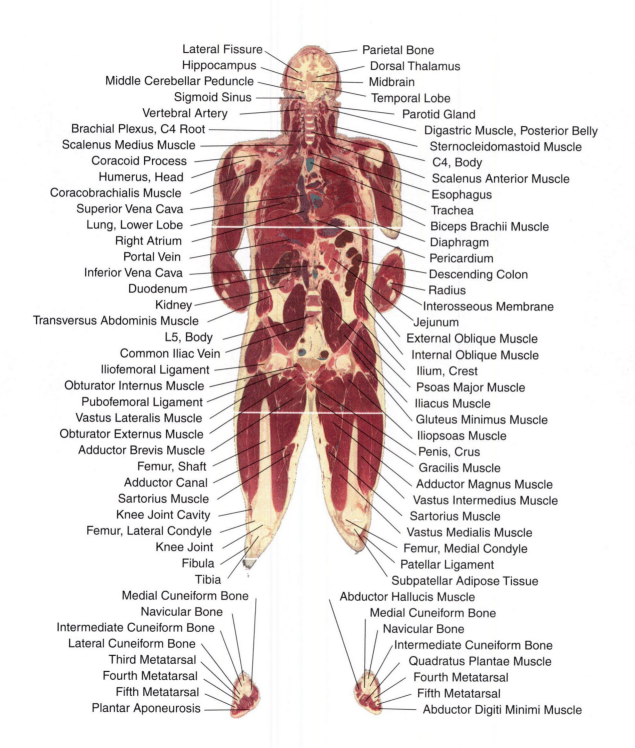

Lateral Fissure — — Parietal Bone
Hippocampus — — Dorsal Thalamus
Middle Cerebellar Peduncle — — Midbrain
Sigmoid Sinus — — Temporal Lobe
Vertebral Artery — — Parotid Gland
Brachial Plexus, C4 Root — — Digastric Muscle, Posterior Belly
Scalenus Medius Muscle — — Sternocleidomastoid Muscle
Coracoid Process — — C4, Body
Humerus, Head — — Scalenus Anterior Muscle
Coracobrachialis Muscle — — Esophagus
Superior Vena Cava — — Trachea
Lung, Lower Lobe — — Biceps Brachii Muscle
Right Atrium — — Diaphragm
Portal Vein — — Pericardium
Inferior Vena Cava — — Descending Colon
Duodenum — — Radius
Kidney — — Interosseous Membrane
Transversus Abdominis Muscle — — Jejunum
L5, Body — — External Oblique Muscle
Common Iliac Vein — — Internal Oblique Muscle
Iliofemoral Ligament — — Ilium, Crest
Obturator Internus Muscle — — Psoas Major Muscle
Pubofemoral Ligament — — Iliacus Muscle
Vastus Lateralis Muscle — — Gluteus Minimus Muscle
Obturator Externus Muscle — — Iliopsoas Muscle
Adductor Brevis Muscle — — Penis, Crus
Femur, Shaft — — Gracilis Muscle
Adductor Canal — — Adductor Magnus Muscle
Sartorius Muscle — — Vastus Intermedius Muscle
Knee Joint Cavity — — Sartorius Muscle
Femur, Lateral Condyle — — Vastus Medialis Muscle
Knee Joint — — Femur, Medial Condyle
Fibula — — Patellar Ligament
Tibia — — Subpatellar Adipose Tissue
Medial Cuneiform Bone — — Abductor Hallucis Muscle
Navicular Bone — — Medial Cuneiform Bone
Intermediate Cuneiform Bone — — Navicular Bone
Lateral Cuneiform Bone — — Intermediate Cuneiform Bone
Third Metatarsal — — Quadratus Plantae Muscle
Fourth Metatarsal — — Fourth Metatarsal
Fifth Metatarsal — — Fifth Metatarsal
Plantar Aponeurosis — — Abductor Digiti Minimi Muscle

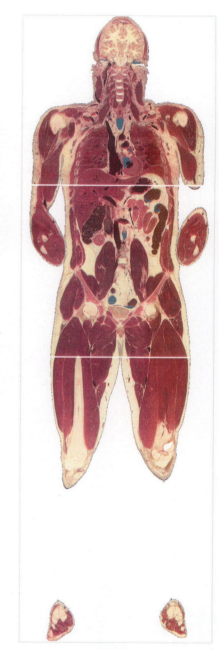

vmc3173

358

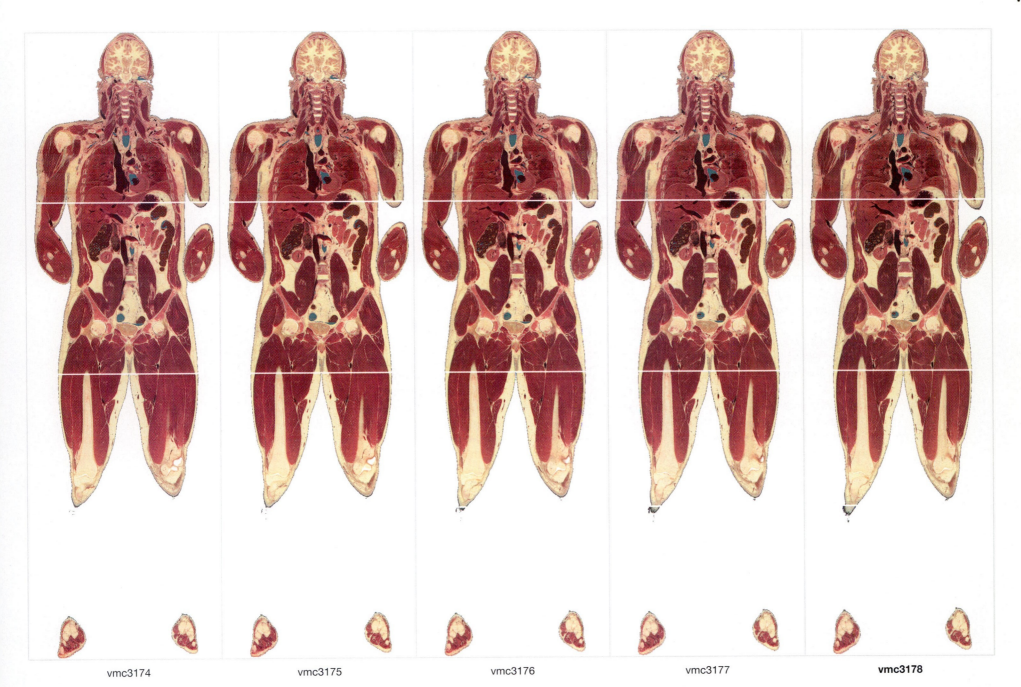

vmc3174   vmc3175   vmc3176   vmc3177   **vmc3178**

Cerebellum — Superior Sagittal Sinus
Middle Cerebellar Peduncle — Parietal Lobe
Obliquus Capitis Inferior Muscle — Lateral Fissure
Sternocleidomastoid Muscle — Medulla Oblongata
Levator Scapulae Muscle — Mastoid Air Cells
Clavicle — Digastric Muscle, Posterior Belly
Scapula, Acromion — Brachial Plexus, C5 Root
Coracoid Process — Humerus
Aorta, Arch — Longus Cervicis Muscle
Pulmonary Artery — Deltoid Muscle
Serratus Anterior Muscle — Coracobrachialis Muscle
Right Ventricle — Esophagus
Inferior Vena Cava — Biceps Brachii Muscle
Liver, Right Lobe — Trachea
Ascending Colon — Lung, Upper Lobe
Inferior Vena Cava — Left Ventricle
Diaphragm, Crus — Stomach
Long Flexor Muscles — Ulna
Kidney — Descending Colon
Transversus Abdominis Muscle — Iliacus Muscle
Psoas Major Muscle — Ilium
Iliofemoral Ligament — Gluteus Medius Muscle
Femur, Neck — Gluteus Minimus Muscle
Obturator Internus Muscle — Tensor Fascia Lata Muscle
Obturator Externus Muscle — Femur, Head
Gracilis Muscle — Sigmoid Colon
Vastus Lateralis Muscle — Bladder
Adductor Canal — Adductor Longus Muscle
Sartorius Muscle — Penis, Crus
Femur, Medial Epicondyle — Vastus Intermedius Muscle
Femur, Lateral Condyle — Femur, Shaft
Knee Joint — Vastus Medialis Muscle
Tibia — Patellar Ligament
Tibialis Anterior Muscle — Tibial Tuberosity
Tibia, Shaft — Femur, Lateral Condyle
Medial Cuneiform Bone — Abductor Hallucis Muscle
Navicular Bone — Navicular Bone
Intermediate Cuneiform Bone — Lateral Cuneiform Bone
Lateral Cuneiform Bone — Quadratus Plantae Muscle
Flexor Digitorum Brevis Muscle — Fourth Metatarsal
Fifth Metatarsal — Fifth Metatarsal
Plantar Aponeurosis — Abductor Digiti Minimi Muscle

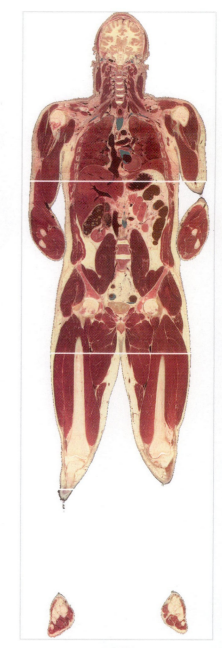

vmc3179

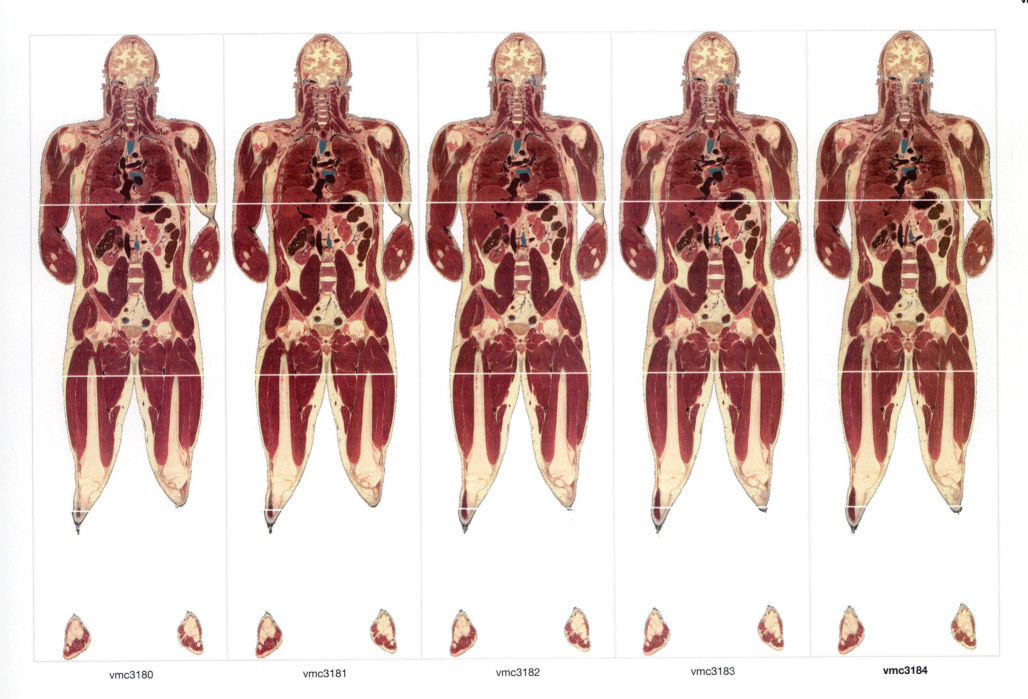

vmc3180   vmc3181   vmc3182   vmc3183   **vmc3184**

Splenium — Parietal Bone
Lateral Ventricle — Superior Sagittal Sinus
Cerebellum — Temporalis Muscle
Sigmoid Sinus — Mastoid Air Cells
Fourth Ventricle — Obliquus Capitis Inferior Muscle
Medulla Oblongata — Sternocleidomastoid Muscle
Levator Scapulae Muscle — Spinal Cord
Shoulder Joint — Humerus, Head
Lung, Upper Lobe — Deltoid Muscle
Subscapularis Muscle — Esophagus
Pulmonary Artery — Aorta, Arch
Right Ventricle — Lung, Oblique Fissure
Triceps Brachii Muscle — Lung, Lower Lobe
Inferior Vena Cava — Stomach
Diaphragm, Crus — Spleen
Radius — Pancreas
Kidney — Descending Colon
Long Flexor Muscles — Renal Vein
L4-5, Intervertebral Disc — Internal Oblique Muscle
Femur, Head — Psoas Major Muscle
Femur, Greater Trochanter — Iliac Crest
Obturator Internus Muscle — Gluteus Minimus Muscle
Iliopsoas Tendon — Gluteus Medius Muscle
Obturator Externus Muscle — Sigmoid Colon
Bladder — Penis, Crus
Pubis, Inferior Ramus — Gracilis Muscle
Adductor Brevis Muscle — Vastus Lateralis Muscle
Gracilis Muscle — Femoral Vessel
Adductor Magnus Muscle — Sartorius Muscle
Adductor Canal — Vastus Intermedius Muscle
Sartorius Muscle — Femur, Shaft
Vastus Medialis Muscle — Knee Joint Cavity
Tibial Collateral Ligament — Knee Joint
Crural Fascia — Tibia, Condyles
Tibialis Anterior Muscle — Tibialis Anterior Muscle
Tibia, Shaft — Abductor Hallucis Muscle
Tibialis Anterior Tendon — Talus Bone
Navicular Bone — Lateral Cuneiform Bone
Medial Cuneiform Bone — Extensor Digitorum Brevis Muscle
Quadratus Plantae Muscle — Cuboid Bone
Fifth Metatarsal — Fourth Metatarsal
Abductor Digiti Minimi Muscle — Fifth Metatarsal

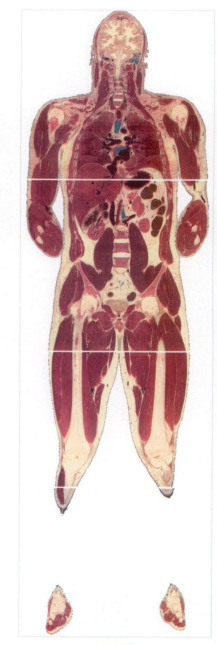

vmc3185

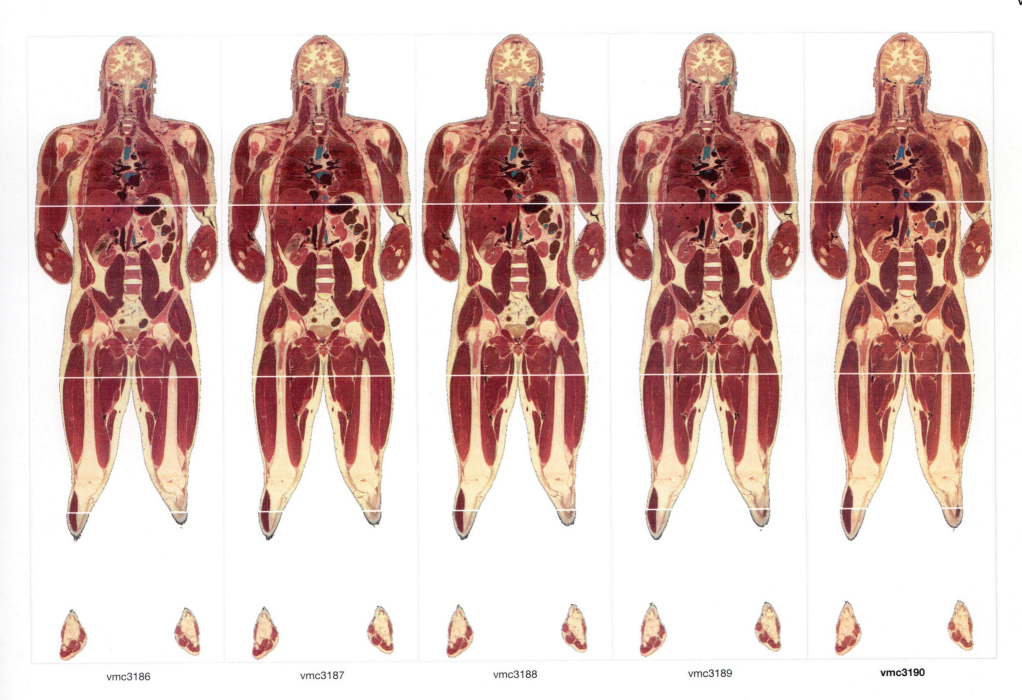

Splenium — Superior Sagittal Sinus
Fornix — Falx Cerebri
Temporal Lobe — Lateral Ventricle
Tentorium — Temporalis Muscle
Transverse Sinus — Cerebellum
Trapezius Muscle — Sternocleidomastoid Muscle
Clavicle — Levator Scapulae Muscle
Scapula, Acromion — Spinal Cord
Supraspinatus Muscle — Humerus
Deltoid Muscle — Trachea, Carina
Subscapularis Muscle — Primary Bronchus
Esophagus — Lung, Oblique Fissure
Triceps Brachii Muscle — Stomach
Liver, Right Lobe — Spleen
Esophagus — Pancreas
Inferior Vena Cava — Radius
Diaphragm, Crus — Ulna
Kidney — Transversus Abdominis Muscle
Psoas Major Muscle — L4-5, Intervertebral Disc
Ilium — Iliacus Muscle
Gluteus Maximus Muscle — Femur, Greater Trochanter
Gluteus Minimus Muscle — Femur, Neck
Bladder — Obturator Internus Muscle
Puborectalis Muscle — Obturator Externus Muscle
Prostate Gland — Obturator Membrane
Bulbospongiosus Muscle — Urogenital Diaphragm
Femur, Linea Aspera — Ischiocavernosus Muscle
Sartorius Muscle — Vastus Lateralis Muscle
Vastus Intermedius Muscle — Gracilis Muscle
Femur, Lateral Condyle — Adductor Magnus Muscle
Anterior Cruciate Ligament — Vastus Medialis Muscle
Tibia, Lateral Condyle — Lateral Meniscus
Tibialis Anterior Muscle — Femur, Medial Condyle
Crural Fascia — Tibia, Medial Condyle
Tibia, Shaft — Tibia, Shaft
Tibialis Anterior Tendon — Crural Fascia
Extensor Hallucis Longus Tendon — Tibia
Talus Bone — Tibia, Medial Malleolus
Navicular Bone — Talus Bone
Extensor Digitorum Brevis Muscle — Extensor Digitorum Brevis Muscle
Abductor Hallucis Muscle — Quadratus Plantae Muscle
Flexor Digiti Minimi Muscle — Flexor Digitorum Brevis Muscle

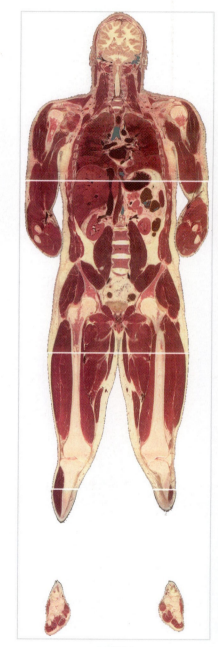

vmc3191

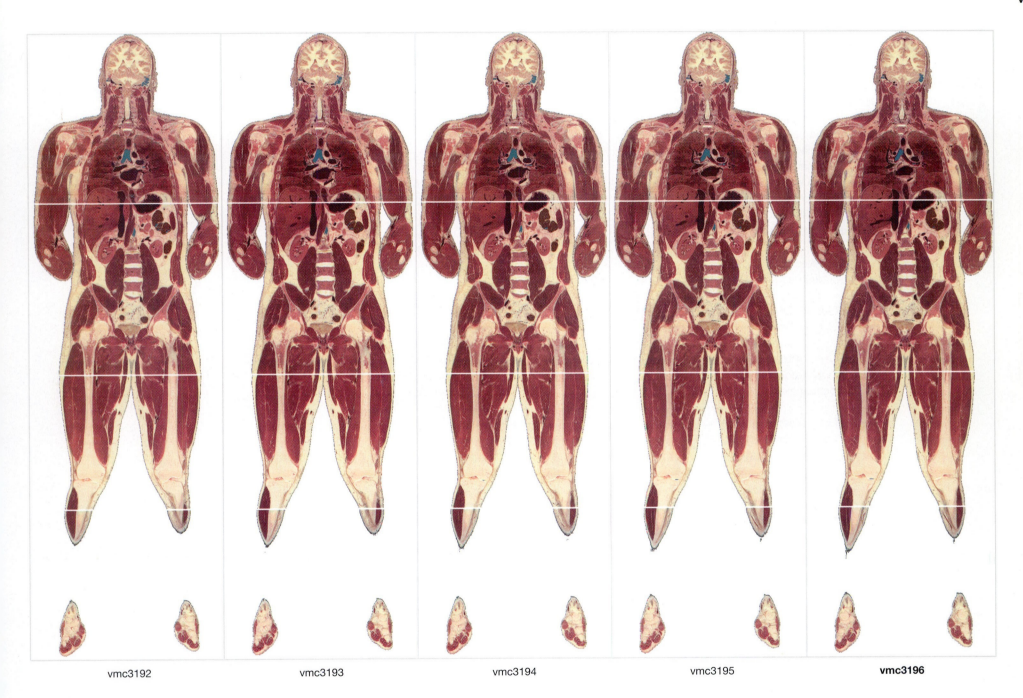

vmc3192   vmc3193   vmc3194   vmc3195   **vmc3196**

Parietal Bone — Superior Sagittal Sinus
Parietal Lobe — Longitudinal Cerebral Fissure
Transverse Sinus — Tentorium
Rectus Capitis Posterior Major Muscle — Longissimus Capitis Muscle
Obliquus Capitis Inferior Muscle — Sternocleidomastoid Muscle
Trapezius Muscle — Semispinalis Cervicis Muscle
Levator Scapulae Muscle — Deltoid Muscle
Subscapularis Muscle — Spinal Cord
Teres Major Muscle — Humerus
Lung, Upper Lobe — Rib 4-5, Intercostal Muscles
Triceps Brachii Muscle — Biceps Brachii Muscle
Liver, Right Lobe — Stomach
Latissimus Dorsi Muscle — Spleen
Kidney — Colon, Splenic Flexure
Psoas Major Muscle — Radius
Flexor Carpi Ulnaris Muscle — Ulna
Ilium — Internal Oblique Muscle
Gluteus Minimus Muscle — Iliacus Muscle
Gluteus Medius Muscle — Sigmoid Colon
Ischium — Femur, Head
Prostate Gland — Femur, Greater Trochanter
Obturator Membrane — Obturator Internus Muscle
Bulbospongiosus Muscle — Obturator Externus Muscle
Fascia Lata — Adductor Brevis Muscle
Adductor Magnus Muscle — Vastus Lateralis Muscle
Sartorius Muscle — Gracilis Muscle
Biceps Femoris Muscle, Short Head — Femur, Shaft
— Vastus Intermedius Muscle
Femur, Medial Condyle — Femur, Medial Epicondyle
Anterior Cruciate Ligament — Femur, Lateral Condyle
Tibia, Medial Condyle — Lateral Meniscus
Tibialis Anterior Muscle — Tibia, Lateral Condyle
Extensor Digitorum Longus Muscle — Medial Meniscus
Tibia, Shaft — Tibia, Shaft
Crural Fascia

Abductor Hallucis Muscle — Tibia, Medial Malleolus
Quadratus Plantae Muscle — Peroneus Tertius Muscle
Extensor Hallucis Brevis Muscle — Flexor Digitorum Longus Tendon
Extensor Digitorum Brevis Muscle — Flexor Hallucis Longus Tendon
Flexor Digitorum Brevis Muscle — Cuboid Bone
Abductor Digiti Minimi Muscle — Extensor Digitorum Brevis Muscle
— Fifth Metatarsal

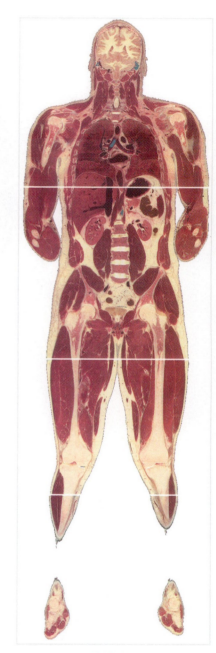

vmc3197

366

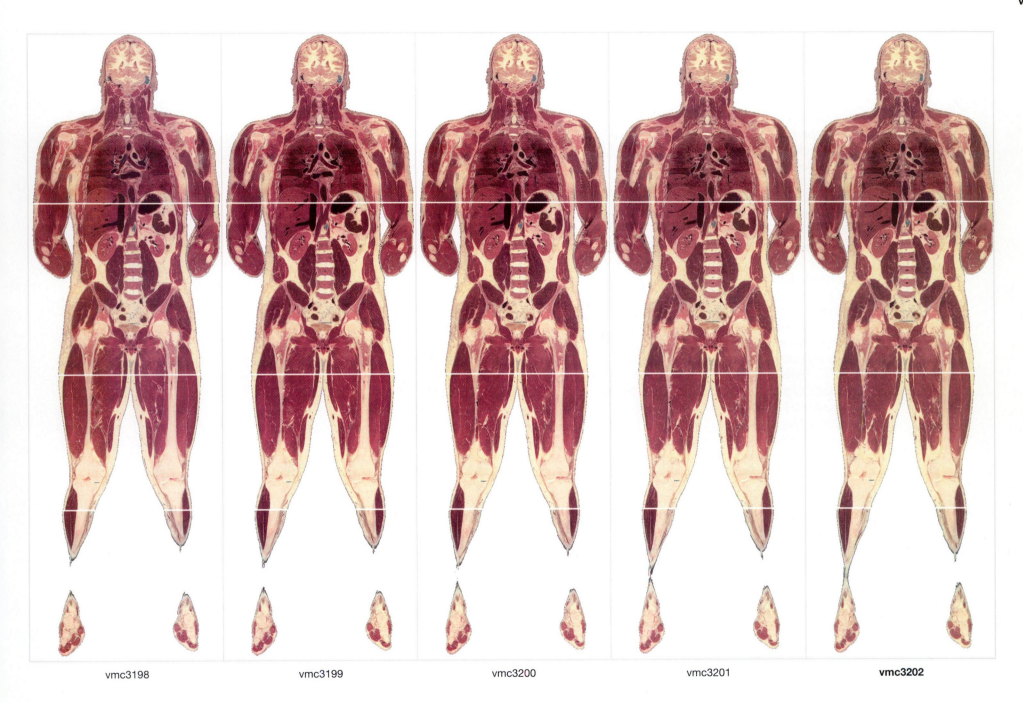

vmc3198   vmc3199   vmc3200   vmc3201   **vmc3202**

Parietal Bone

Central Sulcus

Temporal Lobe

Transverse Sinus

Cerebellum, Vermis

Splenius Capitis Muscle

Trapezius Muscle

Levator Scapulae Muscle

Subscapularis Muscle

Teres Major Muscle

Thoracic Aorta

Latissimus Dorsi Muscle

Liver, Right Lobe

Diaphragm, Crus

Kidney

L2-3, Intervertebral Disc

Flexor Carpi Ulnaris Muscle

Psoas Major Muscle

Common Iliac Vein

Gluteus Minimus Muscle

Gluteus Maximus Muscle

Prostate Gland

Pubococcygeus Muscle

Bulbospongiosus Muscle

Adductor Magnus Muscle

Sartorius Muscle

Popliteal Vessel

Biceps Femoris Muscle

Femur, Medial Condyle

Anterior Cruciate Ligament

Tibia, Medial Condyle

Tibialis Anterior Muscle

Soleus Muscle

Tibia

Tibialis Anterior Tendon

Abductor Hallucis Muscle

Peroneus Tertius Tendon

Quadratus Plantae Muscle

Extensors Hallucis and
Digitorum Brevis Muscles

Flexor Digitorum Brevis Muscle

Abductor Digiti Minimi Muscle

Superior Sagittal Sinus

Falx Cerebri

Longitudinal Cerebral Fissure

Longissimus Capitis Muscle

Sternocleidomastoid Muscle

Semispinalis Cervicis Muscle

Supraspinatus Muscle

Spinal Cord

Pulmonary Artery

Humerus

Bronchus

Biceps Brachii Muscle

Stomach

Pancreas, Tail

Spleen

Radius

Ulna

Transversus Abdominis Muscle

Iliacus Muscle

Ilium

Sigmoid Colon

Femur, Head

Obturator Externus Muscle

Adductor Brevis Muscle

Ischiocavernosus Muscle

Vastus Lateralis Muscle

Gracilis Muscle

Adductor Magnus Muscle

Vastus Intermedius Muscle

Femur, Lateral Condyle

Lateral Meniscus

Medial Meniscus

Gastrocnemius Muscle

Great Saphenous Vein

Extensor Digitorum Longus Muscle

Tibia, Medial Malleolus

Peroneus Tertius Muscle

Flexor Digitorum Longus Tendon

Talus Bone

Extensor Digitorum Brevis Muscle

Cuboid Bone

Fifth Metatarsal

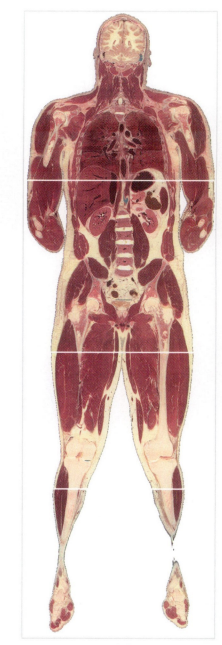

vmc3203

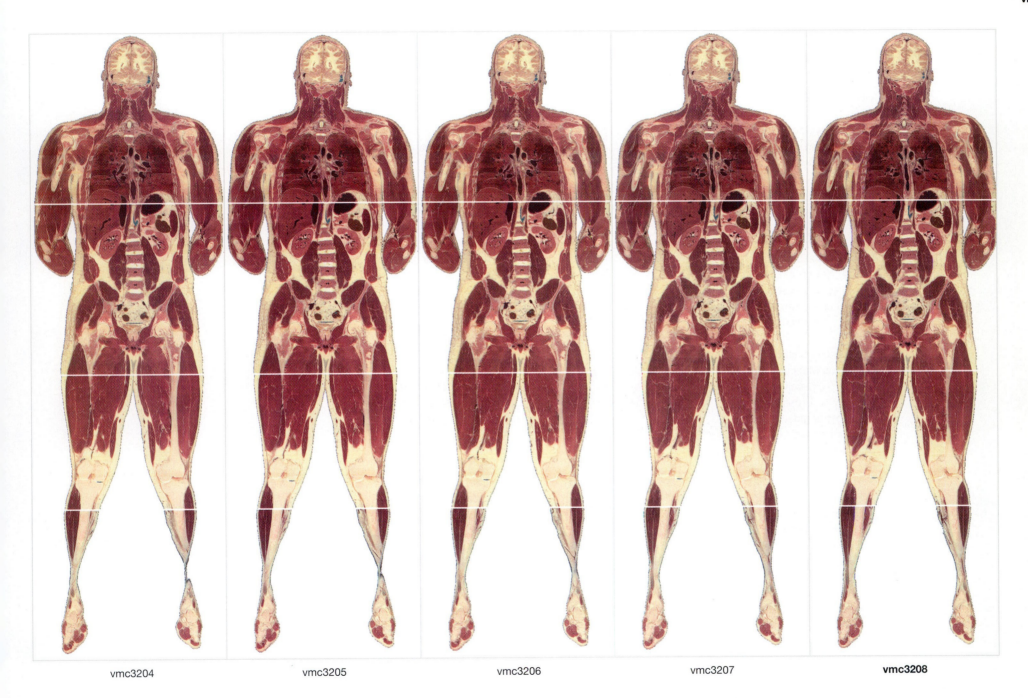

Parietal Bone — Superior Sagittal Sinus
Central Sulcus — Falx Cerebri
Temporal Bone — Tentorium
Rectus Capitis Posterior Major Muscle — Cerebellum, Lateral Hemisphere
Obliquus Capitis Inferior Muscle — Longissimus Capitis Muscle
Semispinalis Capitis Muscle — Sternocleidomastoid Muscle
Trapezius Muscle — Semispinalis Cervicis Muscle
Levator Scapulae Muscle — T1, Transverse Process
Scapula — Lung, Upper Lobe
Teres Minor Muscle — Rib 5-6, Intercostal Muscles
Serratus Anterior Muscle — Pulmonary Vein
Esophagus — Biceps Brachii Muscle
Azygos Vein — Stomach
Brachioradialis Muscle — Pancreas, Tail
T12, Vertebral Body — Spleen
Kidney, Adipose Capsule — Radius
Quadratus Lumborum Muscle — Ulna
Subarachnoid Space — External Oblique Muscle
Sacrum — Sacroiliac Joint
Gluteus Minimus Muscle — Ilium
Gluteus Maximus Muscle — Obturator Internus Muscle
Seminal Vesicle — Femur, Greater Trochanter
Prostate Gland — Obturator Externus Muscle
Bulbospongiosus Muscle — Ischiocavernosus Muscle
Adductor Magnus Muscle — Vastus Lateralis Muscle
Semimembranosus Muscle — Gracilis Muscle
Sartorius Muscle — Adductor Magnus Muscle
Biceps Femoris Muscle — Vastus Intermedius Muscle
Femur, Medial Condyle — Femur, Lateral Condyle
Anterior Cruciate Ligament — Lateral Meniscus
Tibia, Medial Condyle — Medial Meniscus
Tibialis Posterior Muscle — Great Saphenous Vein
Gastrocnemius Muscle — Popliteus Muscle
Extensor Hallucis Longus Muscle — Extensor Digitorum Longus Muscle
Tibialis Anterior Muscle — Peroneus Longus Muscle
Abductor Hallucis Muscle — Tibia
Quadratus Plantae Muscle — Peroneus Tertius Muscle
Fibula, Lateral Malleolus — Flexor Digitorum Longus Tendon
Extensors Hallucis and Digitorum Brevis Muscles — Flexor Hallucis Longus Tendon
Flexor Digitorum Brevis Muscle — Cuboid Bone
Abductor Digiti Minimi Muscle — Fifth Metatarsal

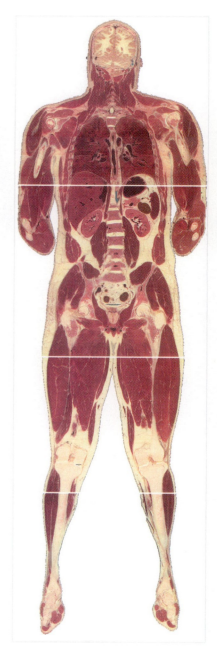

vmc3209

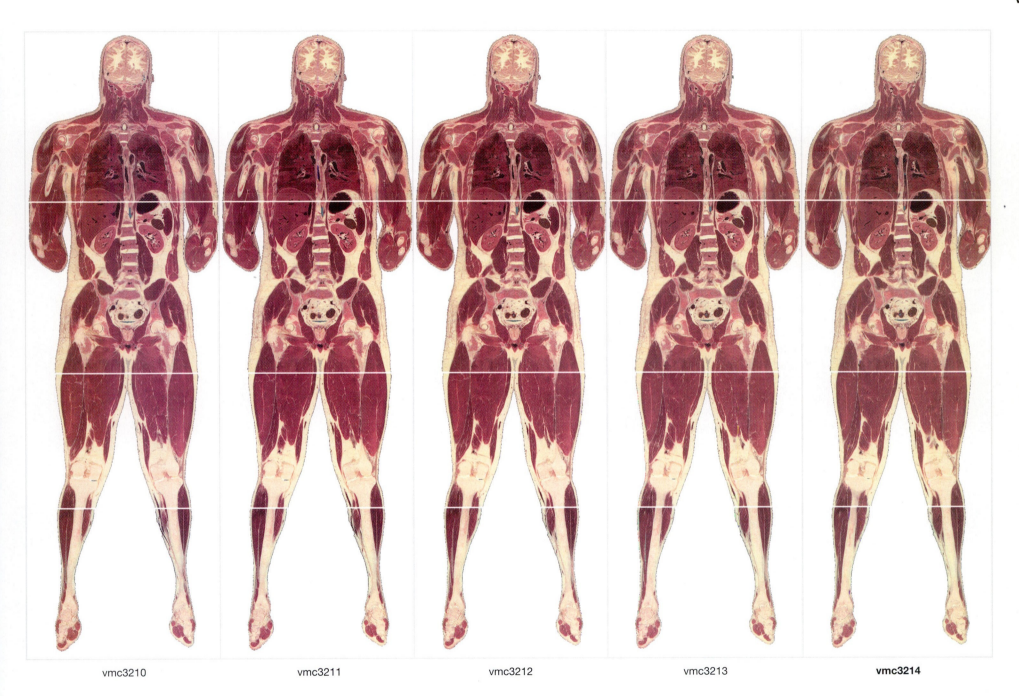

vmc3210   vmc3211   vmc3212   vmc3213   **vmc3214**

Coronal
vmc3220

Temporal Bone
Calcarine Sulcus
Tentorium
Transverse Sinus
Rectus Capitis Posterior Major M.
Trapezius Muscle
Supraspinatus Muscle
Infraspinatus Muscle
Deltoid Muscle
Lung, Upper Lobe
Serratus Anterior Muscle
Latissimus Dorsi Muscle
Diaphragm
Liver, Right Lobe
Brachialis Muscle
Humerus, Capitulum
Radius
L1-2, Intervertebral Disc
Psoas Major Muscle
Quadratus Lumborum Muscle
Sacrum, Lateral Wing
Sacroiliac Joint
Internal Iliac Vein
Sigmoid Colon
Seminal Vesicle
Prostate Gland
Gracilis Muscle
Vastus Lateralis Muscle
Adductor Magnus Muscle
Semimembranosus Muscle
Biceps Femoris Muscle
Gastrocnemius Muscle, Lateral Head
Femur, Lateral Condyle
Tibialis Posterior Muscle
Soleus Muscle
Gastrocnemius Muscle, Medial Head
Peroneus Brevis Muscle
Tibia, Shaft
Peroneus Tertius Muscle
Talus Bone
Abductor Hallucis Muscle
Flexor Digiti Minimi Muscle

Superior Sagittal Sinus
Longitudinal Cerebral Fissure
Occipital Lobe
Cerebellum
Rectus Capitis Posterior Minor M.
Semispinalis Cervicis Muscle
Trapezius Muscle
Spinal Cord
Scapula
Teres Major Muscle
Humerus
Thoracic Aorta
Stomach
Spleen
Kidney
Ulna
Internal Oblique Muscle
Cauda Equina
Gluteus Medius Muscle
Ilium, Body
Gluteus Minimus Muscle
Femur, Greater Trochanter
Pubis
Obturator Externus Muscle
Ischiocavernosus Muscle
Bulbospongiosus Muscle
Vastus Lateralis Muscle
Biceps Femoris Muscle, Short Head
Adductor Magnus Muscle
Gracilis Muscle
Popliteal Vessel
Femur, Lateral Condyle
Posterior Cruciate Ligament
Sartorius Tendon
Peroneus Longus Muscle
Peroneus Brevis Muscle
Tibia, Shaft
Tibia, Medial Malleolus
Ankle Joint
Extensor Digitorum Brevis Muscle
Cuboid Bone
Flexor Digitorum Brevis Muscle

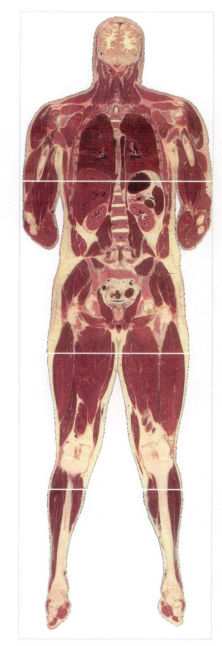

vmc3215

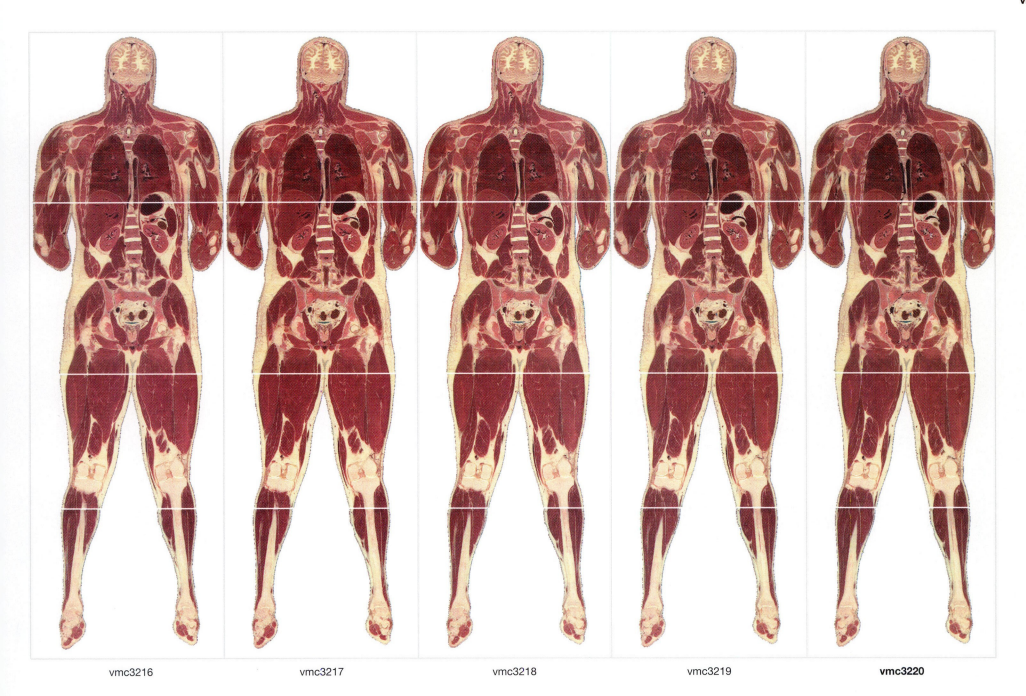

vmc3216   vmc3217   vmc3218   vmc3219   **vmc3220**

Calcarine Sulcus

Tentorium

Transverse Sinus

Cerebellum

Rectus Capitis Posterior Minor M.

Splenius Capitis Muscle

Semispinalis Capitis Muscle

Levator Scapulae Muscle

Infraspinatus Muscle

Subscapularis Muscle

Teres Major Muscle

Latissimus Dorsi Muscle

Serratus Anterior Muscle

Humerus

Diaphragm

Liver, Right Lobe

Renal Fascia

Psoas Major Muscle

L1-2, Intervertebral Disc

Quadratus Lumborum Muscle

Ilium

Sacroiliac Joint

Internal Iliac Vein

Pubis

Levator Ani Muscle

Ischiocavernosus Muscle

Iliotibial Tract

Vastus Lateralis Muscle

Sciatic Nerve

Biceps Femoris Muscle

Semimembranosus Muscle

Lateral Meniscus

Fibula, Head

Popliteus Muscle

Great Saphenous Vein

Tibialis Posterior Muscle

Peroneus Longus Muscle

Tibia, Shaft

Tibia, Medial Malleolus

Ankle Joint

Calcaneus Bone

Plantar Aponeurosis

Superior Sagittal Sinus

Parietal Lobe

Occipital Lobe

Rectus Capitis Posterior Major M.

Trapezius Muscle

Supraspinatus Muscle

Spinal Cord

Deltoid Muscle

Scapula

Subscapularis Muscle

Lung, Upper Lobe

Lung, Oblique Fissure

Thoracic Aorta

Stomach

Spleen

Radius

Ulna

Adrenal Gland

Kidney

External Oblique Muscle

Iliacus Muscle

Gluteus Medius Muscle

Gluteus Minimus Muscle

S1-2, Intervertebral Disc

Sigmoid Colon

Obturator Externus Muscle

Seminal Vesicle

Vastus Lateralis Muscle

Obturator Internus Muscle

Prostatic Urethra

Adductor Magnus Muscle

Gracilis Muscle

Semimembranosus Muscle

Femur, Lateral Condyle

Posterior Cruciate Ligament

Sartorius Muscle

Gastrocnemius Muscle, Medial Head

Talus Bone

Quadratus Plantae Muscle

Extensor Digitorum Brevis Muscle

Flexor Digitorum Brevis Muscle

Flexor Digiti Minimi Muscle

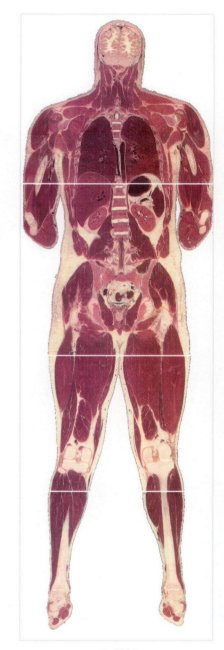

vmc3221

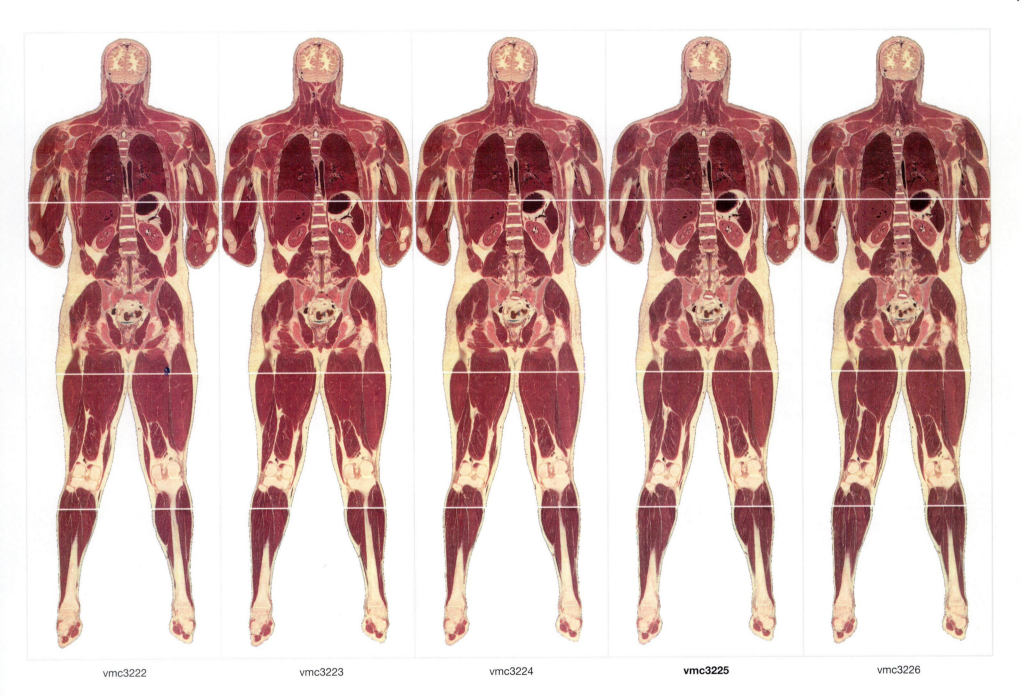

vmc3222   vmc3223   vmc3224   **vmc3225**   vmc3226

Parietal Bone
Calcarine Sulcus
Transverse Sinus
Occipital Bone
Splenius Capitis Muscle
Trapezius Muscle
Supraspinatus Muscle
Deltoid Muscle
Infraspinatus Muscle
Lung, Lower Lobe
Diaphragm
Humerus
Liver, Right Lobe
Brachialis Muscle
Humerus, Trochlea
Kidney
Vertebral Canal
Sacrum, Lateral Wing
Gluteus Medius Muscle
Levator Ani Muscle
Quadratus Femoris Muscle
Adductor Magnus Muscle
Gracilis Muscle
Semimembranosus Muscle
Sciatic Nerve
Biceps Femoris Muscle
Sartorius Muscle
Femur, Medial Condyle
Popliteus Muscle
Fibula, Head
Gastrocnemius Muscle, Lateral Head
Gastrocnemius Muscle, Medial Head
Soleus Muscle
Peroneus Longus Muscle
Flexor Hallucis Longus Muscle
Peroneus Brevis Muscle
Tibia
Talus Bone
Quadratus Plantae Muscle
Calcaneus Bone
Abductor Digiti Minimi Muscle
Plantar Aponeurosis

Superior Sagittal Sinus
Parietal Lobe
Occipital Lobe
Tentorium
Cerebellum
Semispinalis Capitis Muscle
Trapezius Muscle
Scapula, Spine
Subscapularis Muscle
Spinal Cord
Teres Major Muscle
Lung, Oblique Fissure
Lung, Lower Lobe
Serratus Anterior Muscle
Brachioradialis Muscle
Stomach
Spleen
External Oblique Muscle
Kidney
Internal Oblique Muscle
T12-L1, Intervertebral Disc
Subarachnoid Space
Ilium
Gluteus Maximus Muscle
Inferior Gemellus Muscle
Obturator Internus Muscle
Ischiorectal Fossa
Anus
Urogenital Diaphragm
Vastus Lateralis Muscle
Adductor Magnus Muscle
Semimembranosus Muscle
Popliteal Vessel
Tibia, Lateral Condyle
Popliteus Muscle
Gastrocnemius Muscle, Medial Head
Soleus Muscle
Talus Bone
Quadratus Plantae Muscle
Calcaneus Bone
Flexor Digitorum Brevis Muscle
Abductor Digiti Minimi Muscle

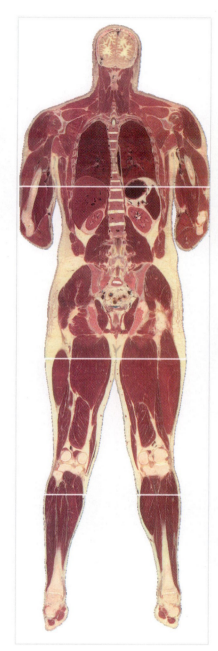

vmc3227

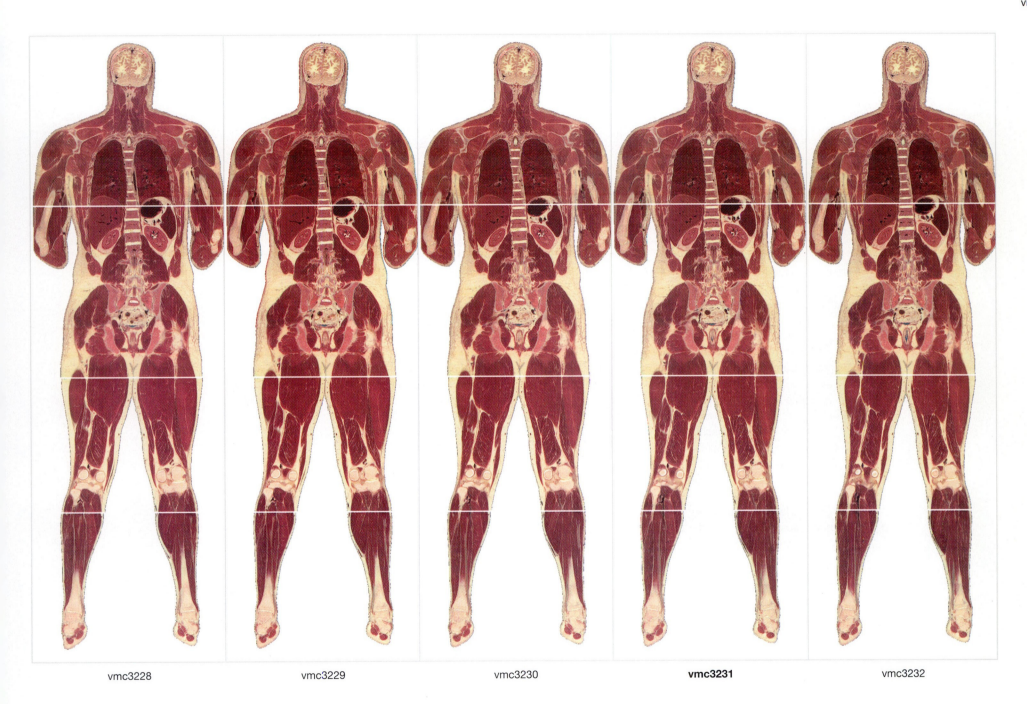

vmc3228   vmc3229   vmc3230   **vmc3231**   vmc3232

Longitudinal Cerebral Fissure
Calcarine Sulcus
Straight Sinus
Transverse Sinus
Cerebellum
Splenius Capitis Muscle
Trapezius Muscle
Supraspinatus Muscle
Infraspinatus Muscle
Spinal Cord
Serratus Anterior Muscle
Lung, Lower Lobe
Diaphragm
Liver, Right Lobe
External Oblique Muscle
Humerus, Trochlea
Ulna, Coronoid Process
Kidney
Psoas Major Muscle
Subarachnoid Space
Quadratus Lumborum Muscle
Spinalis Muscle
Gluteus Medius Muscle
Sacrum
Piriformis Muscle
Obturator Internus Tendon
Obturator Internus Muscle
Vastus Lateralis Muscle
Ischium
Ischiorectal Fossa
Biceps Femoris Muscle, Long Head
Gracilis Muscle
Sartorius Muscle
Small Saphenous Vein
Popliteal Vessels
Gastrocnemius Muscle, Medial Head
Soleus Muscle
Tibia
Fibula, Lateral Malleolus
Talus Bone
Calcaneus Bone
Plantar Aponeurosis

Superior Sagittal Sinus
Parietal Bone
Parietal Lobe
Tentorium
Occipitofrontalis Muscle, Occipital Belly
Semispinalis Capitis Muscle
Scapula, Spine
Deltoid Muscle
Subscapularis Muscle
Teres Minor Muscle
Teres Major Muscle
Triceps Brachii Muscle
Lung, Oblique Fissure
Humerus
Latissimus Dorsi Muscle
Spleen
Stomach
Internal Oblique Muscle
Multifidus Muscle
Ilium
Sacral Foramen
Levator Ani Muscle
Anus
Quadratus Femoris Muscle
Central Tendinous Point
Adductor Magnus Muscle
Semitendinosus Muscle
Semimembranosus Muscle
Biceps Femoris Muscle
Gracilis Muscle
Femur, Lateral Condyle
Tibia, Lateral Condyle
Fibula, Head
Popliteus Muscle
Gastrocnemius Muscle, Lateral Head
Peroneus Longus Muscle
Tibia
Long Saphenous Vein
Fibula, Lateral Malleolus
Quadratus Plantae Muscle
Abductor Digiti Minimi Muscle
Flexor Digitorum Brevis Muscle

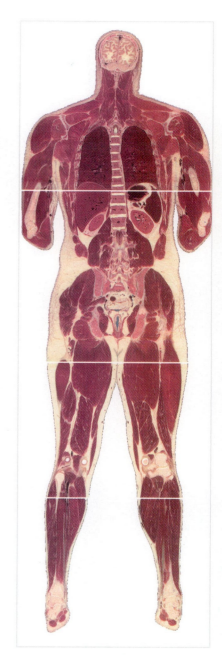

vmc3233

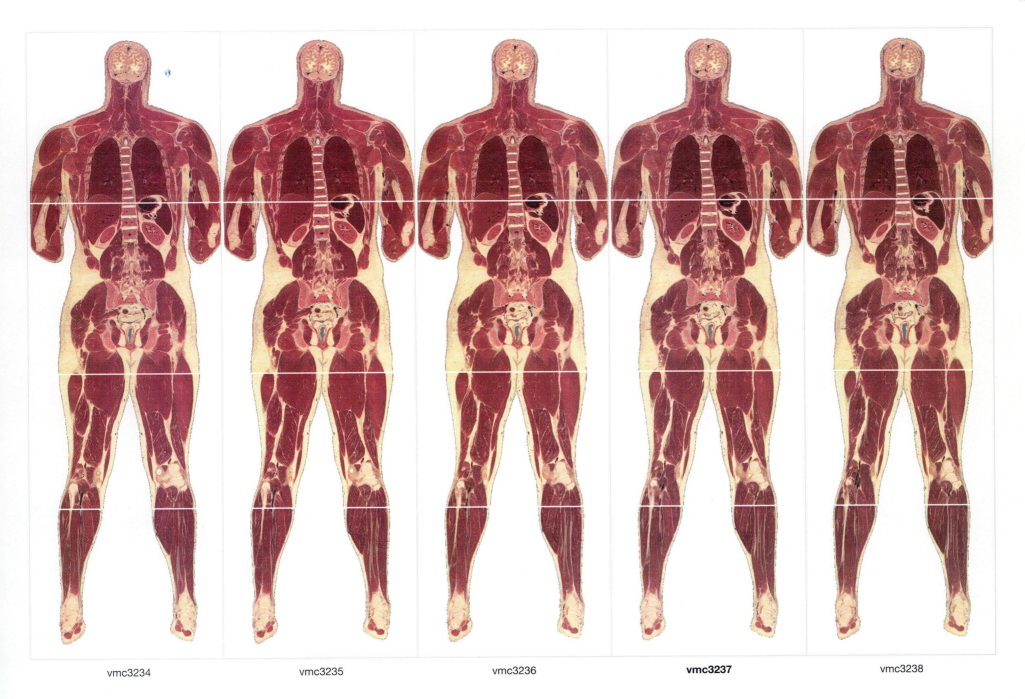

vmc3234   vmc3235   vmc3236   **vmc3237**   vmc3238

Superior Sagittal Sinus — Occipital Bone

Calcarine Sulcus — Longitudinal Cerebral Fissure

Occipitofrontalis Muscle, Occipital Belly — Occipital Lobe

Straight Sinus — Splenius Capitis Muscle

Splenius Capitis Muscle — Semispinalis Capitis Muscle

Levator Scapulae Muscle — Trapezius Muscle

Subscapularis Muscle — Supraspinatus Muscle

Infraspinatus Muscle — Spinal Cord

Deltoid Muscle — Lung, Oblique Fissure

Teres Major Muscle — Lung, Lower Lobe

Latissimus Dorsi Muscle — Latissimus Dorsi Muscle

Triceps Brachii Muscle — Serratus Anterior Muscle

Serratus Anterior Muscle — Diaphragm

Liver, Right Lobe — Spleen

Olecranon Process — Kidney

Humerus, Trochlea — Subarachnoid Space

External Oblique Muscle — L3, Spinous Process

Quadratus Lumborum Muscle — Quadratus Lumborum Muscle

L2, Transverse Process — Multifidus Muscle

Longissimus Muscle — Gluteus Medius Muscle

Ilium — Anus

Gluteus Maximus Muscle — Gluteus Maximus Muscle

Sacrum — Levator Ani Muscle

Sigmoid Colon — Adductor Magnus Muscle

Obturator Internus Muscle — Vastus Lateralis Muscle

Ischiorectal Fossa — Semitendinosus Muscle

Vastus Lateralis Muscle — Semimembranosus Muscle

Biceps Femoris Muscle, Short Head — Biceps Femoris Muscle

Biceps Femoris Muscle, Long Head — Femur, Lateral Condyle

Semitendinosus Muscle — Fibula, Head

Gracilis Muscle — Gastrocnemius Muscle, Lateral Head

Small Saphenous Vein — Gastrocnemius Muscle, Medial Head

Sartorius Muscle — Soleus Muscle

Gastrocnemius Muscle, Medial Head — Flexor Hallucis Longus Muscle

Soleus Muscle — Tibia

Peroneus Longus Muscle — Ankle Joint

Fibula, Shaft — Talus Bone

Flexor Hallucis Longus Muscle — Fibula, Lateral Malleolus

Peroneus Brevis Muscle — Calcaneus Bone

Fibula, Lateral Malleolus — Abductor Hallucis Muscle

Ankle Joint — Abductor Digiti Minimi Muscle

Quadratus Plantae Muscle — Quadratus Plantae Muscle

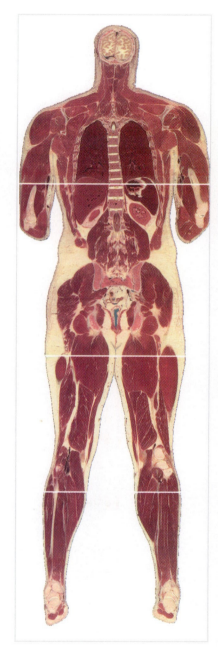

vmc3239

380

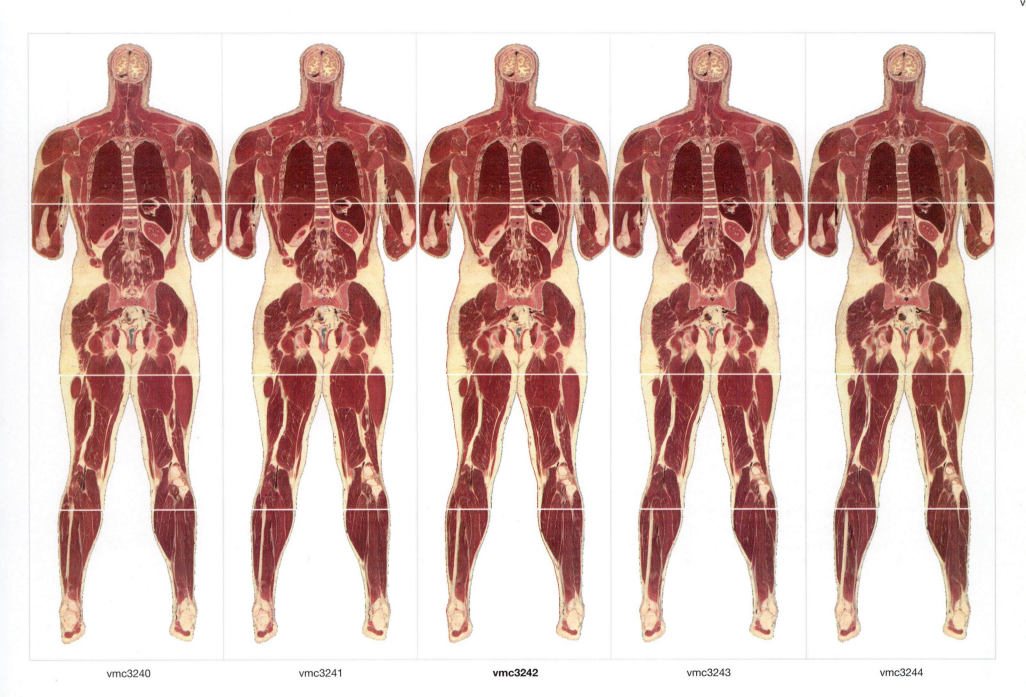

vmc3240   vmc3241   **vmc3242**   vmc3243   vmc3244

Superior Sagittal Sinus

Occipital Bone

Occipital Vessel

Trapezius Muscle

Levator Scapulae Muscle

Multifidus Muscle

Deltoid Muscle

T6-7, Intervertebral Disc

T7, Vertebra

Lung, Lower Lobe

Serratus Anterior Muscle

Triceps Brachii Muscle

Latissimus Dorsi Muscle

Liver, Right Lobe

Humerus, Medial Condyle

Ulna

T12, Vertebra

Spinal Cord

Iliocostalis Muscle

Longissimus Muscle

Gluteus Medius Muscle

Gluteus Maximus Muscle

Sigmoid Colon

Ischial Tuberosity

Ischiorectal Fossa

Anus

Adductor Magnus Muscle

Biceps Femoris Muscle

Gracilis Muscle

Biceps Femoris Tendon

Great Saphenous Vein

Gastrocnemius Muscle, Medial Head

Soleus Muscle

Peroneus Longus Muscle

Peroneus Brevis Muscle

Fibula, Shaft

Flexor Hallucis Longus Muscle

Crural Fascia

Tibia, Medial Malleolus

Ankle Joint

Abductor Digiti Minimi Muscle

Abductor Hallucis Muscle

Sagittal Suture

Occipital Lobe

Semispinalis Capitis Muscle

Trapezius Muscle

Supraspinatus Muscle

Scapula, Spine

Infraspinatus Muscle

Subscapularis Muscle

Spinal Cord

Teres Major Muscle

Rib 6

Rib 6-7, Intercostal Muscles

Diaphragm

Humerus

Basilic Vein

Common Flexor Tendon

Common Extensor Tendon

Kidney

Quadratus Lumborum Muscle

Longissimus Muscle

Ilium

L5, Spinous Process

Piriformis Muscle

Sacrum

Gemelli Muscles

Obturator Internus Muscle

Vastus Lateralis Muscle

Levator Ani Muscle

Semitendinosus Muscle

Semimembranosus Muscle

Small Saphenous Vein

Fibula, Head

Great Saphenous Vein

Gastrocnemius Muscle, Medial Head

Soleus Muscle

Peroneus Longus Muscle

Crural Fascia

Fibula, Shaft

Fibula, Lateral Malleolus

Calcaneus Bone

Abductor Hallucis Muscle

Abductor Digiti Minimi Muscle

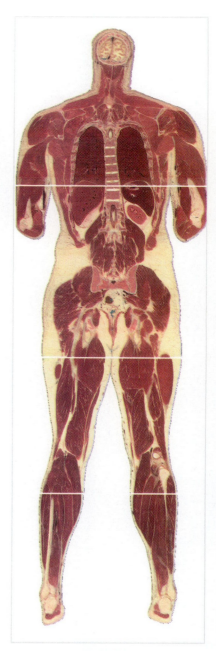

vmc3245

382

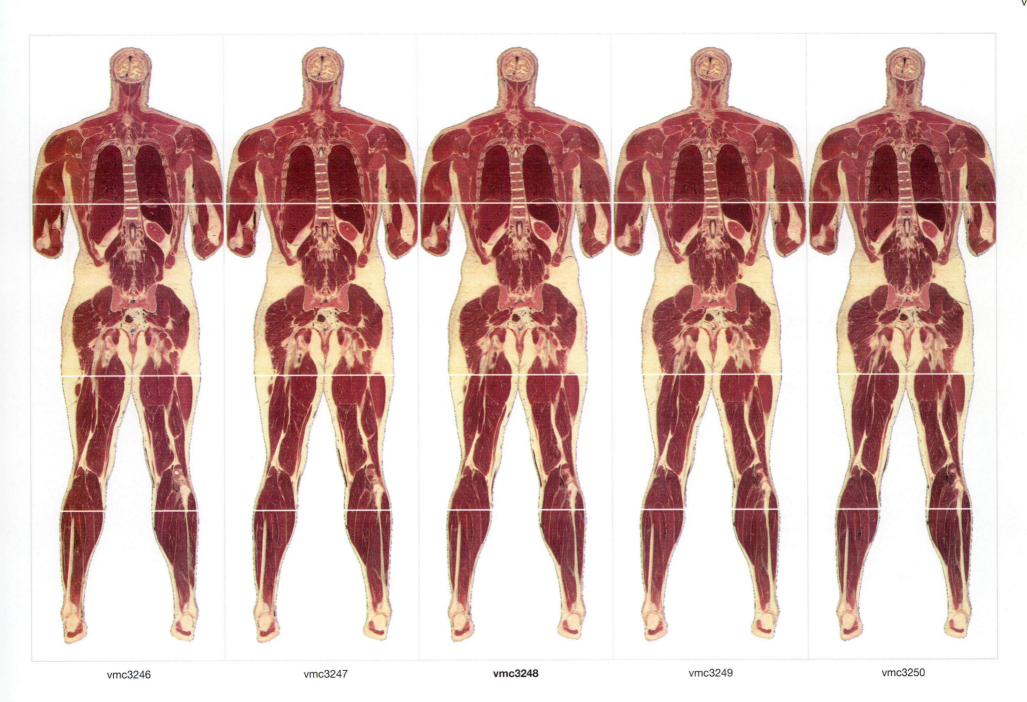

vmc3246   vmc3247   **vmc3248**   vmc3249   vmc3250

Longitudinal Cerebral Fissure

Internal Occipital Protuberance

Occipital Vessel

Trapezius Muscle

Levator Scapulae Muscle

Rhomboid Major Muscle

Scapula, Spine

Multifidus Muscle

Deltoid Muscle

Subscapularis Muscle

Teres Major Muscle

Triceps Brachii Muscle, Long Head

Serratus Anterior Muscle

Latissimus Dorsi Muscle

Diaphragm

Ulna

Humerus, Medial Epicondyle

Liver, Right Lobe

Spinal Cord

L3, Superior Articular Process

Gluteus Medius Muscle

L5, Spinous Process

Ilium

Sacrotuberous Ligament

Ischiorectal Fossa

Anus

Adductor Magnus Muscle

Biceps Femoris Muscle, Long Head

Biceps Femoris Muscle, Short Head

Semimembranosus Muscle

Plantaris Muscle

Gastrocnemius Muscle, Lateral Head

Gastrocnemius Muscle, Medial Head

Soleus Muscle

Crural Fascia

Peroneus Brevis Muscle

Flexor Hallucis Longus Muscle

Fibula, Lateral Malleolus

Calcaneus Bone

Quadratus Plantae Muscle

Abductor Hallucis Muscle

Occipital Bone

Occipital Lobe

Trapezius Muscle

Supraspinatus Muscle

Subscapularis Muscle

Infraspinatus Muscle

Deltoid Muscle

Teres Minor Muscle

Rib 4

Lung, Oblique Fissure

Spinal Cord

Latissimus Dorsi Muscle

Lung, Lower Lobe

Diaphragm

Kidney

Quadratus Lumborum Muscle

External Oblique Muscle

Iliocostalis Muscle

Longissimus Muscle

Sacroiliac Joint

Sacrum

Gluteus Maximus Muscle

Piriformis Muscle

Sigmoid Colon

Ischial Tuberosity

Levator Ani Muscle

Vastus Lateralis Muscle

External Anal Sphincter

Semitendinosus Muscle

Semimembranosus Muscle

Plantaris Muscle

Small Saphenous Vein

Gastrocnemius Muscle, Lateral Head

Great Saphenous Vein

Peroneus Longus Muscle

Soleus Muscle

Fibula, Shaft

Flexor Hallucis Longus Muscle

Fibula, Lateral Malleolus

Calcaneus Bone

Plantar Aponeurosis

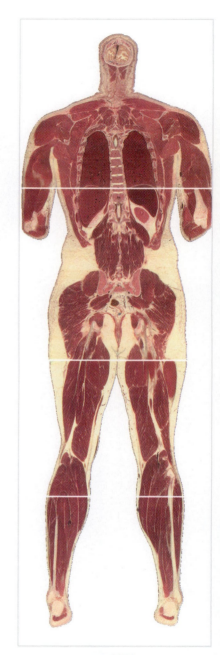

vmc3251

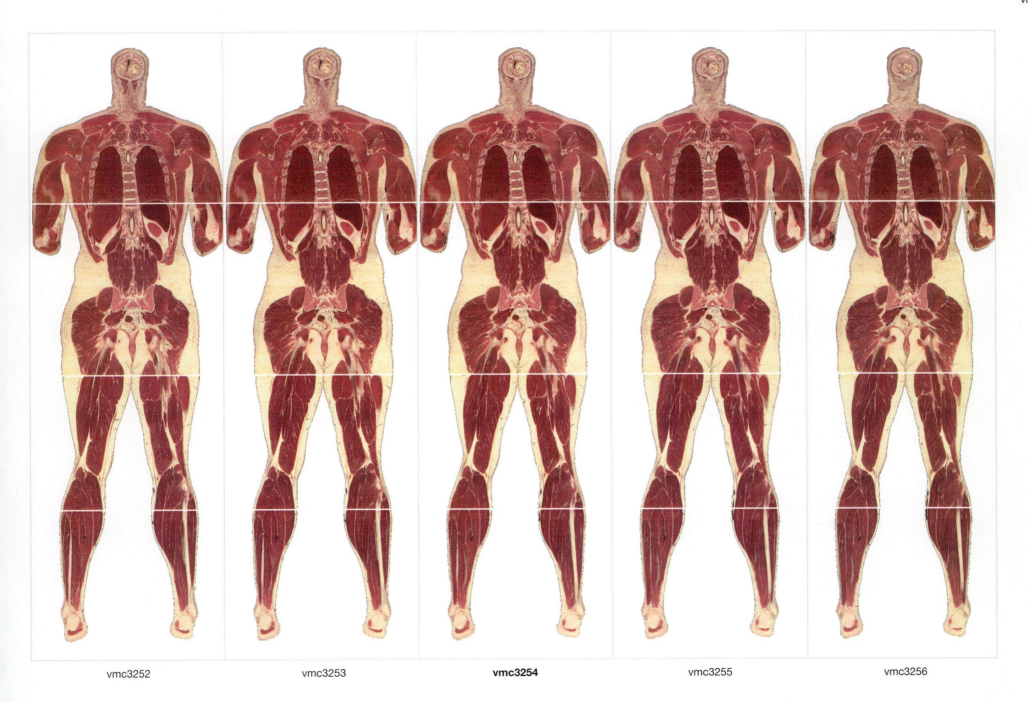

vmc3252   vmc3253   **vmc3254**   vmc3255   vmc3256

Diploe
Outer Table
Occipital Bone
Diploic Vessel
Multifidus Muscle
Supraspinatus Muscle
Deltoid Muscle
Rib 4-5, Intercostal Muscles
Triceps Brachii Muscle, Long Head
Subscapularis Muscle
Spinal Cord
Rib 7
Serratus Anterior Muscle
Lung, Lower Lobe
Triceps Brachii Tendon
Diaphragm
Liver, Right Lobe
Latissimus Dorsi Muscle
Superficial Fascia and Adipose Tissue
Spinalis Muscle
Multifidus Muscle
Ilium
Sacrum
Piriformis Muscle
Ischiorectal Fossa
External Anal Sphincter
Adductor Magnus Muscle
Biceps Femoris Muscle
Fascia Lata
Semitendinosus Muscle
Semimembranosus Muscle
Great Saphenous Vein
Plantaris Muscle
Gastrocnemius Muscle, Medial Head
Gastrocnemius Muscle, Lateral Head
Soleus Muscle
Tibialis Posterior Muscle
Flexor Hallucis Longus Muscle
Peroneus Brevis Muscle
Calcaneus Bone
Quadratus Plantae Muscle
Flexor Digitorum Brevis Muscle

Inner Table
Galea Aponeurotica
Occipital Lobe
Occipitofrontalis Muscle, Occipital Belly
Trapezius Muscle
Supraspinatus Muscle
Scapula, Spine
Infraspinatus Muscle
Deltoid Muscle
Subscapularis Muscle
Scapula
Teres Major Muscle
Triceps Brachii Muscle, Long Head
Lung, Lower Lobe
Diaphragm
Spleen
Quadratus Lumborum Muscle
Longissimus Muscle
Iliocostalis Muscle
Gluteus Medius Muscle
Sacroiliac Joint
Gluteus Maximus Muscle
Internal Iliac Vein
Levator Ani Muscle
Sigmoid Colon
Anus
Iliotibial Tract
Biceps Femoris Muscle, Long Head
Adductor Magnus Muscle
Semitendinosus Muscle
Biceps Femoris Tendon
Semimembranosus Muscle
Small Saphenous Vein
Gastrocnemius Muscle, Lateral Head
Soleus Muscle
Fibula, Shaft
Peroneus Brevis Muscle
Flexor Hallucis Longus Muscle
Fibula, Lateral Malleolus
Peroneus Longus Tendon
Calcaneus Bone
Plantar Aponeurosis

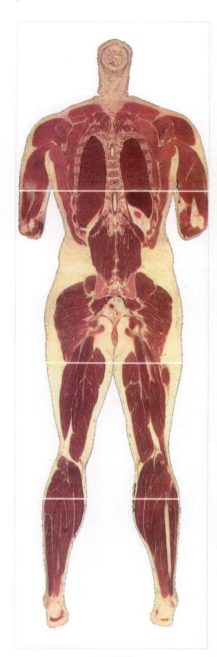

vmc3257

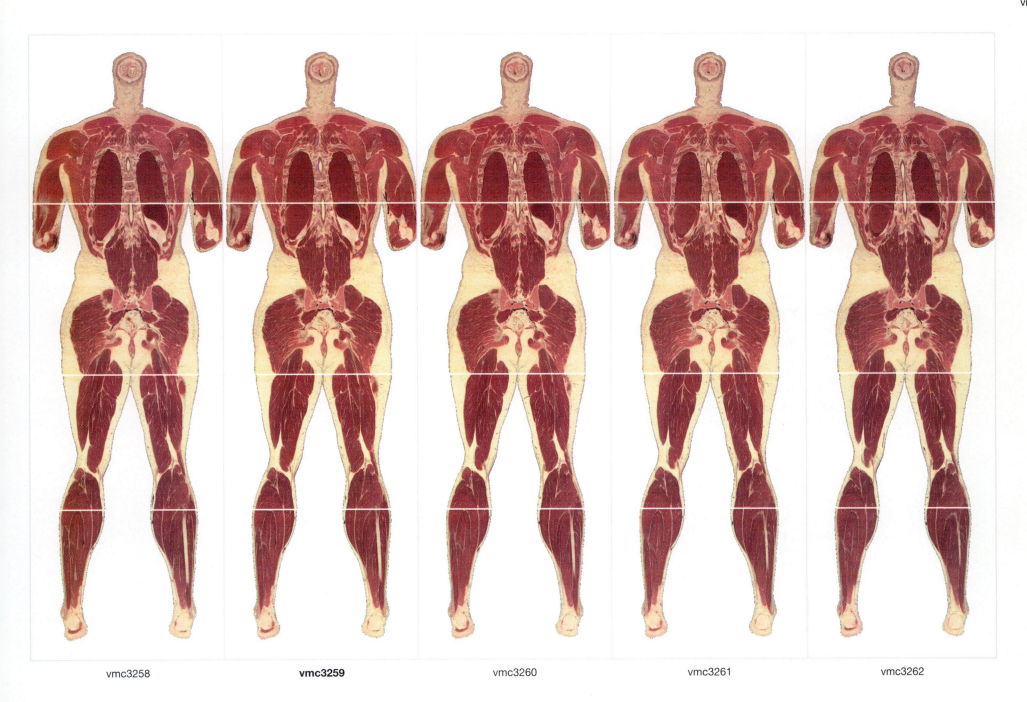

vmc3258   **vmc3259**   vmc3260   vmc3261   vmc3262

Periosteum — Galea Aponeurotica

Occipital Vessels — Occipital Bone

Multifidus Muscle — Trapezius Muscle

Rhomboid Major Muscle — Levator Scapulae Muscle

T3, Spinous Process — Supraspinatus Muscle

Deltoid Muscle — Scapula, Spine

Infraspinatus Muscle — Infraspinatus Muscle

Triceps Brachii Muscle, Long Head — Teres Minor Muscle

Scapula — Semispinalis Cervicis Muscle

Teres Major Muscle — Teres Major Muscle

Lung, Lower Lobe — Serratus Anterior Muscle

Rib 7 — Spinal Cord

Latissimus Dorsi Muscle — Rib 7-8, Intercostal Muscles

Diaphragm — Humerus, Medial Epicondyle

Liver, Right Lobe — Humerus, Trochlea

Rib 11-12, Intercostal Muscle — Spleen

Superficial Fascia and Adipose Tissue — Latissimus Dorsi Muscle

Ilium — Iliocostalis Muscle

Sacroiliac Joint — Spinalis Muscle

Gluteus Maximus Muscle — Longissimus Muscle

Piriformis Muscle — Gluteus Medius Muscle

Middle Rectal Vessel — Sacrum

External Anal Sphincter — Piriformis Muscle

Intergluteal Crease — Levator Ani Muscle

Adductor Magnus Muscle — Anococcygeal Ligament

Biceps Femoris Muscle — Ischiorectal Fossa

Semitendinosus Muscle — Anus

Semimembranosus Muscle — Adductor Magnus Muscle

Fascia Lata — Biceps Femoris Muscle, Long Head

Superficial Fascia — Biceps Femoris Tendon

Small Saphenous Vein — Plantaris Muscle

Plantaris Muscle — Gastrocnemius Muscle, Medial Head

Soleus Muscle — Soleus Muscle

Gastrocnemius Muscle, Medial Head — Peroneus Longus Muscle

Gastrocnemius Muscle, Lateral Head — Peroneus Brevis Muscle

Crural Fascia — Flexor Hallucis Longus Muscle

Calcaneal Subtendinous Space — Calcaneal Subtendinous Space

Calcaneus Bone — Calcaneus Bone

Flexor Digitorum Brevis Muscle — Plantar Aponeurosis

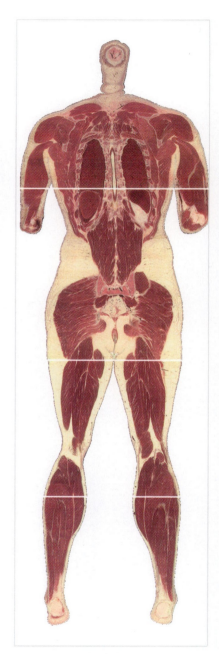

vmc3263

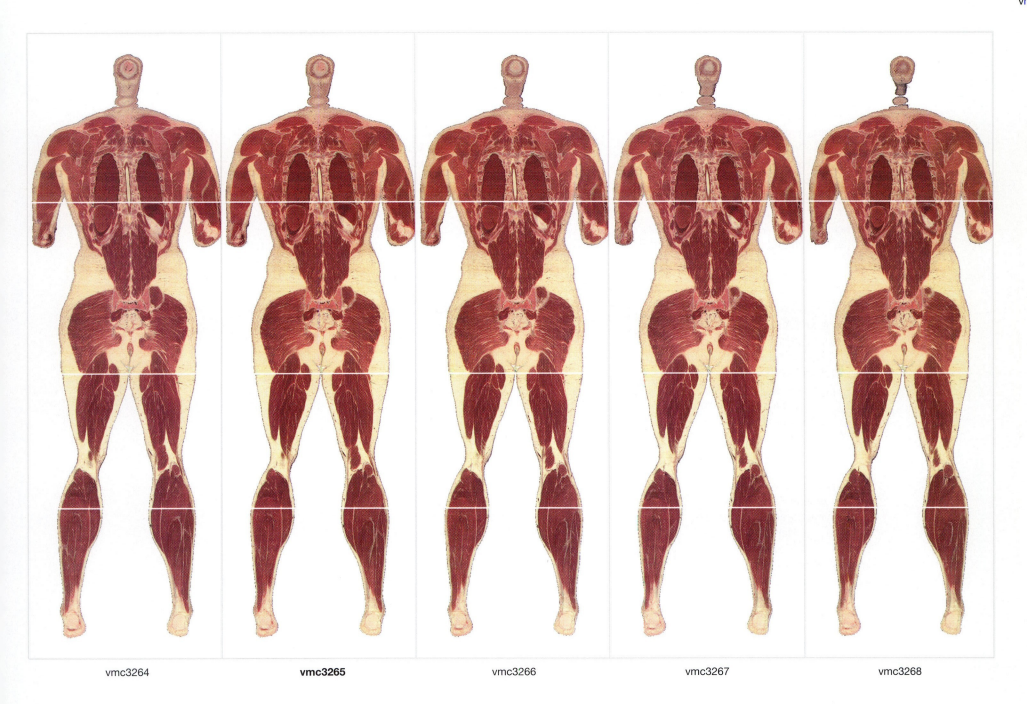

vmc3264   **vmc3265**   vmc3266   vmc3267   vmc3268

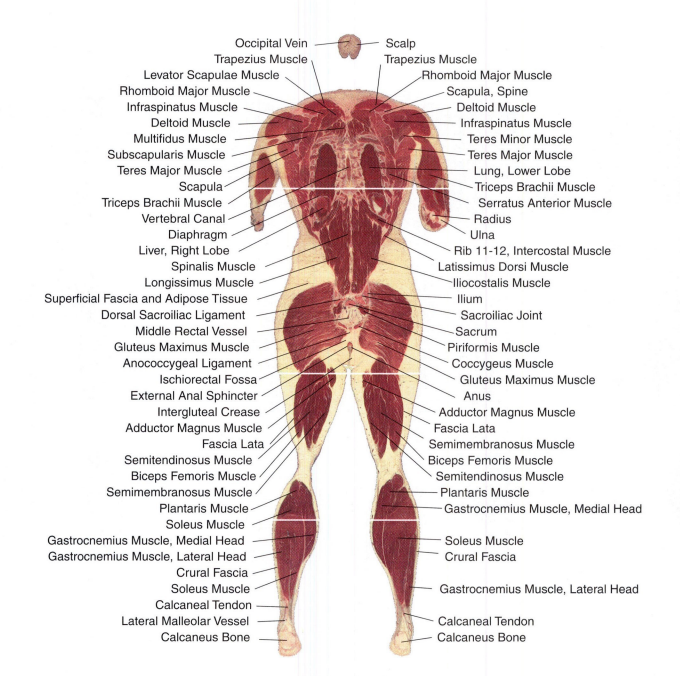

Occipital Vein — Scalp
Trapezius Muscle — Trapezius Muscle
Levator Scapulae Muscle — Rhomboid Major Muscle
Rhomboid Major Muscle — Scapula, Spine
Infraspinatus Muscle — Deltoid Muscle
Deltoid Muscle — Infraspinatus Muscle
Multifidus Muscle — Teres Minor Muscle
Subscapularis Muscle — Teres Major Muscle
Teres Major Muscle — Lung, Lower Lobe
Scapula — Triceps Brachii Muscle
Triceps Brachii Muscle — Serratus Anterior Muscle
Vertebral Canal — Radius
Diaphragm — Ulna
Liver, Right Lobe — Rib 11-12, Intercostal Muscle
Spinalis Muscle — Latissimus Dorsi Muscle
Longissimus Muscle — Iliocostalis Muscle
Superficial Fascia and Adipose Tissue — Ilium
Dorsal Sacroiliac Ligament — Sacroiliac Joint
Middle Rectal Vessel — Sacrum
Gluteus Maximus Muscle — Piriformis Muscle
Anococcygeal Ligament — Coccygeus Muscle
Ischiorectal Fossa — Gluteus Maximus Muscle
External Anal Sphincter — Anus
Intergluteal Crease — Adductor Magnus Muscle
Adductor Magnus Muscle — Fascia Lata
Fascia Lata — Semimembranosus Muscle
Semitendinosus Muscle — Biceps Femoris Muscle
Biceps Femoris Muscle — Semitendinosus Muscle
Semimembranosus Muscle — Plantaris Muscle
Plantaris Muscle — Gastrocnemius Muscle, Medial Head
Soleus Muscle — Soleus Muscle
Gastrocnemius Muscle, Medial Head — Crural Fascia
Gastrocnemius Muscle, Lateral Head —
Crural Fascia — Gastrocnemius Muscle, Lateral Head
Soleus Muscle —
Calcaneal Tendon — Calcaneal Tendon
Lateral Malleolar Vessel —
Calcaneus Bone — Calcaneus Bone

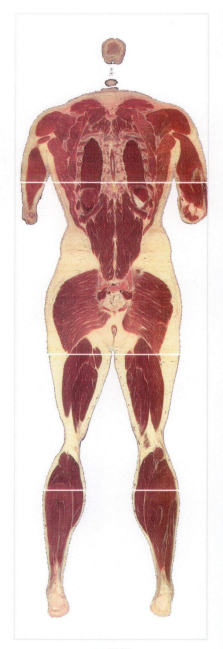

vmc3269

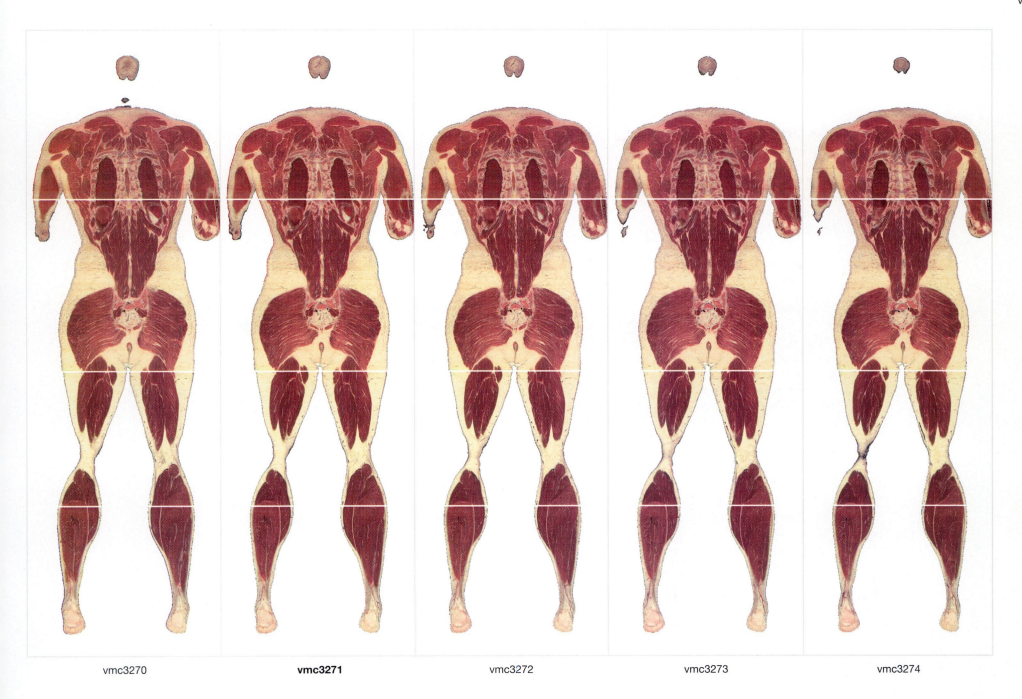

vmc3270 **vmc3271** vmc3272 vmc3273 vmc3274

Rhomboid Minor Muscle

Trapezius Muscle

Levator Scapulae Muscle

Scapula

Semispinalis Cervicis Muscle

Deltoid Muscle

Axillary Vessel

Multifidus Muscle

Teres Major Muscle

Lung, Lower Lobe

T8, Spinous Process

Rib 10

Latissimus Dorsi Muscle

Lumbodorsal Fascia

Superficial Fascia

Sacrum

Median Sacral Vessel

Gluteus Maximus Muscle

Anococcygeal Ligament

Ischiorectal Fossa

Biceps Femoris Muscle

Fascia Lata

Semitendinosus Muscle

Semimembranosus Muscle

Superficial Fascia

Plantaris Muscle

Gastrocnemius Muscle, Medial Head

Soleus Muscle

Crural Fascia

Calcaneal Tendon

Calcaneus Bone

Plantar Aponeurosis

Rhomboid Major Muscle

Trapezius Muscle

Scapula, Spine

Infraspinatus Muscle

Deltoid Muscle

Subscapularis Muscle

Rib 4-5, Intercostal Muscles

Scapula

Teres Major Muscle

Triceps Brachii Muscle

Ligamentum Flavum

Triceps Brachii Tendon

Latissimus Dorsi Muscle

Rib 10-11, Intercostal Muscles

Rib 11

Iliocostalis Muscle

Longissimus Muscle

Spinalis Muscle

Coccygeus Muscle

Coccyx

External Anal Sphincter

Intergluteal Crease

Biceps Femoris Muscle

Semitendinosus Muscle

Semimembranosus Muscle

Fascia Lata

Superficial Fascia

Plantaris Muscle

Gastrocnemius Muscle, Medial Head

Soleus Muscle

Crural Fascia

Calcaneal Tendon

Calcaneus Bone

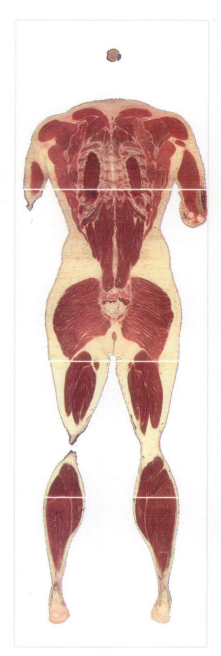

vmc3275

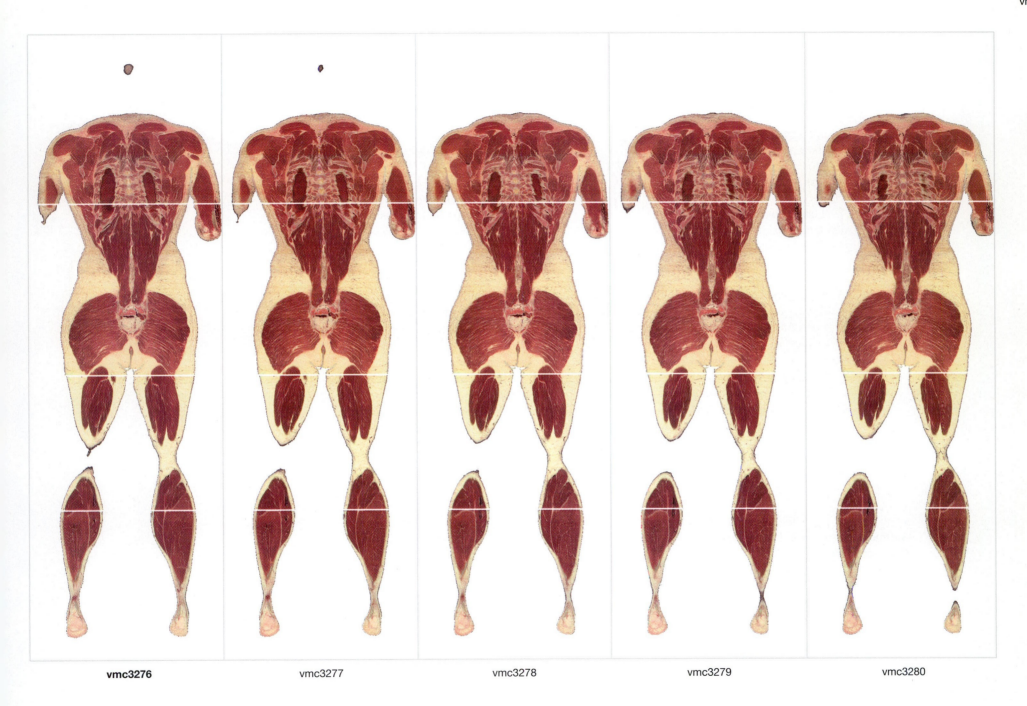

vmc3276   vmc3277   vmc3278   vmc3279   vmc3280

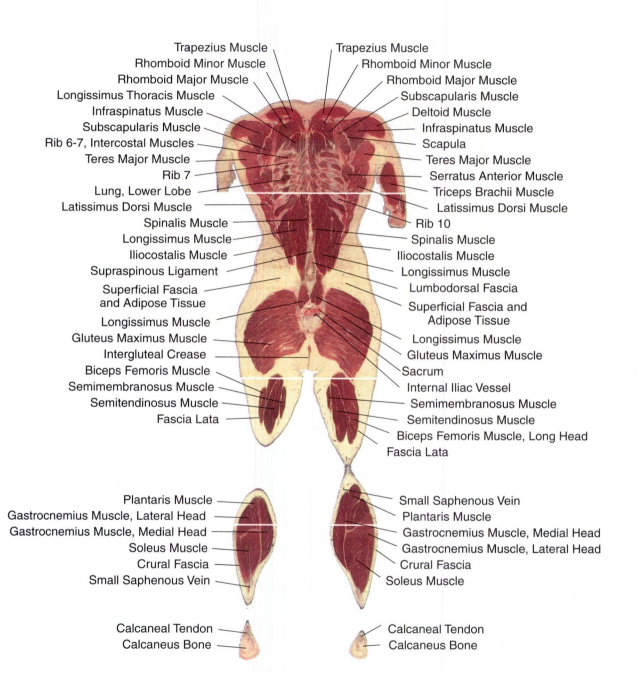

Trapezius Muscle
Rhomboid Minor Muscle
Rhomboid Major Muscle
Longissimus Thoracis Muscle
Infraspinatus Muscle
Subscapularis Muscle
Rib 6-7, Intercostal Muscles
Teres Major Muscle
Rib 7
Lung, Lower Lobe
Latissimus Dorsi Muscle
Spinalis Muscle
Longissimus Muscle
Iliocostalis Muscle
Supraspinous Ligament
Superficial Fascia
and Adipose Tissue
Longissimus Muscle
Gluteus Maximus Muscle
Intergluteal Crease
Biceps Femoris Muscle
Semimembranosus Muscle
Semitendinosus Muscle
Fascia Lata

Trapezius Muscle
Rhomboid Minor Muscle
Rhomboid Major Muscle
Subscapularis Muscle
Deltoid Muscle
Infraspinatus Muscle
Scapula
Teres Major Muscle
Serratus Anterior Muscle
Triceps Brachii Muscle
Latissimus Dorsi Muscle
Rib 10
Spinalis Muscle
Iliocostalis Muscle
Longissimus Muscle
Lumbodorsal Fascia
Superficial Fascia and
Adipose Tissue
Longissimus Muscle
Gluteus Maximus Muscle
Sacrum
Internal Iliac Vessel
Semimembranosus Muscle
Semitendinosus Muscle
Biceps Femoris Muscle, Long Head
Fascia Lata

Plantaris Muscle
Gastrocnemius Muscle, Lateral Head
Gastrocnemius Muscle, Medial Head
Soleus Muscle
Crural Fascia
Small Saphenous Vein

Small Saphenous Vein
Plantaris Muscle
Gastrocnemius Muscle, Medial Head
Gastrocnemius Muscle, Lateral Head
Crural Fascia
Soleus Muscle

Calcaneal Tendon
Calcaneus Bone

Calcaneal Tendon
Calcaneus Bone

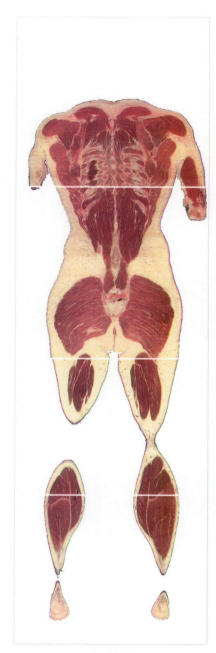

vmc3281

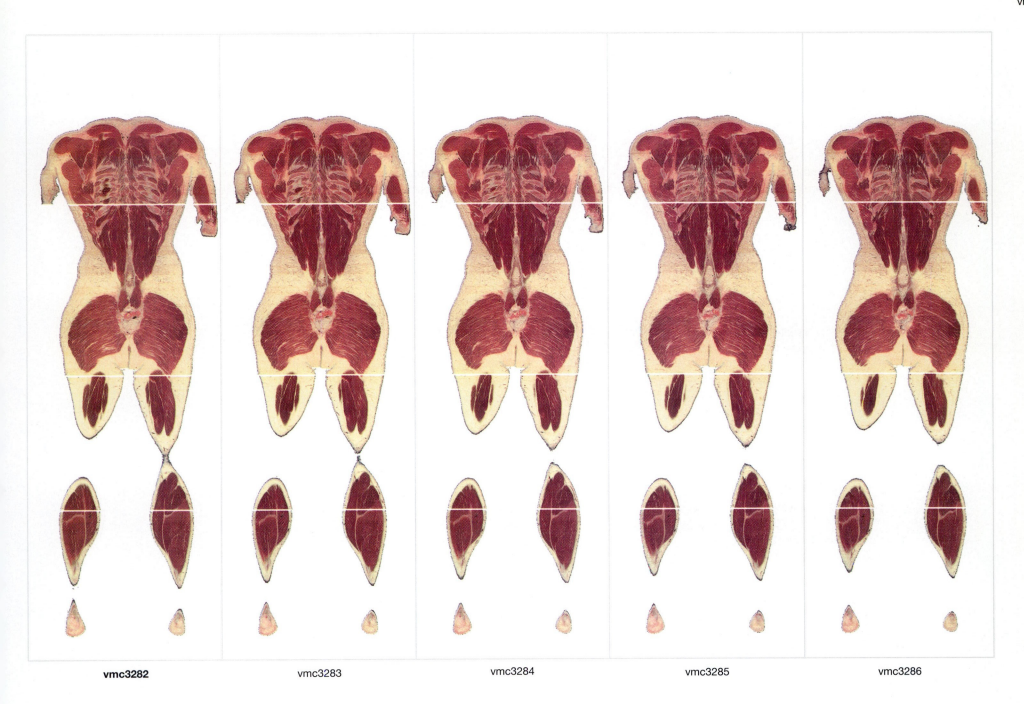

vmc3282   vmc3283   vmc3284   vmc3285   vmc3286

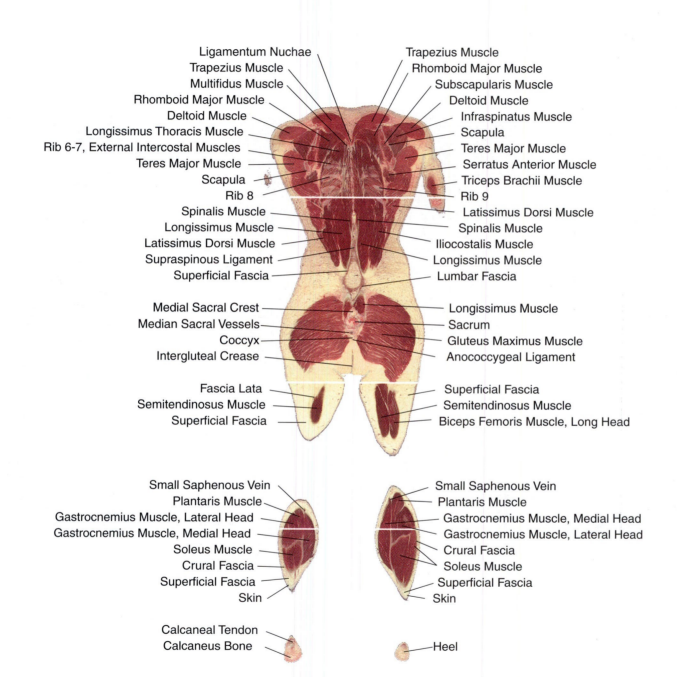

Ligamentum Nuchae
Trapezius Muscle
Multifidus Muscle
Rhomboid Major Muscle
Deltoid Muscle
Longissimus Thoracis Muscle
Rib 6-7, External Intercostal Muscles
Teres Major Muscle
Scapula
Rib 8
Spinalis Muscle
Longissimus Muscle
Latissimus Dorsi Muscle
Supraspinous Ligament
Superficial Fascia

Trapezius Muscle
Rhomboid Major Muscle
Subscapularis Muscle
Deltoid Muscle
Infraspinatus Muscle
Scapula
Teres Major Muscle
Serratus Anterior Muscle
Triceps Brachii Muscle
Rib 9
Latissimus Dorsi Muscle
Spinalis Muscle
Iliocostalis Muscle
Longissimus Muscle
Lumbar Fascia

Medial Sacral Crest
Median Sacral Vessels
Coccyx
Intergluteal Crease

Longissimus Muscle
Sacrum
Gluteus Maximus Muscle
Anococcygeal Ligament

Fascia Lata
Semitendinosus Muscle
Superficial Fascia

Superficial Fascia
Semitendinosus Muscle
Biceps Femoris Muscle, Long Head

Small Saphenous Vein
Plantaris Muscle
Gastrocnemius Muscle, Lateral Head
Gastrocnemius Muscle, Medial Head
Soleus Muscle
Crural Fascia
Superficial Fascia
Skin

Small Saphenous Vein
Plantaris Muscle
Gastrocnemius Muscle, Medial Head
Gastrocnemius Muscle, Lateral Head
Crural Fascia
Soleus Muscle
Superficial Fascia
Skin

Calcaneal Tendon
Calcaneus Bone

Heel

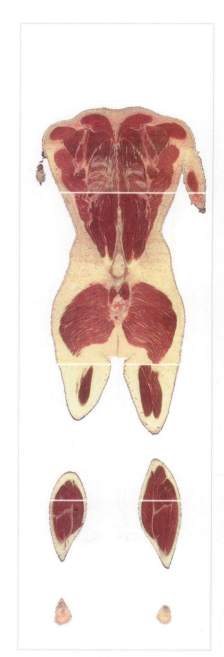

vmc3287

396

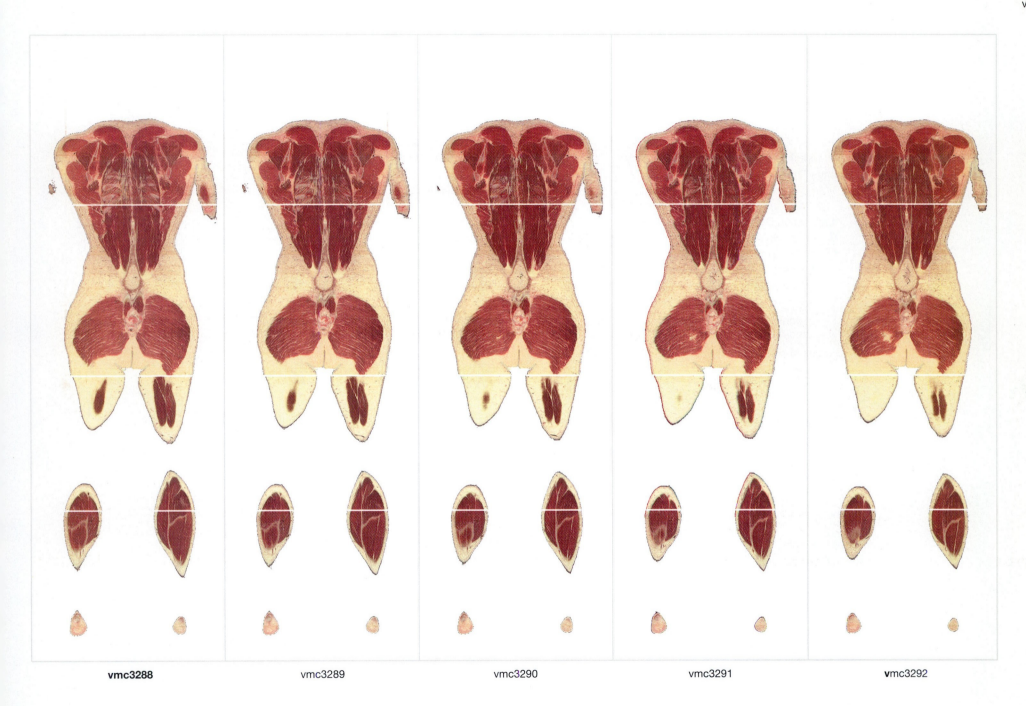

vmc3288   vmc3289   vmc3290   vmc3291   vmc3292

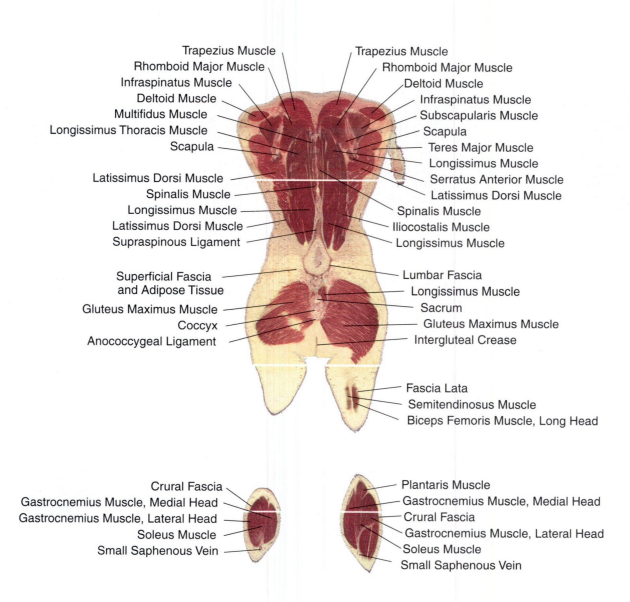

Trapezius Muscle
Trapezius Muscle
Rhomboid Major Muscle
Rhomboid Major Muscle
Infraspinatus Muscle
Deltoid Muscle
Deltoid Muscle
Infraspinatus Muscle
Multifidus Muscle
Subscapularis Muscle
Longissimus Thoracis Muscle
Scapula
Scapula
Teres Major Muscle
Longissimus Muscle
Latissimus Dorsi Muscle
Serratus Anterior Muscle
Spinalis Muscle
Latissimus Dorsi Muscle
Longissimus Muscle
Spinalis Muscle
Latissimus Dorsi Muscle
Iliocostalis Muscle
Supraspinous Ligament
Longissimus Muscle

Superficial Fascia
and Adipose Tissue
Lumbar Fascia
Longissimus Muscle
Gluteus Maximus Muscle
Sacrum
Coccyx
Gluteus Maximus Muscle
Anococcygeal Ligament
Intergluteal Crease

Fascia Lata
Semitendinosus Muscle
Biceps Femoris Muscle, Long Head

Crural Fascia
Plantaris Muscle
Gastrocnemius Muscle, Medial Head
Gastrocnemius Muscle, Medial Head
Gastrocnemius Muscle, Lateral Head
Crural Fascia
Soleus Muscle
Gastrocnemius Muscle, Lateral Head
Small Saphenous Vein
Soleus Muscle
Small Saphenous Vein

Calcaneal Tendon
Calcaneus Bone
Heel

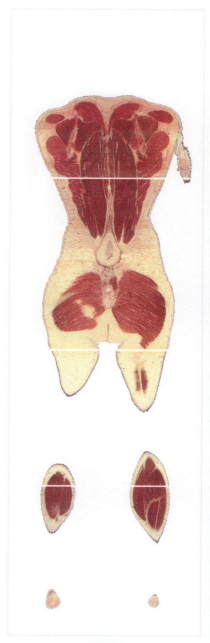

vmc3293

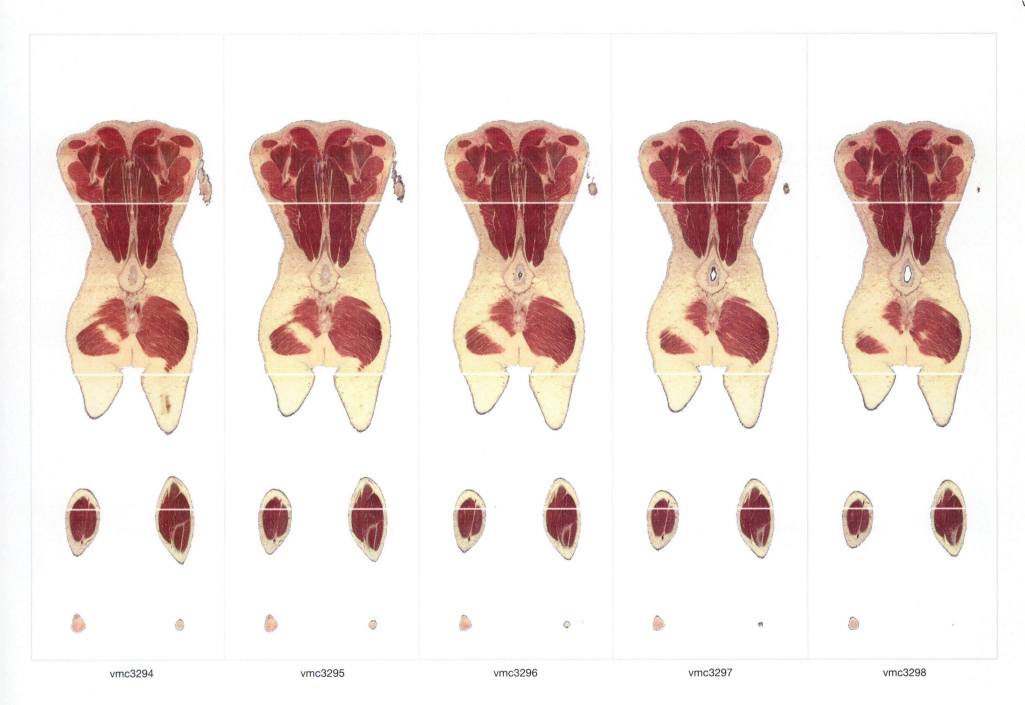

vmc3294   vmc3295   vmc3296   vmc3297   vmc3298

Trapezius Muscle
Multifidus Muscle
Rhomboid Major Muscle
Infraspinatus Muscle
Teres Minor Muscle
Teres Major Muscle
Longissimus Muscle
Spinalis Muscle
Latissimus Dorsi Muscle
Longissimus Muscle
Latissimus Dorsi Muscle
Supraspinous Ligament
Thoracolumbar Fascia
Superficial Fascia and
Adipose Tissue
Buttock
Gluteus Maximus Muscle

Trapezius Muscle
Rhomboid Major Muscle
Deltoid Muscle
Infraspinatus Muscle
Teres Major Muscle
Spinalis Thoracis Muscle
Serratus Anterior Muscle
Latissimus Dorsi Muscle
Spinalis Muscle
Iliocostalis Muscle
Longissimus Muscle
Lumbar Fascia
Superficial Fascia and
Adipose Tissue
Gluteus Maximus Muscle

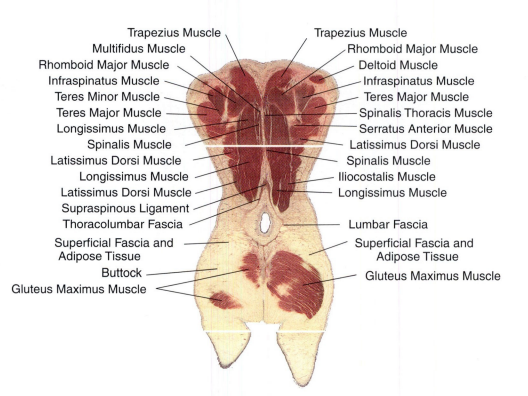

Gastrocnemius Muscle, Medial Head
Gastrocnemius Muscle, Lateral Head
Crural Fascia
Small Saphenous Vein

Plantaris Muscle
Gastrocnemius Muscle, Medial Head
Gastrocnemius Muscle, Lateral Head
Soleus Muscle
Crural Fascia

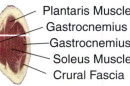

Heel

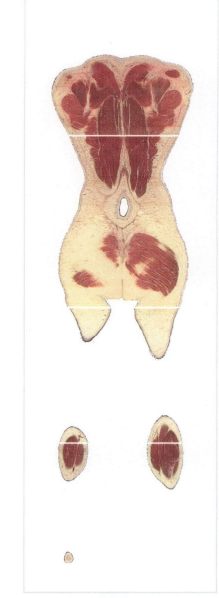

vmc3299

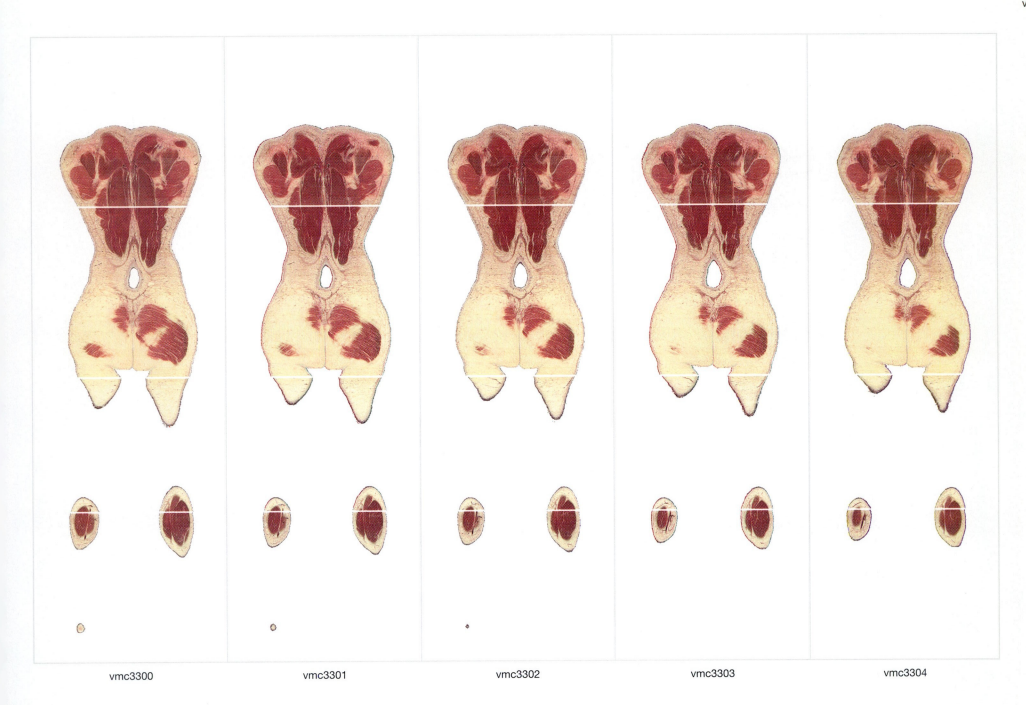

vmc3300   vmc3301   vmc3302   vmc3303   vmc3304

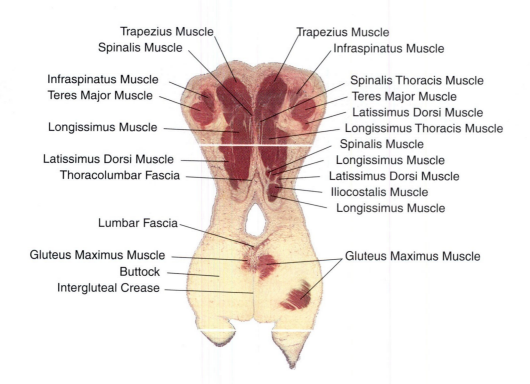

Trapezius Muscle

Trapezius Muscle

Spinalis Muscle

Infraspinatus Muscle

Infraspinatus Muscle

Spinalis Thoracis Muscle

Teres Major Muscle

Teres Major Muscle

Latissimus Dorsi Muscle

Longissimus Muscle

Longissimus Thoracis Muscle

Spinalis Muscle

Latissimus Dorsi Muscle

Longissimus Muscle

Thoracolumbar Fascia

Latissimus Dorsi Muscle

Iliocostalis Muscle

Longissimus Muscle

Lumbar Fascia

Gluteus Maximus Muscle

Gluteus Maximus Muscle

Buttock

Intergluteal Crease

Gastrocnemius Muscle, Lateral Head

Plantaris Muscle

Gastrocnemius Muscle, Lateral Head

Gastrocnemius Muscle, Medial Head

Gastrocnemius Muscle, Medial Head

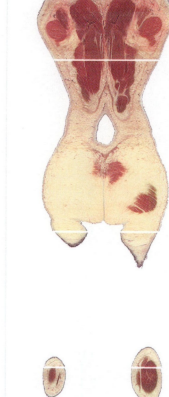

vmc3305

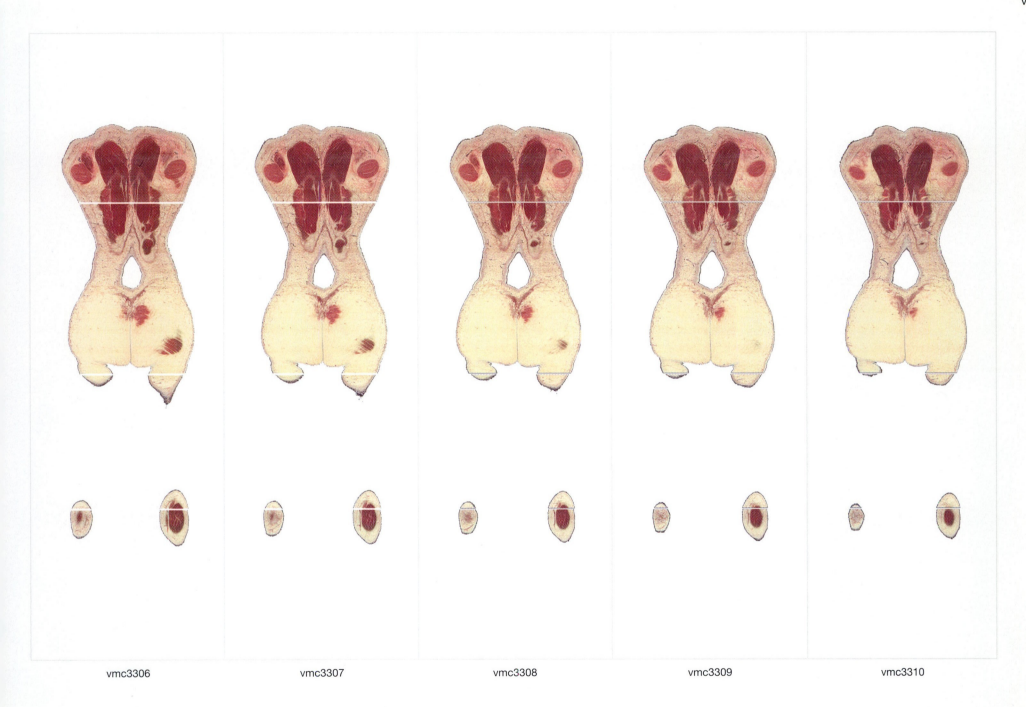

vmc3306   vmc3307   vmc3308   vmc3309   vmc3310

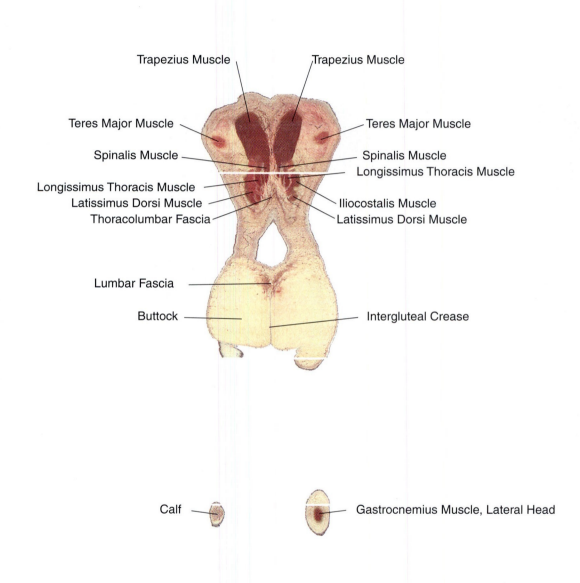

Trapezius Muscle — Trapezius Muscle

Teres Major Muscle — Teres Major Muscle

Spinalis Muscle — Spinalis Muscle
Longissimus Thoracis Muscle

Longissimus Thoracis Muscle — Iliocostalis Muscle
Latissimus Dorsi Muscle — Latissimus Dorsi Muscle
Thoracolumbar Fascia —

Lumbar Fascia

Buttock — Intergluteal Crease

Calf — Gastrocnemius Muscle, Lateral Head

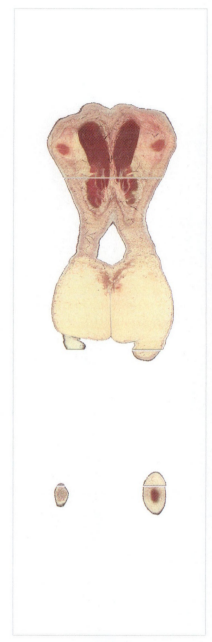

vmc3311

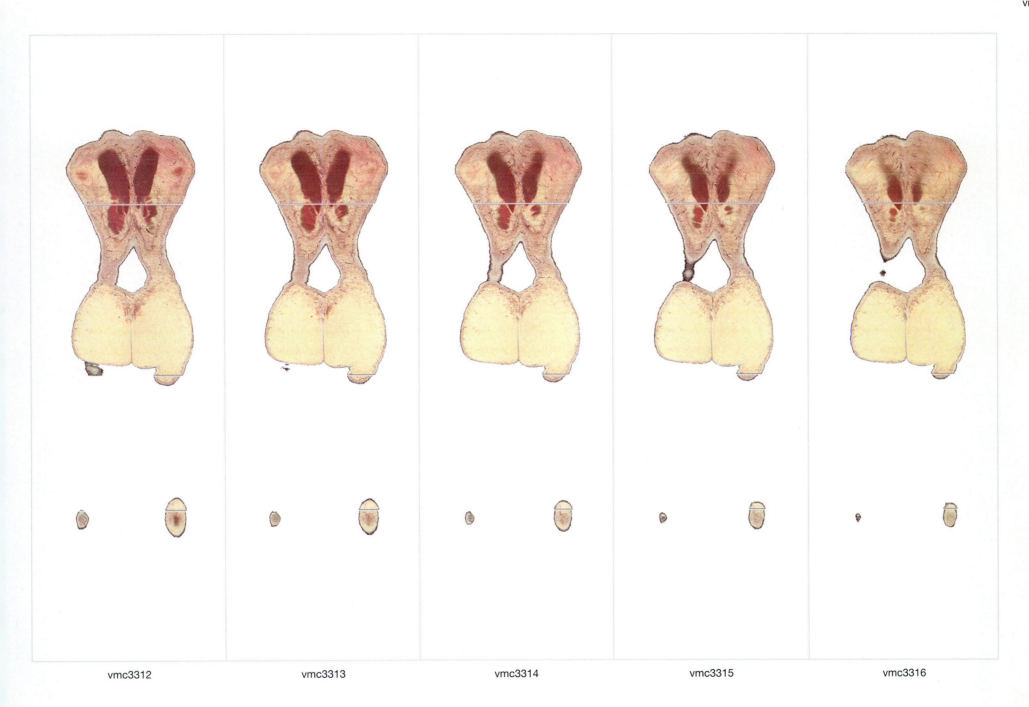

vmc3312 vmc3313 vmc3314 vmc3315 vmc3316

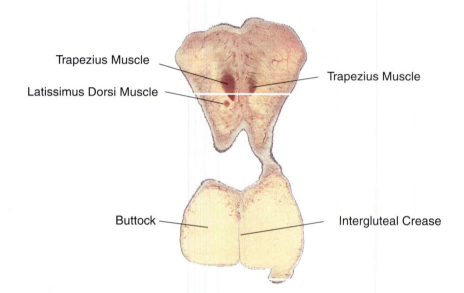

Trapezius Muscle

Trapezius Muscle

Latissimus Dorsi Muscle

Buttock

Intergluteal Crease

 Calf

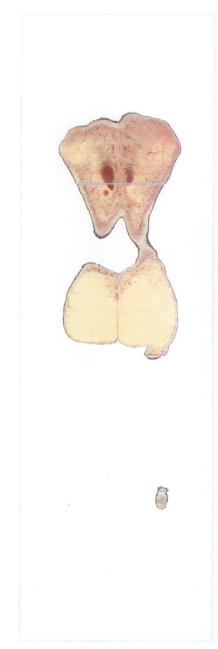

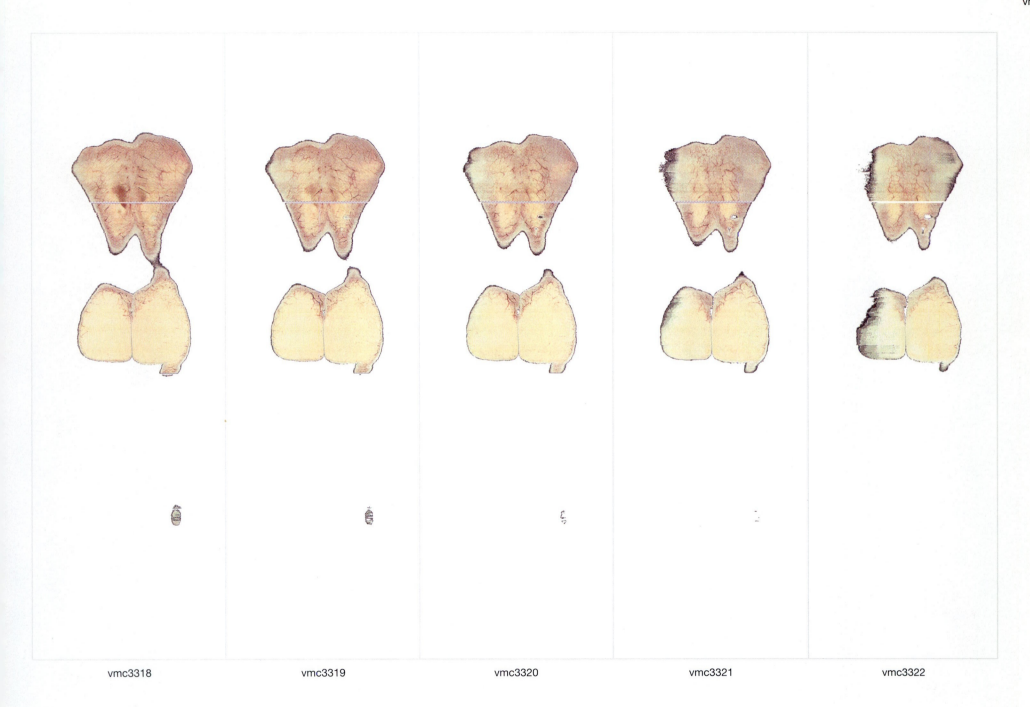

vmc3318   vmc3319   vmc3320   vmc3321   vmc3322

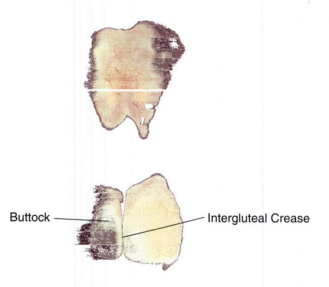

Buttock ——————— ——————— Intergluteal Crease

vmc3323

PART THREE

RECONSTRUCTED IMAGES
Sagittal

The sagittal images are of vertical planes through the long axis of the body parallel to or through that of the skull's sagittal suture. They are standard radiological planes used for direct magnetic resonance and ultrasound imaging. Sagittal clinical images are often reconstructed from transverse computed tomography and single photon emission computed tomography images. A sagittal image best demonstrates the arrangement of our anatomy from front to back. A special sagittal plane, the median sagittal plane, bisects many upaired anatomical structures (nose, mouth, spinal cord, etc.) and is the plane of symmetry for most bilateral anatomical structures (upper- and lower-extremity structures, eyes, kidneys, etc.).

These images, like their coronal counterparts, are virtual slices—the Visible Human Male was never physically cut in sagittal planes. Each sagittal image was computer reconstructed from an edge view of all slices in the transverse image collection as seen from the left lateral perspective and each is of a different depth at one-millimeter intervals through that collection.

The maximum resolution of images reconstructed in the sagittal plane is equivalent to the coronal plane reconstructions and is determined by the one-millimeter thickness of the original transverse sections.

The labeling format for sagittal images is the same as that used for the transverse and coronal images. In the same fashion as the coronal image display, the labeled slice is always repeated from the ten sagittal images on the same and facing pages.

Unlike transverse images, which include all slices, these reconstructed sagittal images begin at a full section through the juntion of the left shoulder and the chest wall and continue through most of the right arm. The scale remains the same for all sagittal images.

Clavicle
Coracoid Process
Coracoclavicular Ligament
Trapezius Muscle
Deltoid Muscle
Supraspinatus Muscle
Coracobrachialis Muscle
Scapula, Spine
Cephalic Vein
Infraspinatus Muscle
Musculocutaneous Nerve
Subscapularis Muscle
Pectoralis Major Muscle
Scapula, Lateral Border
Axillary Vessel
Teres Major Muscle
Brachial Plexus
Latissimus Dorsi Muscle
Axilla
External Oblique Muscle
Serratus Anterior Muscle
Superficial Fascia and
External Oblique Muscle
Adipose Tissue
Rib 8
Gluteus Medius Muscle
Rib 9
Gluteus Minimus Muscle
Rib 9-10, Intercostal M.
Femur, Greater Trochanter
Rib 10
Gluteus Maximus Muscle
Radius
Femur, Shaft
Pronator Quadratus M.
Vastus Lateralis Muscle
Flexor Digitorum
Superficialis Muscle
Flexor Carpi Ulnaris Muscle
Biceps Femoris Muscle,
Tensor Fascia Lata Muscle
Long Head
Rectus Femoris Muscle
Short Head
Vastus Intermedius Muscle
Fascia Lata
Articularis Genu
Popliteal Fossa
Rectus Femoris Tendon
Knee Joint, Cavity
Patella
Femur, Lateral Condyle
Patellar Ligament
Knee Joint
Anterior Cruciate Ligament
Tibia, Lateral Condyle
Tibial Tuberosity
Popliteal Vessel
Tibia, Shaft
Popliteus Muscle
Tibialis Posterior Muscle
Soleus Muscle
Small Saphenous Vein
Crural Fascia
Superficial Fascia
Fifth Metatarsal
Abductor Digiti Minimi Muscle
Fifth Proximal Phalanx
Heel
Fifth Middle Phalanx
Flexor Digiti Minimi Brevis M.
Long Flexor Tendon

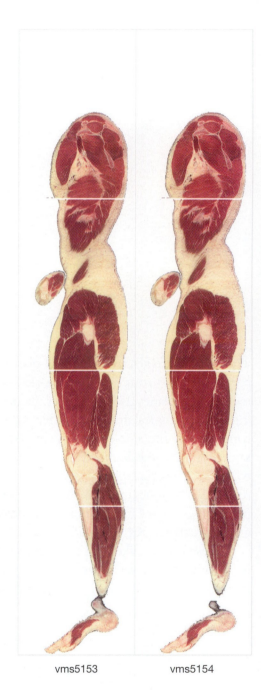

vms5153   vms5154

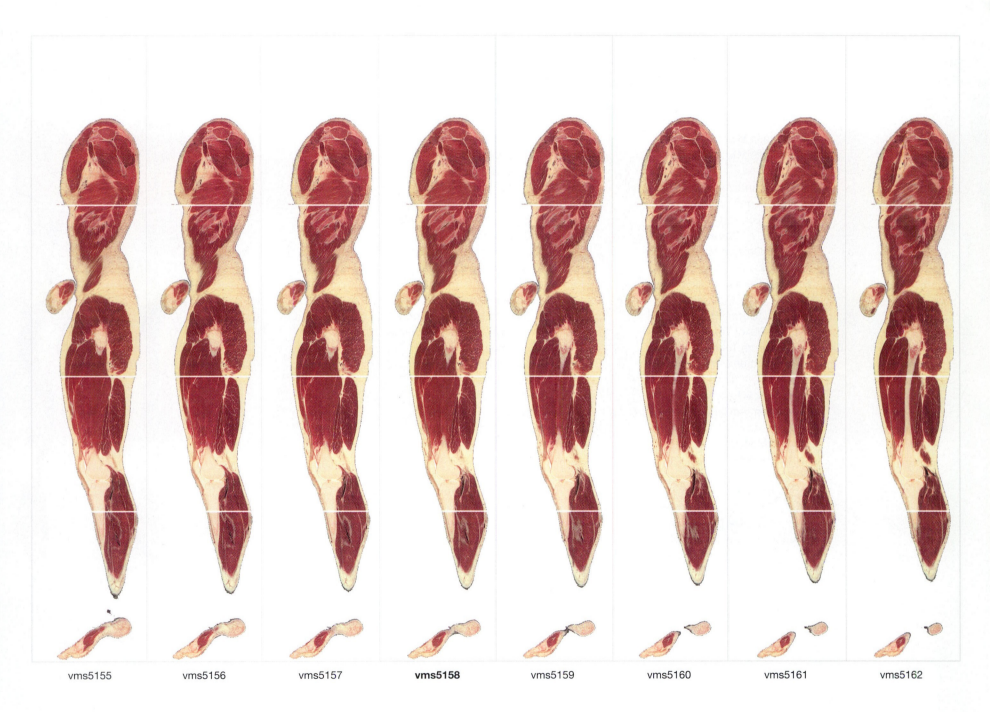

vms5155 vms5156 vms5157 **vms5158** vms5159 vms5160 vms5161 vms5162

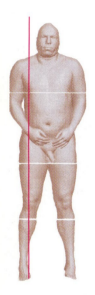

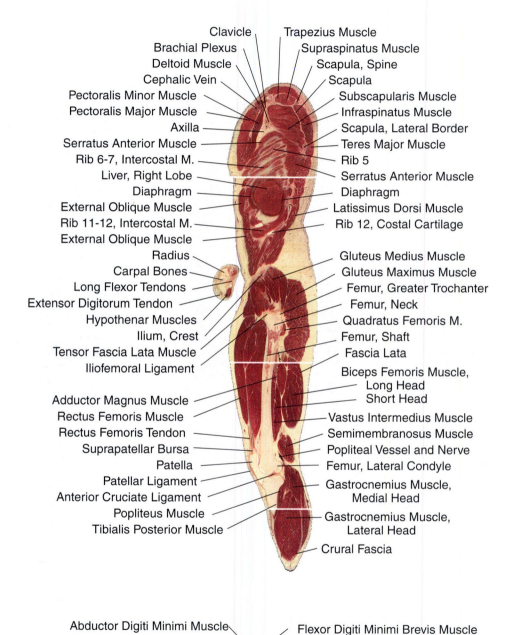

Clavicle
Brachial Plexus
Deltoid Muscle
Cephalic Vein
Pectoralis Minor Muscle
Pectoralis Major Muscle
Axilla
Serratus Anterior Muscle
Rib 6-7, Intercostal M.
Liver, Right Lobe
Diaphragm
External Oblique Muscle
Rib 11-12, Intercostal M.
External Oblique Muscle
Radius
Carpal Bones
Long Flexor Tendons
Extensor Digitorum Tendon
Hypothenar Muscles
Ilium, Crest
Tensor Fascia Lata Muscle
Iliofemoral Ligament
Adductor Magnus Muscle
Rectus Femoris Muscle
Rectus Femoris Tendon
Suprapatellar Bursa
Patella
Patellar Ligament
Anterior Cruciate Ligament
Popliteus Muscle
Tibialis Posterior Muscle

Trapezius Muscle
Supraspinatus Muscle
Scapula, Spine
Scapula
Subscapularis Muscle
Infraspinatus Muscle
Scapula, Lateral Border
Teres Major Muscle
Rib 5
Serratus Anterior Muscle
Diaphragm
Latissimus Dorsi Muscle
Rib 12, Costal Cartilage
Gluteus Medius Muscle
Gluteus Maximus Muscle
Femur, Greater Trochanter
Femur, Neck
Quadratus Femoris M.
Femur, Shaft
Fascia Lata
Biceps Femoris Muscle,
 Long Head
 Short Head
Vastus Intermedius Muscle
Semimembranosus Muscle
Popliteal Vessel and Nerve
Femur, Lateral Condyle
Gastrocnemius Muscle,
 Medial Head
Gastrocnemius Muscle,
 Lateral Head
Crural Fascia

Abductor Digiti Minimi Muscle
Flexor Digiti Minimi Brevis Muscle

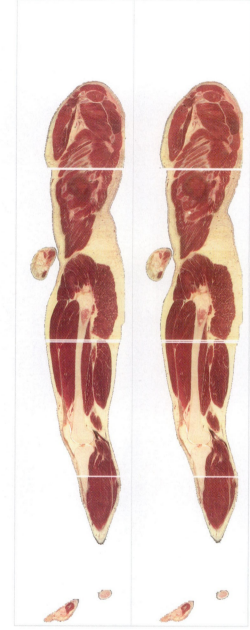

vms5163   vms5164

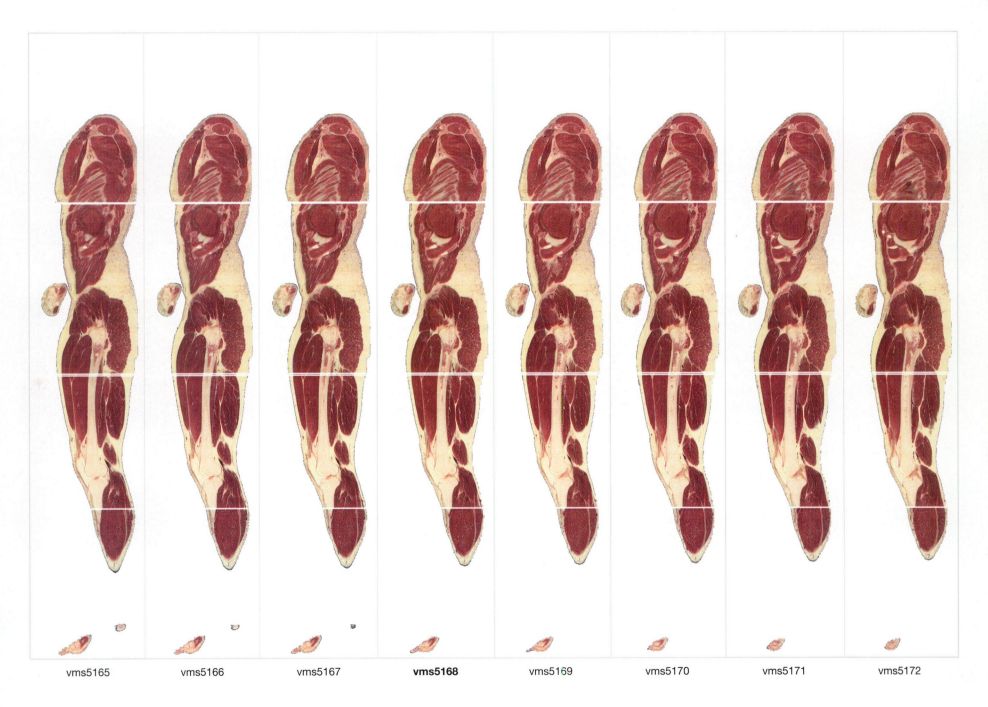

vms5165 vms5166 vms5167 **vms5168** vms5169 vms5170 vms5171 vms5172

Suprascapular Vessel
Clavicle
Subclavius Muscle
Deltoid Muscle
Cephalic Vein
Axillary Vessel
Pectoralis Major Muscle
Brachial Plexus
Pectoralis Minor Muscle
Axilla
Rib 9
Liver, Right Lobe
Intra-abdominal Fat
Transversus Abdominis M.
External Oblique Muscle
Long Flexor Tendons
Carpal Bones
Fourth Metacarpal
Fourth Dorsal Interosseous M.
Fifth Metacarpal
Hypothenar Muscles
Sartorius Muscle
Iliofemoral Ligament
Femur, Neck
Femur, Intertrochanteric Crest
Rectus Femoris Muscle
Vastus Intermedius Muscle
Femur, Shaft
Vastus Medialis Muscle
Rectus Femoris Tendon
Patella
Infrapatellar Fat Pad
Femur, Lateral Condyle
Patellar Ligament
Tibia, Shaft
Popliteus Muscle

Supraspinatus Muscle
Trapezius Muscle
Scapula, Spine
Infraspinatus Muscle
Subscapularis Muscle
Rib 5
Serratus Anterior Muscle
Teres Minor Muscle
Scapula, Inferior Angle
Rib 8
Diaphragm
Rib 11-12, Intercostal M.
Rib 12
Latissimus Dorsi Muscle
External Oblique Muscle
Internal Oblique Muscle
Transversus Abdominis M.
Gluteus Medius Muscle
Ilium, Anterior Inferior
Spine
Obturator Internus Tendon
Gluteus Minimus Muscle
Quadratus Femoris Muscle
Adductor Magnus Muscle
Biceps Femoris Muscle,
Long Head
Short Head
Semimembranosus M.
Posterior Cruciate Ligament
Knee Joint
Tibia
Gastrocnemius Muscle,
Medial Head
Soleus Muscle
Small Saphenous Vein

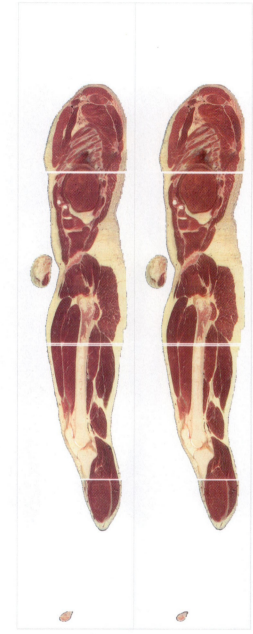

vms5173   vms5174

414

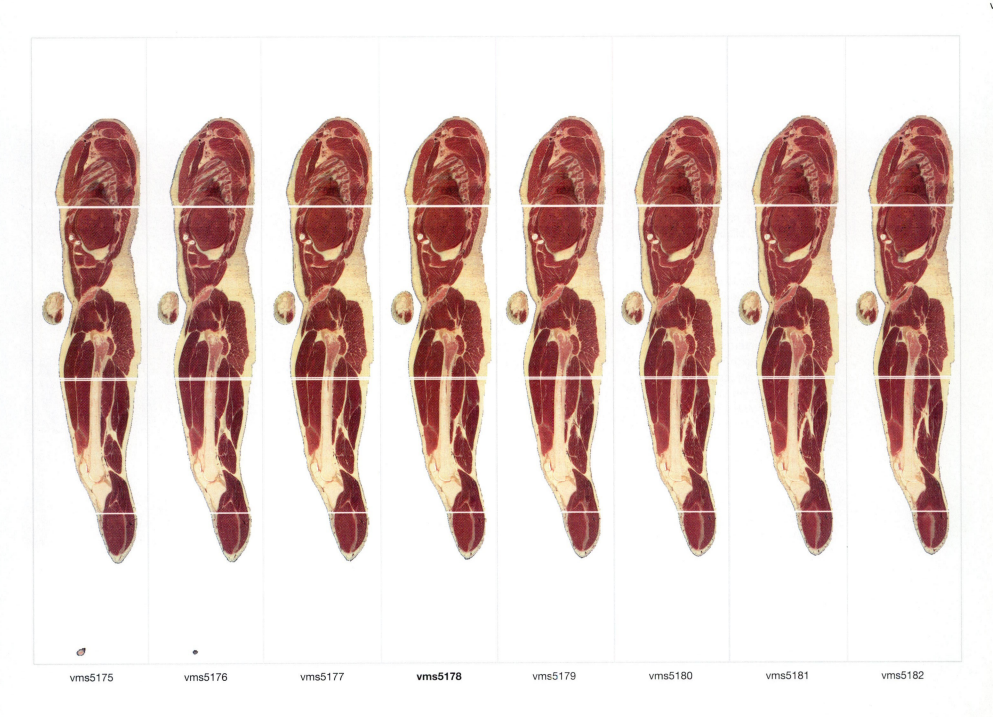

vms5175 vms5176 vms5177 **vms5178** vms5179 vms5180 vms5181 vms5182

Clavicle
Subclavius Muscle
Cephalic Vein
Axilla
Axillary Vessel
Brachial Plexus
Pectoralis Minor Muscle
Pectoralis Major Muscle
Rib 4
Lung, Middle Lobe
External Oblique Muscle
Liver, Right Lobe
Rib 8, Costal Cartilage
External Oblique Muscle
Internal Oblique Muscle
Transversus Abdominis M.
Carpal Bones
Third Dorsal Interosseous
Long Flexor Tendons
Fourth Dorsal Interosseous M.
Fifth Metacarpal
Hypothenar Muscles
Sartorius Muscle
Iliofemoral Ligament
Iliopsoas Muscle
Femur, Neck
Rectus Femoris Muscle
Vastus Intermedius Muscle
Vastus Medialis Muscle
Medial Patellar Retinaculum
Knee Joint
Tibia, Periosteum
Crural Fascia
Soleus Muscle

Trapezius Muscle
Serratus Anterior Muscle
Supraspinatus Muscle
Scapula, Spine
Infraspinatus Muscle
Subscapularis Muscle
Rib 5
Serratus Anterior Muscle
Lung, Upper Lobe
Lung, Oblique Fissure
Lung, Lower Lobe
Diaphragm
Pararenal Fat
Renal Fascia
Ascending Colon
Iliacus Muscle
Ilium
Gluteus Medius Muscle
Gluteus Minimus Muscle
Gluteus Maximus Muscle
Gemelli and Obturator
Internus Tendons
Quadratus Femoris M.
Femur,
Intertrochanteric Crest
Biceps Femoris Muscle,
Long Head
Adductor Magnus Muscle
Semitendinosus Muscle
Femur, Shaft
Semimembranosus M.
Popliteal Vessel
Gastrocnemius Muscle,
Medial Head
Femur, Medial Condyle

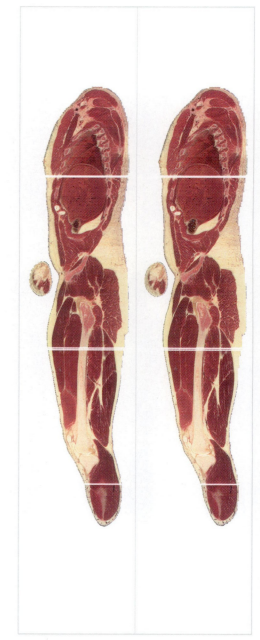

vms5183   vms5184

416

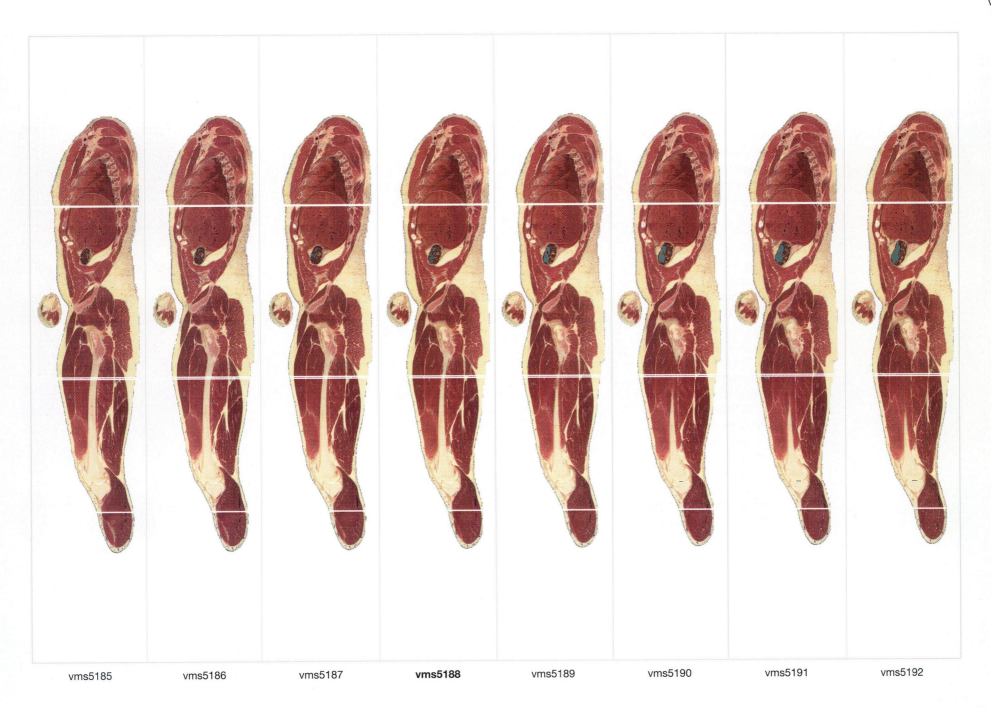

vms5185   vms5186   vms5187   **vms5188**   vms5189   vms5190   vms5191   vms5192

Serratus Anterior Muscle
Clavicle
Subclavius Muscle
Cephalic Vein
Axillary Vessel & Brachial Plexus
Pectoralis Major Muscle
Pectoralis Minor Muscle
Rib 2-3, Intercostal Muscles
Lung, Horizontal Fissure
Lung, Middle Lobe
Rib 4
Liver, Right Lobe
External Oblique Muscle
Transversus Abdominis M.
Ascending Colon
Internal Oblique Muscle
Trapezium Bone
First Dorsal Interosseous M.
Second Metacarpal
Third Metacarpal
Third Dorsal Interosseous M.
Long Flexor Tendons
Fifth Metacarpal
Abductor Digiti Minimi M.
Sartorius Muscle
Rectus Femoris Tendon
Rectus Femoris Muscle
Vastus Intermedius Muscle
Adductor Magnus Muscle
Adductor Canal, Opening
Knee Joint Cavity
Gastrocnemius Muscle,
Medial Head

Omohyoid Muscle, Inferior Belly
Trapezius Muscle
Supraspinatus Muscle
Rib 2
Scapula
Subscapularis Muscle
Trapezius Muscle
Lung, Upper Lobe
Lung, Oblique Fissure
Lung, Lower Lobe
Diaphragm
Iliocostalis Thoracis M.
Rib 10-11, Intercostal M.
Perirenal Fat
Renal Fascia
Latissimus Dorsi Muscle
Internal Oblique Muscle
Transversus Abdominis M.
Ilium, Crest
Iliacus Muscle
Gluteus Medius Muscle
Gluteus Maximus Muscle
Ilium, Body
Femur, Head
Quadratus Femoris Muscle
Femur, Lesser Trochanter
Biceps Femoris Muscle,
Long Head
Semitendinosus Muscle
Semimembranosus Muscle
Femur, Medial Condyle
Knee Joint
Tibia, Medial Condyle

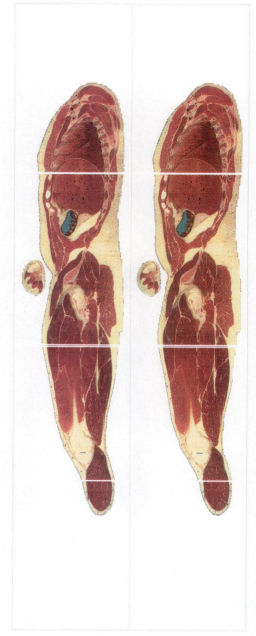

vms5193   vmsv5194

418

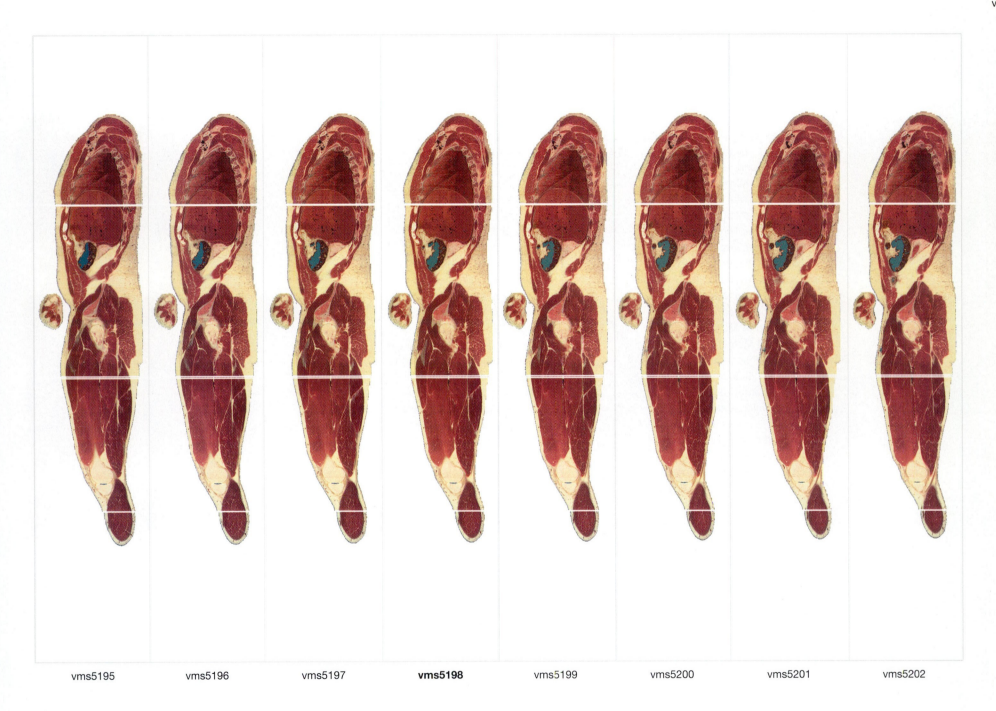

vms5195   vms5196   vms5197   **vms5198**   vms5199   vms5200   vms5201   vms5202

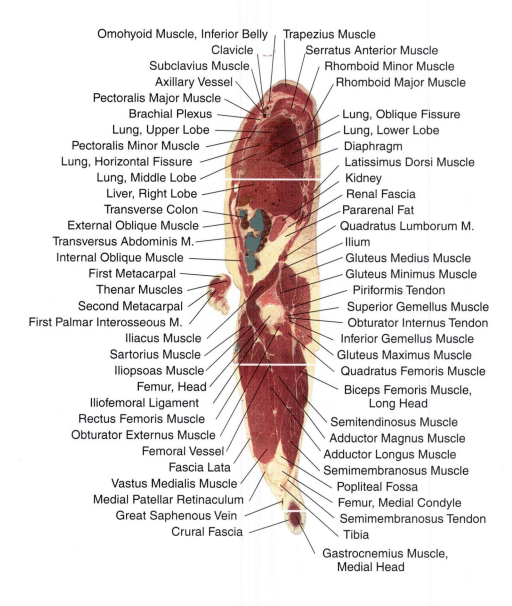

Omohyoid Muscle, Inferior Belly

Trapezius Muscle

Clavicle

Serratus Anterior Muscle

Subclavius Muscle

Rhomboid Minor Muscle

Axillary Vessel

Rhomboid Major Muscle

Pectoralis Major Muscle

Brachial Plexus

Lung, Oblique Fissure

Lung, Upper Lobe

Lung, Lower Lobe

Pectoralis Minor Muscle

Diaphragm

Lung, Horizontal Fissure

Latissimus Dorsi Muscle

Lung, Middle Lobe

Kidney

Liver, Right Lobe

Renal Fascia

Transverse Colon

Pararenal Fat

External Oblique Muscle

Quadratus Lumborum M.

Transversus Abdominis M.

Ilium

Internal Oblique Muscle

Gluteus Medius Muscle

First Metacarpal

Gluteus Minimus Muscle

Thenar Muscles

Piriformis Tendon

Second Metacarpal

Superior Gemellus Muscle

First Palmar Interosseous M.

Obturator Internus Tendon

Iliacus Muscle

Inferior Gemellus Muscle

Sartorius Muscle

Gluteus Maximus Muscle

Iliopsoas Muscle

Quadratus Femoris Muscle

Femur, Head

Biceps Femoris Muscle, Long Head

Iliofemoral Ligament

Rectus Femoris Muscle

Semitendinosus Muscle

Obturator Externus Muscle

Adductor Magnus Muscle

Femoral Vessel

Adductor Longus Muscle

Fascia Lata

Semimembranosus Muscle

Vastus Medialis Muscle

Popliteal Fossa

Medial Patellar Retinaculum

Femur, Medial Condyle

Great Saphenous Vein

Semimembranosus Tendon

Crural Fascia

Tibia

Gastrocnemius Muscle, Medial Head

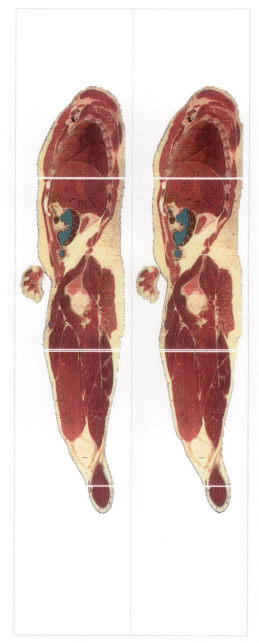

vms5203

vms5204

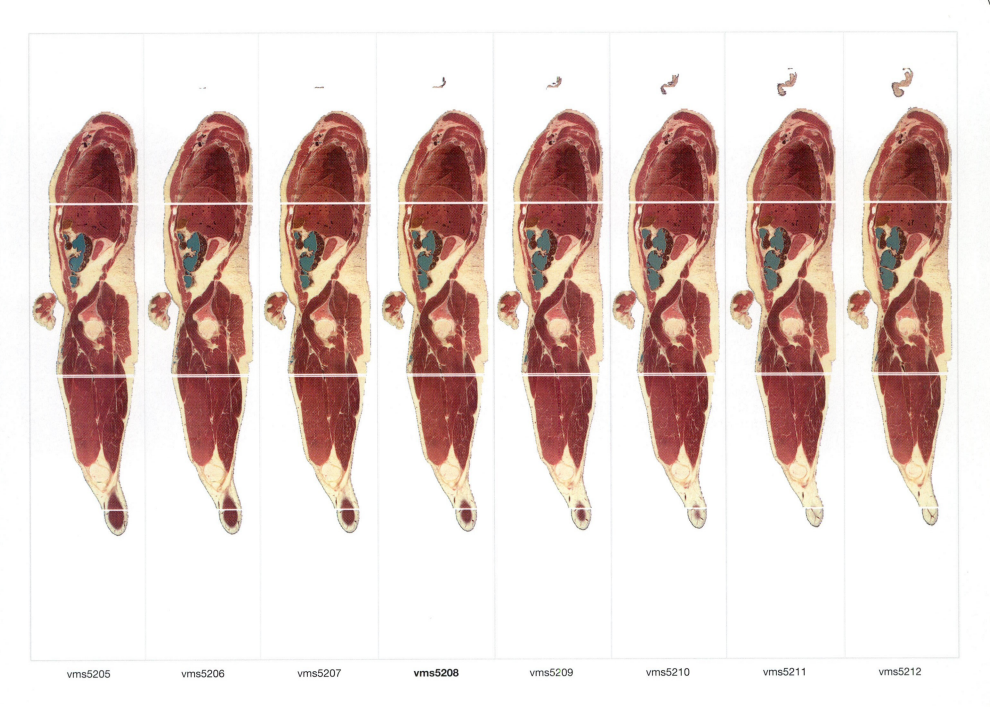

vms5205 vms5206 vms5207 **vms5208** vms5209 vms5210 vms5211 vms5212

Superficial Fascia
Parotid Gland
Transverse Cervical Artery
Omohyoid M., Inferior Belly
Subclavian Artery
Clavicle
Subclavian Vein
Pectoralis Major Muscle
Liver, Right Lobe
Pectoralis Major Muscle
External Oblique Muscle
Transverse Colon
Rectus Abdominis Muscle
Internal Oblique Muscle
Tendinous Intersection
Ascending Colon
First Metacarpal
Thenar Muscles
First Dorsal
Interosseous Muscle
Second Metacarpal
Third Metacarpal
Internal Oblique Muscle
Transversus Abdominis M.
Sartorius Muscle
Femur, Head
Obturator Externus Muscle
Adductor Magnus Muscle
Femoral Vessel
Vastus Medialis Muscle
Medial Patellar Retinaculum
Great Saphenous Vein

External Acoustic Meatus
Auricular Cartilage
Superficial Temporal Vein
Trapezius Muscle
Levator Scapulae Muscle
Serratus Anterior Muscle
Rib 3
Rhomboid Major Muscle
Lung, Upper Lobe
Lung, Oblique Fissure
Lung, Lower Lobe
Diaphragm
Latissimus Dorsi Muscle
Iliocostalis Muscle
Renal Fascia
Kidney
Quadratus Lumborum M.
Gluteus Medius Muscle
Iliacus Muscle
Ilium
Piriformis Muscle
Obturator Internus
Tendon
Gluteus Maximus Muscle
Ischial Tuberosity
Semitendinosus Muscle
Semimembranosus Muscle
Femur, Medial Condyle
Gracilis Tendon
Sartorius Tendon
Small Saphenous Vein

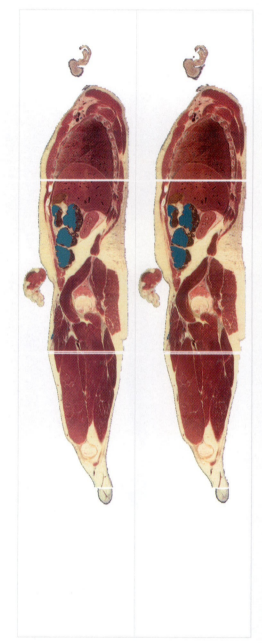

vms5213   vms5214

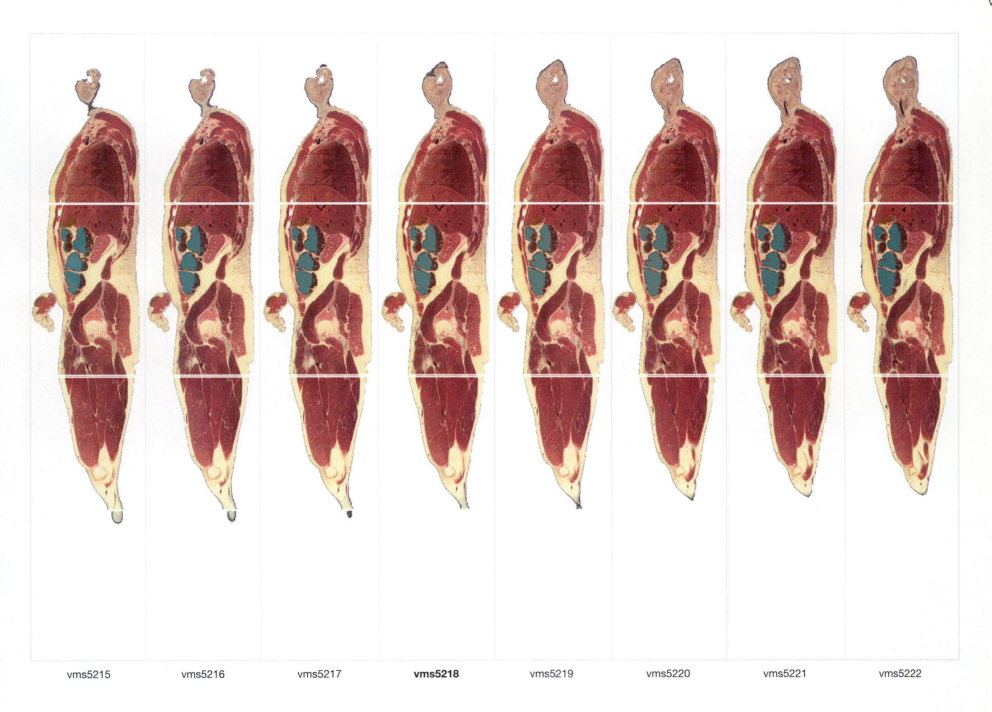

vms5215   vms5216   vms5217   **vms5218**   vms5219   vms5220   vms5221   vms5222

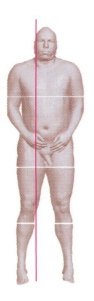

Temporalis Muscle — Galea Aponeurotica
Frontal Vein — Skull, Periosteum
Superficial Temporal Vein — Occipitofrontalis Muscle,
Masseter Muscle — Occipital Belly
External Jugular Vein — External Acoustic Meatus
Platysma Muscle — Auricular Cartilage
Sternocleidomastoid Muscle — Trapezius Muscle
Omohyoid M., Inferior Belly — Levator Scapulae Muscle
Clavicle — Scalenus Medius Muscle
Pectoralis Major Muscle — Rib 3
Subclavian Vein — Subclavian Artery
Lung, Oblique Fissure — Lung, Upper Lobe
Liver, Right Lobe — Lung, Lower Lobe
Transverse Colon — Diaphragm
Ascending Colon — Latissimus Dorsi Muscle
Rectus Abdominis Muscle — Gall Bladder
Tendinous Intersection — Kidney
First Metacarpal — Quadratus Lumborum M.
Thenar Muscles — Psoas Major Muscle
First Dorsal — Gluteus Medius Muscle
Interosseous Muscle — Iliacus Muscle
Cecum — Ilium
Inguinal Canal — Piriformis Muscle
Internal Oblique and — Gluteus Maximus Muscle
Transversus Abdominis M. — Femur, Head
Sartorius Muscle — Ischial Tuberosity
Obturator Externus Muscle — Semitendinosus Muscle
Adductor Magnus Muscle — Semimembranosus Muscle
Vastus Medialis Muscle — Gracilis Muscle
Femoral Vessel — Sartorius Muscle
Great Saphenous Vein

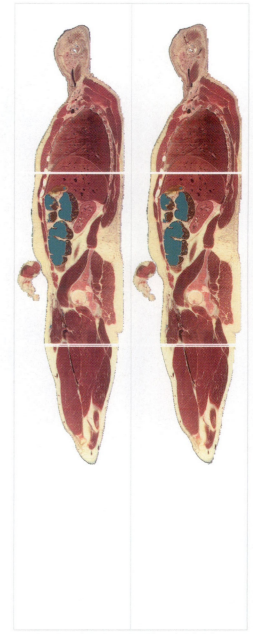

vms5223   vms5224

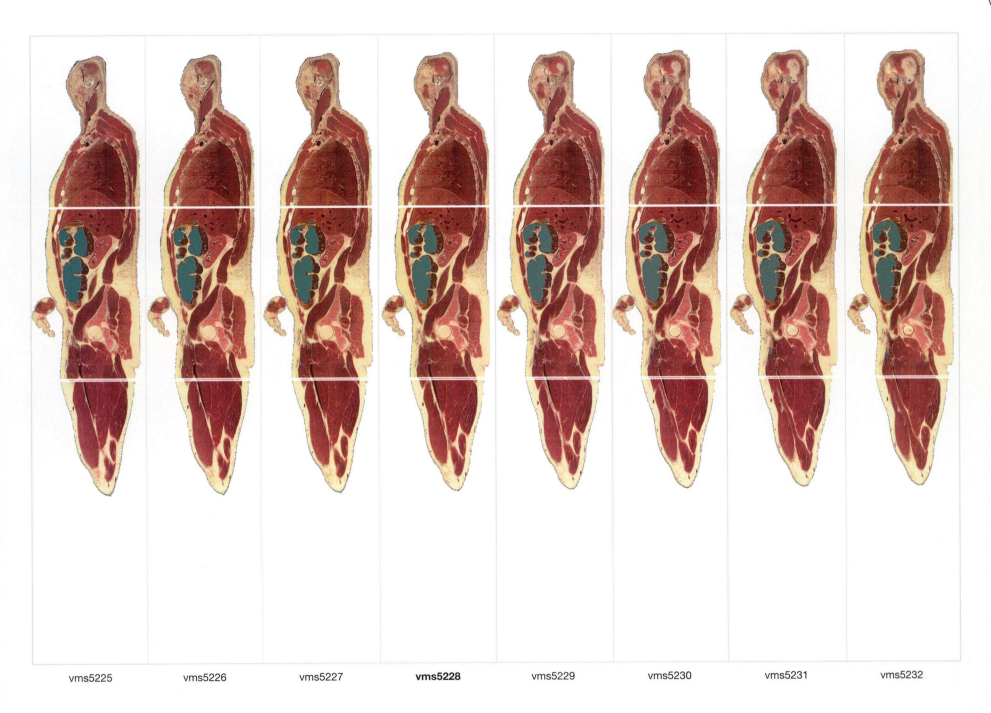

vms5225   vms5226   vms5227   **vms5228**   vms5229   vms5230   vms5231   vms5232

Frontal Bone
Temporalis Muscle
Orbicularis Oculi Muscle
Superficial Temporal Vein
Masseter Muscle
Facial Vein
Platysma Muscle
Sternocleidomastoid Muscle
Communicating Vein
Scalenus Medius Muscle
Brachial Plexus
Subclavian Vein
Pectoralis Major Muscle
Diaphragm
Colon, Hepatic Flexure
Portal Vein
Ascending Colon
Tendinous Intersection
Rectus Abdominis Muscle
First Metacarpal, Head
Second Metacarpal
Thenar Muscles
External Oblique Aponeurosis
Internal Oblique Muscle
Femoral Vein
Iliacus Muscle
Pubis
Pectineus Muscle
Great Saphenous Vein
Adductor Longus Muscle
Sartorius Muscle
Vastus Medialis Muscle
Great Saphenous Vein

Parietal Bone
Lateral Fissure
Occipitofrontalis Muscle,
    Occipital Belly
Temporal Lobe
External Acoustic Meatus
Auricular Cartilage
Splenius Capitis Muscle
Trapezius Muscle
Levator Scapulae Muscle
Rhomboid Major Muscle
Trapezius Muscle
Lung, Upper Lobe
Iliocostalis Thoracis M.
Liver, Caudate Lobe
Kidney
Latissimus Dorsi Muscle
Iliocostalis Muscle
Quadratus Lumborum M.
Gluteus Medius Muscle
Psoas Major Muscle
Ilium
Piriformis Muscle
Gluteus Maximus Muscle
Obturator Internus Muscle
Ischium
Semitendinosus Muscle
Obturator Externus Muscle
Adductor Brevis Muscle
Semimembranosus Muscle
Adductor Magnus Muscle
Gracilis Muscle

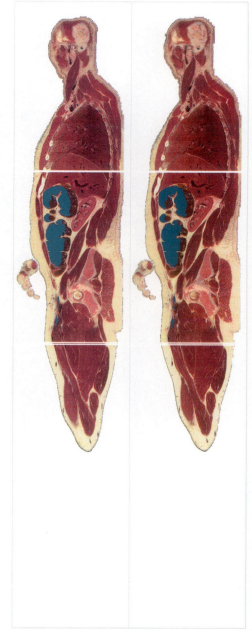

vms5233   vms5234

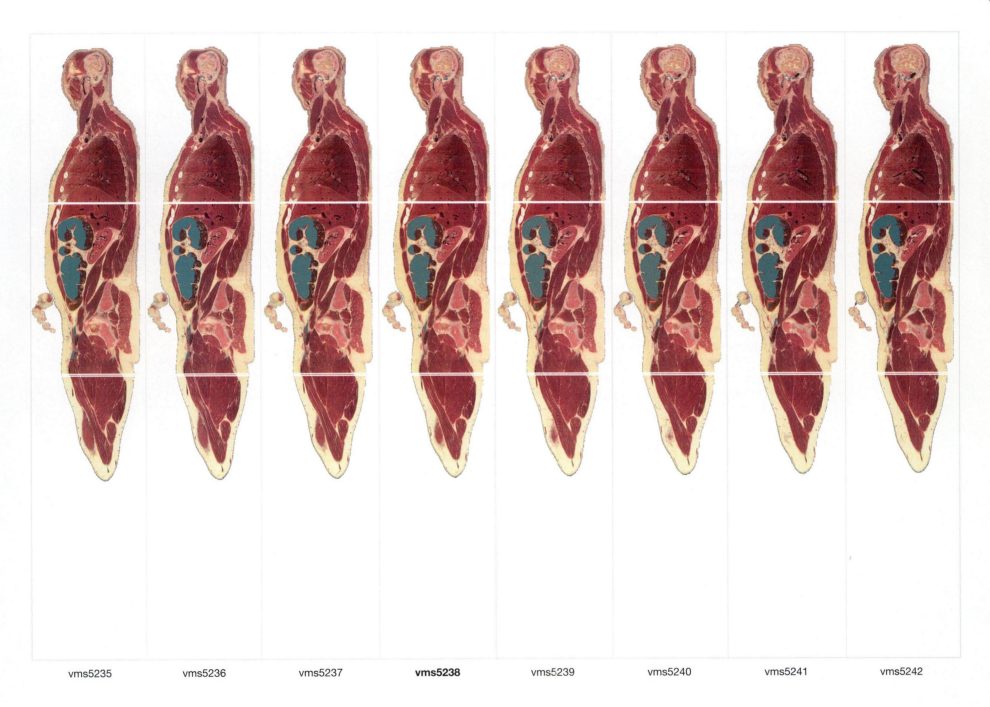

vms5235 vms5236 vms5237 **vms5238** vms5239 vms5240 vms5241 vms5242

Frontal Bone — Galea Aponeurotica
Frontal Lobe — Parietal Bone
Temporalis Muscle — Parietal Lobe
Lateral Pterygoid Muscle — Lateral Fissure
Masseter Muscle — Transverse Sinus
Facial Vein — Cerebellum
Digastric M., Posterior Belly — Semispinalis Capitis M.
Internal Jugular Vein — Trapezius Muscle
Sternocleidomastoid M. — Levator Scapulae Muscle
Scalenus Anterior Muscle — Scalenus Posterior Muscle
Clavicle — Brachial Plexus
Pectoralis Major Muscle — Lung, Upper Lobe
Lung, Middle Lobe — Pulmonary Artery
Liver, Right Lobe — Diaphragm
Colon, Hepatic Flexure — Adrenal Gland
Ascending Colon — Kidney
Tendinous Intersection — Duodenum
Rectus Abdominis Muscle — Longissimus Muscle
Cecum — Psoas Major Muscle
Inguinal Canal, Internal Ring — Ilium
Spermatic Cord — Sacrum, Lateral Wing
Femoral Vein — Piriformis Muscle
Femoral Vessel — Obturator Vessel
Pubis — Gluteus Maximus Muscle
Pectineus Muscle — Obturator Internus Muscle
Adductor Brevis Muscle — Ischium
Adductor Longus Muscle — Obturator Externus Muscle
Sartorius Muscle — Semimembranosus Muscle
Superficial Fascia — Adductor Magnus Muscle
Great Saphenous Vein — Gracilis Muscle

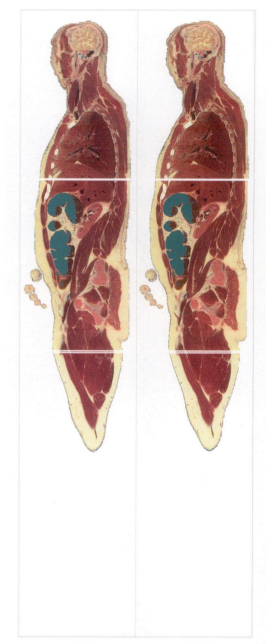

vms5243   vms5244

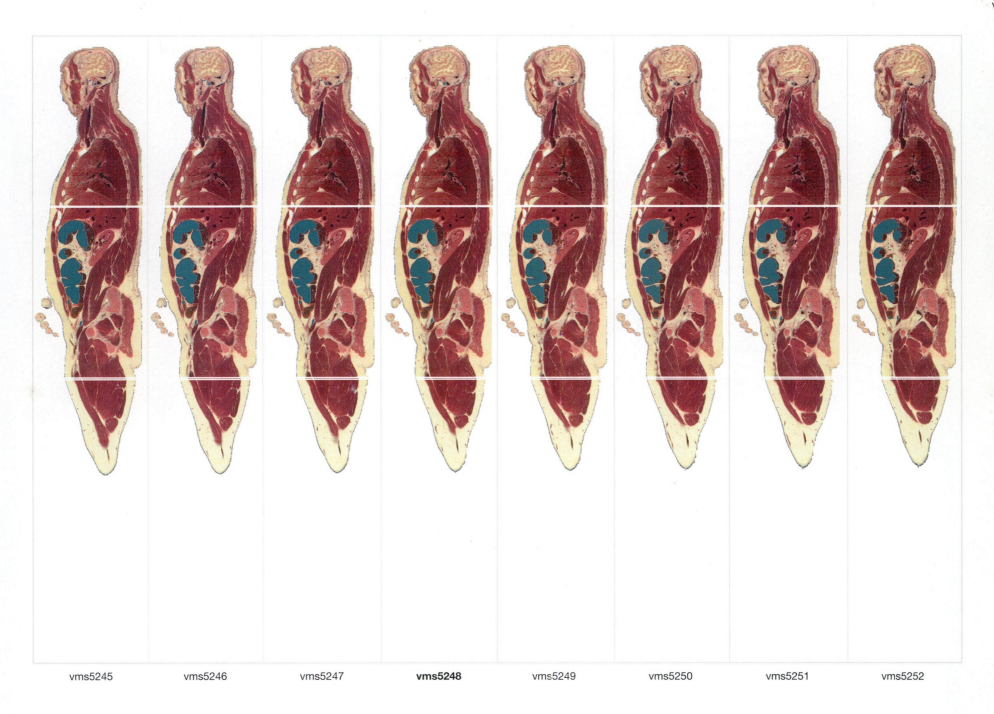

vms5245 vms5246 vms5247 **vms5248** vms5249 vms5250 vms5251 vms5252

Frontal Bone — Galea Aponeurotica
Frontal Lobe — Parietal Bone
Frontal Sinus — Parietal Lobe
Eye Globe — Lateral Fissure
Orbicularis Oculi Muscle — Transverse Sinus
Maxillary Sinus — Tentorium
Temporal Lobe — Cerebellum
Orbicularis Oris Muscle — Sigmoid Sinus
Buccinator Muscle — Obliquus Capitis Inferior M.
Mandible — Trapezius Muscle
Medial Pterygoid Muscle — Lateral Pterygoid Muscle
Common Carotid Artery — Styloglossus Muscle
Sternocleidomastoid M. — Levator Scapulae Muscle
Brachiocephalic Vein — Lung, Upper Lobe
Lung, Middle Lobe — Pulmonary Artery
Right Atrium — Diaphragm
Secondary Bronchus — Adrenal Gland
Liver, Right Lobe — Kidney
Transverse Colon — Duodenum
Ascending Colon — Iliocostalis Muscle
Ileum — Psoas Major Muscle
Rectus Abdominis Muscle — Ilium
Spermatic Cord — Sacroiliac Joint
Pubis, Superior Ramus — Sacrum, Lateral Wing
Pectineus Muscle — Gluteus Maximus Muscle
Adductor Longus Muscle — Obturator Internus Muscle
Adductor Brevis Muscle — Ischium
Vastus Medialis Muscle — Obturator Externus Muscle
Great Saphenous Vein — Adductor Magnus Muscle
Superficial Fascia — Gracilis Muscle

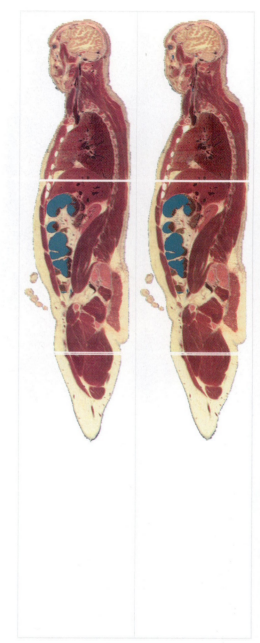

vms5253   vms5254

430

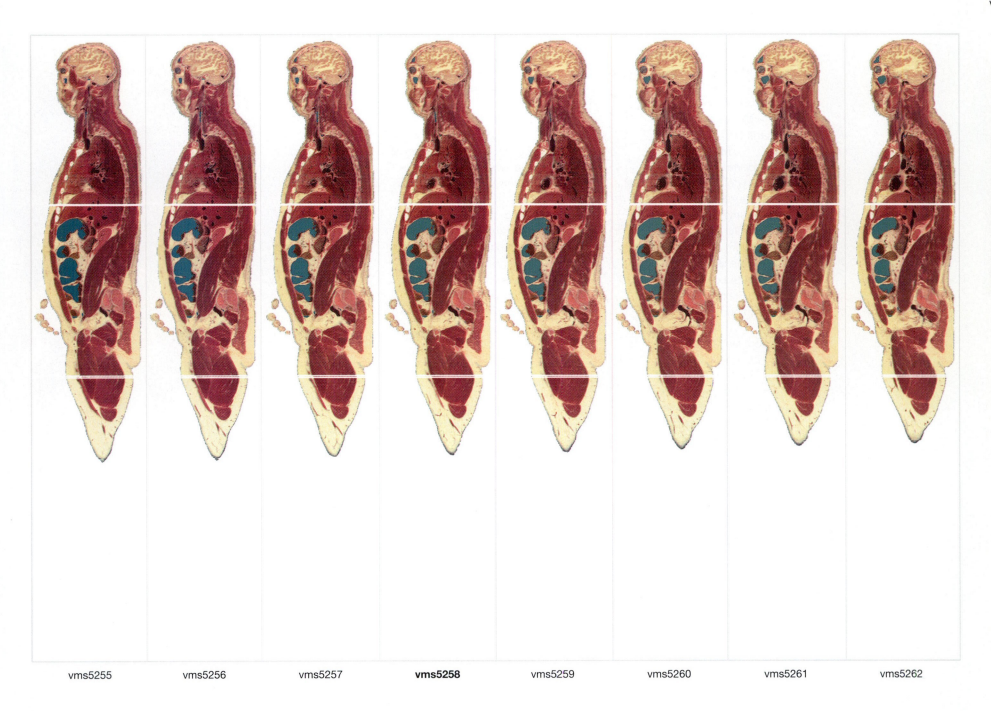

vms5255   vms5256   vms5257   **vms5258**   vms5259   vms5260   vms5261   vms5262

Frontal Bone — Galea Aponeurotica
Frontal Lobe — Insula
Frontal Sinus — Lateral Fissure
Eye Globe — Transverse Sinus
Orbicularis Oculi Muscle — Tentorium
Maxillary Sinus — Cerebellum
Orbicularis Oris Muscle — Sigmoid Sinus
Mandible — Obliquus Capitis Inferior M.
Buccinator Muscle — Trapezius Muscle
Medial Pterygoid Muscle — Semispinalis Cervicis M.
Common Carotid Artery — Thyroid Gland
Sternocleidomastoid M. — Lung, Upper Lobe
Superior Vena Cava — Primary Bronchus
Right Atrium — Pulmonary Artery
External Oblique Muscle — Diaphragm
Liver, Quadrate Lobe — Adrenal Gland
Colon, Hepatic Flexure — Portal Vein
Ascending Colon — Inferior Vena Cava
Ileum — Longissimus Muscle
Rectus Abdominis Muscle — Duodenum
Spermatic Cord — Ilium
Pubis, Superior Ramus — Psoas Major Muscle
Pectineus Muscle — Piriformis Muscle
Adductor Brevis Muscle — Internal Iliac Vein
Adductor Longus Muscle — Gluteus Maximus Muscle
Fascia Lata — Bladder
Superficial Fascia — Adductor Magnus Muscle
Gracilis Muscle

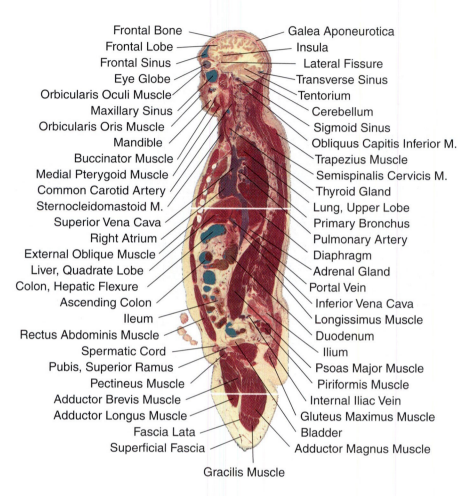

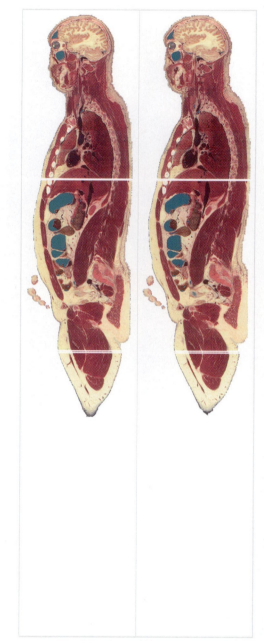

vms5263   vms5264

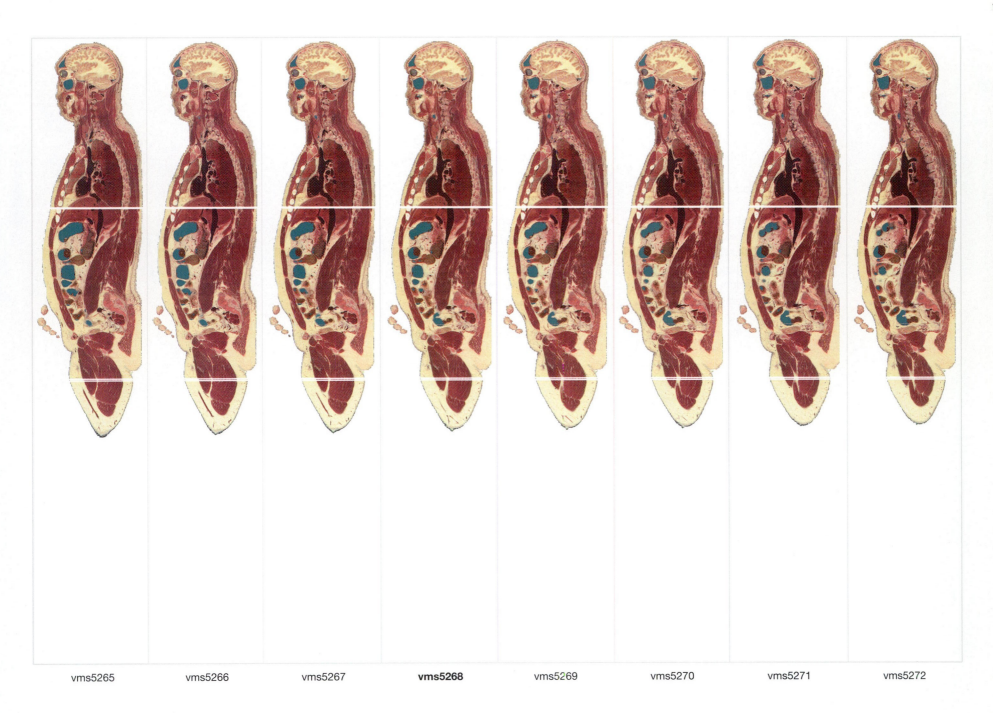

vms5265 vms5266 vms5267 **vms5268** vms5269 vms5270 vms5271 vms5272

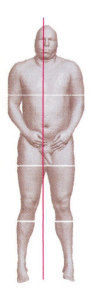

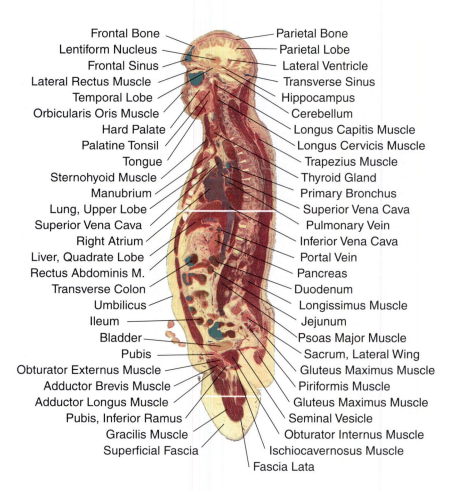

Frontal Bone — Parietal Bone
Lentiform Nucleus — Parietal Lobe
Frontal Sinus — Lateral Ventricle
Lateral Rectus Muscle — Transverse Sinus
Temporal Lobe — Hippocampus
Orbicularis Oris Muscle — Cerebellum
Hard Palate — Longus Capitis Muscle
Palatine Tonsil — Longus Cervicis Muscle
Tongue — Trapezius Muscle
Sternohyoid Muscle — Thyroid Gland
Manubrium — Primary Bronchus
Lung, Upper Lobe — Superior Vena Cava
Superior Vena Cava — Pulmonary Vein
Right Atrium — Inferior Vena Cava
Liver, Quadrate Lobe — Portal Vein
Rectus Abdominis M. — Pancreas
Transverse Colon — Duodenum
Umbilicus — Longissimus Muscle
Ileum — Jejunum
Bladder — Psoas Major Muscle
Pubis — Sacrum, Lateral Wing
Obturator Externus Muscle — Gluteus Maximus Muscle
Adductor Brevis Muscle — Piriformis Muscle
Adductor Longus Muscle — Gluteus Maximus Muscle
Pubis, Inferior Ramus — Seminal Vesicle
Gracilis Muscle — Obturator Internus Muscle
Superficial Fascia — Ischiocavernosus Muscle
Fascia Lata

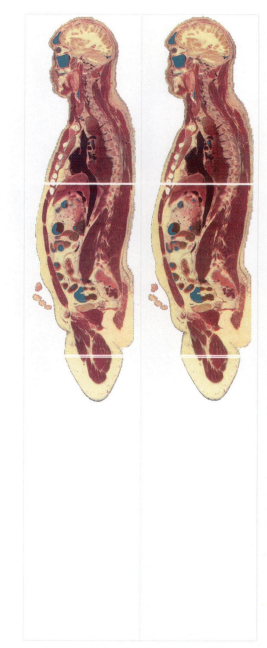

vms5273   vms5274

434

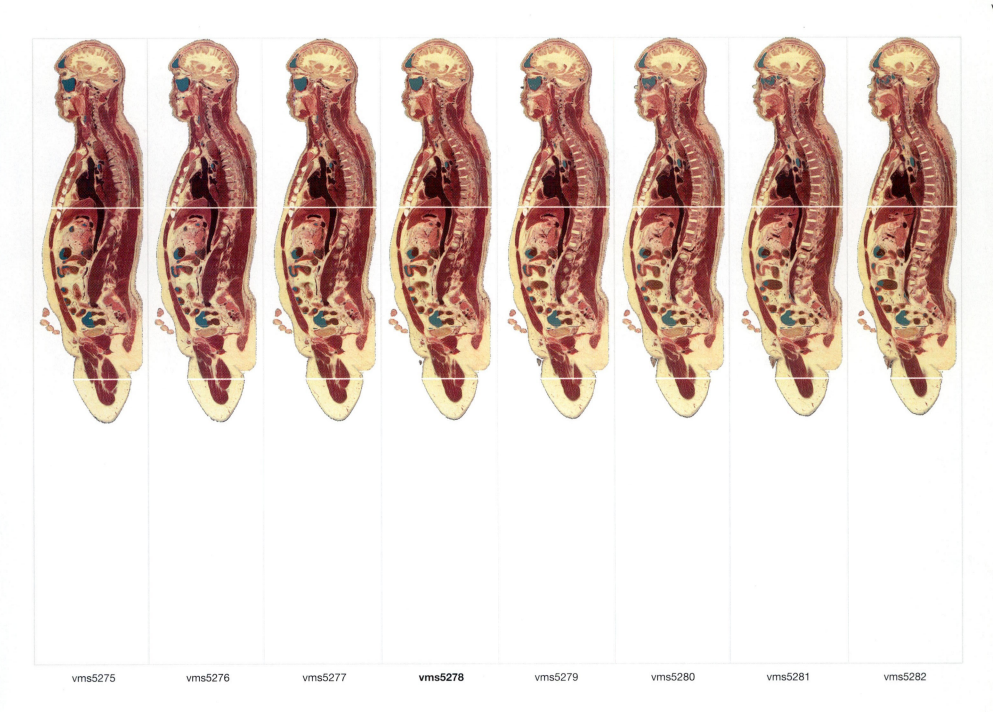

vms5275 vms5276 vms5277 **vms5278** vms5279 vms5280 vms5281 vms5282

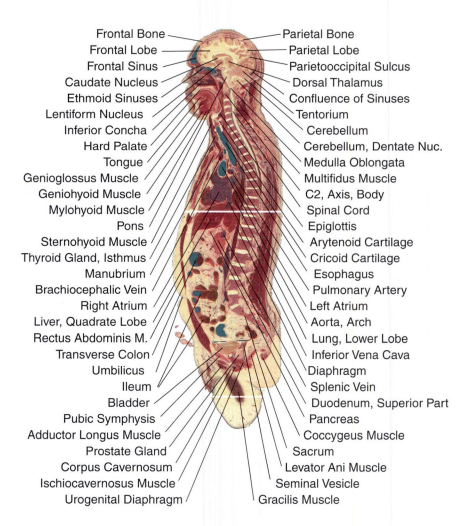

Frontal Bone — — Parietal Bone
Frontal Lobe — — Parietal Lobe
Frontal Sinus — — Parietooccipital Sulcus
Caudate Nucleus — — Dorsal Thalamus
Ethmoid Sinuses — — Confluence of Sinuses
Lentiform Nucleus — — Tentorium
Inferior Concha — — Cerebellum
Hard Palate — — Cerebellum, Dentate Nuc.
Tongue — — Medulla Oblongata
Genioglossus Muscle — — Multifidus Muscle
Geniohyoid Muscle — — C2, Axis, Body
Mylohyoid Muscle — — Spinal Cord
Pons — — Epiglottis
Sternohyoid Muscle — — Arytenoid Cartilage
Thyroid Gland, Isthmus — — Cricoid Cartilage
Manubrium — — Esophagus
Brachiocephalic Vein — — Pulmonary Artery
Right Atrium — — Left Atrium
Liver, Quadrate Lobe — — Aorta, Arch
Rectus Abdominis M. — — Lung, Lower Lobe
Transverse Colon — — Inferior Vena Cava
Umbilicus — — Diaphragm
Ileum — — Splenic Vein
Bladder — — Duodenum, Superior Part
Pubic Symphysis — — Pancreas
Adductor Longus Muscle — — Coccygeus Muscle
Prostate Gland — — Sacrum
Corpus Cavernosum — — Levator Ani Muscle
Ischiocavernosus Muscle — — Seminal Vesicle
Urogenital Diaphragm — — Gracilis Muscle

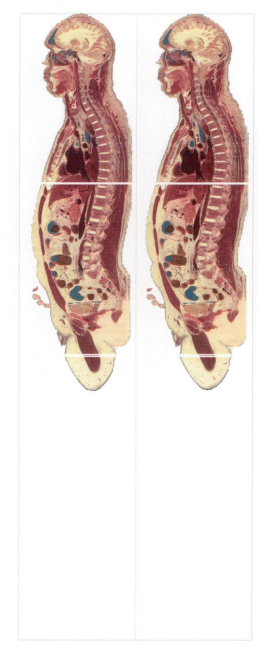

vms5283   vms5284

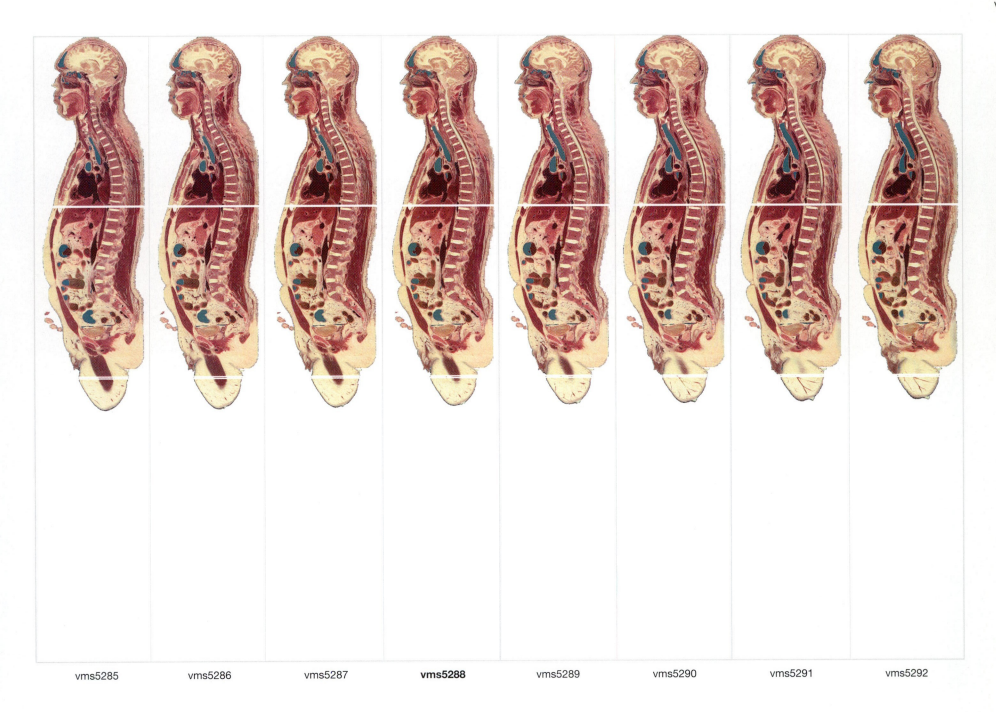

vms5285   vms5286   vms5287   **vms5288**   vms5289   vms5290   vms5291   vms5292

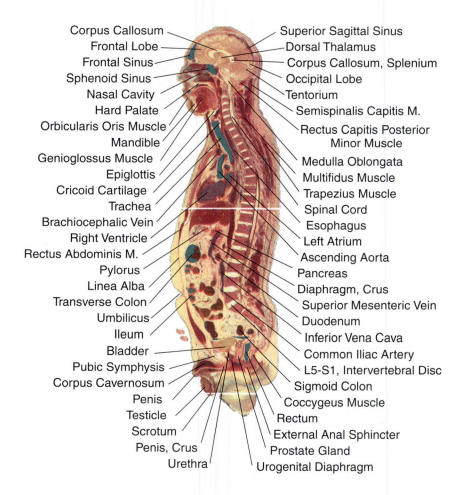

Corpus Callosum

Frontal Lobe

Frontal Sinus

Sphenoid Sinus

Nasal Cavity

Hard Palate

Orbicularis Oris Muscle

Mandible

Genioglossus Muscle

Epiglottis

Cricoid Cartilage

Trachea

Brachiocephalic Vein

Right Ventricle

Rectus Abdominis M.

Pylorus

Linea Alba

Transverse Colon

Umbilicus

Ileum

Bladder

Pubic Symphysis

Corpus Cavernosum

Penis

Testicle

Scrotum

Penis, Crus

Urethra

Superior Sagittal Sinus

Dorsal Thalamus

Corpus Callosum, Splenium

Occipital Lobe

Tentorium

Semispinalis Capitis M.

Rectus Capitis Posterior
Minor Muscle

Medulla Oblongata

Multifidus Muscle

Trapezius Muscle

Spinal Cord

Esophagus

Left Atrium

Ascending Aorta

Pancreas

Diaphragm, Crus

Superior Mesenteric Vein

Duodenum

Inferior Vena Cava

Common Iliac Artery

L5-S1, Intervertebral Disc

Sigmoid Colon

Coccygeus Muscle

Rectum

External Anal Sphincter

Prostate Gland

Urogenital Diaphragm

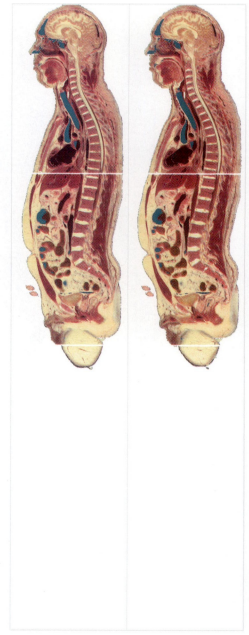

vms5293   vms5294

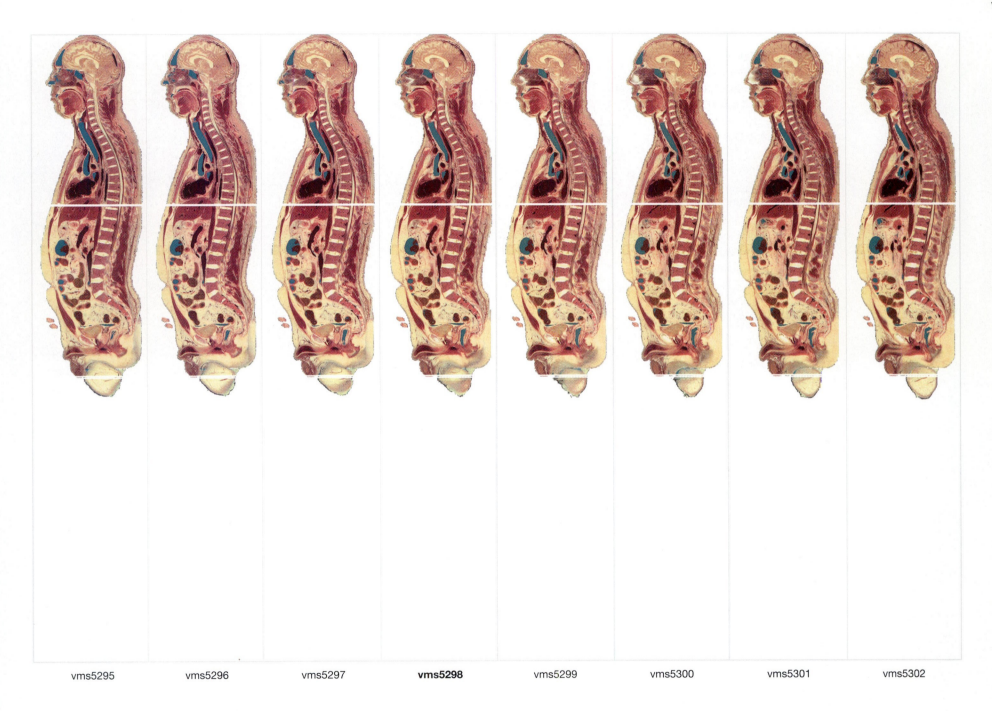

vms5295 vms5296 vms5297 **vms5298** vms5299 vms5300 vms5301 vms5302

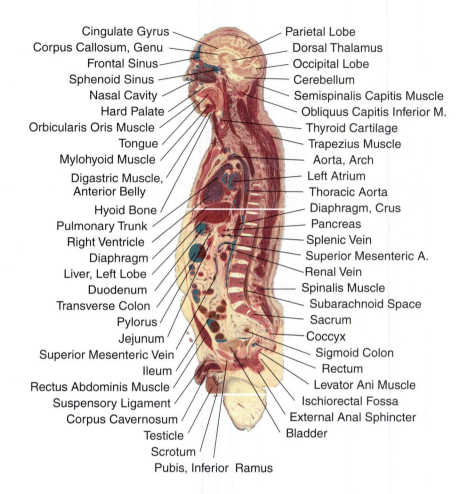

Cingulate Gyrus

Corpus Callosum, Genu

Frontal Sinus

Sphenoid Sinus

Nasal Cavity

Hard Palate

Orbicularis Oris Muscle

Tongue

Mylohyoid Muscle

Digastric Muscle, Anterior Belly

Hyoid Bone

Pulmonary Trunk

Right Ventricle

Diaphragm

Liver, Left Lobe

Duodenum

Transverse Colon

Pylorus

Jejunum

Superior Mesenteric Vein

Ileum

Rectus Abdominis Muscle

Suspensory Ligament

Corpus Cavernosum

Testicle

Scrotum

Pubis, Inferior Ramus

Parietal Lobe

Dorsal Thalamus

Occipital Lobe

Cerebellum

Semispinalis Capitis Muscle

Obliquus Capitis Inferior M.

Thyroid Cartilage

Trapezius Muscle

Aorta, Arch

Left Atrium

Thoracic Aorta

Diaphragm, Crus

Pancreas

Splenic Vein

Superior Mesenteric A.

Renal Vein

Spinalis Muscle

Subarachnoid Space

Sacrum

Coccyx

Sigmoid Colon

Rectum

Levator Ani Muscle

Ischiorectal Fossa

External Anal Sphincter

Bladder

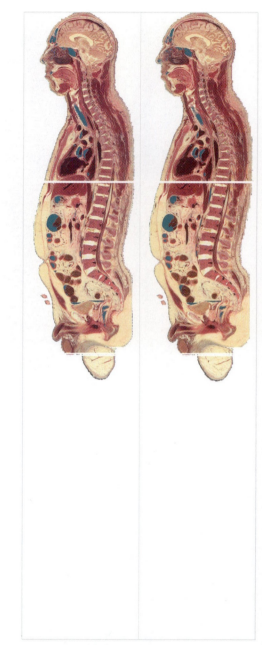

vms5303

vms5304

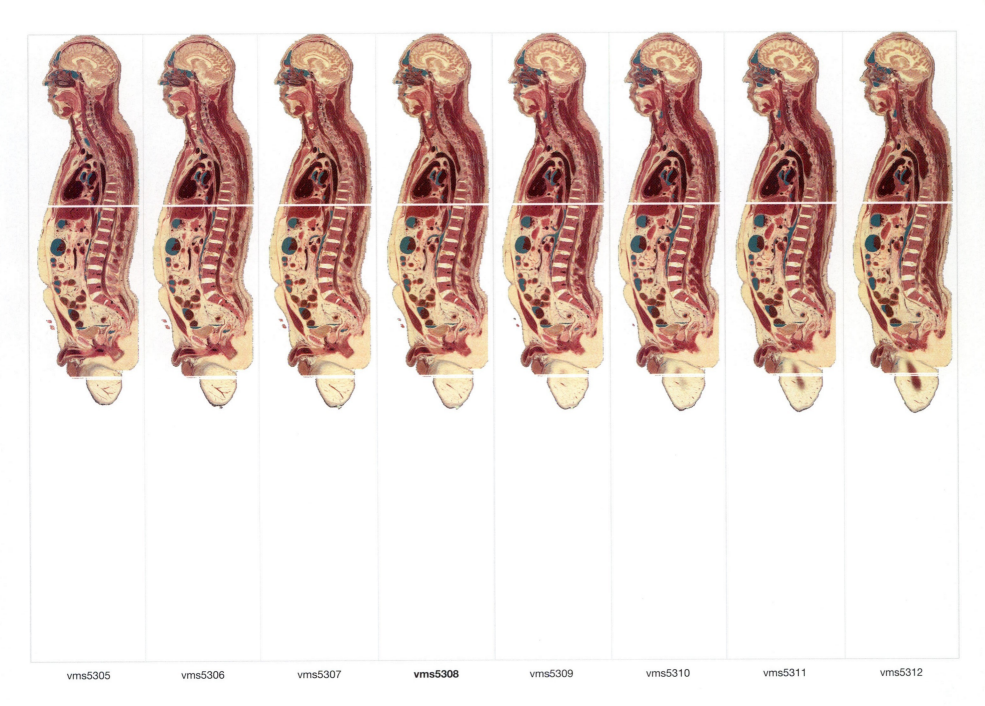

vms5305 vms5306 vms5307 **vms5308** vms5309 vms5310 vms5311 vms5312

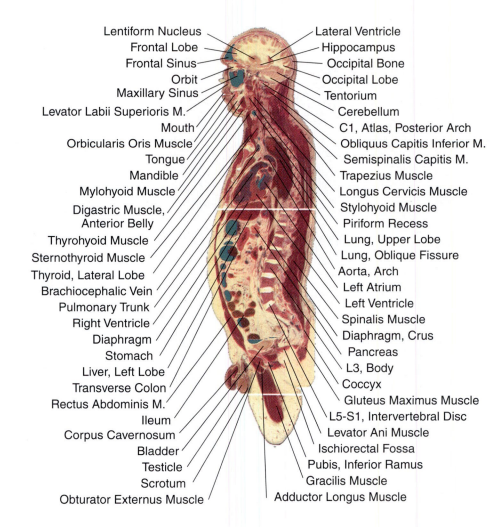

Lentiform Nucleus — Lateral Ventricle
Frontal Lobe — Hippocampus
Frontal Sinus — Occipital Bone
Orbit — Occipital Lobe
Maxillary Sinus — Tentorium
Levator Labii Superioris M. — Cerebellum
Mouth — C1, Atlas, Posterior Arch
Orbicularis Oris Muscle — Obliquus Capitis Inferior M.
Tongue — Semispinalis Capitis M.
Mandible — Trapezius Muscle
Mylohyoid Muscle — Longus Cervicis Muscle
Digastric Muscle, Anterior Belly — Stylohyoid Muscle
Thyrohyoid Muscle — Piriform Recess
Sternothyroid Muscle — Lung, Upper Lobe
Thyroid, Lateral Lobe — Lung, Oblique Fissure
Brachiocephalic Vein — Aorta, Arch
Pulmonary Trunk — Left Atrium
Right Ventricle — Left Ventricle
Diaphragm — Spinalis Muscle
Stomach — Diaphragm, Crus
Liver, Left Lobe — Pancreas
Transverse Colon — L3, Body
Rectus Abdominis M. — Coccyx
Ileum — Gluteus Maximus Muscle
Corpus Cavernosum — L5-S1, Intervertebral Disc
Bladder — Levator Ani Muscle
Testicle — Ischiorectal Fossa
Scrotum — Pubis, Inferior Ramus
Obturator Externus Muscle — Gracilis Muscle
Adductor Longus Muscle

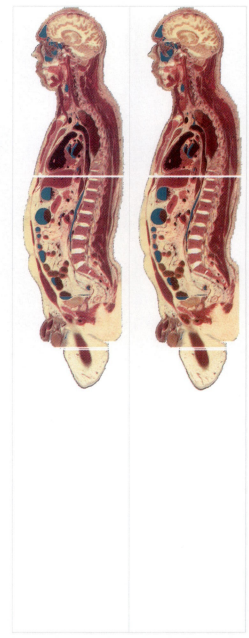

vms5313   vms5314

442

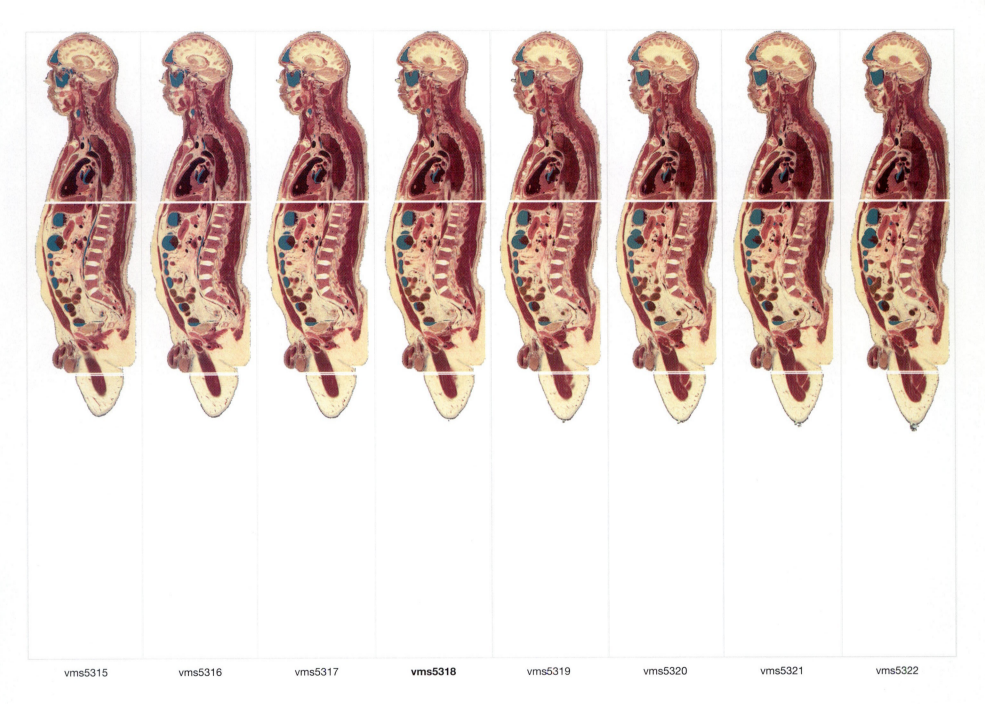

vms5315   vms5316   vms5317   **vms5318**   vms5319   vms5320   vms5321   vms5322

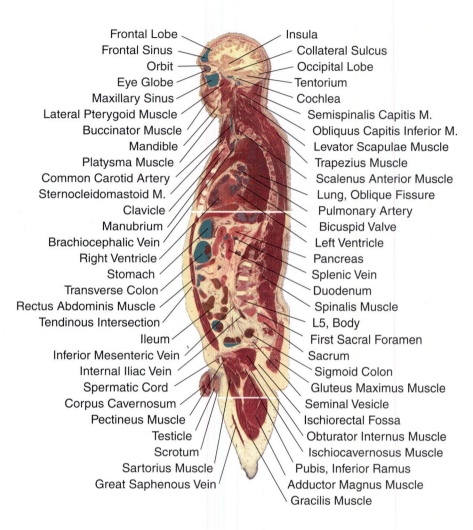

Frontal Lobe — Insula
Frontal Sinus — Collateral Sulcus
Orbit — Occipital Lobe
Eye Globe — Tentorium
Maxillary Sinus — Cochlea
Lateral Pterygoid Muscle — Semispinalis Capitis M.
Buccinator Muscle — Obliquus Capitis Inferior M.
Mandible — Levator Scapulae Muscle
Platysma Muscle — Trapezius Muscle
Common Carotid Artery — Scalenus Anterior Muscle
Sternocleidomastoid M. — Lung, Oblique Fissure
Clavicle — Pulmonary Artery
Manubrium — Bicuspid Valve
Brachiocephalic Vein — Left Ventricle
Right Ventricle — Pancreas
Stomach — Splenic Vein
Transverse Colon — Duodenum
Rectus Abdominis Muscle — Spinalis Muscle
Tendinous Intersection — L5, Body
Ileum — First Sacral Foramen
Inferior Mesenteric Vein — Sacrum
Internal Iliac Vein — Sigmoid Colon
Spermatic Cord — Gluteus Maximus Muscle
Corpus Cavernosum — Seminal Vesicle
Pectineus Muscle — Ischiorectal Fossa
Testicle — Obturator Internus Muscle
Scrotum — Ischiocavernosus Muscle
Sartorius Muscle — Pubis, Inferior Ramus
Great Saphenous Vein — Adductor Magnus Muscle
Gracilis Muscle

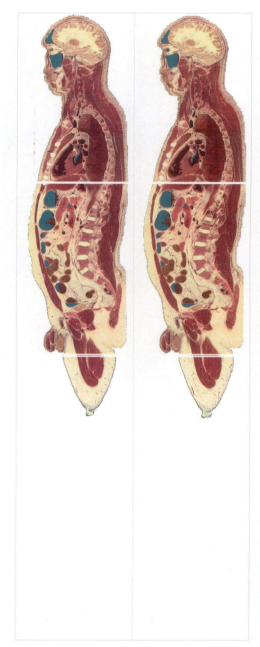

vms5323   vms5324

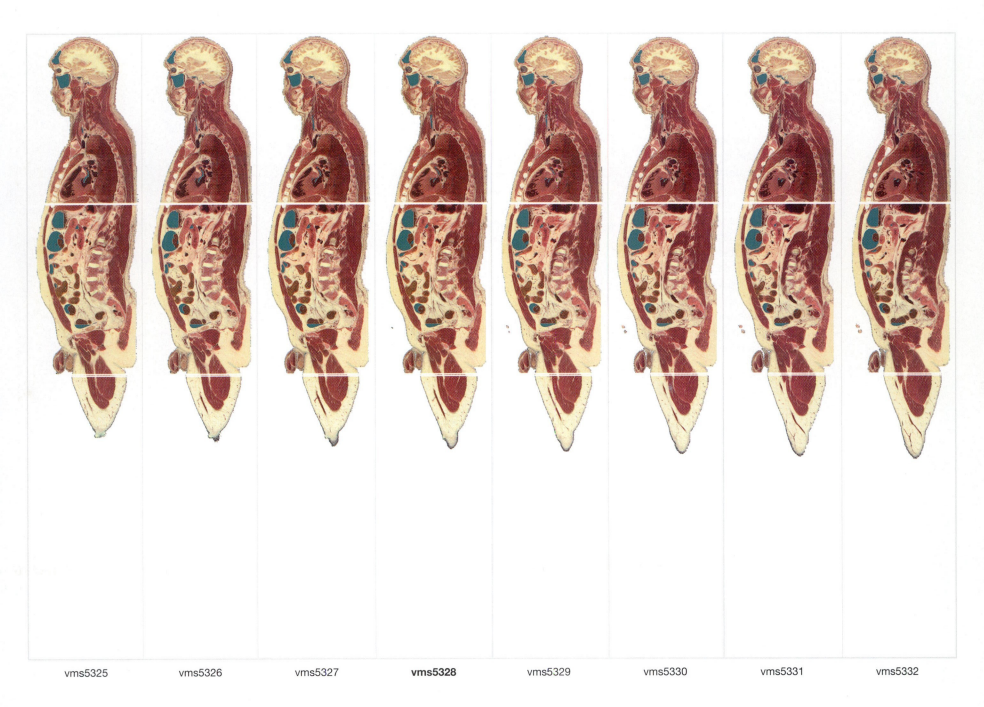

vms5325 vms5326 vms5327 **vms5328** vms5329 vms5330 vms5331 vms5332

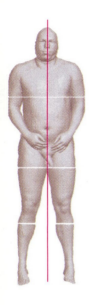

Frontal Bone — Parietal Lobe
Frontal Sinus — Lateral Fissure
Orbit — Tentorium
Temporalis Muscle — Cerebellum
Maxillary Sinus — Mastoid Air Cells
Masseter Muscle — Splenius Capitis Muscle
Submandibular Gland — Digastric Muscle,
Posterior Belly
Sternocleidomastoid Muscle —
Scalenus Anterior Muscle — Levator Scapulae Muscle
Bronchus — Trapezius Muscle
Left Ventricle — Scalenus Medius and
Posterior Muscles
Right Ventricle —
Stomach — Brachial Plexus
Transverse Colon — Pulmonary Artery
Jejunum — Diaphragm
Rectus Abdominis Muscle — Spleen
Ileum — Pancreas
Descending Colon — Kidney
Sigmoid Colon — Splenic Vein
Pubis, Superior Ramus — Psoas Major Muscle
Sacrum
Spermatic Cord — Piriformis Muscle
Pectineus Muscle — Internal Iliac Vein
Corpus Cavernosum — Sacrotuberous Ligament
Adductor Longus Muscle — Gluteus Maximus Muscle
Obturator Internus Muscle — Ischiorectal Fossa
Vastus Medialis Muscle — Ischial Tuberosity
Sartorius Muscle — Semimembranosus Muscle
Gracilis Muscle — Adductor Magnus Muscle
Superficial Fascia

Great Saphenous Vein

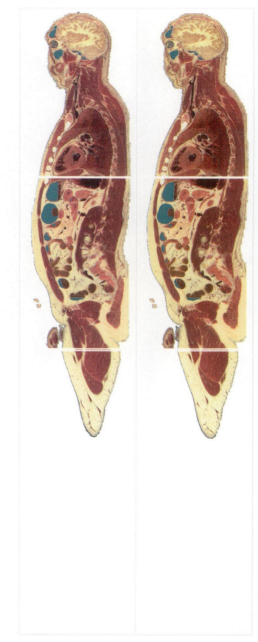

vms5333   vms5334

446

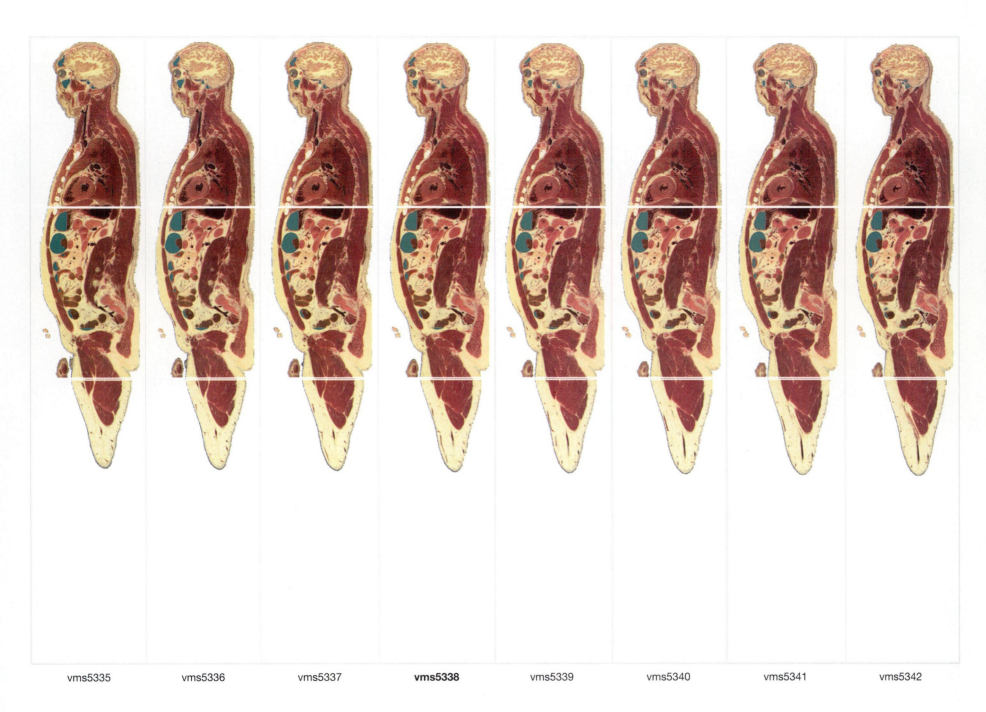

vms5335　　vms5336　　vms5337　　**vms5338**　　vms5339　　vms5340　　vms5341　　vms5342

Frontal Bone — Galea Aponeurotica
Orbit — Parietal Bone
Temporalis Muscle — Parietal Lobe
Zygomatic Bone — Lateral Fissure
Masseter Muscle — Mastoid Air Cells
External Acoustic Meatus — Splenius Capitis Muscle
Parotid Gland — Semispinalis Capitis M.
Retromandibular Vein — Trapezius Muscle
Platysma Muscle — Levator Scapulae Muscle
Sternocleidomastoid Muscle — Scalenus Medius and
Scalenus Anterior Muscle — Posterior Muscles
Clavicle — Brachial Plexus
Subclavian Vein — Lung, Oblique Fissure
Pectoralis Major Muscle — Lung, Lower Lobe
Lung, Lingula — Left Ventricle
Colon, Splenic Flexure — Spleen
Jejunum — Diaphragm
Rectus Abdominis Muscle — Stomach
Ileum — Latissimus Dorsi Muscle
Descending Colon — Pancreas
Sigmoid Colon — Kidney
Spermatic Cord — Multifidus Muscle
Pubis, Superior Ramus — Psoas Major Muscle
— Ilium
Pectineus Muscle — Sacrum
Adductor Longus Muscle — Gluteus Maximus Muscle
Vastus Medialis Muscle — Obturator Internus Muscle
Sartorius Muscle — Ischial Tuberosity
Gracilis Muscle — Obturator Externus Muscle
Superficial Fascia — Adductor Magnus Muscle
Great Saphenous Vein — Semimembranosus Muscle

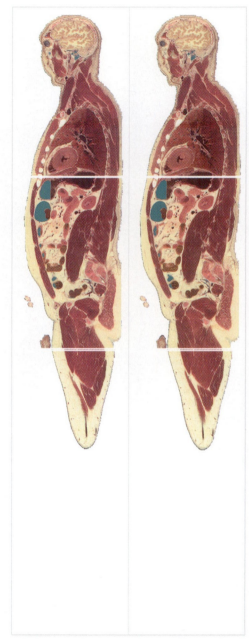

vms5343   vms5344

448

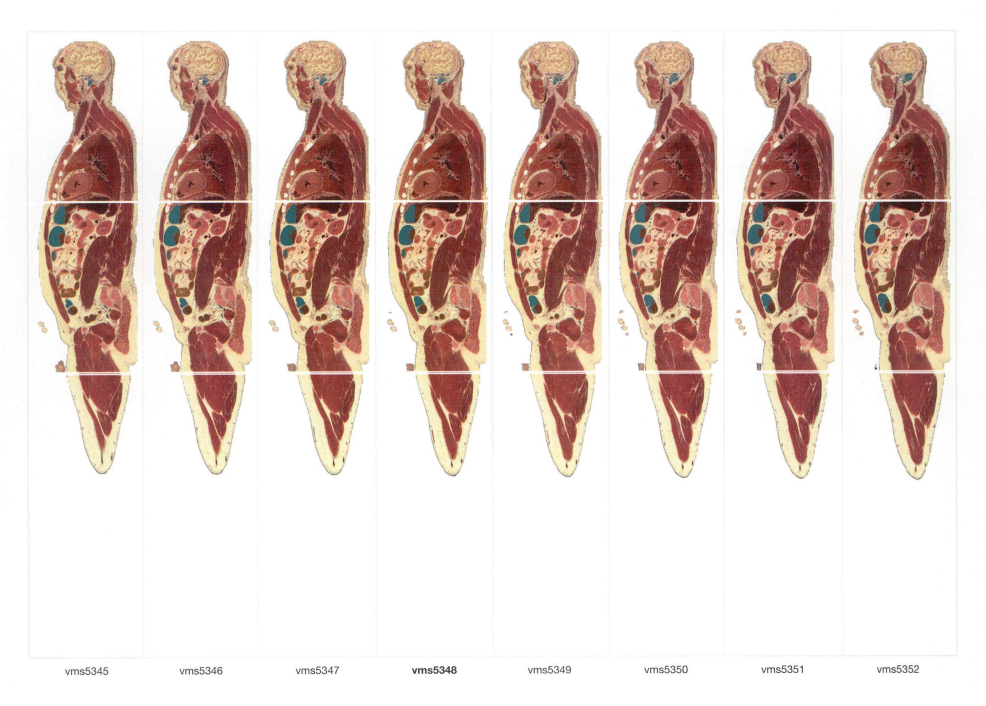

vms5345 vms5346 vms5347 **vms5348** vms5349 vms5350 vms5351 vms5352

Parietal Bone

Frontal Bone

Temporalis Muscle

Zygomatic Bone

Masseter Muscle

External Acoustic Meatus

Parotid Gland

Platysma Muscle

Sternocleidomastoid Muscle

Omohyoid M., Inferior Belly

Clavicle

Subclavian Vein

Pectoralis Major Muscle

Lung, Upper Lobe

Transverse Colon

Jejunum

Rectus Abdominis Muscle

Tendinous Intersection

Ileum

Cecum

Spermatic Cord

Pubis, Superior Ramus

Pectineus Muscle

Obturator Internus Muscle

Adductor Brevis Muscle

Adductor Longus Muscle

Sartorius Muscle

Gracilis Muscle

Vastus Medialis Muscle

Superficial Fascia

Great Saphenous Vein

Lateral Fissure

Temporal Lobe

Occipitofrontalis Muscle,
Occipital Belly

Mastoid Air Cells

Communicating Vein

Levator Scapulae Muscle

Trapezius Muscle

Scalenus Medius and
Posterior Muscles

Rhomboid Major Muscle

Brachial Plexus

Lung, Oblique Fissure

Left Ventricle

Spleen

Stomach

Latissimus Dorsi Muscle

Pancreas

Kidney

Ilium

Psoas Major Muscle

Sacrum

Piriformis Muscle

Gluteus Maximus Muscle

Ischium

Ischial Tuberosity

Obturator Externus Muscle

Fascia Lata

Adductor Magnus Muscle

Semimembranosus Muscle

Great Saphenous Vein

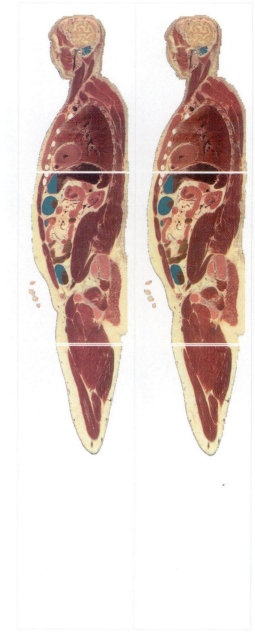

vms5353   vms5354

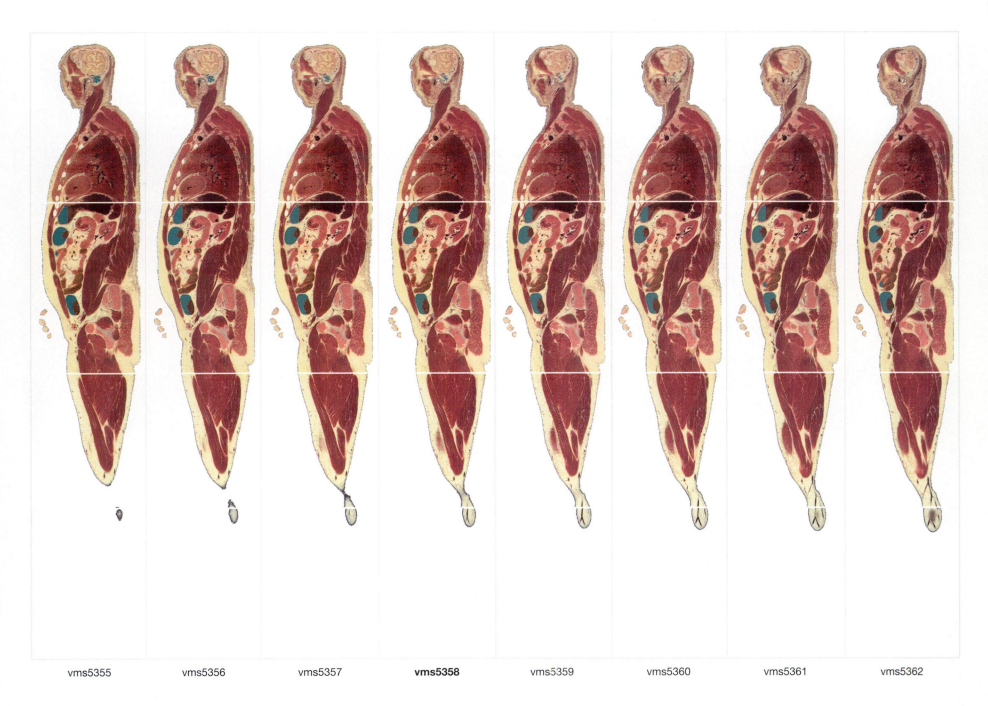

vms5355   vms5356   vms5357   **vms5358**   vms5359   vms5360   vms5361   vms5362

Temporalis Muscle
Superficial Temporal Vessel
External Acoustic Meatus
Superficial Facial Vessel
Parotid Gland
Platysma Muscle
Omohyoid M., Inferior Belly
Clavicle
Axillary Vessel and
Brachial Plexus
Axillary Vein
Rib 1-2, Intercostal M.
Pectoralis Major Muscle
Rib 3
Pericardium
Pylorus
Transverse Colon
Jejunum
Rectus Abdominis Muscle
Tendinous Intersection
Intestinal Vessel
Ileum
Cecum
Internal Oblique Muscle
Inguinal Canal
Femoral Vein
Femur, Head
Sartorius Muscle
Obturator Externus Muscle
Ischium
Femoral Vessel
Vastus Medialis Muscle
Popliteal Fossa
Medial Patellar Retinaculum
Great Saphenous Vein
Sartorius Tendon

Parietal Scalp Vessel
Periosteum
Galea Aponeurotica
Auricular Cartilage
Posterior Auricular Vessel
Trapezius Muscle
Supraspinatus Muscle
Scapula
Serratus Anterior Muscle
Subscapularis Muscle
Rhomboid Major Muscle
Lung, Upper Lobe
Lung, Oblique Fissure
Diaphragm
Left Ventricle
Spleen
Stomach
Splenic Vein
Pancreas, Tail
Latissimus Dorsi Muscle
Renal Vein
Quadratus Lumborum M.
Gluteus Medius Muscle
Ilium, Body
Gluteus Minimus Muscle
Inferior Gemellus Muscle
Gluteus Maximus Muscle
Superior Gemellus Muscle
Biceps Femoris Muscle,
Long Head
Semitendinosus Muscle
Adductor Magnus Muscle
Semimembranosus Muscle
Gracilis Tendon
Crural Fascia
Gastrocnemius Muscle,
Medial Head

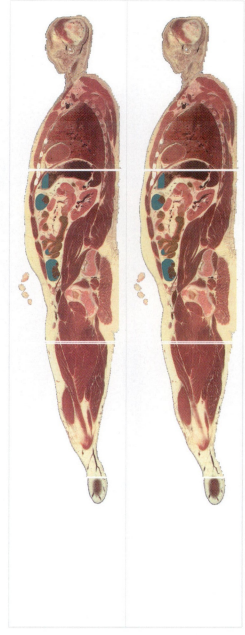

vms5363   vms5364

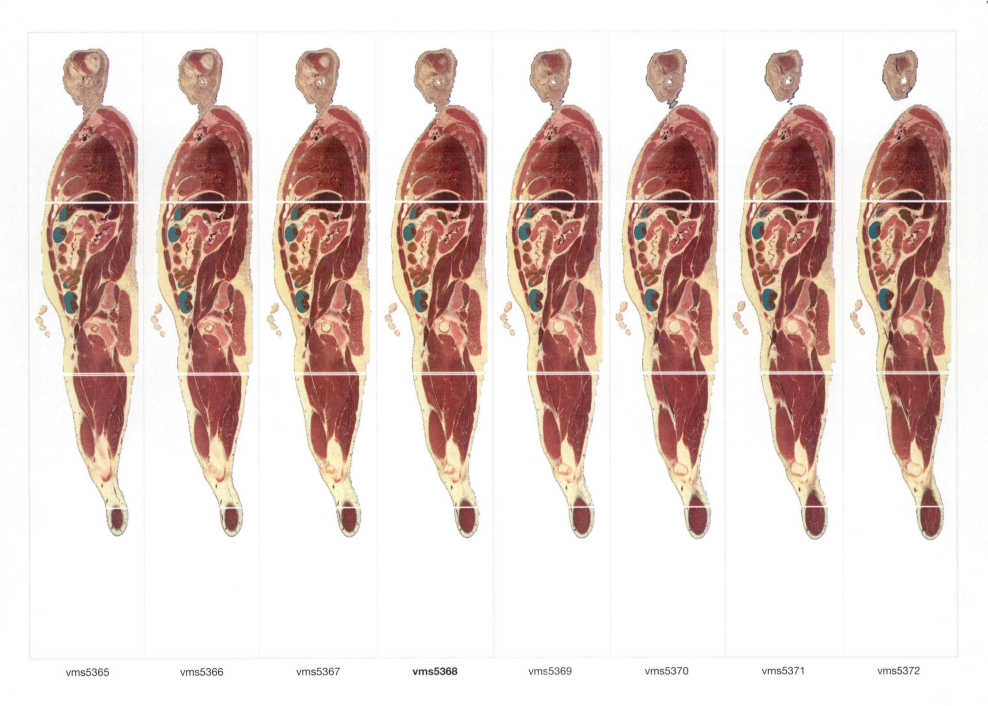

vms5365   vms5366   vms5367   **vms5368**   vms5369   vms5370   vms5371   vms5372

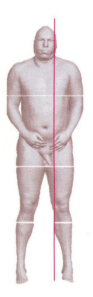

Scalp | Auricular Cartilage

Omohyoid Muscle, Inferior Belly | Trapezius Muscle
Clavicle | Serratus Anterior Muscle
Brachial Plexus | Supraspinatus Muscle
Subclavius Muscle | Subscapularis Muscle
Axillary Vein | Scapula
Pectoralis Major Muscle | Trapezius Muscle
Pectoralis Minor Muscle | Lung, Upper Lobe
Rib 2-3, Intercostal Muscles | Rhomboid Major Muscle
Rib 3 | Lung, Oblique Fissure
Pericardium | Lung, Lower Lobe
Left Ventricle | Diaphragm
Rib 6, Costal Cartilage | Stomach
Transverse Colon | Spleen
Jejunum | Latissimus Dorsi Muscle
Rectus Abdominis Muscle | Splenic Vein
Tendinous Intersection | Colon, Splenic Flexure
Mesentery | Kidney
Ileum | Kidney, Hilus
First Proximal Phalanx | Iliocostalis Muscle
Second Middle Phalanx | Quadratus Lumborum M.
Third Middle Phalanx | Ilium, Crest
Descending Colon | Gluteus Medius Muscle
Internal Oblique Muscle | Psoas Major Muscle
Inguinal Canal, Internal Ring | Gluteus Maximus Muscle
Sartorius Muscle | Iliacus Muscle
Femoral Vein | Ilium, Body
Great Saphenous Vein | Quadratus Femoris Muscle
Pectineus Muscle | Biceps Femoris Muscle
Femur, Head | Obturator Externus Muscle
Adductor Longus Muscle | Semitendinosus Muscle
Vastus Medialis Muscle | Adductor Magnus Muscle
Knee Joint Capsule | Semimembranosus Muscle
Tibial Collateral Ligament | Femur, Medial Condyle
Gastrocnemius Muscle, Medial Head | Crural Fascia
Great Saphenous Vein | Semimembranosus Tendon

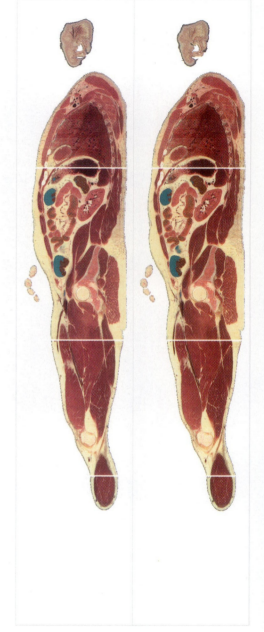

vms5373   vms5374

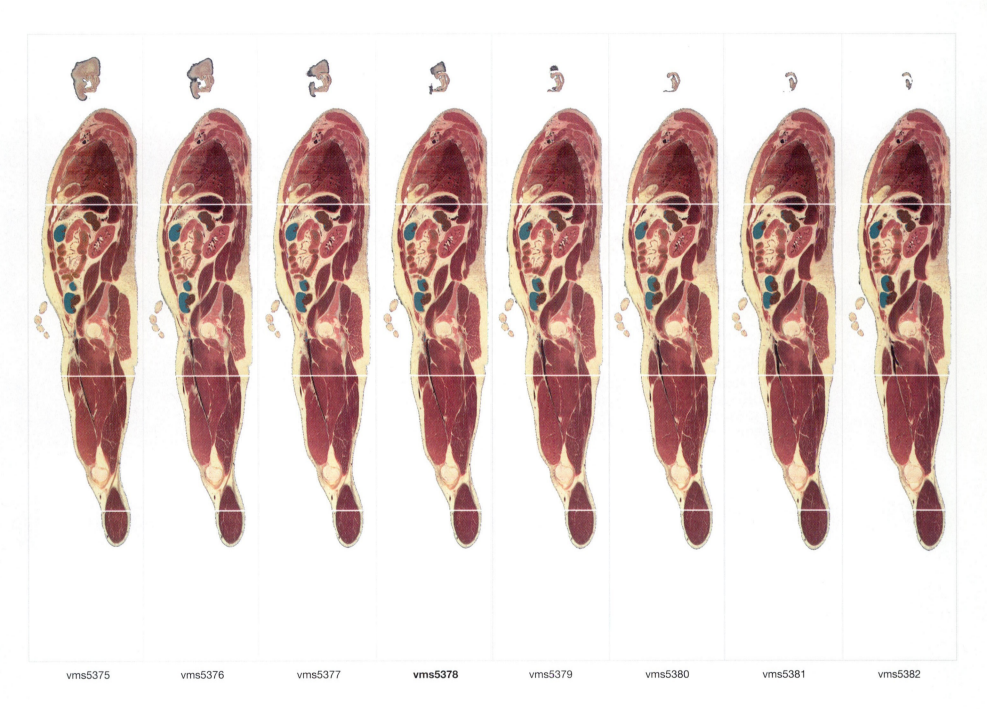

vms5375   vms5376   vms5377   **vms5378**   vms5379   vms5380   vms5381   vms5382

Omohyoid Muscle, Inferior Belly

Brachial Plexus

Clavicle

Subclavius Muscle

Axillary Vessel

Rib 2

Pectoralis Minor Muscle

Pectoralis Major Muscle

Rib 3

Lung, Upper Lobe

Lung, Lingula

Diaphragm

Colon, Splenic Flexure

Transverse Colon

Tendinous Intersection

Mesentery

Ileum

Linea Semilunaris

First Metacarpal

Abductor Pollicis Muscle

Rectus Abdominis M.

Descending Colon

Internal Oblique Muscle

Transversus Abdominis M.

Sartorius Muscle

Iliofemoral Ligament

Femur, Head

Iliopsoas Tendon

Great Saphenous Vein

Pectineus Muscle

Femoral Vein

Adductor Longus Muscle

Vastus Medialis Muscle

Femur, Medial Epicondyle

Tibia, Medial Condyle

Knee Joint

Great Saphenous Vein

Serratus Anterior Muscle

Trapezius Muscle

Supraspinatus Muscle

Scapula, Spine

Infraspinatus Muscle

Subscapularis Muscle

Serratus Anterior Muscle

Lung, Oblique Fissure

Lung, Lower Lobe

Latissimus Dorsi Muscle

Stomach

Splenic Vessels

Spleen

Iliocostalis Muscle

Jejunum

Latissimus Dorsi Muscle

Kidney

Renal Fascia

Quadratus Lumborum M.

Ilium, Crest

Gluteus Medius Muscle

Gluteus Minimus Muscle

Piriformis Muscle

Gluteus Maximus Muscle

Superior Gemellus Muscle

Inferior Gemellus Muscle

Quadratus Femoris Muscle

Biceps Femoris Muscle

Obturator Externus Muscle

Semitendinosus Muscle

Adductor Magnus Muscle

Semimembranosus M.

Femur, Medial Condyle

Semimembranosus Tendon

Gastrocnemius Muscle, Medial Head

Soleus Muscle

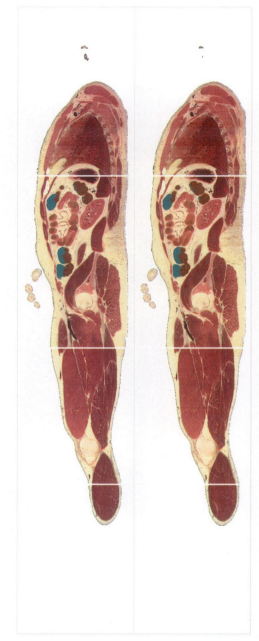

vms5383   vms5384

456

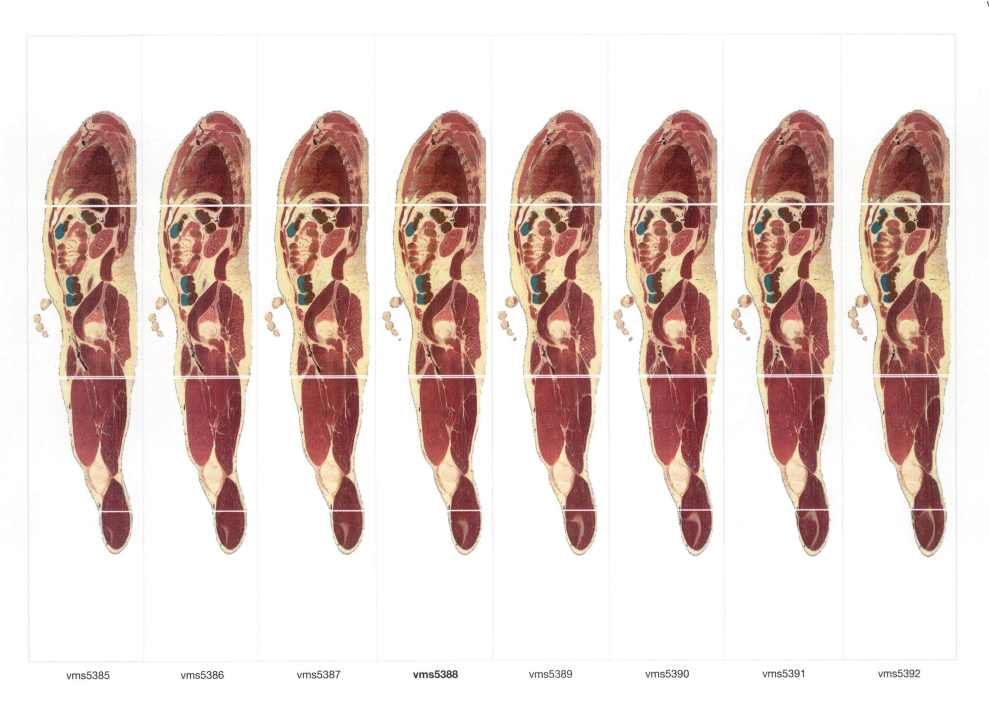

vms5385   vms5386   vms5387   **vms5388**   vms5389   vms5390   vms5391   vms5392

Axilla
Trapezius Muscle
Clavicle
Supraspinatus Muscle
Serratus Anterior Muscle
Subscapularis Muscle
Subclavius Muscle
Scapula, Spine
Cephalic Vein
Infraspinatus Muscle
Axillary Vessel and
Brachial Plexus
Lung, Oblique Fissure
Serratus Anterior Muscle
Pectoralis Minor Muscle
Lung, Lower Lobe
Pectoralis Major Muscle
Scapula, Inferior Angle
Lung, Upper Lobe
Diaphragm
Transverse Colon
Colon, Splenic Flexure
Rib 8, Costal Cartilage
Spleen
Jejunum
External Oblique Muscle
Transversus Abdominis M.
Kidney
Descending Colon
Quadratus Lumborum M.
Internal Oblique Muscle
Ilium, Crest
Linea Semilunaris
Gluteus Medius Muscle
First Metacarpal
Ilium, Body
Thenar Muscles
Gluteus Minimus Muscle
Second Metacarpal
Acetabulum
Third Metacarpal
Femur, Head
Transversus Abdominis M.
Superior Gemellus Muscle
Iliofemoral Ligament
Obturator Internus Tendon
Iliopsoas Muscle
Gluteus Maximus Muscle
Rectus Femoris Muscle
Inferior Gemellus Muscle
Femoral Vessel
Quadratus Femoris M.
Sartorius Muscle
Pectineus Muscle
Perforating Vessel
Biceps Femoris Muscle,
Long Head
Great Saphenous Vein
Adductor Magnus Muscle
Adductor Longus Muscle
Vastus Medialis Muscle
Semitendinosus Muscle
Femur, Shaft
Semimembranosus M.
Knee Joint Cavity
Popliteal Vessel
Articular Cartilage
Small Saphenous Vein
Tibia, Medial Condyle
Popliteal Fossa
Tibia, Periosteum
Femur, Medial Condyle
Gastrocnemius Muscle,
Medial Head
Gastrocnemius Muscle,
Lateral Head
Crural Fascia
Knee Joint
Great Saphenous Vein
Soleus Muscle

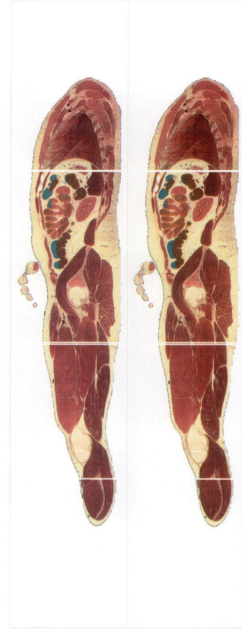

vms5393   vms5394

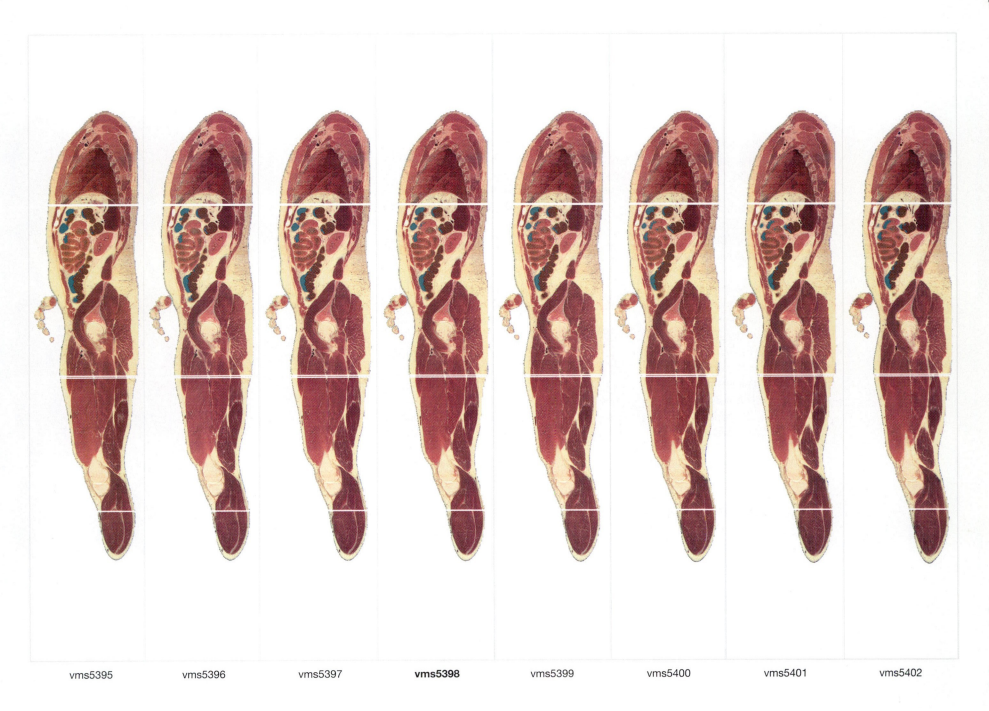

vms5395   vms5396   vms5397   **vms5398**   vms5399   vms5400   vms5401   vms5402

Clavicle — Trapezius Muscle
Subclavius Muscle — Supraspinatus Muscle
Axilla — Infraspinatus Muscle
Cephalic Vein — Scapula, Spine
Serratus Anterior Muscle — Subscapularis Muscle
Axillary Vessel — Scapula
Pectoralis Major Muscle — Rib 5
Pectoralis Minor Muscle — Lung, Oblique Fissure
Lung, Upper Lobe — Scapula, Inferior Angle
Rib 4 — Teres Major Muscle
Rib 4-5, Intercostal Muscles — Serratus Anterior Muscle
Lung, Lingula — Lung, Lower Lobe
Diaphragm — Latissimus Dorsi Muscle
Transverse Colon — Spleen
Jejunum — Colon, Splenic Flexure
Transversus Abdominis M. — Renal Fascia
External Oblique Muscle — Kidney
Internal Oblique Muscle — Serratus Post. Inferior M.
First Metacarpal — Descending Colon
First Dorsal Interosseous — Quadratus Lumborum M.
Abductor Pollicis Muscle — Iliacus Muscle
Third Metacarpal — Gluteus Medius Muscle
Flexor Pollicis Brevis Muscle — Piriformis Muscle
Sartorius Muscle — Gluteus Maximus Muscle
Iliofemoral Ligament — Superior Gemellus Muscle
Femur, Head — Inferior Gemellus Muscle
Rectus Femoris Muscle — Quadratus Femoris Muscle
Iliopsoas Muscle — Femur, Intertrochanteric Crest
Femur, Shaft
Vastus Medialis Muscle — Femur, Neck
Vastus Medialis Muscle — Biceps Femoris Muscle
Femur, Shaft — Adductor Magnus Muscle
Medial Patellar Retinaculum — Sciatic Nerve
Knee Joint Cavity — Semimembranosus M.
Femur, Medial Condyle — Popliteal Vein
Knee Joint — Gastrocnemius Muscle, Lateral Head
Tibia, Shaft
Popliteus Muscle — Gastrocnemius Muscle, Medial Head
Posterior Cruciate Ligament
Soleus Muscle

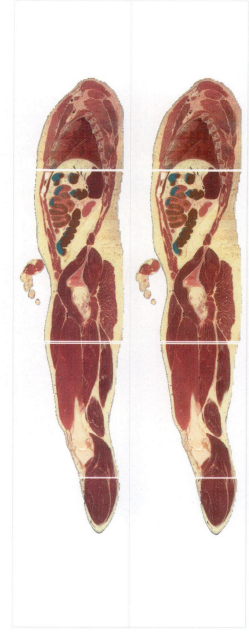

vms5403   vms5404

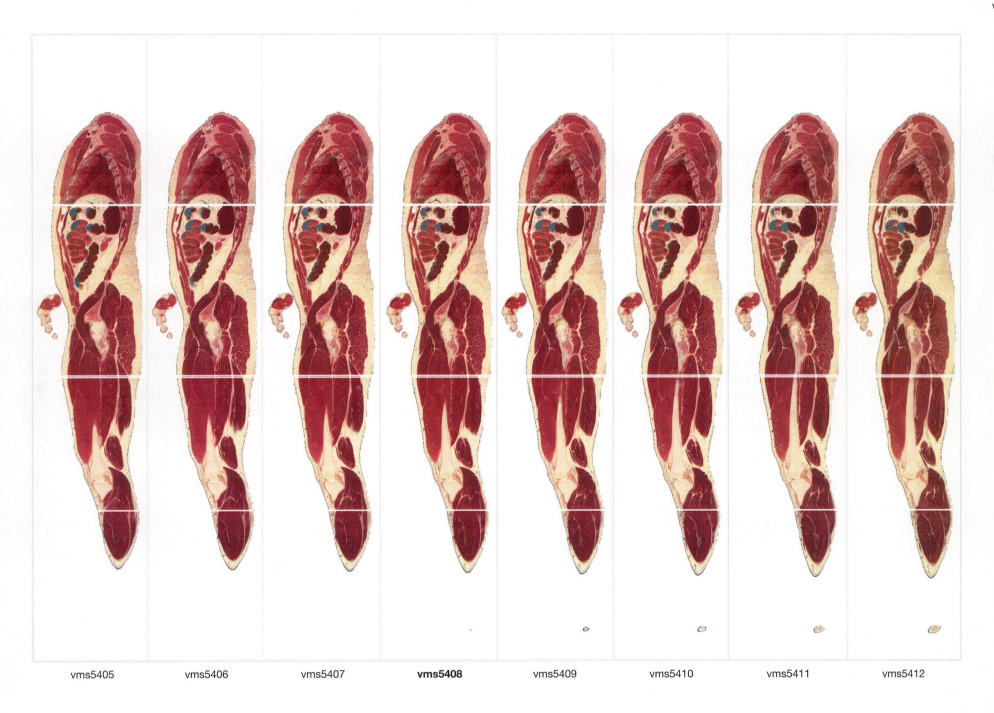

vms5405 vms5406 vms5407 **vms5408** vms5409 vms5410 vms5411 vms5412

Clavicle

Trapezius Muscle

Deltoid Muscle

Supraspinatus Muscle

Axillary Vessels and Nerve

Scapula, Spine

Cephalic Vein

Infraspinatus Muscle

Pectoralis Minor Muscle

Subscapularis Muscle

Serratus Anterior Muscle

Serratus Anterior Muscle

Pectoralis Major Muscle

Scapula, Inferior Angle

Rib 4

Teres Major Muscle

Lung, Upper Lobe

Lung, Lower Lobe

Lung, Oblique Fissure

Diaphragm

Transverse Colon

Latissimus Dorsi Muscle

Jejunum

Spleen

External Oblique Muscle

Rib 11

Internal Oblique Muscle

Rib 11-12, Intercostal M.

Transversus Abdominis M.

External Oblique Muscle

First Metacarpal

Internal Oblique Muscle

First Dorsal Interosseous

Iliacus Muscle

Opponens Pollicis M.

Ilium

Second Metacarpal

Gluteus Medius Muscle

Second Dorsal Interosseous

Gluteus Maximus Muscle

Third Metacarpal

Gluteus Minimus Muscle

Sartorius Muscle

Piriformis Tendon

Rectus Femoris Tendon

Obturator Internus Tendon

Iliopsoas Muscle

Quadratus Femoris M.

Iliofemoral Ligament

Femur,
Intertrochanteric Crest

Femoral Vessel

Vastus Intermedius Muscle

Femur, Shaft

Rectus Femoris Muscle

Biceps Femoris Muscle,
Long Head

Femur, Neck

Vastus Medialis Muscle

Adductor Magnus Muscle

Rectus Femoris Tendon

Biceps Femoris Muscle,
Short Head

Suprapatellar Bursa

Femur, Lateral Condyle

Popliteal Nerve

Tibia, Lateral Condyle

Semimembranosus Muscle

Anterior Cruciate Ligament

Popliteal Vein

Tibia, Shaft

Medial Meniscus

Soleus Muscle

Popliteus Muscle

Tibialis Posterior Muscle

Soleus Tendon

Feet, Superficial Fascia

Great Saphenous Vein

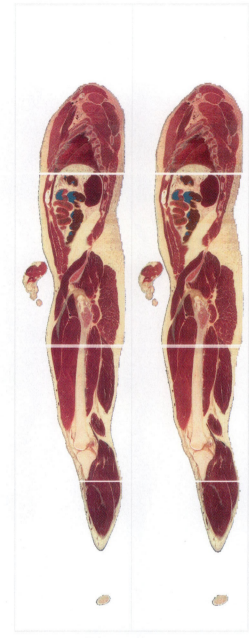

vms5413   vms5414

462

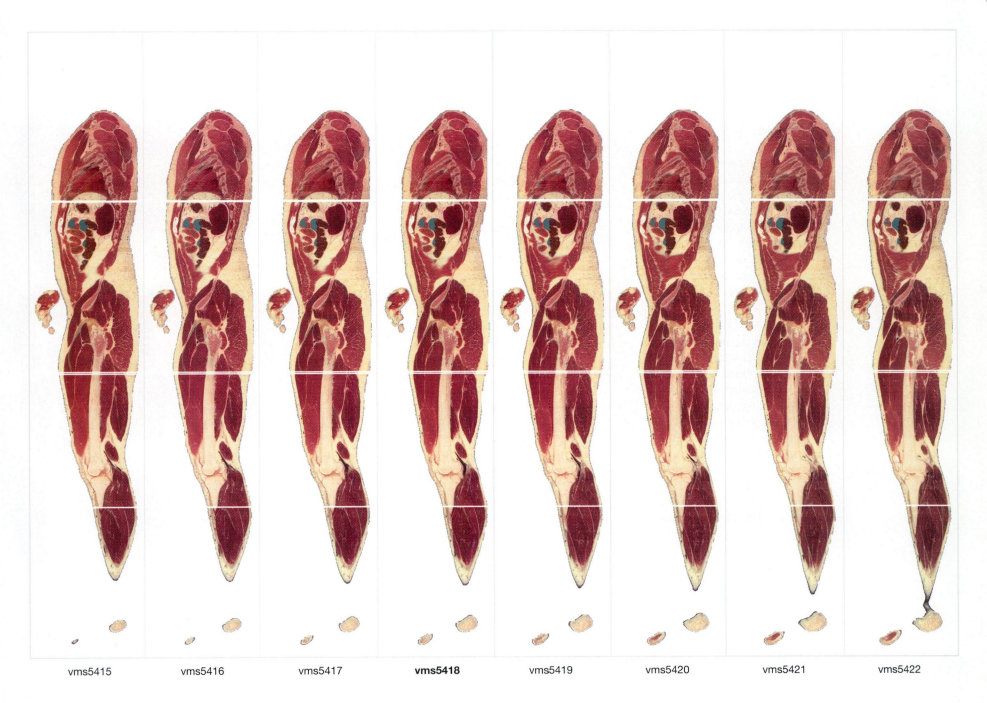

vms5415   vms5416   vms5417   **vms5418**   vms5419   vms5420   vms5421   vms5422

Clavicle
Scapula
Subscapularis Muscle
Coracobrachialis Muscle
Axillary Vessel and Brachial Plexus
Axilla
Pectoralis Minor Muscle
Rib 5
Descending Colon
External Oblique Muscle
Descending Colon
Rib 9, Costal Cartilage
Transversus Abdominis M.
External Oblique Muscle
First Metacarpal
First Palmar Interosseous M.
Second Metacarpal
Second Palmar
Interosseous Muscle
Third Metacarpal
Fourth Metacarpal
Fifth Metacarpal
Sartorius Muscle
Rectus Femoris Muscle
Iliofemoral Ligament
Vastus Intermedius Muscle
Rectus Femoris Tendon
Patella
Patellar Ligament
Knee Joint
Tibia, Medial Condyle
Tibia, Shaft
Tibialis Posterior Muscle
Ankle Joint
Flexor Digitorum Brevis Muscle
Adductor Hallucis Muscle
Plantar Aponeurosis
Abductor Hallucis Muscle
First Metacarpal, Head
First Proximal Phalanx

Trapezius Muscle
Supraspinatus Muscle
Scapula, Spine
Infraspinatus Muscle
Deltoid Muscle
Rib 5-6, Intercostal M.
Serratus Anterior Muscle
Teres Major Muscle
Rib 7
Latissimus Dorsi Muscle
Diaphragm
Spleen
Diaphragm
Rib 12
External Oblique Muscle
Superficial Fascia and
Adipose Tissue
Internal Oblique Muscle
Transversus Abdominis M.
Iliacus Muscle
Gluteus Medius Muscle
Ilium
Femur, Greater Trochanter
Gluteus Maximus Muscle
Gluteus Minimus Muscle
Femur, Neck
Femur, Shaft
Vastus Lateralis Muscle
Biceps Femoris Muscle,
Long Head
Short Head
Popliteal Fossa
Gastrocnemius Muscle,
Lateral Head
Soleus Muscle
Flexor Digitorum Longus
Muscle
Fibula
Calcaneal Tendon
Calcaneus Bone
Quadratus Plantae Muscle

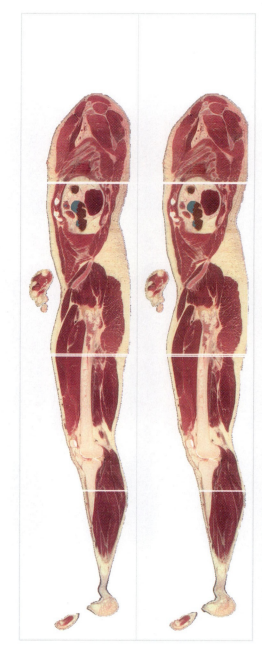

vms5423   vms5424

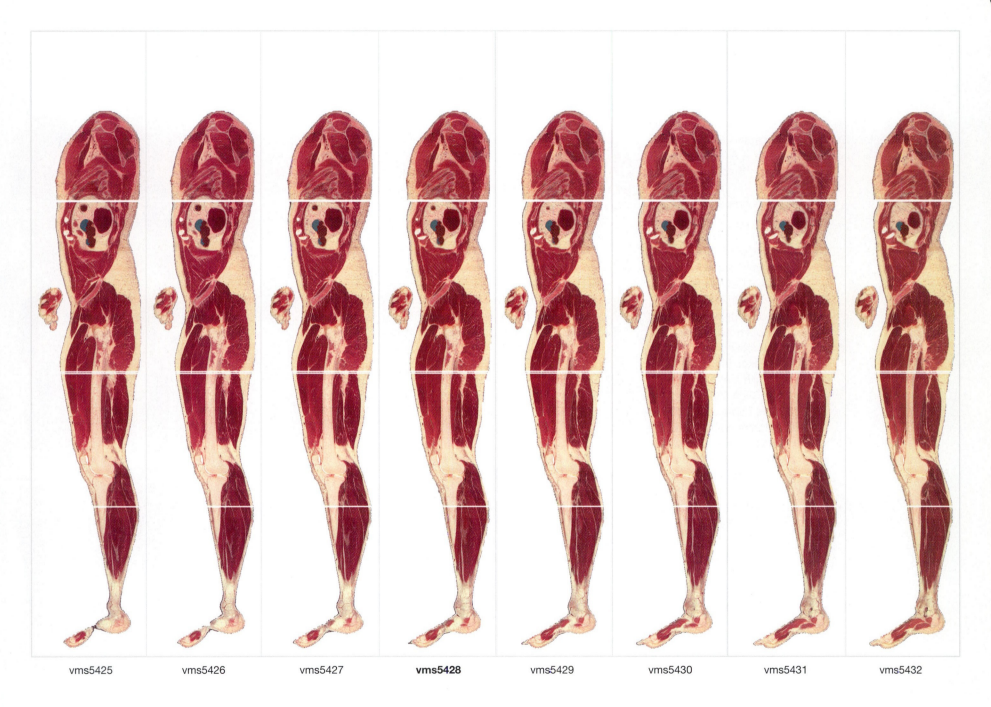

vms5425   vms5426   vms5427   **vms5428**   vms5429   vms5430   vms5431   vms5432

Clavicle
Coracoacromial Ligament
Deltoid Muscle
Coracoid Process
Coracobrachialis Muscle
Pectoralis Major Muscle
Axillary Vessel
Brachial Plexus
Axilla
Serratus Anterior Muscle
Rib 5
Diaphragm
Rib 10, Costal Cartilage
External Oblique Muscle
Internal Oblique Muscle
First Metacarpal
Second Metacarpal
Second Dorsal Interosseous
First Palmar Interosseous M.
Long Flexor Tendons
Fourth Metacarpal
Fifth Metacarpal
Rectus Femoris Muscle
Vastus Intermedius Muscle
Articularis Genu Muscle
Rectus Femoris Tendon
Suprapatellar Bursa
Patella
Femur, Lateral Condyle
Patellar Ligament
Tibial Tuberosity
Tibia, Lateral Condyle
Tibia, Shaft
Great Saphenous Vein
Tibia, Medial Malleolus
Tibialis Posterior Tendon
Abductor Hallucis Muscle
Flexor Hallucis Longus Tendon
Flexor Hallucis Brevis Muscle
First Metatarsal
First Proximal Phalanx

Trapezius Muscle
Supraspinatus Muscle
Scapula, Spine
Subscapularis Muscle
Infraspinatus Muscle
Scapula
Teres Major Muscle
Serratus Anterior Muscle
Latissimus Dorsi Muscle
Diaphragm
Descending Colon
Rib 12, Costal Cartilage
External Oblique Muscle
Internal Oblique Muscle
Ilium
Gluteus Medius Muscle
Buttock
Femur, Greater Trochanter
Gluteus Maximus Muscle
Femur, Shaft
Vastus Intermedius M.
Biceps Femoris Muscle,
Long Head
Short Head
Femur, Shaft
Popliteal Fossa
Knee Joint
Gastrocnemius Muscle,
Lateral Head
Soleus Muscle
Crural Fascia
Tibialis Posterior Muscle
Flexor Digitorum Longus M.
Small Saphenous Vein
Flexor Hallucis Longus M.
Subtendinous Space
Calcaneal Tendon
Calcaneus Bone
Quadratus Plantae Muscle
Flexor Digiti Minimi Brevis
Plantar Aponeurosis

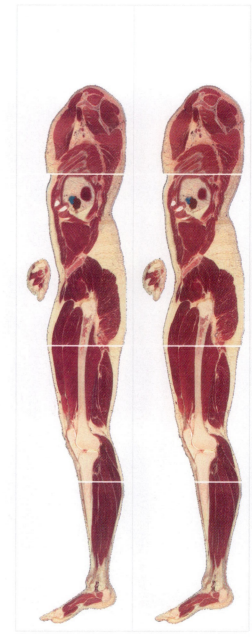

vms5433   vms5434

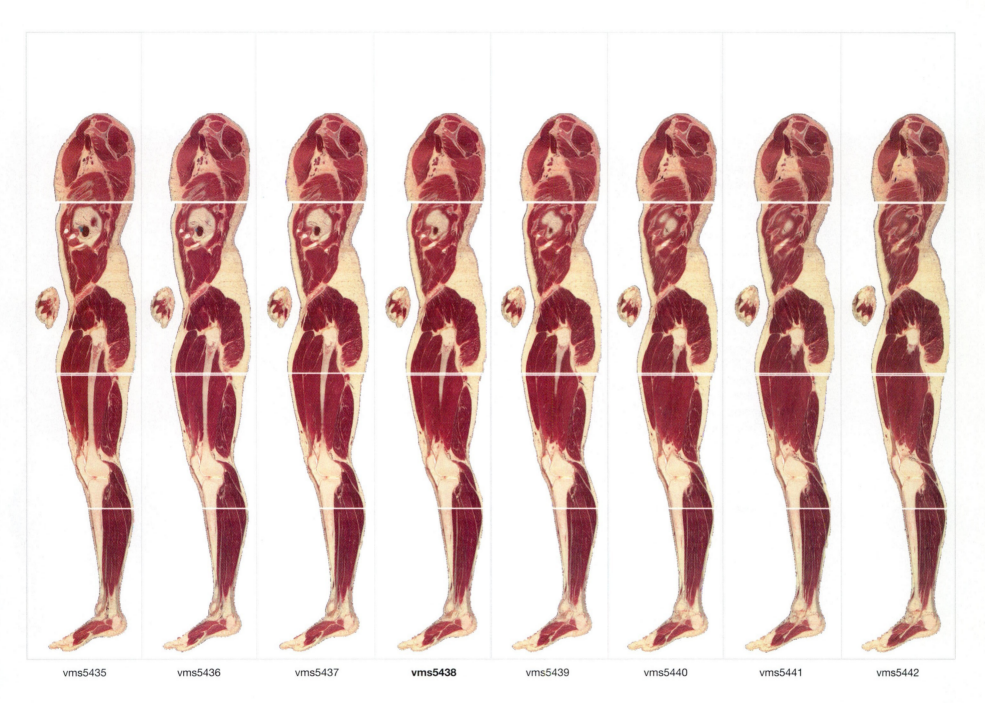

vms5435   vms5436   vms5437   **vms5438**   vms5439   vms5440   vms5441   vms5442

Clavicle

Shoulder Joint

Deltoid Muscle

Glenohumeral Ligament

Deltoid Muscle

Scapula

Coracobrachialis Muscle

Pectoralis Major Muscle

Axillary Vessel and
Brachial Plexus

Thoracolumbar Vein

Rib 9-10, Intercostal Muscles

Rib 10

Internal Oblique Muscle

External Oblique Muscle

Radius

Long Flexor Tendons

Carpal Bones

Fourth Metacarpal

Fourth Dorsal Interosseous M.

Fifth Metacarpal

Hypothenar Muscles

Tensor Fascia Lata Muscle

Vastus Lateralis Muscle

Rectus Femoris Muscle

Rectus Femoris Tendon

Articularis Genu Muscle

Suprapatellar Bursa

Patella

Patellar Ligament

Tibia, Periosteum

Tibia, Shaft

Crural Fascia

Tibia, Medial Malleolus

Medial Cuneiform Bone

First Metatarsal

Adductor Hallucis Muscle

Flexor Digitorum
Longus Tendon

Trapezius Muscle

Supraspinatus Muscle

Scapula, Spine

Deltoid Muscle

Infraspinatus Muscle

Scapula, Neck

Teres Minor Muscle

Teres Major Muscle

Subscapularis Muscle

Latissimus Dorsi Muscle

Rib 11

External Oblique Muscle

Superficial Fascia

Gluteus Medius Muscle

Gluteus Minimus Muscle

Femur, Greater Trochanter

Gluteus Maximus Muscle

Buttock

Vastus Intermedius M.

Fascia Lata

Biceps Femoris Muscle,
Long Head
Short Head

Femur, Lateral Condyle

Knee Joint

Lateral Meniscus

Tibia, Lateral Condyle

Gastrocnemius Muscle,
Lateral Head

Tibialis Posterior Muscle

Flexor Digitorum
Longus Muscle

Flexor Hallucis Longus M.

Talus Bone

Quadratus Plantae Muscle

Calcaneus Bone

Flexor Digiti Minimi Brevis M.

Abductor Digiti Minimi M.

Plantar Aponeurosis

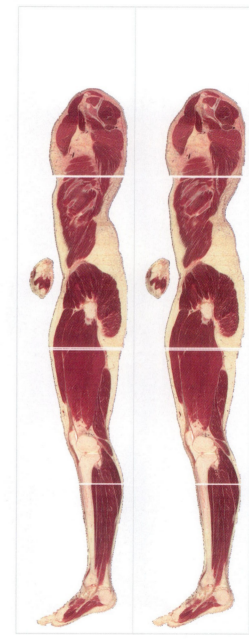

vms5443   vms5444

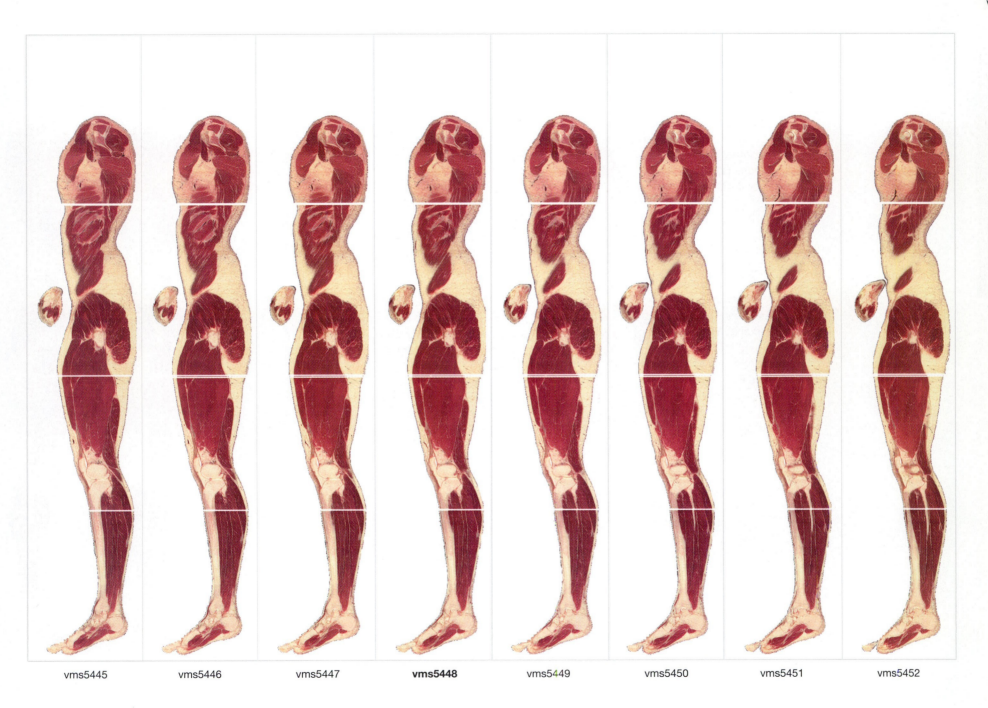

vms5445   vms5446   vms5447   **vms5448**   vms5449   vms5450   vms5451   vms5452

Humerus, Head

Humerus, Lesser Tubercle

Shoulder Joint, Capsule

Deltoid Muscle

Coracobrachialis Muscle

Biceps Brachii Muscle

Pectoralis Major Muscle

Clavicle

Trapezius Muscle

Scapula, Acromion

Supraspinatus Muscle

Deltoid Muscle

Infraspinatus Muscle

Teres Minor Muscle

Subscapularis Tendon

Teres Major Muscle

Latissimus Dorsi Muscle

External Oblique Muscle

Long Flexor Tendons

Pronator Quadratus Muscle

Radius

Carpal Bones

Long Extensor Tendons

Hypothenar Muscles

Superficial Fascia and Adipose Tissue

External Oblique Muscle

Gluteus Medius Muscle

Gluteus Maximus Muscle

Buttock

Tensor Fascia Lata Muscle

Rectus Femoris Muscle

Vastus Lateralis Muscle

Fascia Lata

Vastus Intermedius Muscle

Biceps Femoris Muscle, Short Head

Knee Joint Cavity

Patella

Patellar Ligament

Lateral Meniscus

Tibia, Lateral Condyle

Tibialis Anterior Muscle

Extensor Digitorum Longus M.

Fibula, Shaft

Tibia

Navicular Bone

Medial Cuneiform Bone

Plantar Aponeurosis

Adductor Hallucis Muscle

First Dorsal Interosseous M.

First Metatarsal

Fibular Collateral Ligament

Biceps Femoris Tendon, Long Head

Fibula, Head

Crural Fascia

Gastrocnemius Muscle, Lateral Head

Fibula, Shaft

Peroneus Brevis Muscle

Ankle Joint

Talus Bone

Quadratus Plantae Muscle

Calcaneus Bone

Abductor Digiti Minimi M.

Flexor Digitorum Brevis M.

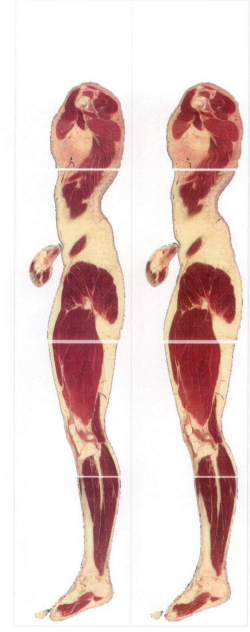

vms5453   vms5454

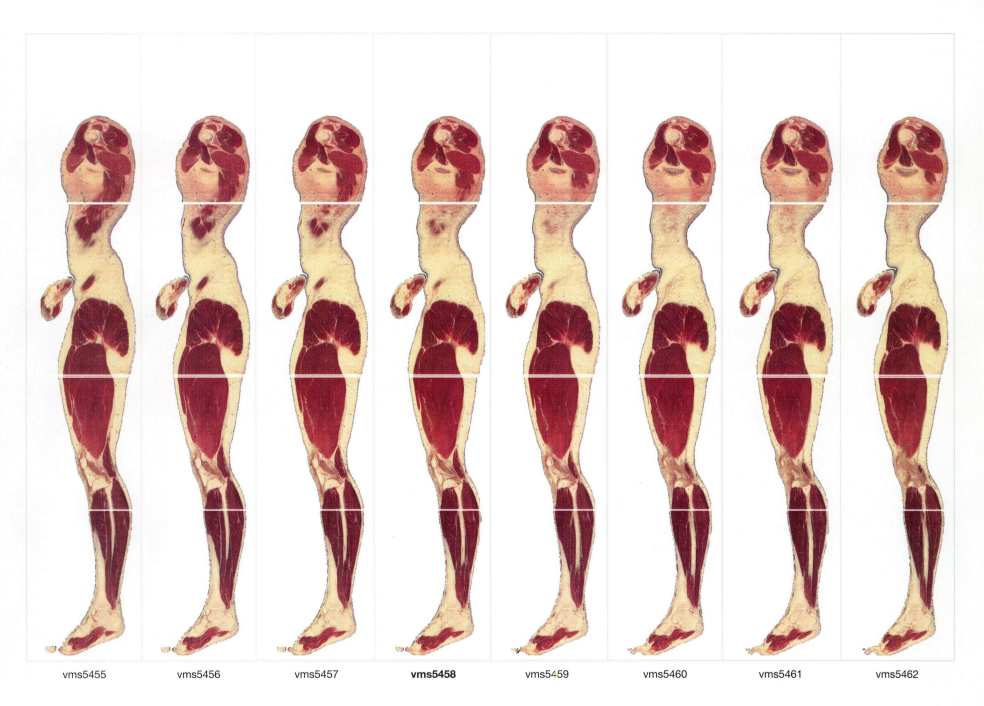

vms5455   vms5456   vms5457   **vms5458**   vms5459   vms5460   vms5461   vms5462

Acromioclavicular Ligament

Humerus, Head

Deltoid Muscle

Glenohumeral Ligament

Humerus, Neck

Biceps Brachii Muscle,
Long Head

Anterior Axillary Fold

Superficial Fascia
and Adipose Tissue

Scapula, Acromion

Supraspinatus Muscle

Deltoid Muscle

Infraspinatus Muscle

Articular Capsule

Teres Major Muscle

Posterior Humeral
Circumflex Vessel

Posterior Axillary Fold

Flexor Digitorum
Profundus Muscle

Radius

Scaphoid Bone

Extensor Carpi Radialis
Brevis Tendon

Carpal Bones

Tensor Fascia Lata Muscle

Vastus Lateralis Muscle

Rectus Femoris Muscle

Vastus Intermedius Muscle

Lateral Patellar Retinaculum

Tibialis Anterior Muscle

Extensor Hallucis Longus Muscle

Tibialis Anterior Tendon

Tibia

Talus Bone

Navicular Bone

Quadratus Plantae Muscle

Second Metatarsal

First Dorsal Interosseous Muscle

Adductor Hallucis Muscle

Plantar Aponeurosis

Second Proximal Phalanx

Flexor Digitorum
Superficialis Muscle

Flexor Carpi
Ulnaris Muscle

Hypothenar Muscles

Gluteus Maximus Muscle

Gluteus Medius Muscle

Superficial Fascia and
Adipose Tissue

Fascia Lata

Lateral Genicular Vessel

Biceps Femoris Tendon

Fibula, Head

Gastrocnemius Muscle,
Lateral Head

Extensor Digitorum
Longus Muscle

Crural Fascia

Fibula, Shaft

Peroneus Brevis Muscle

Lateral Cuneiform Bone

Calcaneus Bone

Cuboid Bone

Abductor Digiti Minimi M.

Peroneus Longus Tendon

Flexor Digitorum Brevis M.

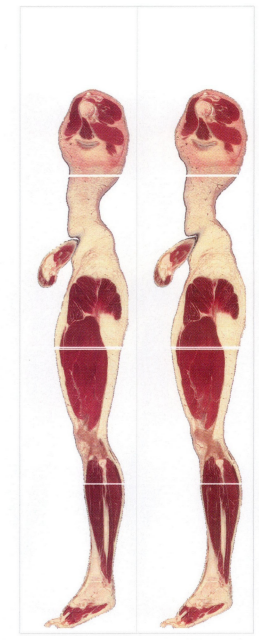

vms5463   vms5464

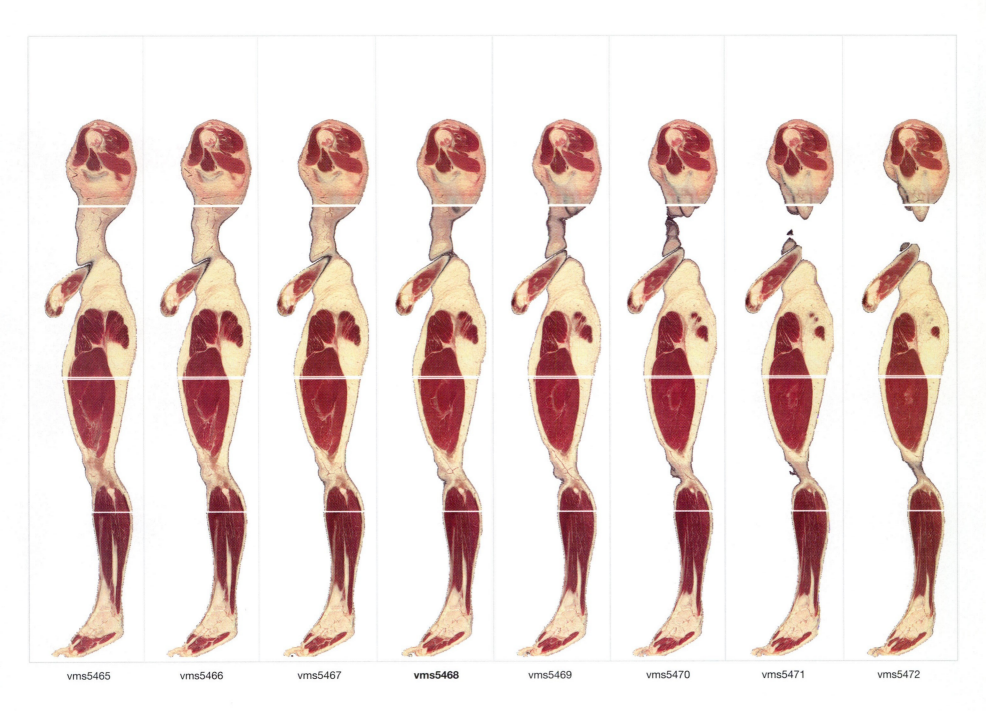

vms5465   vms5466   vms5467   **vms5468**   vms5469   vms5470   vms5471   vms5472

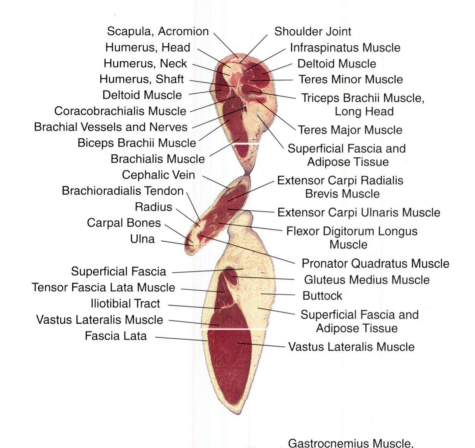

Scapula, Acromion — Shoulder Joint
Humerus, Head — Infraspinatus Muscle
Humerus, Neck — Deltoid Muscle
Humerus, Shaft — Teres Minor Muscle
Deltoid Muscle — Triceps Brachii Muscle,
Coracobrachialis Muscle — Long Head
Brachial Vessels and Nerves — Teres Major Muscle
Biceps Brachii Muscle — Superficial Fascia and
Brachialis Muscle — Adipose Tissue
Cephalic Vein — Extensor Carpi Radialis
Brachioradialis Tendon — Brevis Muscle
Radius — Extensor Carpi Ulnaris Muscle
Carpal Bones — Flexor Digitorum Longus
Ulna — Muscle
— Pronator Quadratus Muscle
Superficial Fascia — Gluteus Medius Muscle
Tensor Fascia Lata Muscle — Buttock
Iliotibial Tract — Superficial Fascia and
Vastus Lateralis Muscle — Adipose Tissue
Fascia Lata — Vastus Lateralis Muscle

Tibialis Anterior Muscle — Gastrocnemius Muscle,
Extensor Digitorum Longus Muscle — Lateral Head
Extensor Hallucis Longus Muscle — Peroneus Brevis Muscle
Tibialis Anterior Tendon — Crural Fascia
Intermediate Cuneiform Bone — Talus Bone
Third Metatarsal — Fibula
Second Dorsal Interosseous — Peroneus Longus Tendon
Muscle — Fibula, Lateral Malleolus
Third Dorsal Interosseous M. — Cuboid Bone
Plantar Aponeurosis — Calcaneus Bone
— Fourth Metatarsal
— Flexor Digitorum Brevis M.

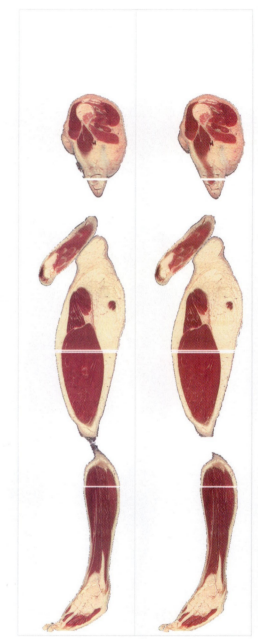

vms5473   vms5474

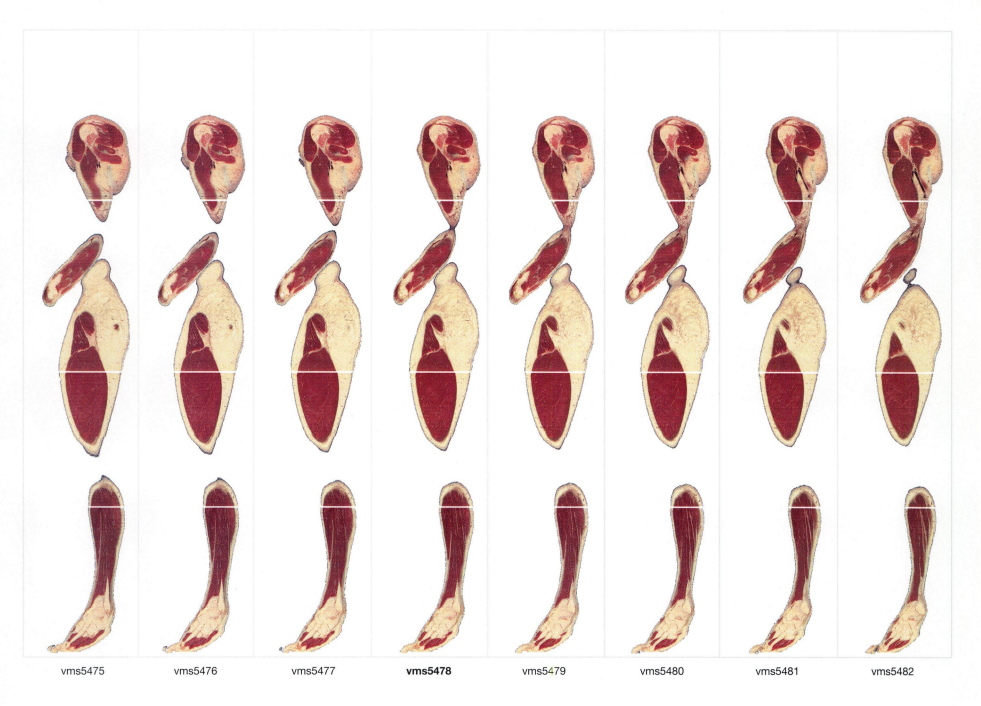

vms5475   vms5476   vms5477   **vms5478**   vms5479   vms5480   vms5481   vms5482

Rotator Cuff | Scapula, Acromion
Humerus, Head | Infraspinatus Muscle
Humerus, Neck | Deltoid Muscle
Humerus, Shaft | Teres Minor Muscle
Deltoid Muscle | Triceps Brachii Muscle,
Coracobrachialis Muscle | Long Head
Median and Ulnar Nerves | Medial Head
Biceps Brachii Muscle | Brachial Vessel

Extensor Carpi Radialis
Longus Muscle | Brachial Vein

Extensor Carpi Radialis
Brevis Muscle | Flexor Digitorum
Superficialis Muscle

Brachioradialis Muscle | Flexor Digitorum
Radius | Profundus Muscle

Extensor Pollicis
Longus Muscle | Flexor Carpi Ulnaris Muscle

Extensor Digitorum Muscle | Buttock
Ulna | Superficial Fascia and
Adipose Tissue

Vastus Lateralis Muscle | Fascia Lata
Skin

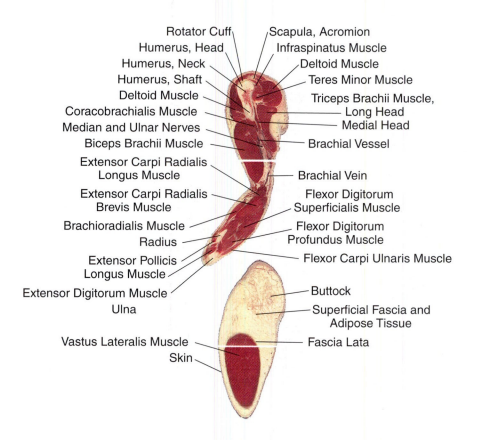

Extensor Digitorum Longus Muscle | Peroneus Longus Muscle
Anterior Crural | Crural Fascia
Intermuscular Septum | Fifth Metatarsal
Lateral Cutaneous Vein
Extensor Hallucis Longus Muscle | Fibula
Extensor Digitorum Brevis Muscle | Fibula, Lateral Malleolus
Intermediate Cuneiform Bone | Calcaneus Bone
Third Metatarsal | Cuboid Bone
Third Dorsal Interosseous Muscle | Plantar Aponeurosis
Fourth Metatarsal | Fourth Dorsal Interosseous M.
Fourth Proximal Phalanx | Flexor Digiti Minimi Brevis M.

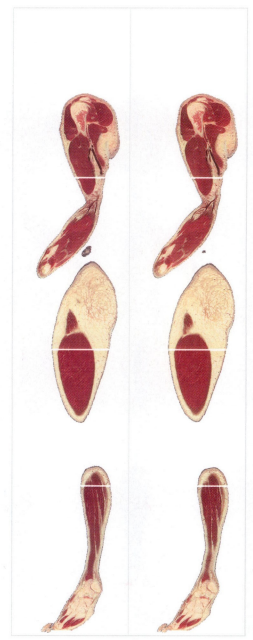

vms5483   vms5484

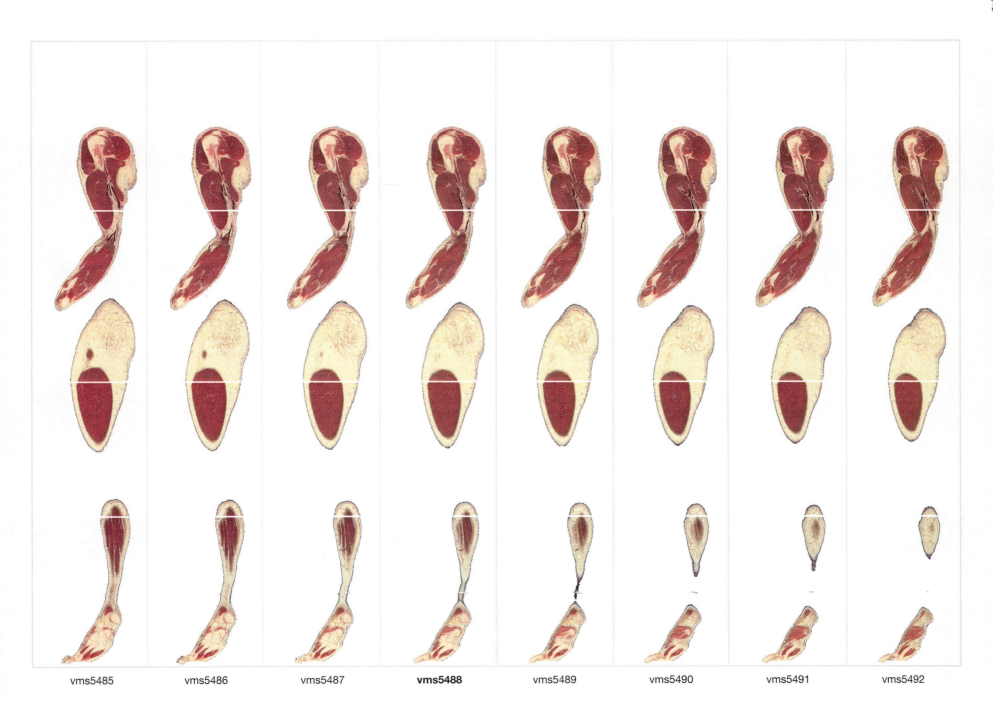

vms5485   vms5486   vms5487   **vms5488**   vms5489   vms5490   vms5491   vms5492

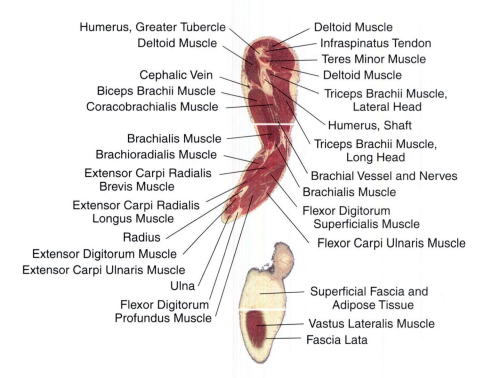

Humerus, Greater Tubercle

Deltoid Muscle

Cephalic Vein

Biceps Brachii Muscle

Coracobrachialis Muscle

Brachialis Muscle

Brachioradialis Muscle

Extensor Carpi Radialis
Brevis Muscle

Extensor Carpi Radialis
Longus Muscle

Radius

Extensor Digitorum Muscle

Extensor Carpi Ulnaris Muscle

Ulna

Flexor Digitorum
Profundus Muscle

Deltoid Muscle

Infraspinatus Tendon

Teres Minor Muscle

Deltoid Muscle

Triceps Brachii Muscle,
Lateral Head

Humerus, Shaft

Triceps Brachii Muscle,
Long Head

Brachial Vessel and Nerves

Brachialis Muscle

Flexor Digitorum
Superficialis Muscle

Flexor Carpi Ulnaris Muscle

Superficial Fascia and
Adipose Tissue

Vastus Lateralis Muscle

Fascia Lata

Fifth Middle Phalanx

Fifth Distal Phalanx

Extensor Digitorum Brevis Muscle

Abductor Digiti Minimi Muscle

Flexor Digiti Minimi Brevis M.

vms5493

vms5494

478

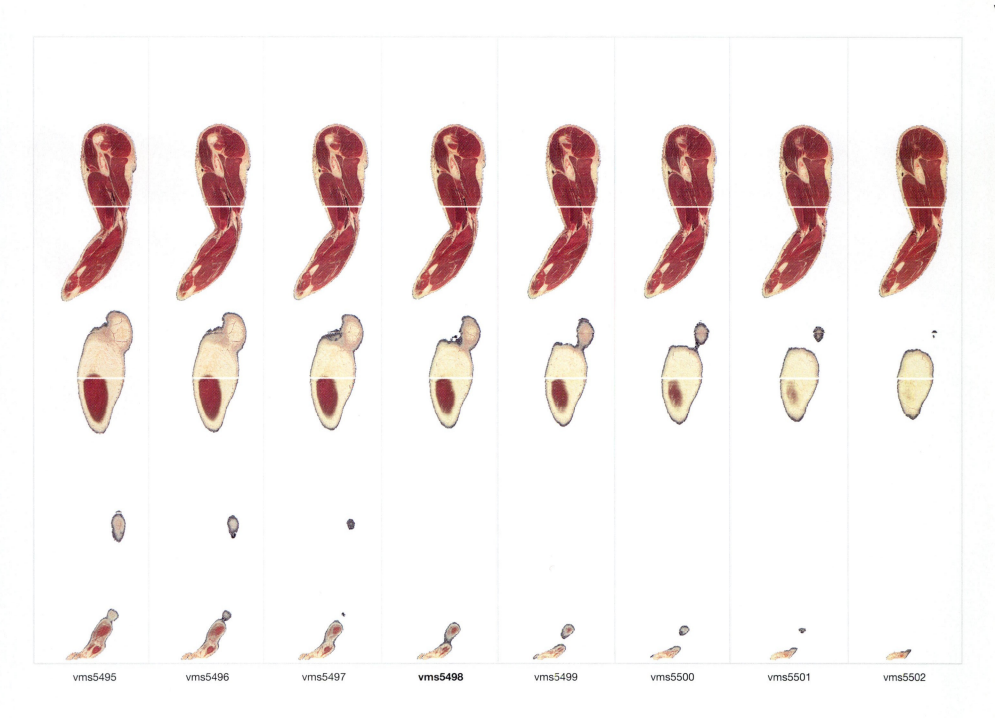

vms5495 vms5496 vms5497 **vms5498** vms5499 vms5500 vms5501 vms5502

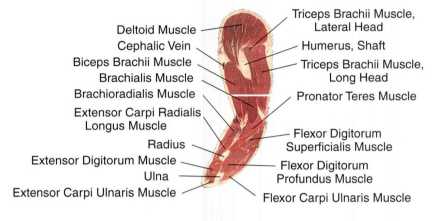

Deltoid Muscle

Cephalic Vein

Biceps Brachii Muscle

Brachialis Muscle

Brachioradialis Muscle

Extensor Carpi Radialis
Longus Muscle

Radius

Extensor Digitorum Muscle

Ulna

Extensor Carpi Ulnaris Muscle

Triceps Brachii Muscle,
Lateral Head

Humerus, Shaft

Triceps Brachii Muscle,
Long Head

Pronator Teres Muscle

Flexor Digitorum
Superficialis Muscle

Flexor Digitorum
Profundus Muscle

Flexor Carpi Ulnaris Muscle

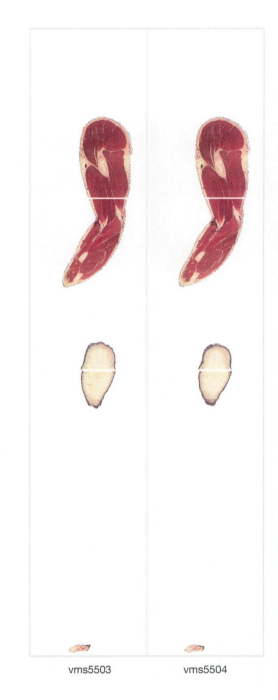

vms5503   vms5504

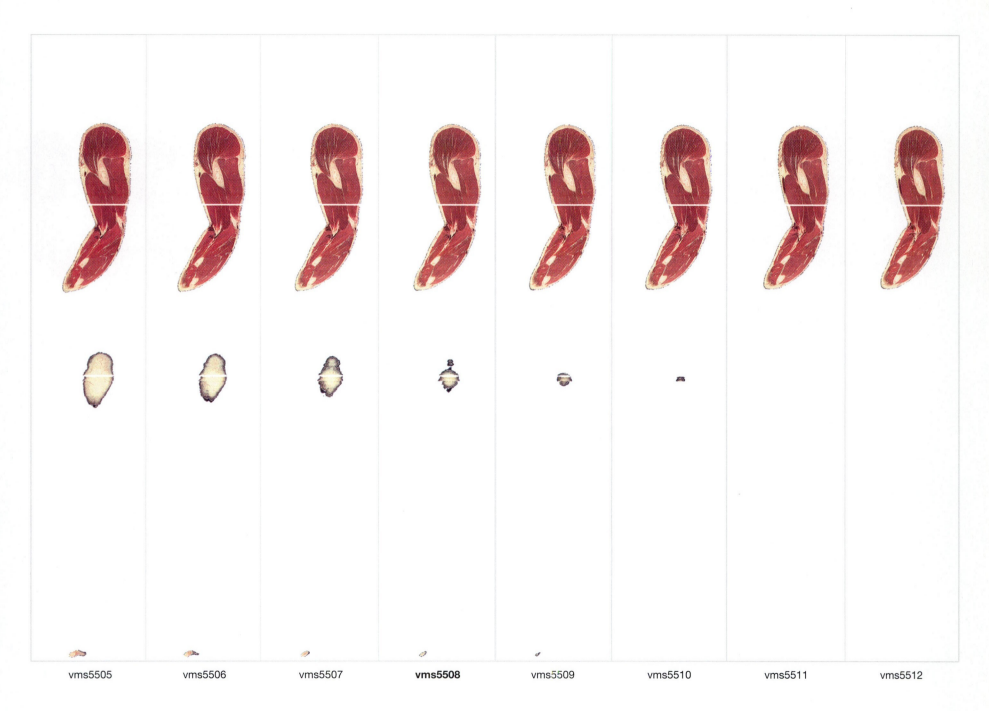

vms5505   vms5506   vms5507   **vms5508**   vms5509   vms5510   vms5511   vms5512

Deltoid Muscle

Triceps Brachii Muscle,
Lateral Head

Humerus, Deltoid Tuberosity

Biceps Brachii Muscle

Humerus, Shaft

Brachialis Muscle

Triceps Brachii Muscle,
Long Head

Brachioradialis Muscle

Extensor Carpi Radialis
Longus Muscle

Pronator Teres Muscle

Flexor Digitorum
Superficialis Muscle

Radius

Extensor Digitorum Muscle

Flexor Digitorum
Profundus Muscle

Ulna

Extensor Carpi Ulnaris Muscle

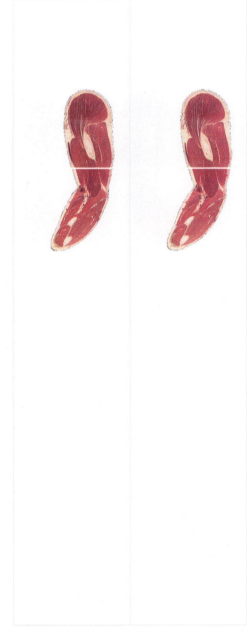

vms5513   vms5514

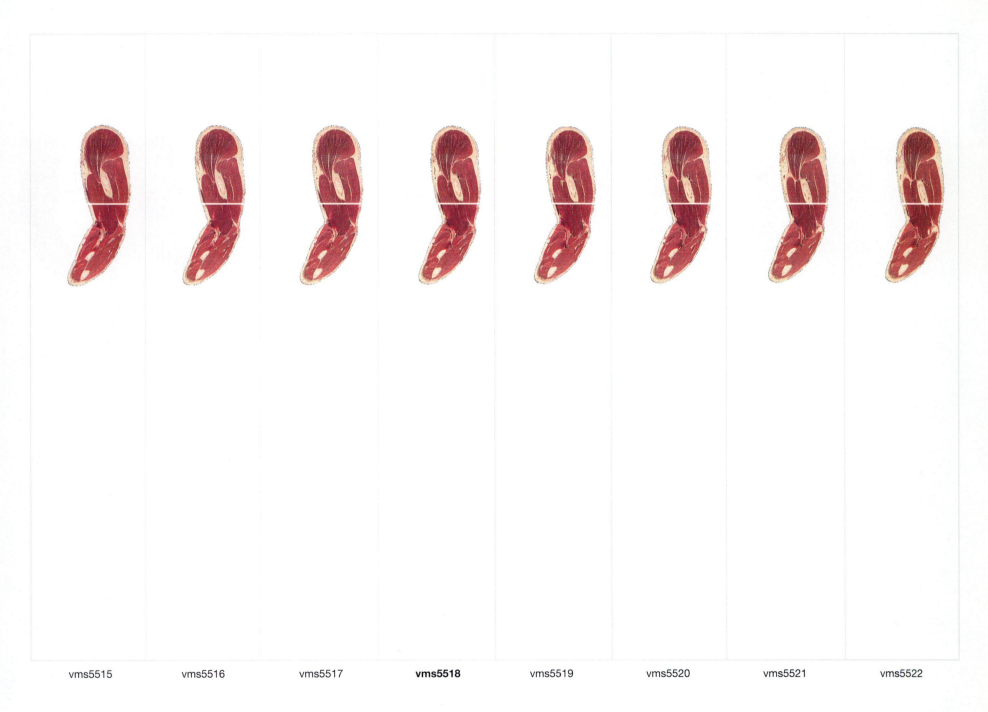

vms5515   vms5516   vms5517   **vms5518**   vms5519   vms5520   vms5521   vms5522

Brachialis Muscle
Biceps Brachii Muscle
Cephalic Vein
Brachioradialis Muscle
Extensor Carpi Radialis
Longus Muscle
Radius
Extensor Carpi Radialis
Brevis Muscle
Extensor Digitorum
Longus Muscle
Ulna

Deltoid Muscle
Triceps Brachii Muscle,
Long Head
Triceps Brachii Muscle,
Lateral Head
Humerus, Shaft
Humerus,
Medial Epicondyle
Humerus, Trochlea
Flexor Digitorum
Superficialis Muscle

vms5523   vms5524

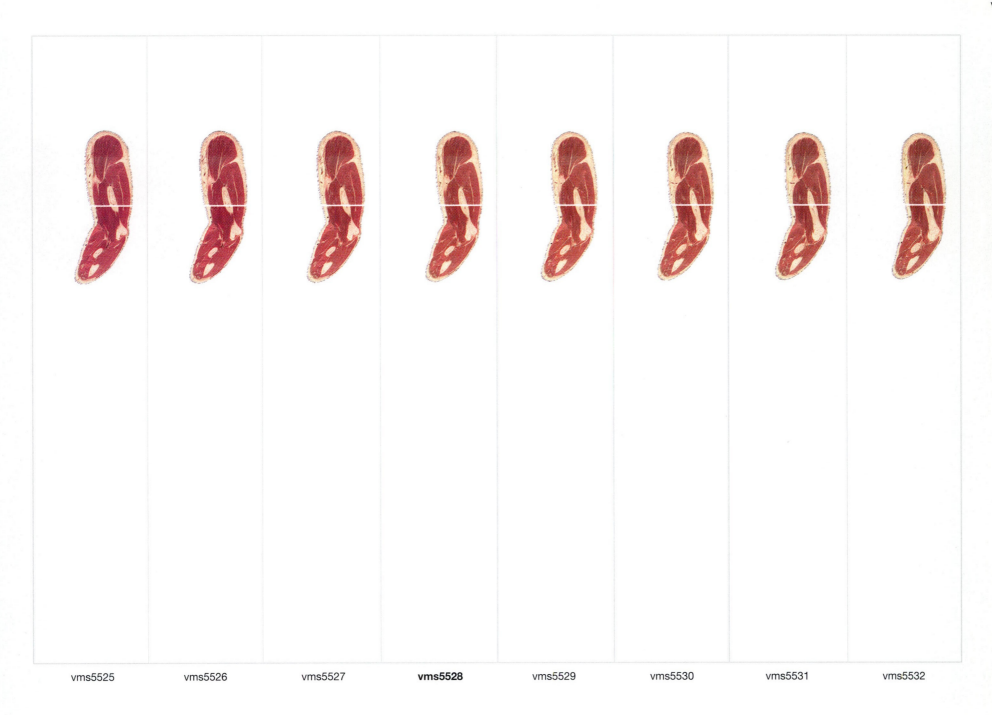

vms5525   vms5526   vms5527   **vms5528**   vms5529   vms5530   vms5531   vms5532

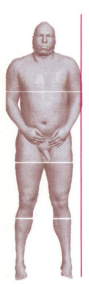

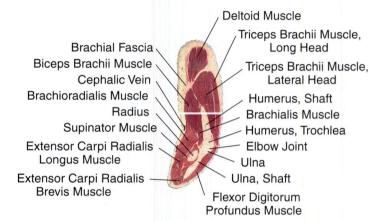

Deltoid Muscle

Triceps Brachii Muscle, Long Head

Brachial Fascia

Biceps Brachii Muscle

Triceps Brachii Muscle, Lateral Head

Cephalic Vein

Brachioradialis Muscle

Humerus, Shaft

Radius

Brachialis Muscle

Supinator Muscle

Humerus, Trochlea

Extensor Carpi Radialis Longus Muscle

Elbow Joint

Ulna

Extensor Carpi Radialis Brevis Muscle

Ulna, Shaft

Flexor Digitorum Profundus Muscle

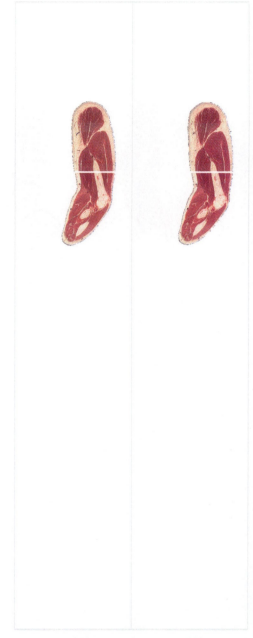

cms5533   vms5534

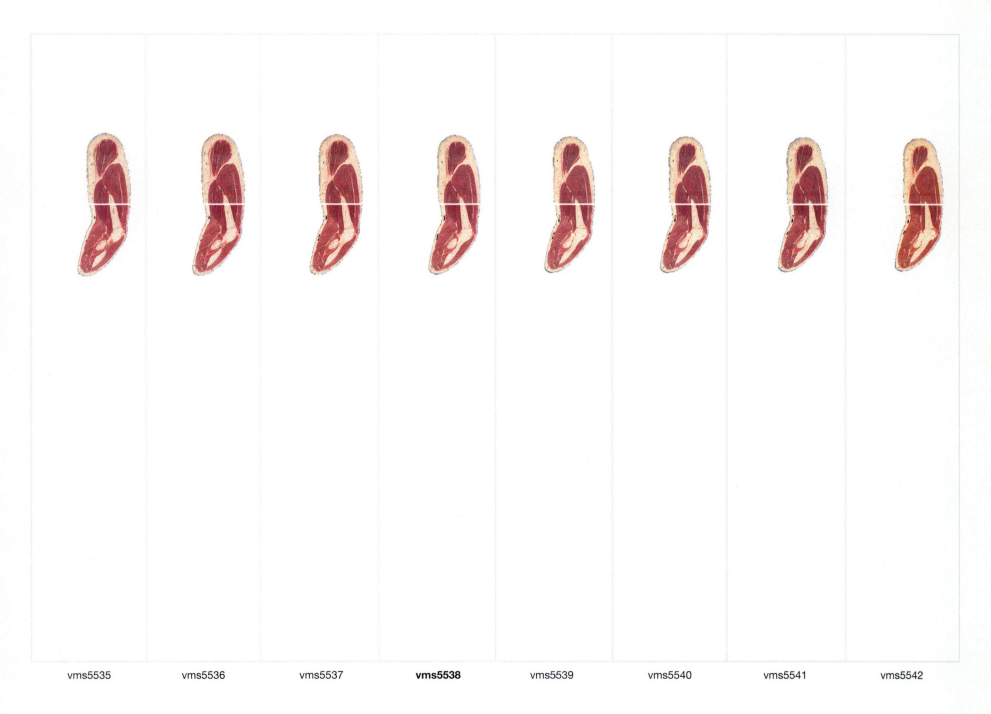

vms5535   vms5536   vms5537   **vms5538**   vms5539   vms5540   vms5541   vms5542

PART FOUR

Image Gallery

The following twenty pages contain a potpourri of some exciting views and applications of the Visible Human Male dataset. The first display is of images from the transverse, coronal, and sagittal parts. These images are particularly interesting slices because they show collections of anatomical structures not easily visualized in conventional displays (e.g., a transverse section through the optic chiasma where both optic nerves are visualized in the same slice, the median sagittal plane, and a coronal image through the entire length of body). These images are displayed here without labels and leader lines and at maximum size. Next, there are three "slices" that are not of any of the standard planes. The first of these is cut in the Frankel imaging plane, which is widely used in clinical computed tomography to examine the base of the skull. The other two images are not planes at all but rather curved surfaces taken through the center of the vertebral column and through the center of the spinal cord. These reconstructions show how a volume of computed image data can be manipulated to demonstrate entire lengths of anatomical structures (i.e., the spinal cord or vertebral column).

This part also contains ray-traced renderings of the classified data of the Visible Human Project. The 3D effect has been achieved by illuminating the surfaces of the visualized structures and analyzing the reflected light. The classified data used to create these pictures is courtesy of Gold Standard Multimedia, Inc. These pictures are unique in that their color and/or texture are attribuites of the original cross-sectional data. Some of the images demonstrate that each structure can be uniquely identified by assignment of different colors to distinct structures.

The part concludes with three pages displaying developments at the University of Colorado Center for Human Simulation. These images demonstrate how the classified data of volume-based image data is being utilized for simulating medical porocedures. In addition to anatomical tissue type classification, each voxel has also been designated with physical characteristics such that the user can now "feel" the data as well as see the cut planes and 3D renderings.

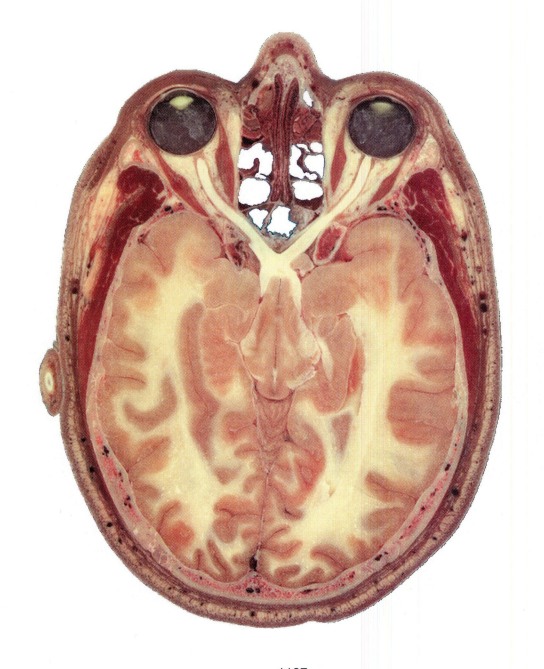

a_vm1107

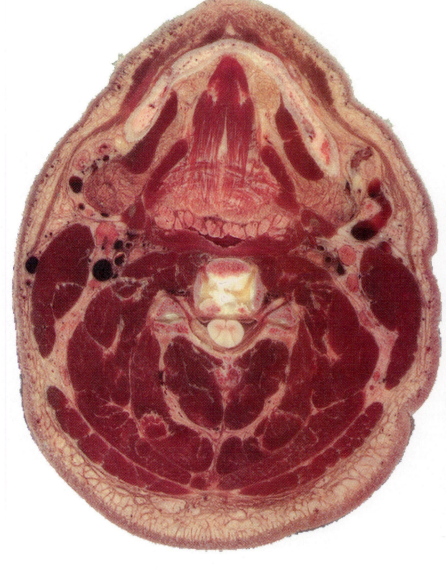

a_vm1212

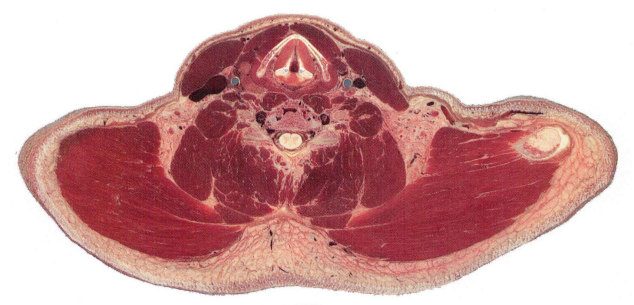

a_vm1260

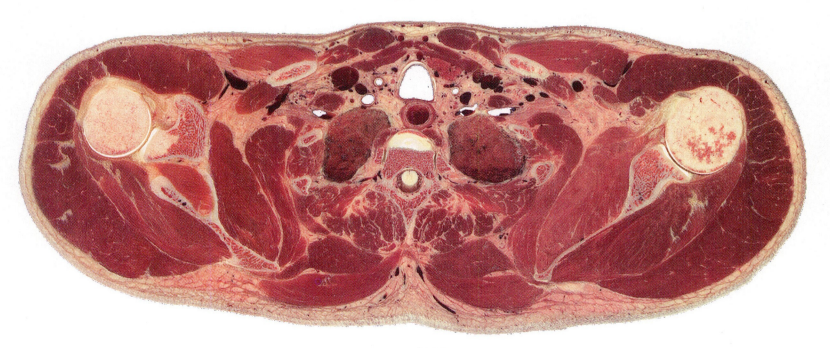

a_vm1307

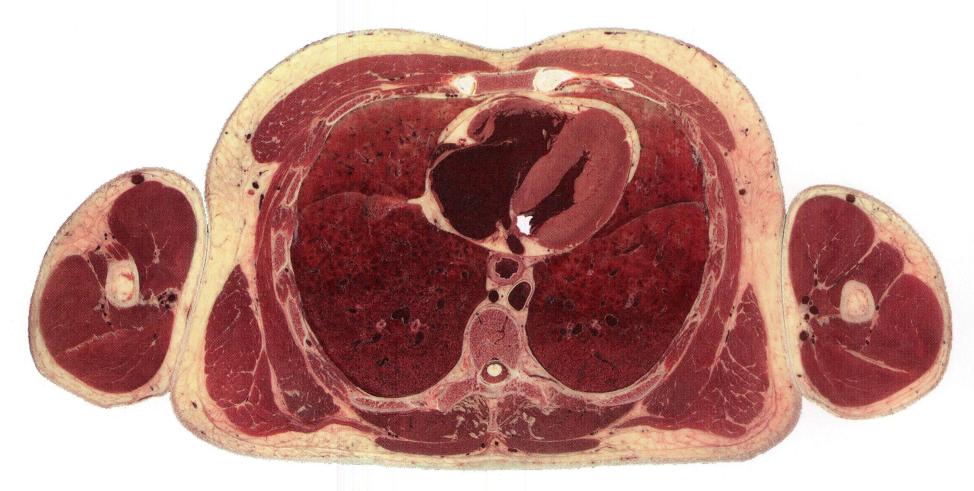

a_vm1452

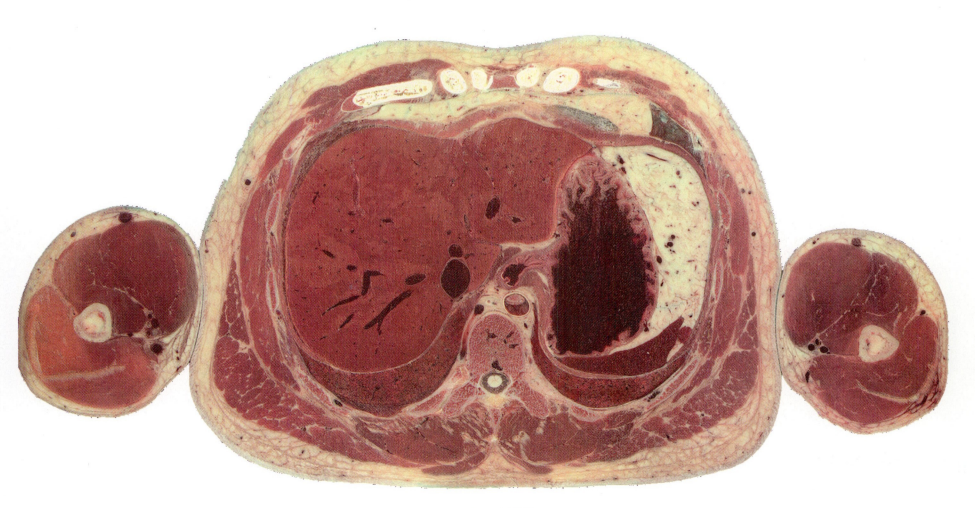

a_vm1510

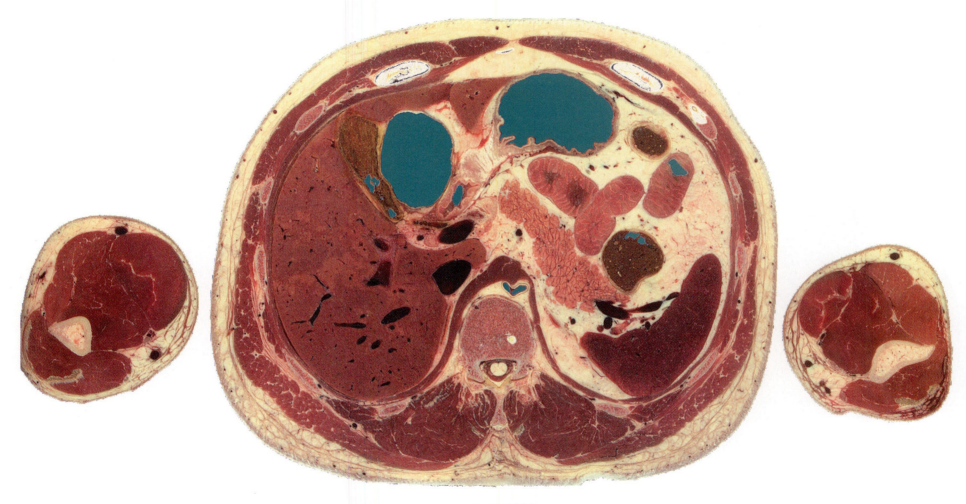

a_vm1555

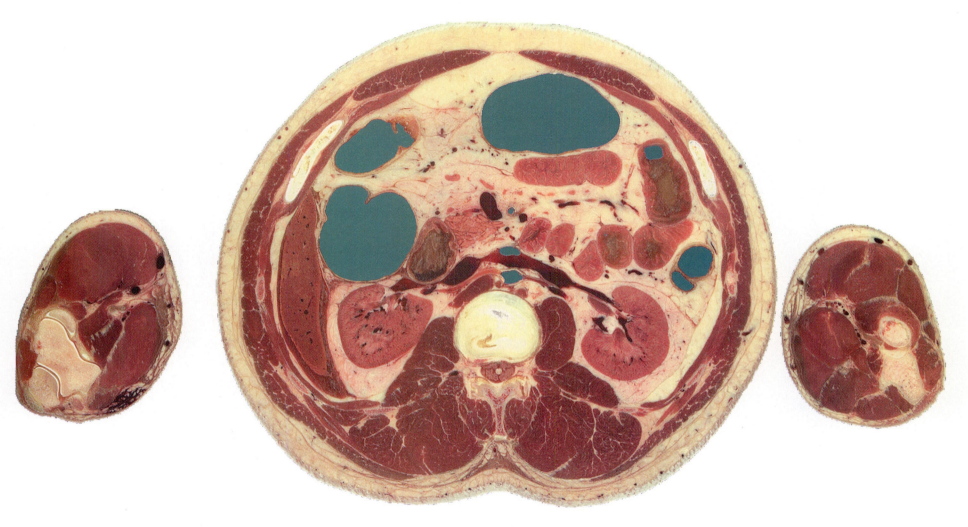

a_vm1616

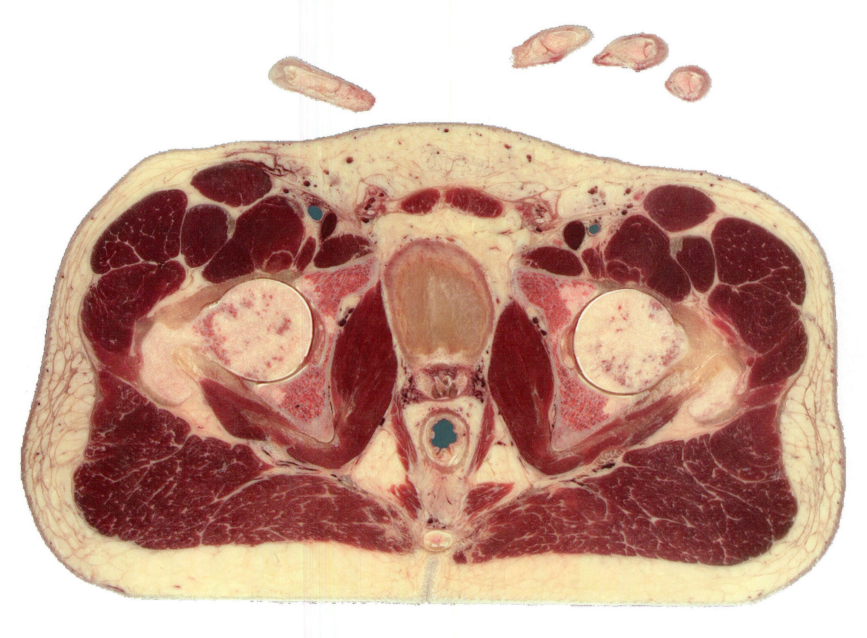

a_vm1886

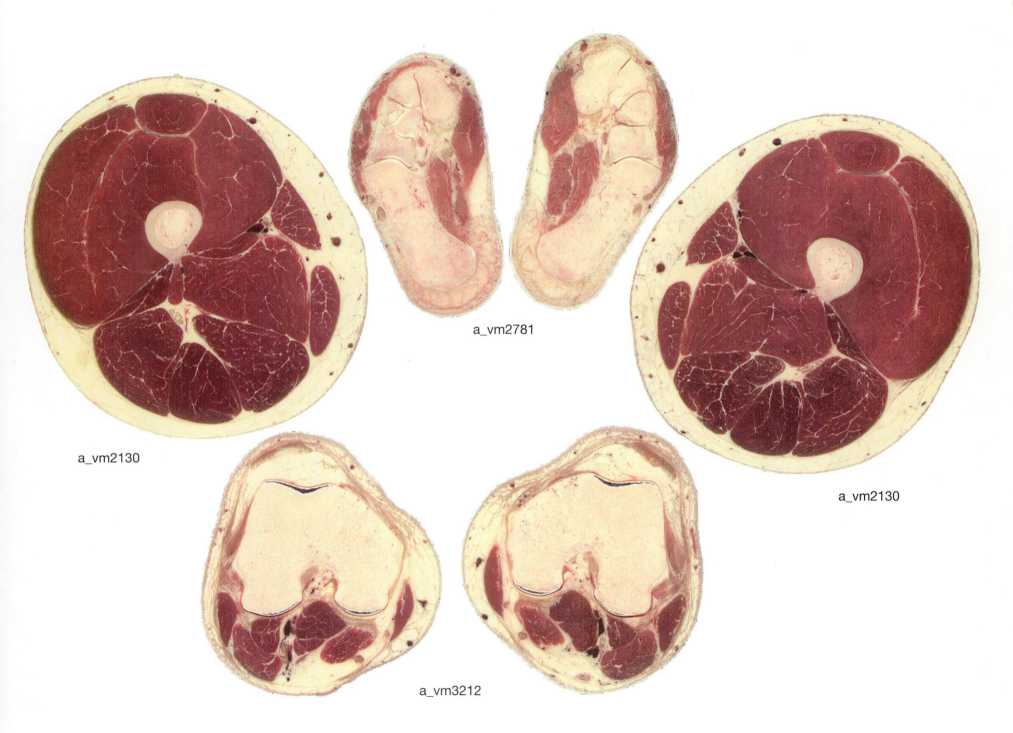

a_vm2781

a_vm2130

a_vm2130

a_vm3212

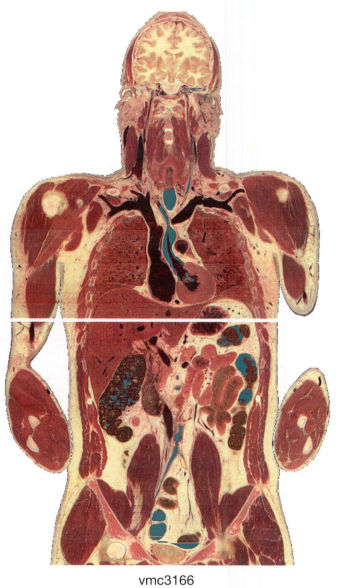

vmc3166

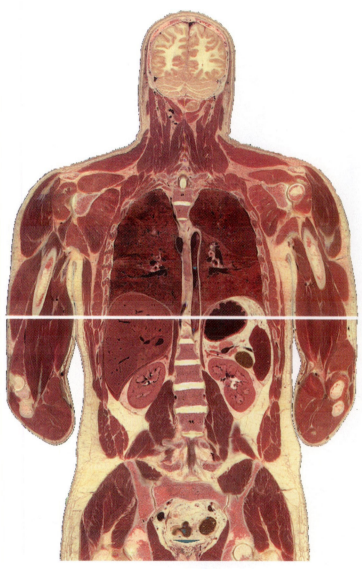

vmc3215

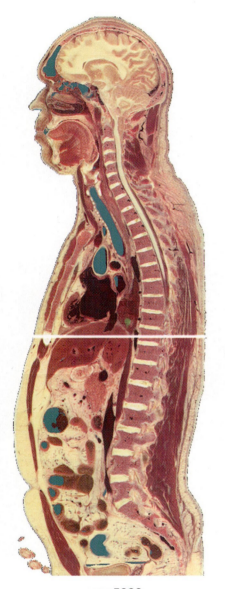

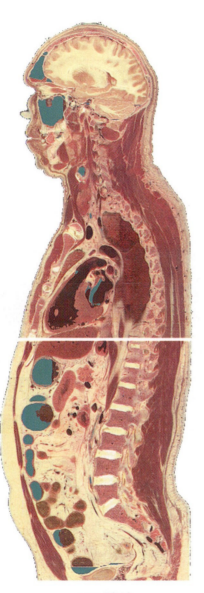

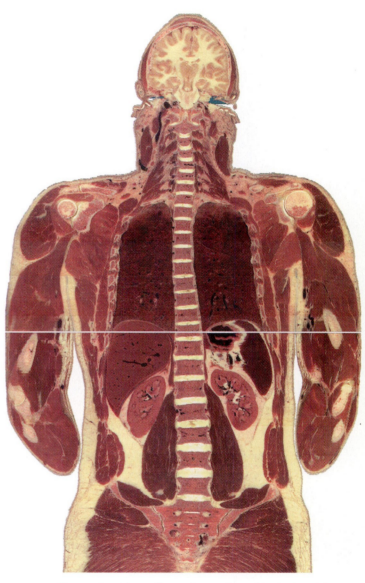

vms5288

vms5318

Curved surface through center of spinal cord

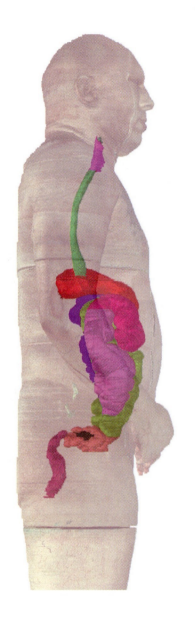
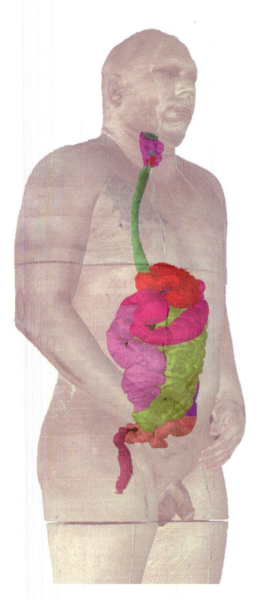
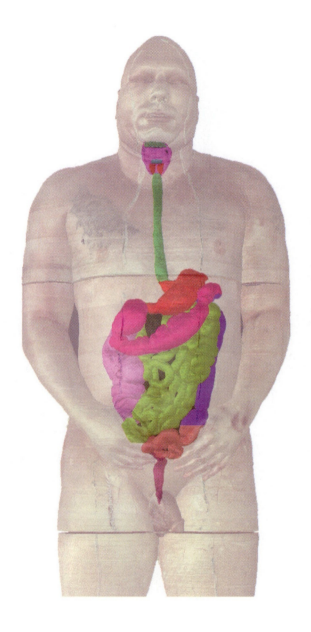

Digestive System

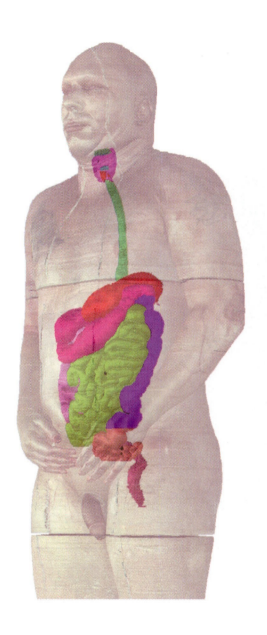

Digestive System

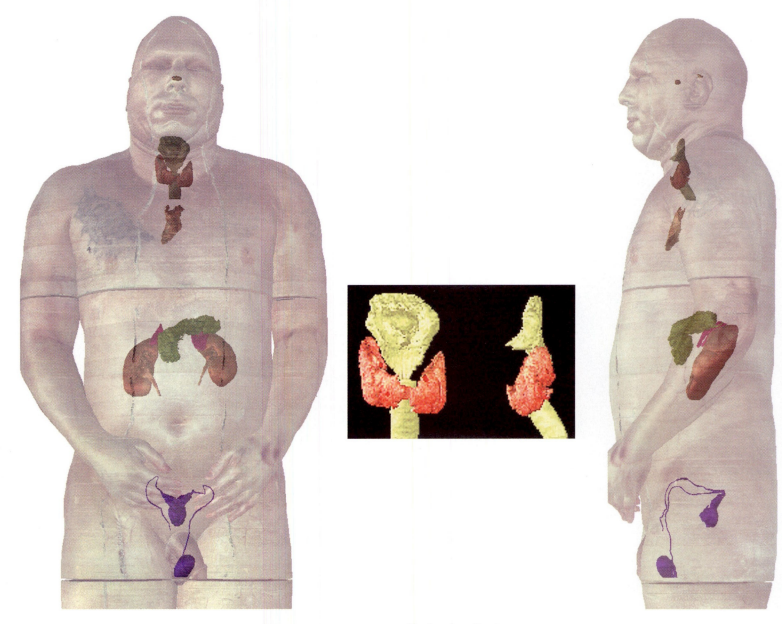

Endocrine System

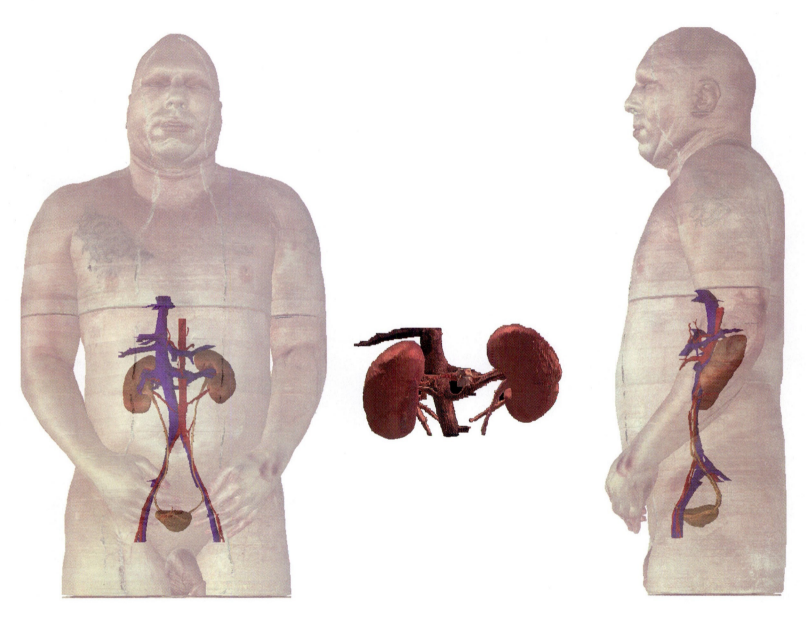

Urinary System

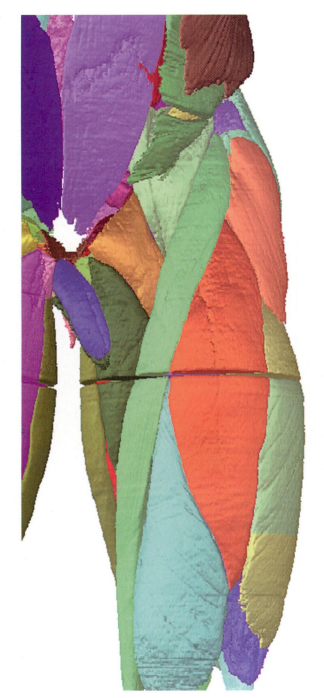

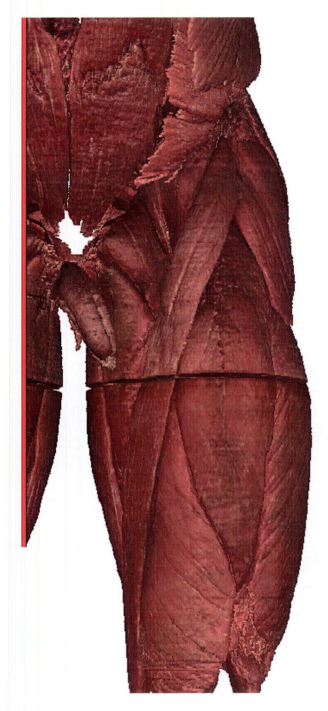

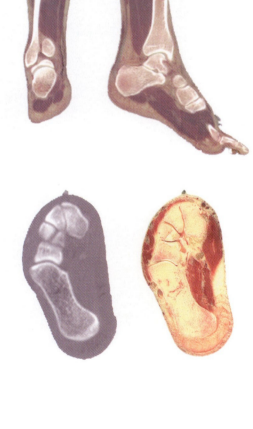

Musculoskeletal System

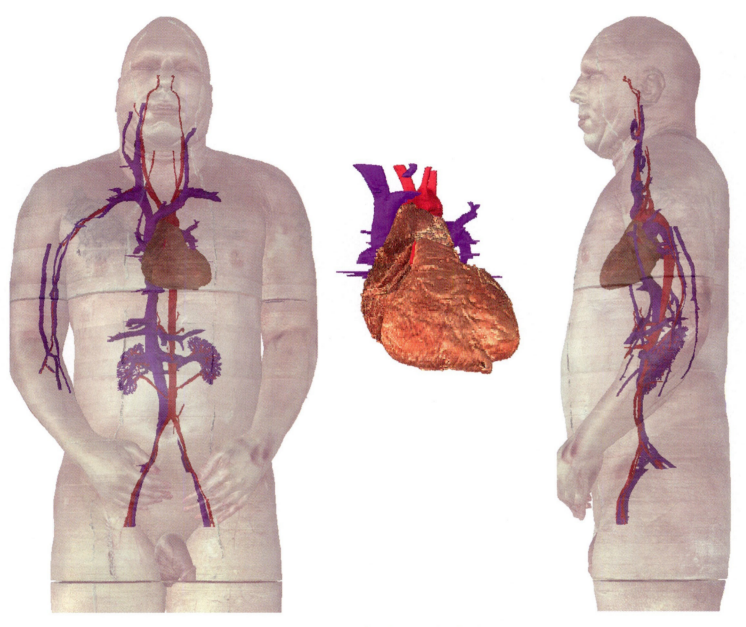

Cardiovascular System

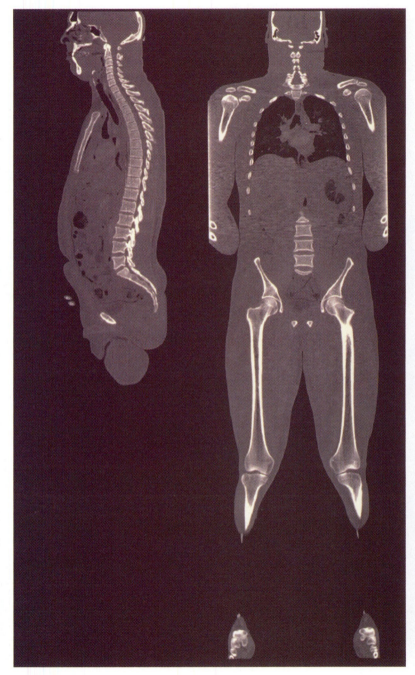

CT Skeletal Images

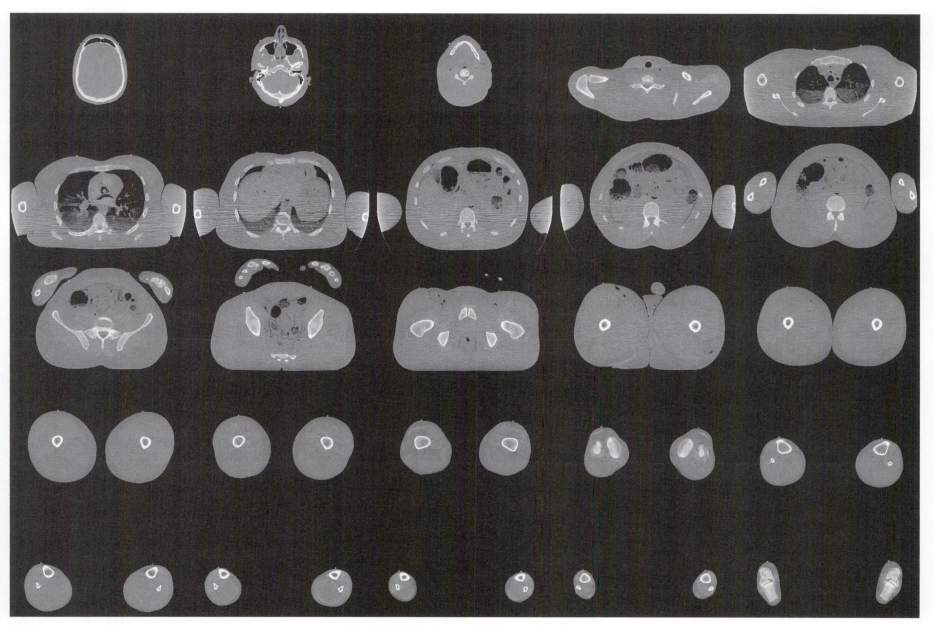

The 1871 Transverse CT Cross-Sections

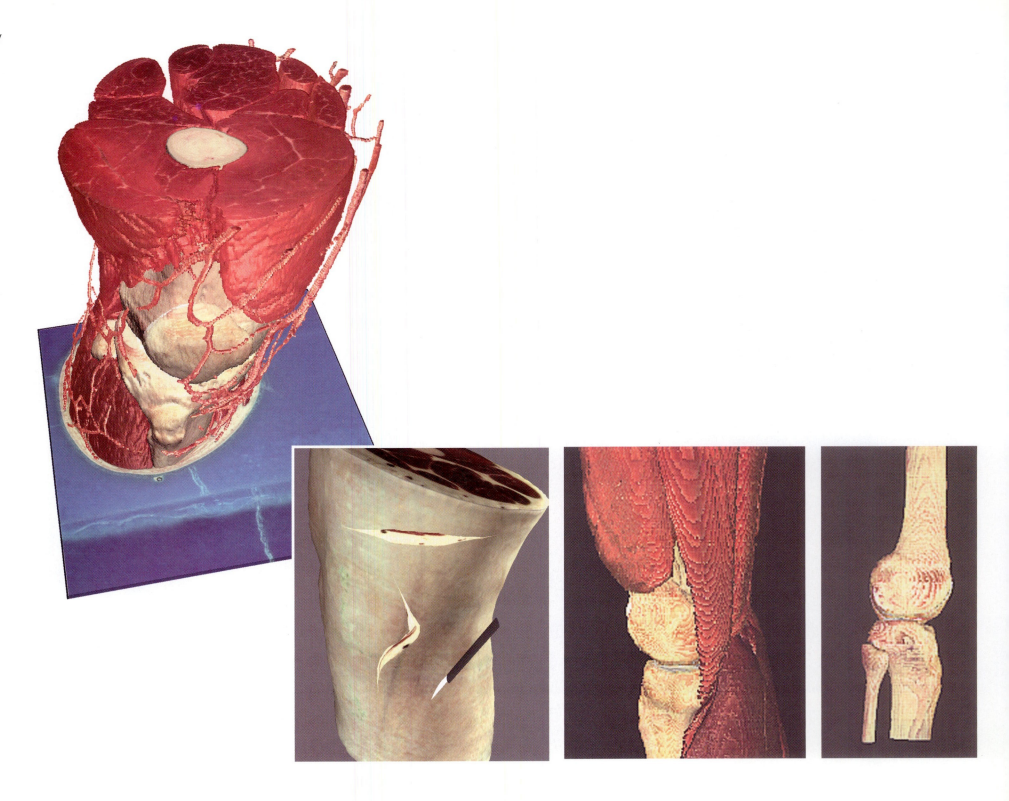

INDEX